Cultural Heritage Science

More information about this series at http://www.springer.com/series/13104

Francesca Casadio • Katrien Keune • Petria Noble
Annelies Van Loon • Ella Hendriks
Silvia A. Centeno • Gillian Osmond
Editors

Metal Soaps in Art

Conservation and Research

Foreword by Taco Dibbits

Springer

Editors
Francesca Casadio
The Art Institute of Chicago
Chicago, IL, USA

Petria Noble
Conservation Department
Rijksmuseum
Amsterdam, The Netherlands

Ella Hendriks
Programme Conservation and
Restoration of Cultural Heritage
University of Amsterdam
Amsterdam, The Netherlands

Gillian Osmond
Queensland Art Gallery/
Gallery of Modern Art
Brisbane, Australia

Katrien Keune
Conservation Department
Rijksmuseum
Van't Hoff Institute for Molecular Sciences
University of Amsterdam
Amsterdam, The Netherlands

Annelies Van Loon
Conservation Department
Rijksmuseum
Amsterdam, The Netherlands

Silvia A. Centeno
Department of Scientific Research
The Metropolitan Museum of Art
New York, NY, USA

ISSN 2366-6226 ISSN 2366-6234 (electronic)
Cultural Heritage Science
ISBN 978-3-319-90616-4 ISBN 978-3-319-90617-1 (eBook)
https://doi.org/10.1007/978-3-319-90617-1

Library of Congress Control Number: 2018953888

This Springer imprint is published by the registered company Springer Nature Switzerland AG.
The registered company address is: Gewerbestrasse 11, 6330 Cham, Switzerland

Foreword

It was in the late 1990s that so-called metal soap “protrusions” were first observed in the Netherlands. At the time, it was thought that these were just a few isolated cases, but over the course of the last two decades, the problems associated with metal soaps have become more extensive and affected paintings are still being discovered. As a result, the subject of “metal soaps” has evolved into one of the most prominent topics of painting conservation research, and these protrusions are now acknowledged as a threat to the condition of paintings.

It was therefore important and a great honor for the Rijksmuseum, together with the University of Amsterdam and The Art Institute of Chicago, to organize within the framework of the Netherlands Institute for Conservation, Art History and Science (NICAS), a conference on 14–15 March 2016 dedicated to the subject of “metal soaps.” The International Scientific Committee consisting of Katrien Keune, Annelies Van Loon, Robert van Langh, and Petria Noble (Rijksmuseum, Amsterdam), along with Ella Hendriks and Maartje Witlox (University of Amsterdam, Amsterdam), Klaas Jan Van den Berg (Cultural Heritage Agency of the Netherlands, Amsterdam), Francesca Casadio (The Art Institute of Chicago and NU-ACCESS, Chicago), Silvia A. Centeno (The Metropolitan Museum of Art, New York), Gillian Osmond (Queensland Art Gallery/Gallery of Modern Art, Brisbane, Australia), Caroline Tokarski (University of Bordeaux, France), and Marika Spring (National Gallery, London), needs to be especially thanked for its collective vision and energy in organizing such a successful conference that attracted more than 220 participants from 20 countries around the globe. I hope the results of the conference lead to a better understanding of metal soap-related degradation phenomena in paintings, and will bring all of us in the paintings and museum communities a step closer to answering the many questions surrounding these phenomena.

As a museum, it is our responsibility to present paintings in optimal condition and to ensure their preservation for future generations. We strive to treat and maintain paintings in the best possible manner. Metal soap-related degradation is a good example of the changes that paintings can undergo over time. It is therefore important for conservators to know how to deal with this. Should you treat it? Can you stabilize it? Can we ignore it?

In order to better understand and minimize such changes, it is of utmost importance to invest in technologies that further our knowledge of the condition of paintings, to develop our knowledge of artists' materials and techniques, and to carry out research into the preservation and treatment histories of paintings. In the Rijksmuseum, and in the framework of NICAS, we are developing this knowledge through the collaboration of multidisciplinary teams of scientists, art historians, and conservators in an academic and museum environment to answer both applied and fundamental questions.

It gives me great pleasure to present this volume, which was compiled as a result of the conference mentioned above. With great commitment and care, the international team of editors has brought together a collection of 24 peer-reviewed interdisciplinary articles by experts in the field that are of the highest quality and that comply with current standards in scientific publishing. Here you will find articles that highlight the latest insights into metal soap degradation in paintings from different perspectives. It is invaluable that this compilation of current knowledge and expertise relating to metal soaps is now bundled into a single volume, and I am convinced that this book will be an important reference for anyone concerned with the future preservation of paintings in the museum field and beyond.

Amsterdam, The Netherlands

Taco Dibbits
(General Director, Rijksmuseum)

Preface

Our knowledge of the phenomenological manifestations of the reaction of metals in pigments or driers with oil-based binding media and varnishes in painted surfaces has increased substantially in the past two decades. Commonly termed "metal soaps," these organometallic compounds can sometimes contribute to stabilizing paints; however, with increasing frequency they are implicated in degradation phenomena in paintings. For a long time, conservators, curators, and other museum professionals have observed increased transparency of paints; formation of insoluble whitish surface hazes and crusts; tiny aggregates deforming the paint surface and eventually leading to pin-point paint losses; paint flaking and interlayer delamination; and water sensitive and dripping paints. It was not until the early 2000s though that researchers started to recognize that these wide-ranging chemical, physical, and visual changes were but some examples of paint defects connected to the formation of metal soaps. While these changes are now acknowledged as a serious threat to the condition of painted surfaces, the potential consequences for art historical interpretation of the works of art are not yet widely recognized.

In recent years, numerous case studies have emerged and significant advances have been made in the detection and scientific characterization of metal soaps, both in actual works of art and in model systems. The formation, migration, and treatment of metal soaps have been explored and theoretical and experimental frameworks for their formation in paint films proposed. Developing from the initial studies in the early 2000s, which focused on case studies and basic characterization of the different types of metal soaps, more detailed insights of the underlying mechanisms of metal soap-related degradation in paints have recently been gained thanks to advances in scientific techniques that are able to probe reaction boundaries and structures at the micro- and nanoscales, as well as noninvasive mapping techniques that allow epidemiological studies of large groups of paintings. Armed with this knowledge, researchers are also starting to explore the role of anthropogenic causes that are extrinsic to the paints, such as the influence of conservation treatments and indoor climate on the formation of metal soaps, and how these external factors affect molecular and supramolecular dynamics and transport mechanisms.

Since metal soaps were first described and characterized in the scientific and conservation literature at the end of the last century, two topical conferences were held: the De Mayerne Day in the Netherlands in 2005 and the Paintings Specialty Session at the Annual Meeting of the American Institute for Conservation of Historic and Artistic Works held in the United States in 2006. A decade later, the need to capture the increased body of new research carried out in recent years culminated in the organization of the conference "Metal Soaps in Art," held in Amsterdam in 2016. The conference brought to the fore the realization that much of this emerging research is scattered in various scientific publications that are not easily accessible by the conservation community, or is contained in conservation publications not widely available and often out of print. Furthermore, the pressing need to link the robust findings of the scientific community to the practical issues faced by conservators of painted surfaces worldwide became strongly apparent. Outside of the comfort zone of tightly controlled experimental conditions in scientific laboratories, conservators are confronted with paintings and painted objects that exhibit metal soap-related degradation phenomena, and are wondering how solvents, heat, and moisture introduced during cleaning or structural treatments affect metal soaps. Furthermore, they are searching for suitable treatment options and optimal display and storage conditions. Current pressure to relax temperature and relative humidity parameters for museum collections raises further concerns about the effect of environmental conditions on the rate of formation of metal soaps and other preventive conservation issues related to metal soap degradation. It thus became clear that a platform for sharing knowledge between scientists and conservators was needed to jointly translate present scientific knowledge into best practices for the treatment and display of painted surfaces.

This book gathers state-of-the-art research on metal soap formation concerning both scientific analysis and the implications for conservation and preservation. The diverse geographical and professional origin of the authors and the guest editorial board – including conservators, conservation scientists, and scientists from museums, academia, and in private practice, working in Europe, Australia, and North America – reflect the need for an interdisciplinary and global approach to this subject. With sections that cover the history of the study of metal soaps in paintings, research on historical sources related to manufacturing processes, and detailed scientific investigations on the formation, migration, and effect of environmental conditions, this volume gathers studies performed with cutting-edge techniques and simulation tools, complemented by case studies from old masters paintings to twentieth-century art. The book also includes a section introducing some practical considerations that can inform conservation strategies and initial indications to help guide treatment decisions.

In editing the book and guiding its authors we have striven to produce essays with introductions that are approachable for an audience with varied backgrounds (conservators, students of conservation and conservation science, academics, and their students interested in the chemistry of art) but which also have enough technical details to be stimulating for those active in this area of research, with a balanced approach between reviewing the state of the art in the field and presenting new findings. The result is highly interdisciplinary, spanning different analytical techniques and 20 years of research, as well as dealing with the practical aspects of metal soap-related degradation, preventive conservation, and treatment.

This volume represents the strong commitment of the field of conservation and conservation science to the dissemination of research. In addition to thanking all the authors for their valuable contributions and the book editors for the countless hours they voluntarily gave to bring the project to fruition, we are very grateful to the many anonymous peer reviewers that have ensured all chapters are of the highest quality and up to the most current standards in scientific publishing. We also gratefully acknowledge AkzoNobel, sponsor of the Rijksmuseum, for generously supporting this publication and the 2016 "Metal Soaps in Art" conference.

It can be confidently stated that a vast majority of oil-based paintings and painted objects in museum collections are affected by some form of metal soap-related phenomena, making metal soap degradation one of the defining issues in the conservation of painted surfaces. In order to find viable solutions to the issues of metal soaps in paintings, scientists and conservators need to join forces and set future goals and research priorities. While capturing a panoramic overview of today's state of knowledge, and aspiring to become the go-to reference for the next decade, this volume also critically examines what the open questions are and casts a glance onto the future. It is our hope that this book will constitute a useful reference for practitioners in the field and become a springboard for future research efforts.

Chicago, IL, USA Francesca Casadio
Amsterdam, The Netherlands Katrien Keune
Amsterdam, The Netherlands Petria Noble
Amsterdam, The Netherlands Annelies Van Loon
Amsterdam, The Netherlands Ella Hendriks
New York, NY, USA Silvia A. Centeno
Brisbane, Australia Gillian Osmond

Contents

Contributors

Chiara Anselmi Istituto CNR di Scienze e Tecnologie Molecolari (CNR-ISTM), Perugia, Italy

Centro di Eccellenza SMAArt, Università degli Studi di Perugia, Perugia, Italy

Wendy Baker Canadian Conservation Institute, Ottawa, ON, Canada

Ludovic Bellot-Gurlet Sorbonne Universités, UPMC Université Pierre et Marie Curie – Paris 6, MONARIS "de la Molécule aux Nano-objets: Réactivité, Interactions et Spectroscopies", UMR 8233, UPMC/CNRS, Paris, France

Victòria Beltran Departamento d'Enginyeria Química, EPSEVG, and Center for Research in Nano-Engineering, Universitat Politècnica de Catalunya, Barcelona, Spain

Barbara H. Berrie National Gallery of Art, Washington, DC, USA

Nancy E. Binnie Canadian Conservation Institute, Ottawa, ON, Canada

Madeleine Bisschoff Independent paintings conservator, Amsterdam, The Netherlands

Zahira Velíz Bomford The Museum of Fine Arts Houston, Houston, TX, USA

Brunetto Giovanni Brunetti Istituto CNR di Scienze e Tecnologie Molecolari (CNR-ISTM), Perugia, Italy

Centro di Eccellenza SMAArt, Università degli Studi di Perugia, Perugia, Italy

Aviva Burnstock Department of Conservation and Technology, Courtauld Institute of Art, London, UK

Salvador Butí Departamento d'Enginyeria Química, EPSEVG and Center for Research in Nano-Engineering, Universitat Politècnica de Catalunya, Barcelona, Spain

David Buti CATS-SMK (Centre for Art Technological Studies and Conservation, Statens Museum for Kunst), Copenhagen, Denmark

Laura Cartechini Istituto CNR di Scienze e Tecnologie Molecolari (CNR-ISTM), Perugia, Italy

Centro di Eccellenza SMAArt, Università degli Studi di Perugia, Perugia, Italy

Francesca Casadio The Art Institute of Chicago, Chicago, IL, USA

Jaclyn Catalano Montclair State University, Montclair, NJ, USA

Silvia A. Centeno Department of Scientific Research, The Metropolitan Museum of Art, New York, NY, USA

Emilie Checroun Checroun RCPM, Magny-le-Hongre, France

Epitopos, Strasbourg, France

Oscar Chiantore Dipartimento di Chimica, Università degli Studi di Torino, Torino, Italy

Gianfelice Cinque Diamond Light Source, Oxfordshire, UK

Oliver Cossairt Electrical Engineering and Computer Science, Northwestern University, Evanston, IL, USA

Marine Cotte European Synchrotron Radiation Facility, Grenoble, France

Laboratoire d'archéologie moléculaire et structurale (LAMS), Sorbonne Université, CNRS, UMR 8220, Paris, France

Fabio Cucinotta School of Chemistry, Newcastle University, Newcastle Upon Tyne, UK

Hannie Diependaal Independent paintings conservator, Amsterdam, The Netherlands

Laurence De Viguerie Laboratoire d'archéologie moléculaire et structurale (LAMS), Sorbonne Université, CNRS, UMR 8220, Paris, France

Brenda Doherty Istituto CNR di Scienze e Tecnologie Molecolari (CNR-ISTM), Perugia, Italy

Centro di Eccellenza SMAArt, Università degli Studi di Perugia, Perugia, Italy

Cecil Dybowski University of Delaware, Newark, DE, USA

Alejandro Fernández-Martínez CNRS, ISTerre, Université Grenoble Alpes, Grenoble, France

Élisabeth Forest Centre de conservation de Québec, Québec, QC, Canada

Francesca Gabrieli Istituto CNR di Scienze e Tecnologie Molecolari (CNR-ISTM), Perugia, Italy

Centro di Eccellenza SMAArt, Università degli Studi di Perugia, Perugia, Italy

Muriel Geldof Cultural Heritage Agency of the Netherlands (RCE), Amsterdam, The Netherlands

Laura Eva Hartman Royal Picture Gallery Mauritshuis, The Hague, The Netherlands

Ralph Haswell Shell Global Solutions International, Amsterdam, The Netherlands

Pauline Hélou-de La Grandière Paintings Conservator, Nîmes, France

Kate Helwig Canadian Conservation Institute, Ottawa, ON, Canada

Joen J. Hermans Van't Hoff Institute for Molecular Sciences, University of Amsterdam, Amsterdam, The Netherlands

Catherine Higgitt The National Gallery, London, UK

Sally Higgs Conservator in Private Practice, London, UK

Ruth Hoppe Gemeentemuseum Den Haag, The Hague, The Netherlands

Piet D. Iedema Van't Hoff Institute for Molecular Sciences, University of Amsterdam, Amsterdam, The Netherlands

Koen Janssens Department of Chemistry – AXES group, University of Antwerp, Antwerp, Belgium

Jordi Juanhuix CELLS-ALBA Synchrotron, Cerdanyola del Vallès, Barcelona, Spain

Aggelos K. Katsaggelos Electrical Engineering and Computer Science, Northwestern University, Evanston, IL, USA

Katrien Keune Conservation Department, Rijksmuseum, Van't Hoff Institute for Molecular Sciences, University of Amsterdam, Amsterdam, The Netherlands

Lizet Klaassen Royal Museum of Fine Arts Antwerp, Antwerp, Belgium

Rick P. Kramer Eindhoven University of Technology, Eindhoven, Netherlands

Dale Kronkright Georgia O'Keeffe Museum, Santa Fe, NM, USA

Maite Leal The Museum of Fine Arts Houston, Houston, TX, USA

Margje Leeuwestein Kröller-Müller Museum, Otterlo, The Netherlands

Stijn Legrand Department of Chemistry – AXES group, University of Antwerp, Antwerp, Belgium

Fengqiang Li Electrical Engineering and Computer Science, Northwestern University, Evanston, IL, USA

Francisco J. Martínez-Casado MAX IV Laboratory, Lund University, Lund, Sweden

Departamento Química-Física I. Facultad de Ciencias Químicas, Universidad Complutense de Madrid, Madrid, Spain

Costanza Miliani Istituto CNR di Scienze e Tecnologie Molecolari (CNR-ISTM), Perugia, Italy

Centro di Eccellenza SMAArt, Università degli Studi di Perugia, Perugia, Italy

Elizabeth Moffatt Canadian Conservation Institute, Ottawa, ON, Canada

Letizia Monico Istituto CNR di Scienze e Tecnologie Molecolari (CNR-ISTM), Perugia, Italy

Centro di Eccellenza SMAArt, Università degli Studi di Perugia, Perugia, Italy

Anna Murphy University of Delaware, Newark, DE, USA

Marco Nervo Centro Conservazione e Restauro "La Venaria Reale", Piazza Della Repubblica, Venaria Reale (TO), Italy

Petria Noble Conservation Department, Rijksmuseum, Amsterdam, The Netherlands

Gillian Osmond Queensland Art Gallery/Gallery of Modern Art, Brisbane, Australia

Céline Paris Sorbonne Universités, UPMC Université Pierre et Marie Curie – Paris 6, MONARIS "de la Molécule aux Nano-objets: Réactivité, Interactions et Spectroscopies", UMR 8233, UPMC/CNRS, Paris, France

Anna Piccirillo Centro Conservazione e Restauro "La Venaria Reale", Piazza Della Repubblica, Venaria Reale (TO), Italy

Tommaso Poli Dipartimento di Chimica, Università degli Studi di Torino, Torino, Italy

Marie Postec Independent Painting Conservator, Brussels, Belgium

Carol Pottasch Royal Picture Gallery Mauritshuis, The Hague, The Netherlands

Jennifer Poulin Canadian Conservation Institute, Ottawa, ON, Canada

Trinitat Pradell Center for Research in Nano-Engineering and Departament de Física, Universitat Politècnica de Catalunya, Barcelona, Spain

Matthieu Réfrégiers Synchrotron SOLEIL, Gif-sur-Yvette, France

Miguel Ramos-Riesco Departamento Química-Física I. Facultad de Ciencias Químicas, Universidad Complutense de Madrid, Madrid, Spain

Laura E. Raven Conservation Department, Rijksmuseum, Amsterdam, The Netherlands

Conservation & Restoration, University of Amsterdam, Amsterdam, The Netherlands

María Isabel Redondo-Yélamos Departamento Química-Física I. Facultad de Ciencias Químicas, Universidad Complutense de Madrid, Madrid, Spain

José A. Rodríguez-Cheda Departamento Química-Física I. Facultad de Ciencias Químicas, Universidad Complutense de Madrid, Madrid, Spain

Dawn V. Rogala Paintings Conservator, Museum Conservation Institute, Smithsonian Institution, Washington, DC, USA

Corina E. Rogge The Museum of Fine Arts Houston, Houston, TX, USA

Francesca Rosi Istituto CNR di Scienze e Tecnologie Molecolari (CNR-ISTM), Perugia, Italy

Centro di Eccellenza SMAArt, Università degli Studi di Perugia, Perugia, Italy

Nati Salvadó Departamento d'Enginyeria Química, EPSEVG and Center for Research in Nano-Engineering, Universitat Politècnica de Catalunya, Barcelona, Spain

Johanna Salvant Centre de recherche et restauration des musées de France, Paris, France

Michael Schilling The Getty Conservation Institute, Los Angeles, CA, USA

Marika Spring The National Gallery, London, UK

Suzanne Stangier Restauratie Atelier Stangier, Guttecoven, Netherlands

Maartje Stols-Witlox Conservation & Restoration, University of Amsterdam, Amsterdam, The Netherlands

Jean Susini European Synchrotron Radiation Facility, Grenoble, France

Mathieu Thoury IPANEMA, CNRS, ministère de la Culture et de la Communication, Université de Versailles Saint-Quentin-en-Yvelines, Muséum National d'Histoire Naturelle, USR 3461, Université Paris-Saclay, Gif-sur-Yvette, France

Aimie Turcotte Centre de conservation de Québec, Québec, QC, Canada

Manuela Vagnini Associazione Laboratorio di Diagnostica per i Beni Culturali, Spoleto, Italy

Margriet H. van Eikema Hommes Delft University of Technology, Delft, Netherlands

Cultural Heritage Agency of Netherlands, Amsterdam, Netherlands

Annelies Van Loon Conservation Department, Rijksmuseum, Amsterdam, The Netherlands

Klaas Jan van den Berg Cultural Heritage Agency of the Netherlands, Amsterdam, The Netherlands

University of Amsterdam, Amsterdam, The Netherlands

Julia van den Burg Royal Picture Gallery Mauritshuis, The Hague, The Netherlands

Geert van der Snickt Department of Chemistry – AXES group, University of Antwerp, Antwerp, Belgium

Conservation Studies, University of Antwerp, Antwerp, Belgium

Frederik Vanmeert Department of Chemistry – AXES group, University of Antwerp, Antwerp, Belgium

Philippe Walter Laboratoire d'archéologie moléculaire et structurale (LAMS), Sorbonne Université, CNRS, UMR 8220, Paris, France

Marc Walton Materials Science and Engineering Department, Northwestern University, Evanston, IL, USA

Yao Yao University of Delaware, Newark, DE, USA

Chia-Kai Yeh Electrical Engineering and Computer Science, Northwestern University, Evanston, IL, USA

Nicholas Zumbulyadis Independent Researcher, Rochester, NY, USA

About the Editors

Francesca Casadio joined the Art Institute of Chicago in July 2003 as the museum's first A. W. Mellon senior conservation scientist. Since January 2018, she is the Grainger Executive Director of Conservation and Science. With a PhD in chemistry from the University of Milan, Italy, Casadio is also the founder and co-director of the Northwestern University/Art Institute of Chicago Center for Scientific Studies in the Arts (NU-ACCESS). Her research interests are interdisciplinary studies mainly focused on artistic materials and practices of the nineteenth and twentieth century.

Katrien Keune is a research scientist at the Rijksmuseum Amsterdam, The Netherlands. She is responsible for the scientific research in the conservation studios. She also holds an appointment as associate professor of chemistry at the University of Amsterdam (UvA) and contributes to the Netherlands Institute for Conservation, Art and Science (NICAS) at a scientific and organizational level. She received a PhD in chemistry from FOM Institute AMOLF, UvA. Her expertise lies in the ageing, deterioration and migration processes in oil paintings, especially related to pigment-binding medium interactions.

Petria Noble is senior paintings conservator and Head of Paintings Conservation at the Rijksmuseum in Amsterdam, a position she has held since 2014. Prior to this, she worked for 18 years at the Mauritshuis in The Hague. As an expert in the material aspects and conservation of seventeenth-century Dutch paintings, she has published widely in conservation and scientific journals. She is a strong advocate for scientific investigations of paintings as a key to understanding changes in appearance. Her current research activities include the application of noninvasive imaging techniques for the study of late Rembrandt paintings.

Annelies Van Loon trained as a chemist and a paintings conservator, before specializing in the materials analysis and ageing processes of Old Master Paintings. She received a PhD in chemistry from FOM Institute AMOLF, University of Amsterdam on the topic of 'Color changes and chemical reactivity in seventeenth-century oil paintings.' From 2012 to 2016, she was project leader of the Science4Arts *PAinT* Project focusing on aspects of metal soaps and pigment-binder interactions. Currently she holds a position as paintings research scientist at the Rijksmuseum Amsterdam.

Ella Hendriks is full professor of conservation and restoration of moveable cultural heritage at the University of Amsterdam. She holds a bachelor's degree in art history (University of Manchester, UK), a postgraduate diploma in the conservation of easel paintings (Hamilton Kerr Institute, University of Cambridge, UK) and a PhD in technical art history/conservation (University of Amsterdam). From 1987 to 1999, she was head of conservation at the Frans Hals Museum in Haarlem and, from 1999 to 2016, senior paintings conservator at the Van Gogh Museum in Amsterdam.

Silvia A. Centeno is a research scientist at the Metropolitan Museum of Art where her main responsibilities include the investigation of artists' materials and techniques and deterioration processes in paintings, photographs, and works of art on paper. She has published and lectured on a number of topics, including soap deterioration in oil paintings, pigment- and platinum-based photographic processes, daguerreotypes, early lithographic inks, and nineteenth-century and Renaissance paintings. She received a PhD in chemistry from Universidad Nacional de La Plata, Argentina.

Gillian Osmond is a paintings conservator at the Queensland Art Gallery/Gallery of Modern Art in Australia with three decades of industry experience. She has a long-standing interest in the technical examination of paintings and paint cross sections and has been studying incidences of zinc soaps in paintings for more than 15 years. In 2014, Gillian was awarded a PhD from the University of Queensland for her research on zinc oxide-centred deterioration in artists' oil paints and consequences for paintings.

Chapter 1
A Brief History of Metal Soaps in Paintings from a Conservation Perspective

Petria Noble

Abstract Drawing on past research and the pioneering case studies of metal soap degradation published more than 15 years ago, this paper reviews the early findings of metal soap-related degradation in seventeenth-century oil paintings carried out by the MOLART and De Mayerne group of researchers in the Netherlands. The various manifestations of lead soap alterations identified are described: aggregates and associated pinpoint losses and texture alterations, insoluble efflorescent crusts, and darkening due to increased transparency from saponification of lead white paints. Mention is made of degradation phenomena associated with other metal soaps, potassium, calcium, copper, aluminum, and especially zinc soaps. A brief history of metal soaps is also presented. Finally, the essay looks at the implications for conservation in the light of recent advances in our understanding of the underlying mechanisms of metal soap formation, and discusses the challenges currently faced in treating and caring for works of art affected by metal soap alterations.

1.1 Introduction

Although conservators have long observed and treated deterioration problems in paintings, little have they known that metal carboxylates, or metal soaps as they are known, are to blame for many seemingly disparate but related phenomena.[1] It is more than 20 years ago that I first became aware of metal soaps when in 1996–1998, during the examination and treatment of Rembrandt van Rijn's *The Anatomy Lesson of Dr Nicolaes Tulp* (Mauritshuis, The Hague), the paint surface was found to

[1]Interview with Sarah Everts during the Metal Soaps in Art conference: "Art conservators struggle with microscopic eruptions in masterpieces," May 23, 2016. http://cen.acs.org/articles/94/i21/Art-conservationists-struggle-microscopic-eruptions.html

P. Noble (✉)
Conservation Department, Rijksmuseum, Amsterdam, The Netherlands
e-mail: p.noble@rijksmuseum.nl

F. Casadio et al. (eds.), *Metal Soaps in Art*, Cultural Heritage Science,
https://doi.org/10.1007/978-3-319-90617-1_1

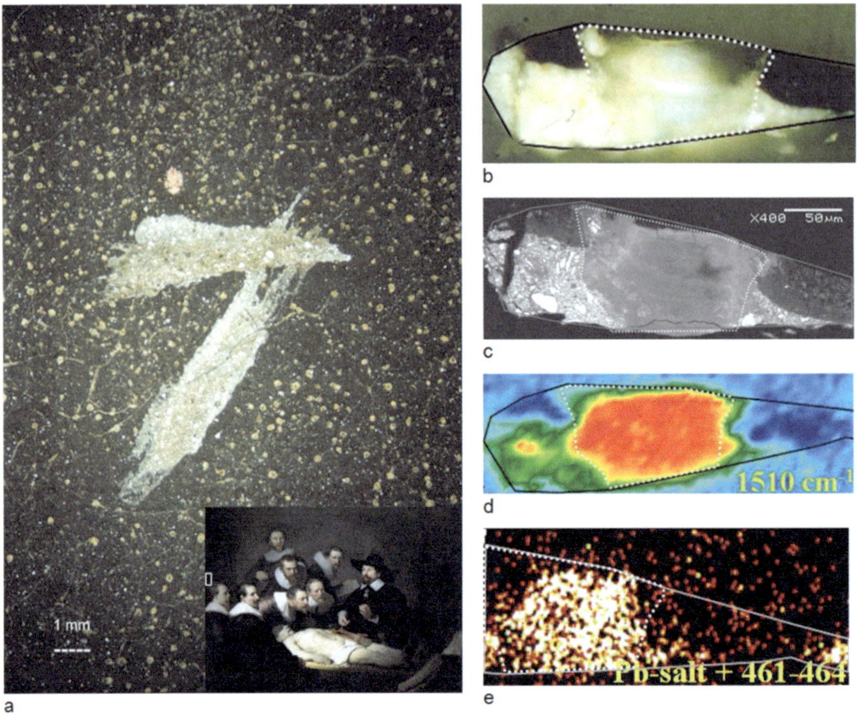

Fig. 1.1 Rembrandt van Rijn (1606–1669), *The Anatomy Lesson of Dr Nicolaes Tulp*, 1632, canvas, 169 × 216.5 cm, Mauritshuis, inv. 146. Here protruding lead soap aggregates have resulted in minuscule paint losses over the entire surface. Overall (insert) and microscope detail of brown background at upper left (**a**). Light microscopy and chemical imaging made it possible to identify the lead soaps in the paint cross sections. Cross sections of the lumps revealed an origin in the upper light grey ground layer (**b**). Corresponding SEM-BSE image demonstrates a striking difference in granularity between aggregate mass and surrounding ground and paint layers (**c**). FTIR imaging in reflection mode shows strong absorption for lead carboxylates at 1510 cm^{-1} (**d**), and secondary ion mass spectroscopy (SIMS) image at *m/z* 461–464, corresponds to the $CH_3(CH_2)_{14}COO\text{-}Pb^+$ ion of lead palmitate (**e**) (Heeren et al. 1999)

be riddled with minute crater-like holes (100–200 µm in diameter)[2] (Fig. 1.1a). The systematic study of the painting that followed made clear that the holes were caused by the formation and subsequent loss/abrasion of lumps of lead soap aggregates that had protruded up through the paint surface causing minuscule paint losses. Over time, as the protruding lumps were lost or abraded, crater-like structures containing residual lead soap is all that remains. In 2002, we coined the term "protrusions" to describe this phenomenon, which is still the most common term used today by

[2]Appearing as light spots on the paint surface. When darkened varnish accumulates in the crater-like holes, they can appear as dark spots.

conservators to denote metal soap aggregates that break through the surface of the paint (Noble et al. 2002: note 4).

Drawing on past research and our many published case studies, this essay seeks to address the importance of scientific research for conservation practice. In the last two decades, the results of numerous research initiatives concerning metal soaps have significantly advanced our understanding of the phenomena and chemistry of metal soap-related degradation in oil paintings, which can visually manifest itself in various ways including aggregates, efflorescent crusts, crystalline exudates, increased transparency, as well as pinpoint losses, delamination, and cracking. Such defects have been characterized in thousands of paintings and are now understood as part of the dynamics of the aging of oil paint that form from the interaction between carboxylic groups on fatty acids and other organic constituents in the oil binder, and metal ions from the pigments and driers. Disfiguring lumps of metal carboxylates have also been found to form in egg tempera paints and varnishes, due to the presence of fatty acids in the binder and acid groups in resins. Defects associated with both lead and zinc soaps are the most common, although degradation phenomena associated with potassium, calcium, copper, and aluminum soaps are increasingly being identified. Although oil-based paintings and painted objects from all periods can be affected by some form of metal soap-related degradation, modern paintings are by far the worst affected.

1.2 Past History

According to historical sources, metal soaps have been known and used since antiquity (see Cotte et al. 2019, 85–106). The earliest use of the term metallic soap appears to be in the second half of the eighteenth century. The French chemist Michel Eugène Chevreul (1786–1889), whose pioneering investigations constituted a significant scientific contribution to the chemistry of fats and oils (Chevreul and France 1850), is credited with first describing the process of saponification in 1823 (Corkery 1998). Toward the middle of the nineteenth century, the use of lead white and particularly lead-based driers was suspected of playing a role in the degradation of paintings. Carlyle discusses a number of nineteenth-century writers, including Jean Francois Leonor Mérimée (1757–1836), whose comprehensive book *The Art of Painting in Oil* from 1830 (translated into English in 1839) describes problems caused by the practice of adding litharge to ground layers to assist with drying:

> . . . it has been discovered that pictures painted on these sorts of preparations, in a few years have their surfaces covered with a multitude of little grains like sand . . . (Mérimée 1839: 219).

Likewise, George Field (1777–1854), an English color manufacturer and chemist, stated in his popular treatise first published in 1835 that an excess of driers renders oil "saponaceous" and that this "will ultimately effloresce on the surface and throw off the color in sandy spots . . . " (Field 1869). Arthur Herbert Church in his chapter on *Siccatives or Dryers* also described a phenomenon that appears consistent with what is now known as lead soap aggregates (Carlyle 2001:9–13, 41–54, 172).

> . . . I have seen one of the results of this comingling of sugar lead [lead acetate drier] with the medium or the paint in the production of an immense number of small spots in the picture, sometimes appearing through the surface-varnish in the form of a white efflorescence. . . . When oil is left in contact with them [lead driers], and especially when heat is applied to the mixture, some of the lead dissolves, forming, with the fatty acids of the oil, lead soaps. (Church 1890:93–94).

By the end of the nineteenth century, numerous authors had expressed concerns about the reactivity of lead white and lead driers and possible degradation problems. In the 1892 book by H.C. Standage entitled *The Use and Abuse of Colours and Mediums in Oil Painting*, it is stated that the lead ground is liable to react chemically with paints applied over it. As a result, it was advised that lead white films should be covered with a layer of zinc white or varnish, and precautions were taken to prevent the profusion of air pollutants from behind the canvas (Carlyle 2001:45, 260–61). It was due to these concerns, as well as toxicity issues, that over the course of the late nineteenth and twentieth centuries, lead white was gradually superseded by zinc white.

In the early twentieth century, the formation of lead soaps was also blamed for the increased transparency in oil paints. This was the conclusion of the German chemist Alexander Eibner, who in 1907 became Director of the Research Institute and Information Office for Painting Techniques (*Versuchsanstalt und Auskunftsstelle für Maltechnik*) in Munich, later known as the Doerner Institute (Eibner 1909:121). The British chemist A.P. Laurie (1861–1949), who also carried out experiments on this subject in the 1920s and 1930s, came to the conclusion that the phenomenon was due to an increase in the refractive index of the oil binding medium with age (Laurie 1926:110, 140–55). Interestingly in the classic book, *Painting Materials: A Short Encyclopaedia*, first published in 1942 by Rutherford J. Gettens and George L. Stout from the Fogg Art Museum, Eibner's conclusion is cited in this regard (Gettens and Stout 1966:175). Unfortunately, it was Laurie's interpretation that prevailed in most of the English language conservation literature right up until the end of the twentieth century when our renewed research on metal soaps demonstrated that saponification of the paint was largely responsible for the increased transparency in artists' paints (Noble et al. 2005, 2008b; Shimadzu et al. 2008; Van Loon 2008).

Metal soaps were also discussed in the industrial paint literature at the beginning of the twentieth century. They became essential ingredients in paint formulations to aid in the grinding and dispersion of pigments (Tumosa 2001). Defects were also reported: the phenomena of "pinholing" in paint films from lead soaps and the embrittlement of paint associated with zinc soap formation are discussed for instance, by Hess. He also describes the anticorrosive effect of metal soaps formed with alkaline pigments in oil media (Hess 1965).

In the 1970s Léopold Kockaert, a conservation scientist at KIK-IRPA in Belgium, investigated minuscule fluorescent lumps (100–200 μm) in lead-tin yellow paint layers of the Flemish primitives – including works by Jan van Eyck. Using microchemical staining tests, the lumps were erroneously interpreted as proteinaceous in nature, suggesting the use of an oil-egg emulsion binding medium (Spring and Higgitt 2006). It was only with the development of sophisticated analytical techniques in the 1980s, particularly Fourier transform infrared (FTIR) microscopy, that it became possible to analyze such microscale features. In the late 1990s, similar fluorescent lumps in the ground of a seventeenth-century painting that had been previously interpreted as proteinaceous lumps, were identified with FTIR microscopy as metal carboxylates (soaps) (Plahter 1999). The lumps observed by Kockaert in paints in early Netherlandish paintings were also later identified as lead soap carboxylates (Higgitt and Plater 2004). Scientists at the National Gallery in London also discovered that the presence of metal soaps can give rise to false-positive results with staining tests for protein and can also alter the fatty acid ratios traditionally used for the identification of oils and proteins leading to misinterpretation of the nature of the binding medium (Higgitt et al. 2003; Spring et al. 2005; Spring and Higgitt 2006).

In the 1980s and 1990s, a number of significant publications reported on the identification of fatty acids and metal soaps in whitish crystalline efflorescence on the inside of the glass of glazed paintings, as well as on the surface of paintings, painted sculptures, and a set of modern wood-block prints (oil paint on paper), dating from the late nineteenth through the twentieth centuries from the collections of the Museum of Modern Art in New York and the Los Angeles County Museum of Art (Williams 1989; Ordonez and Twilley 1998). Several hypotheses were put forth at the time to explain the migration of materials out of the paint that ranged from phase separation to the low glass transition temperature (Tg) of the oil paint. These studies were an important step forward revealing further research was needed to better understand the degradation phenomena and the mechanisms involved. In 1999 a study of various manifestations of whitish surface deposits showed that a wide range of different organic and inorganic compounds could be identified on paintings (Skaliks 1999).

1.3 Lead Soap Aggregates

Although the tiny crater-like holes that we observed in Rembrandt's *The Anatomy Lesson of Dr Nicolaes Tulp* were just a few hundred microns in diameter, at the time we were concerned that they could negatively influence the cleaning of the painting and that they would eventually undermine the picture's stability. Before research was undertaken, various causes were entertained including an earlier treatment with copaiba balsam. Close inspection of the X-radiograph showed that the crater-like holes and lumps corresponded with tiny black spots spread over the entire surface pointing to an origin in the ground layer.[3] The holes had been observed in the 1970s, but at that time it was concluded that they were caused by the painting's exposure to heat from the numerous linings or possibly the fire in the surgeons' guild chamber of 1723 (De Vries et al. 1978:89). Comparison with paintings known to have been damaged by this fire, however, made clear that the nature and order of magnitude of the holes in *The Anatomy Lesson of Dr Nicolaes Tulp* was very different (Noble et al. 2002:49).

Thanks to the MOLART (Molecular Aspects of Aging in Painted Works of Art) (1995–2002) and De Mayerne research programs (2002–2006) sponsored by the Netherlands Organization for Scientific Research (NWO), and especially the contribution of professor emeritus Dr. Jaap Boon (FOM Institute AMOLF) who masterminded and spearheaded both research programs, we went on to analyze and characterize the holes as being caused by lead soap aggregates formed as a result of saponification of the lead-containing paint. The application of (at the time) novel imaging FTIR in reflection mode made it possible to identify and localize the carboxylate functional group associated with lead soaps *in the paint cross section* by its characteristic IR absorption in the 1510–50 cm^{-1} region. Summarized results appeared in the 1998 exhibition catalogue (Middelkoop et al. 1998) that accompanied the presentation of the newly restored Rembrandt painting, and full scientific results were presented at *Art et Chimie*, *La Couleur* conference in Paris in 1998 (Noble et al. 2000) and *ICOM-CC* conference in Lyon (Heeren et al. 1999). Our push for a scientific explanation meant we had put an end to the speculation of what the holes were. The schematic drawing in Fig. 1.2, presented at the ICOM-CC conference in Rio de Janeiro in 2002, summarized our observations on lead soap formation (Boon et al. 2002).

Together with Jaap Boon and the team of MOLART researchers, we went on to investigate many other paintings that exhibited the same phenomena. From our initial studies, we concluded that the presence of lead soap aggregates in works by different artists with different provenances and material histories pointed to commonalities in the composition of the paint or the ground (Noble et al. 2002). In this period, in addition to imaging FTIR, another advanced analytical imaging technique also became available: secondary ion mass spectroscopy (SIMS). Combined with elemental information obtained with scanning electron microscopy with energy dispersive X-ray analyses (SEM-EDX), these chemical imaging techniques made it possible to identify and localize the metal (SEM-EDX, SIMS), metal carboxylate groups (imaging FTIR), and associated fatty acids (SIMS) of the metal soaps in paint cross sections. SIMS revealed that the organic fraction of lead soap masses inside the agglomerates mainly consisted of palmitic and stearic acids (saturated fatty acids), and in a few cases also azelaic acid, clearly indicating an origin from the oil binding medium of the paint (Keune 2005).

Around the same time, the Canadian Conservation Institute (CCI) also identified protruding lumps of lead soaps in nineteenth-century paintings (Robinet and Corbeil

[3]Aggregates can appear as either dark or light spots in X-radiographs depending on the radio-absorbency of the layer in which they are formed.

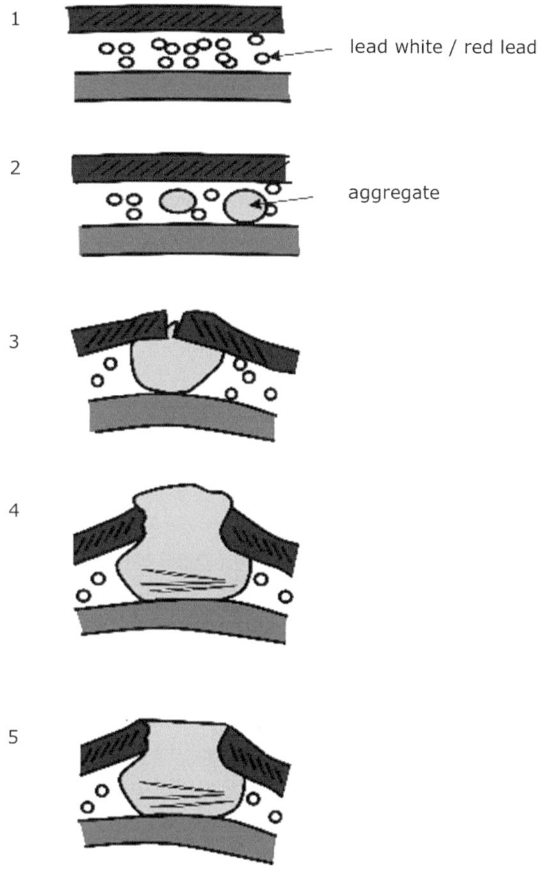

Fig. 1.2 Schematic diagram showing the various stages of lead soap aggregate formation in an intermediate layer. Stages 1–5: (1) intact paint, (2) early stage with small aggregates, (3) expansion of the mature aggregate leading to eruption through the paint surface, (4) protruding aggregate with remineralization, and (5) late stage with mature aggregate decapitated after repeated cleaning. (From Boon et al. 2002:405)

2003). At the Centre for Research and Restoration of the Museums of France (C2RMF) in France, metal soaps in ancient cosmetics were also being studied. *Lithargeage*, the phenomenon of small inclusions protruding out of the paint, has been observed by restorers in France since the 1970s and is traditionally attributed to the use of lead-based driers (Cotte et al. 2007, 2017). Scientists at the National Gallery in London also began a systematic study of metal soap aggregates in red lead- and lead-tin yellow-containing paints (Higgitt et al. 2003; Spring et al. 2005).

1.3.1 International Survey

"Protrusions in oil paintings are a worldwide problem" was the headline of the press release from the Netherlands Organization for Scientific Research after we presented our findings at the "Highlights of MOLART 1995-2002" symposium that took place on the June 19, 2003 (Clarke and Boon 2003). During the follow-up

De Mayerne research program (2002–2006), an international survey we sent out to museums around the world brought to light thousands of examples of paintings with metal soap-related degradation problems by different artists, from different periods, on all kinds of supports including canvas, panel, and copper, with a wide range of conservation and exhibition histories including both treated and untreated works. Only a small proportion of reported cases were analyzed, but in each of the cases studied, metal soap aggregates were identified as having originated in lead- or zinc-containing grounds or paint layers. Research carried out in the National Gallery in London also showed that lead soap aggregates were frequently encountered in red lead-containing preparatory layers of Medieval wall paintings (Higgitt et al. 2003). Similarly, results of the survey showed that in old master paintings, aggregate formation was found to occur most often in preparatory layers and in paint layers containing lead-tin yellow type 1 and red lead, and to a lesser extent in layers containing lead white. Interestingly, lead soap aggregates are not found in pure, densely packed lead white used for instance, for highlights. And where large aggregates are formed in lead white layers, it is still unclear whether this is due to the presence of lead driers/red lead, or the ready availability of fatty acids. In lead-tin yellow surface paints, lead soap aggregate formation was found to cause an otherwise opaque yellow paint to have a pronounced bumpy texture and appear translucent and patchy. Metal soap aggregates were also encountered in dark, oil-rich paint layers containing pigments such as chalk, earth pigments organic lakes, and bone black, pointing to the presence of lead-based driers, particularly red lead, or to the migration of lead from lead-rich underlayers, the oil-rich layer providing a source of mobile reactive fatty acids. From this it was concluded that the paint layer buildup plays an important role in the availability and quantity of fatty acids. Only a few occurrences of lead soaps in early Italian paintings in egg tempera were recorded, also likely connected to the availability of fatty acids. Many examples of zinc soap-related degradation in nineteenth- and twentieth-century paintings came to light, including several works by Vincent van Gogh where the extrusion and eruption of highly mobile zinc soaps severely embrittled the paint (Van der Weerd et al. 2003). In many of the reported paintings, metal soap aggregates, increased transparency, and efflorescence were found to occur simultaneously (Noble et al. 2005).

In addition to protruding aggregates and associated paint loss, examples of extreme texture alterations were reported that had sometimes been misinterpreted in the past as deliberately textured grounds or paints. A notable example is John Singer Sargent's *Madame X*, 1883–1884 (The Metropolitan Museum of Art, New York), where lead soap aggregates had formed under the paint surface of the black gown creating an unintentional surface texture. Analyses of a paint cross section revealed that the bumpy texture was due to the presence of globules of lead soaps in the ground layer (Mahon and Centeno 2005). Another important example is Johannes Vermeer's *View of Delft*, c. 1665 (Mauritshuis, The Hague), where for years it was thought that the whitish lumps in the red paint of the rooftops were the result of the inclusion of sand or coarse lead white particles added by the artist for a deliberate textural effect. This notion, which was purely based on visual inspection, made

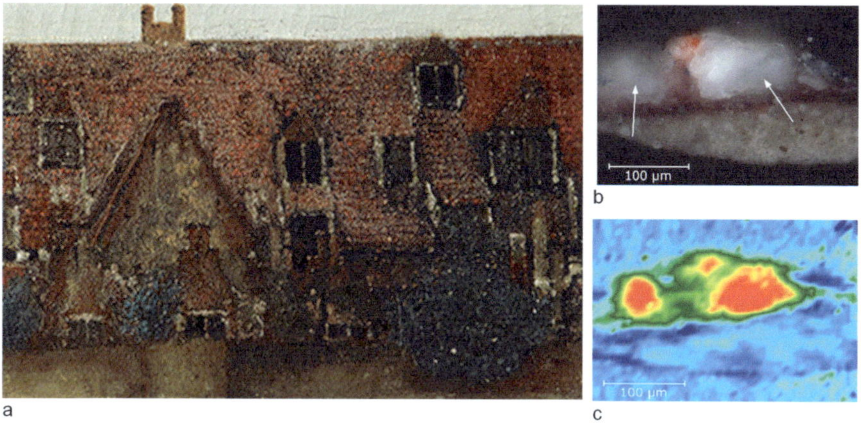

Fig. 1.3 Detail of the red rooftops, Johannes Vermeer, *View of Delft*, c. 1660–61, canvas, 96.5 × 115.7 cm, Mauritshuis, inv. 92, showing protruding lead soap aggregates. In the past, these were interpreted as sand or coarse lumps of lead white added by the artist for a deliberate textural effect (**a**). Paint cross section from red roof showing beige ground and red lake paint of the roof in which the lead soap masses have formed and erupted through the overlying green blue paint of the foliage (**b**). That the lumps (primarily) consist of lead soaps was demonstrated with FTIR imaging in reflection mode of the same cross section. The red areas in the false color FTIR map correspond to areas of high intensity of the lead carboxylate $\nu_{COO_{as}}$ vibration (**c**) (Van Loon 2009)

newspaper headlines in the *NRC Handelsblad* (April 2, 1994) when the painting was treated, thereafter appearing in much of the art historical literature.

> ... The meticulous way that Vermeer worked on this masterpiece is shown by the fact that he mixed grains of sand into some of his paint to achieve a certain texture... (Bailey 1995:60–62).

In 2008, the picture was brought to the conservation studio. From the characteristic appearance, and deformation of the paint surface, microscopic examination made clear that the protruding white lumps in the red rooftops were lead soap aggregates (Fig. 1.3). The identification of lead soaps was subsequently confirmed by light microscopic, SEM-EDX, and FTIR analyses of a paint cross section (Noble et al. 2008a; Van Loon 2009).

1.4 Other Lead Soap Manifestations

Efflorescent Hazes and Crusts

In 2005 during the treatment of two other Rembrandt paintings in the Mauritshuis we went on to identify another form of metal soap-related degradation: efflorescent hazes and crusts. An insoluble whitish haze consisting of fine dense particles of complex salt mixtures could be discerned with the naked eye in the reddish black

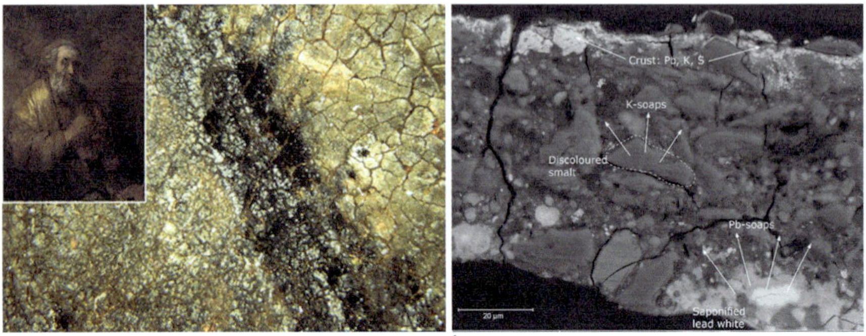

a b

Fig. 1.4 Rembrandt van Rijn (1606–1669), *Homer*, 1663, canvas, 107 × 82 cm, Mauritshuis, inv. 584. Overall (insert) and microscope detail of the insoluble crust in black area of drapery at lower edge (**a**). SEM backscatter image of paint cross section (**b**) showing the dissolution and saponification of lead white in the upper ground layer, depletion of potassium from smalt particles, formation of potassium soaps, and deposition of highly scattering Pb-S-K-rich deposits at the paint surface (Van Loon et al. 2011, 2012a). In this painting the insoluble crust has formed at and below the paint surface and cannot be removed without damage to the original paint

curtain in the upper right of Rembrandt's *Simeon's Song of Praise*, 1631 (not illustrated). It was only at 100–200× magnification that the white crystals associated with this haze could be resolved (Noble and Van Loon 2007). A year later, when investigating Rembrandt's *Homer* from 1663, disturbing whitish surface deposits were found to cover most of the smalt- and lake-rich dark yellow/brown surface paint. SEM-EDX analysis, supplemented with FTIR and Raman, demonstrated that the insoluble crust consists of lead-potassium-sulfate salt(s), and calcium oxalates. Here it was postulated that potassium and lead ions from the smalt-rich paint and the lead white-containing (upper) ground had migrated to the paint surface where they underwent further reactions over time with sulfur from atmospheric compounds. Importantly, BSE images showed that the crust had formed at and below the paint surface and that it could not be removed without risk of damage to the original paint (Keune and Boon 2007; Spring et al. 2005; Van Loon et al. 2011, 2012a; Van Loon 2008) (Fig. 1.4). Since then, various degradation products, such as lead-potassium-sulfates, lead sulfates and lead and calcium oxalates, have been regularly identified on the surface of historical paintings (Spring and Higgitt 2006; Boon and Oberthaler 2010; Van Loon et al. 2011, 2012a, 2019a, 285–297; Klaassen et al. 2019, 265–282). Several degradation processes are now considered to be involved in the formation of these alteration products, including metal soap formation.

Increased Transparency

Examination and treatment of several other paintings in the Mauritshuis also led to important insights regarding the relationship of increased transparency and metal soaps. In Roelandt Savery's *Orpheous enchanting the animals with his music*, c.1620 (Mauritshuis, The Hague), it was shown that saponification of the lead white particles in the brown foreground paint had caused a visible color shift from opaque medium brown to dark brown, distorting the light/dark contrasts in the painting to the extent that many of the small animals and birds in the foreground are no longer visible. Here it was concluded that the loss of opacity of lead white is most dramatic in medium to dark paint layers leading to an overall darkening in tone that can have an enormous visual impact on the appearance of paintings (Noble and Boon 2007; Noble et al. 2005; Van Loon 2008; Van Loon et al. 2012b).

We also came to realize that increased transparency can manifest itself in many ways depending on the physical structure of the support. Aert van der Neer's *River Landscape*, c. 1650 (Mauritshuis, The Hague), is an example of selective darkening associated with the ground-filled deep channels of the early wood of oak panels (Fig. 1.5). Using light microscopy and SEM-EDX, we demonstrated that the disturbing lines correspond to localized saponification of the overlying lead white-rich beige priming layer. Lead soaps having a lower refractive index, closer to that of oil than that of lead white, explains the increased transparency/darkening observed (Robinet and Corbeil 2003). The expansion of affected areas also resulted in loss of the overlying surface paint. In this painting, the free fatty acids required for saponification are provided by the oil-soaked chalk ground present in the deep pores of the wood grain. The resulting loss of reflectance of the chalk ground together with increased transparency of the saponified priming layer creates the localized darkening effect since light can penetrate more deeply in these areas (Fig. 1.6). That this phenomenon is frequently observed in Dutch panel paintings is linked to the use of coarse-grained oak panels, the lead white-rich priming layers, and an economical painting technique involving one or two paint layers. In the past, such darkened streaks have been variously interpreted as abrasion or as a deliberate part of the pictorial design (Noble et al. 2005; Noble et al. 2008b:76; Van Loon 2008; Van Loon et al. 2012a).

A comparable phenomenon related to the physical structure of the painting support, referred to as "ground staining" or "variable translucency," was observed by Zucker in numerous nineteenth-century canvas paintings, including those by the American nineteenth-century artist, Frederic Church (1826–1900). In many of these paintings, dark vertical lines and patchiness have produced disturbing patterns never intended by the artists. In Church's oil sketches, both lead soap aggregates and efflorescence were identified. Considering the lead white in the ground is intact (and not saponified), the degradation can be attributed to the excessive use of a lead drier (lead acetate), unfortunately no longer detectable since it would have reacted away (Boon et al. 2007; Keune and Boon 2007; Zucker and Boon 2007).

These studies, which were presented at the American Institute for Conservation (AIC) annual meeting held in Providence, Rhode Island in 2006 and subsequently published in the Paintings Working Group postprints, helped create awareness of metal soap-related degradation within the conservation community (Boon et al. 2007; Jones et al. 2007; Noble and Boon 2007; Zucker and Boon 2007). Moreover, they show how the different factors, the composition of the paint, the paint layer buildup, and the physical structure of the support, along with conservation history and environmental conditions, contribute to different manifestations of metal soaps.

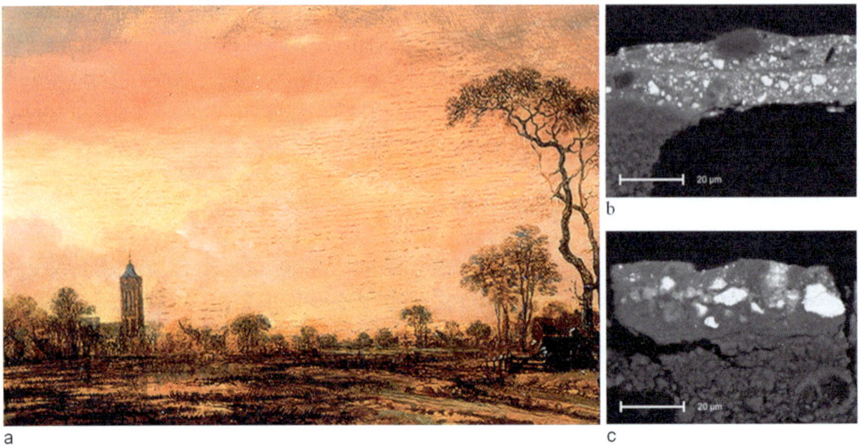

Fig. 1.5 Aert van der Neer, *River Landscape*, c.1650, panel, 44.8 × 63 cm, Mauritshuis, inv. 912. Detail from center right sky showing dark streaks due to selective darkening associated with the wood grain (**a**). SEM backscatter images of paint cross section from unaffected sky (**b**) and from dark streak where the sky paint layer has been lost due to volume increase of saponified priming layer below (**c**). The bottom layer in both cross sections is the chalk ground and the layer above it, the saponified lead white containing beige priming layer (Noble et al. 2005)

1.5 Zinc, Potassium, Calcium, Copper, and Aluminum Soap Manifestations

Zinc soaps Zinc-containing pigments and driers readily react with fatty acids in oil-based paints to form zinc carboxylates. Severe cleavage, paint loss, disfiguring lumps, increased transparency, and surface efflorescence are just some of the problems associated with zinc soaps. Detailed case studies include paintings from the late nineteenth and twentieth centuries (Maines et al. 2011; Osmond 2014, 2019, 25–43; Rogala et al. 2010; Rogala 2019, 317–330; Shimadzu et al. 2008; Van der Weerd et al. 2003). As they crystallize zinc soaps can cause serious delamination – both between and inside paint layers – resulting in the loss of structural integrity of the paint (Van Loon et al. 2019b). Recently Hermans has demonstrated two different phases of zinc soap formation: amorphous zinc soaps, considered to be an early or intermediate stage, and crystalline zinc soaps, the final stage in the degradation process that causes delamination and increased brittleness of the paint (Hermans et al. 2015, 2019, 47–65; Raven et al. 2019, 345–357).

Potassium soaps Are known to form in the oil binding medium of smalt-rich paint layers as a result of the leaching of potassium from the smalt pigment particles, and are largely responsible for the brown appearance of the paint matrix and surface whitening effects. Comparable degradation products have been identified on the surface of historic glass (Spring et al. 2005).

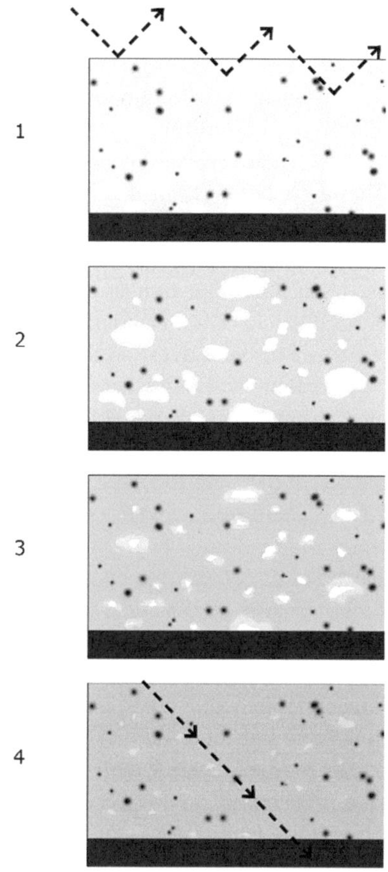

Fig. 1.6 Schematic diagram of a paint cross section explaining increased transparency (darkening) due to saponification of lead white in the paint. Stages 1–4: (1) intact paint, light is scattered; (2) early stage, the small particles are reacting away, and edges of the large particles are dissolving; (3) mature stage, the small particles have reacted away, and large particles are dissolving; and (4) final stage, nearly all lead white has dissolved, and light is absorbed (Van Loon 2008)

Calcium soaps Although more rarely identified in oil paint than lead or zinc soaps, calcium soaps can form aggregates that disrupt the paint surface, or result in a brown discoloration (Ferreira et al. 2011; Helwig et al. 2019, 299–312).

Copper soaps The formation of copper soaps due to the reaction of fatty acids in oil paint with copper pigments and the copper support has long been known. Copper carboxylates are always present for instance, in verdigris in oil and in the green reaction layer that forms between fatty acids in oil paint with the copper support (Horovitz 1999; Gunn et al. 2002; Pavlopoulou and Watkinson 2006).

Aluminum soaps The addition of metal soaps (aluminum, zinc, and magnesium stearates) included by manufacturers in modern paints as dispersion agents, stabilizers, and extenders, together with those formed in the paint by the reaction of the pigments and fatty acids is considered to contribute to the problem of water sensitivity in unvarnished modern paintings (Burnstock and Van den Berg 2014; Osmond 2019, 25–46; Van den Berg et al. 2019, 331–343).

1.6 Time Frame

An important question frequently asked by conservators is the time frame in which metal soaps aggregates form as this may help date later interventions, such as overpaint and varnish. In some cases, degradation phenomena can provide insight. The presence of abraded protrusions below a paint or varnish layer, for instance, implies that the overlying varnish or paint layer cannot be original. In a recent study of a painting by Matthijs Maris in the Rijksmuseum, the presence of abraded lumps established that the varnish, initially suspected of having been applied by the artist himself, could not be original (Iddi 2015). How quickly such defects form, however, remains a difficult question to answer as this depends on several factors.

Research involving model systems that mimic lead soap degradation processes in oil paintings has shown that lead soap aggregates can form within 1 year in oil paint reconstructions (Van Loon et al. 2016). Cotte also demonstrated that depending on the reactive species involved, aggregates can form in a few weeks (Cotte et al. 2007, 2017). The speed at which this occurs, however, depends on several factors: the composition/nature of the paint, the paint layer buildup (availability of fatty acids), as well as environmental conditions.

The degree of (re)mineralization within the aggregates is also informative. In Rembrandt's *The Anatomy Lesson of Dr Nicolaes Tulp*, the identification of lead carbonates, chlorides, and oxides within the lead soap aggregates, visible in BSE images as lamellar precipitation bands, indicates the lumps have formed over a long period of time and that the process has reached a stable end stage. The nature of the remineralization will vary from painting to painting, and possibly area to area, depending on the composition of the paint and environmental pollutants. The remineralized compounds also account for the relative hardness and opacity of the lead soap aggregates and are the reason why lead soaps are more stable in comparison with the softer and more mobile zinc soaps. Depending on the environmental conditions, the aggregates can remain active. There are examples where they have broken through later additions, including the tiny numbers added in the eighteenth century to Rembrandt's *The Anatomy Lesson of Dr Nicolaes Tulp* (Fig. 1.1a) and areas of modern retouching in Berchem's *Allegory of Summer*, c.1680 (Fig. 1.7). External factors in the case of *The Anatomy Lesson of Dr Nicolaes Tulp* include the numerous treatments (including five linings) the painting has undergone in the past involving the use of water, solvents, and heat. The picture has also been subjected to temperature gradients in the galleries, especially since the 1980s when air-conditioning was introduced. In the past, heating of the galleries with coal stoves would have had a similar effect in the winter months, also exposing the painting to atmospheric pollutants, in particular sulfur dioxide.

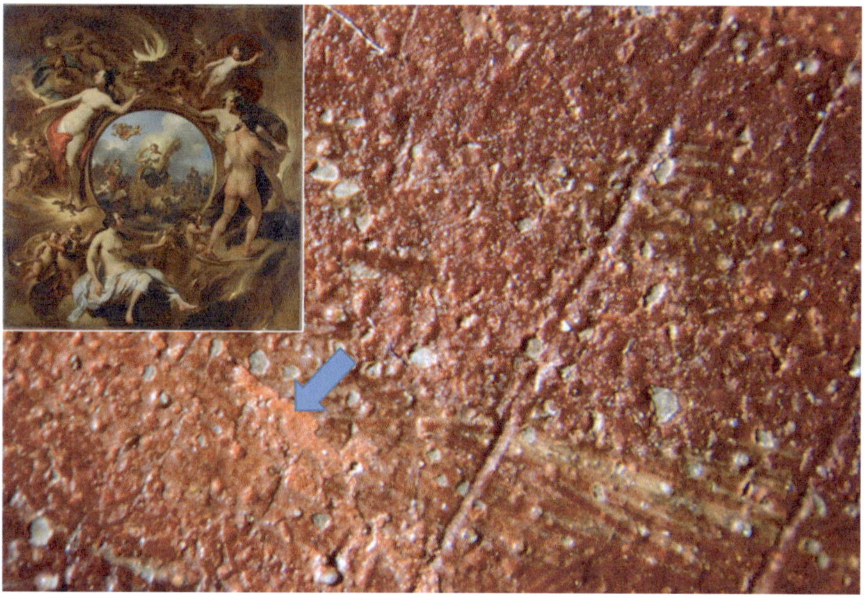

Fig. 1.7 Nicolaes Berchem (1620–1683), *Allegory of Summer*, c.1680, canvas, 94 × 88 cm, Mauritshuis, inv. 1091. Here lead soap aggregates, 100–200 μm in diameter, have broken through two layers of modern overpaint (indicated by arrow), causing masses of pinpoint paint losses. Overall (insert) and microscope detail from red drapery of standing figure left of center (Noble et al. 2005)

1.7 Conservation Perspectives

Given the enormous variation in the visual manifestations, conservators should be alert that even commonly encountered phenomena such as pinpoint losses (Fig. 1.7), delamination, and even cracking may be related to metal soap degradation. Certainly, defects associated with modern paintings present the greatest challenges. Dripping paintings and water-sensitive oil paint (in some cases), are now also thought to be associated with metal soaps (Burnstock 2019, 245–262).

The necessity for diagnostic procedures and chemical analyses is therefore much greater than ever before. Microscopic examination of the paint surface, prior to and during cleaning, is recommended, and where possible, analyses of paint cross sections with optical microscopy and other analytical/chemical imaging techniques, particularly scanning electron microscopy (SEM) and ATR-FTIR.

The practical and ethical challenges conservators currently face are huge. Should the reduction or removal of efflorescent surface crusts be carried out knowing they have formed from the original painting materials? What is the best method for consolidation of inter- and intralamellar delamination caused by zinc soaps? Will the use of heat- and aqueous-based adhesives trigger further degradation? Should paintings with a history of metal soap degradation be stored and displayed under stricter environmental conditions? Should paintings have restricted viewing like other vulnerable works of art, such as photographs or watercolors? What additional preventative measures should be taken to prevent the risk of further deterioration? Another important question is whether the removal of metal soap crusts will trigger the formation of new metal soap phases. Hermans suggests this is not the case since the conversion of amorphous phase metal soaps to a crystalline phase is not an equilibrium reaction. This means that the removal of crystalline material will not affect the equilibrium in the paint between free carboxylate groups, metals and amorphous metal soaps (Hermans 2017; Hermans et al. 2019, 47–65).

Treatment considerations and a range of approaches are discussed in the case studies presented by Burnstock (2019, 245–262), Higgs and Burnstock (2019, 123–140), Klaassen et al. (2019, 265–282), and Van Loon et al. (2019a, 285–297). Metal soap efflorescent hazes and crusts, which are some of the visually most disturbing and challenging, can sometimes be thinned or removed with an aqueous gel, but this will depend on the nature of the surface deposits and the degree to which they have become imbibed in the paint surface, as well as the sensitivity of the underlying paint. Great caution must be exercised in the choice and use of cleaning agents. In the case of inorganic crusts that have become intimately bound with the original paint, it must be realized that such layers cannot be removed without risk of damage to the original paint. This underscores the necessity of FTIR (or Raman) and SEM-EDX analyses of paint samples to characterize the surface deposits and to determine the nature of the interface between the deposits and the paint surface. Test panels can be prepared to gain more information and to help develop a safe working method (Van Loon et al. 2019a). For some paintings, mechanical removal may be possible. A film consisting of zinc soaps was successfully reduced using vinyl and electric erasers (Keune and Boéve-Jones 2014). In all cases, the risk of damage to the original paint versus aesthetic gain and long-term stability needs to be carefully weighed.

Both inter- and intralamellar delaminations due to zinc soap formation, are serious problems that in some paintings are still ongoing. In these cases standard approaches to consolidation have to be reconsidered as recent research suggests both aqueous- and solvent-based adhesives should be avoided (Raven et al. 2019). It has also been speculated whether innovative nano-techniques in the future will provide viable treatment options that could help stabilize such paint systems. For dripping paintings, radical measures, such as the addition of metal ions and exposure to ionizing radiation, may provide treatment options in the future (Boon and Hoogland 2014; Bronken et al. 2015).

Conservators should be aware that treatments involving the use of heat, solvents, and penetrating coatings and varnishes may only temporarily mask problems and could actually do more damage in the long term. Potentially high-risk treatments may even contribute to an acceleration of the degradation process. For some paintings treatment is simply not possible, and appearance can be improved by retouching.

Attention to environmental control is important, since changes in humidity are considered sufficient to trigger the crystallization of metal soaps (Hermans et al.

2016). Humidity also influences the rate of hydrolysis of the drying oil, whereby more free fatty acids are released and become available to react with metal ions in the ground and painting layers. Moreover, it is now known that moisture drives the mobility of fatty acids and that water plays a crucial role in the saponification of metal ions and fatty acids. This is important considering that recent studies have shown that metal ions are able to migrate through the paint, moving from one carboxylate group to another, and that the rate at which this occurs increases with oil polymer mobility from the swelling of paint films due to the use of solvents or heat (Hermans et al. 2015, 2019, 47–65). The knowledge that the majority of historical paintings have undergone numerous treatments in the past, involving solvents, moisture, and heat, reinforces the generally accepted wisdom that exposure of paintings to these elements should be minimized as much as possible.

1.8 Outlook

Metal soap research is certainly one of – if not the most – significant topics of painting conservation research in the last 20 years. It is thanks to the numerous research initiatives, including the MOLART, De Mayerne and Science4Arts PAinT project (2012–2016) sponsored by the Netherlands Organization for Scientific Research, that we are now starting to understand why metal soap behavior in multilayer oil paint systems is so complex.

Questions remain with respect to the conditions that favor the formation of metal soap-related degradation phenomena, their origin inside the paint layers, the processes involved, and the rate at which they form. Accurate identification of the starting materials presents challenges, for example, when two different lead species, such as a lead-based pigment and a lead drier, are present. Here it is anticipated that the further development of advanced analytical techniques will provide breakthroughs in the characterization of metal soaps and the mechanisms involved.

For conservation scientists, it is important to pursue research directions that are of direct relevance for the practicing conservator. More fundamental knowledge is required to determine and understand the consequences of the loss of fatty acids from the paint, the influence of heat and moisture, and the role of solvents. The importance of strict environmental control is also borne out by the knowledge that moisture and water play a crucial role, both in the saponification and hydrolysis processes.

Close collaboration between scientists and conservators is necessary if we hope to develop a more comprehensive understanding of the aging of paint and to develop scientifically supported conservation strategies. Only in this way can we make informed decisions regarding guidelines for treatment and environmental conditions for the display and storage of paintings to ensure their preservation for future generations. Continued research into artists' materials and techniques, along with treatment history and conservation materials used in the past, will also inform this.

Acknowledgments The important contribution of Jaap Boon, Annelies Van Loon and Katrien Keune who carried out much of the analyses reported here is gratefully acknowledged. I am sincerely thankful for their stimulating and productive collaboration throughout the years. I would also like to thank other researchers who took part in the MOLART and De Mayerne programs, particularly Leslie Carlyle, Ron Heeren, and Jaap van der Weerd. The Mauritshuis and the Rijksmuseum are also acknowledged for their support. The research described here was financed by the Netherlands Organization for Scientific Research (NWO, The Hague) through the MOLART (1998–2002) and De Mayerne research (2002–2006) programs.

References

Bailey M (1995) Vermeer. Phaidon, London, pp 60–62

Boon JJ, Hoogland FG (2014) Investigating fluidizing dripping pink commercial paint on van Hemert's seven-series works from 1990–1995. In: Van den Berg KJ, Burnstock A, de Tagle A, de Keijzer M, Heydenreich G, Krueger J, Learner T (eds) Issues in contemporary oil paints. Springer, Cham, pp 227–246

Boon JJ, Oberthaler E (2010) Mechanical weakness and chemical reactivity observed in the paint structure and surface of 'the art of painting'. In: Vermeer, "Die Malkunst": Spurensicherung an einem Meisterwerk, [exh. cat.]. Kunsthistorisches Museum, Vienna, pp 235–253 and 328-335

Boon JJ, Van der Weerd J, Keune K, Noble P, Wadum J (2002) Mechanical and chemical changes in Old Master paintings: dissolution, metal soap formation and remineralisation crystallization processes in lead pigmented ground/intermediate paint layers of 17th century paintings. In: Vontobel R (ed) Preprints of the ICOM committee for conservation 13th triennial meeting, Rio de Janeiro, 22–27 Sept 2002, pp 401–406

Boon JJ, Hoogland F, Keune K (2007) Chemical processes in aged oil paints affecting metal soap migration and aggregation. In: Mar Parkin H (ed) AIC paintings specialty group postprints, Providence, 16–19 June 2006, vol 19, pp 16–23

Bronken I, Boon JJ, Vila A (2015) In search of a solution for softening and weeping paints in contemporary paintings by Tal R and Frank van Hemert. In: Bone L, Clarricoates R, Dowding H (eds) Current technical challenges in the conservation of paintings. Archetype, London

Burnstock A (2019) Taking different forms: metal soaps in paintings, diagnosis of condition and issues for treatment. In: Casadio F, Keune K, Noble P, Van Loon A, Hendriks E, Centeno S, Osmond G (eds) Metal soaps in art: conservation and research. Springer, Cham, pp 245–262

Burnstock A, Van den Berg KJ (2014) The Interface between science and conservation and the challenges for modern oil paint research. In: Van den Berg KJ, Burnstock A, de Tagle A, de Keijzer M, Heydenreich G, Krueger J, Learner T (eds) Issues in contemporary oil paints. Springer, Cham, pp 1–20

Carlyle L (2001) The Artist's assistant, oil painting instruction manuals and handbooks in Britain 1800–1900 with reference to selected eighteenth-century sources. Archetype, London

Chevreul ME, France IND (1850) Recherches expérimentales sur la peinture à l'huile, F. Didot frères

Church AH (1890) The chemistry of paints and painting, 1st edn. Seeley: Service and Co. Limited, London. https://archive.org/stream/chemistrypaints01churgoog#page/n5/mode/2up. Accessed 26 May 2017

Clarke M, Boon JJ (2003) A multidisciplinary NWO PRIORITEIT project on molecular aspects of ageing in painted works of art: final report and highlights 1995–2002. FOM Institute AMOLF, Amsterdam

Corkery RW (1998) Artificial biomineralisation and metallic soaps. Ph.D. thesis, Australian National University, Canberra

Cotte M, Checroun E, Susini J, Walter P (2007) Micro-analytical study of interactions between oil and lead compounds in paintings. Appl Phys A Mater Sci Process 89(4):841–848

Cotte M, Checroun E, De Nolf W, Taniguchi Y, De Viguerie L, Burghammer M, Walter P, Rivard C, Salomé M, Janssens K, Susini J (2017) Lead soaps in paintings, friends or foes? Stud Conserv 62:2–23

Cotte M, De Viguerie L, Checroun E, Susini J, Walter P (2019) Historical evolutions of lead-fat/oil formula from antiquity to modern times in a set of European pharmaceutical and painting treatises. In: Casadio F, Keune K, Noble P, Van Loon A, Hendriks E, Centeno S, Osmond G (eds) Metal soaps in art: conservation and research. Springer, Cham, pp 85–106

De Vries AB, Tóth-Ubbens M, Froentjes W (1978) Rembrandt in the Mauritshuis. Alphen aan de Rijn

Eibner A (1909) Malmaterialenkunde als Grundlage de Maltechnik. Verlag von Julius Springer, Berlin

Everts S (2016) Art conservators struggle with microscopic eruptions in masterpieces. 94(21):21–23. 23 May 2016. http://cen.acs.org/articles/94/i21/Art-conservationists-struggle-microscopic-eruptions.html

Ferreira ESB, Boon JJ, Stampanoni M, Marone F (2011) Study of the mechanism of formation of calcium soaps in an early 20th-century easel painting with correlative 2D and 3D microscopy. Preprints of the ICOM committee for conservation 16th triennial meeting, Lisbon, 26–30 Sept 2011, pp 1604–1608

Field G (1869) Chromatography or treatise on colours and pigments as used by artists. In: Salter T (ed) vol 52. London. http://www.gutenberg.org/files/20915/20915-h/20915-h.htm. Accessed 26 May 2017

Gettens RJ, Stout GL (1966) Painting materials: a short encyclopaedia. Dover Publications, New York

Gunn MG, Chottard G, Rivière E, Girerd JJ, Chottard JC (2002) Chemical reactions between copper pigments and Oleoresinous media. Stud Conserv 47(1):12–23

Heeren RMA, Boon JJ, Noble P, Wadum J (1999) Integrating imaging FTIR and secondary ion mass spectrometry for the analysis of embedded paint cross-sections. In: Bridgland J (ed) Preprints of the ICOM committee for conservation 12th triennial meeting, Lyon, pp 228–233

Helwig K, Forest E, Turcotte A, Baker W, Binnie NE, Moffatt E, Poulin J (2019) The formation of calcium fatty acid salts in oil paint: two case studies. In: Casadio F, Keune K, Noble P, Van Loon A, Hendriks E, Centeno S, Osmond G (eds) Metal soaps in art: conservation and research. Springer, Cham, pp 299–312

Hermans JJ (2017) Metal soaps in oil paint: structure, mechanisms and dynamics. Ph.D. thesis, University of Amsterdam

Hermans JJ, Keune K, Van Loon A, Iedema PD (2015) An infrared spectroscopic study of the nature of zinc carboxylates in oil paintings. J Anal At Spectrom 30:1600–1608

Hermans JJ, Keune K, Van Loon A, Iedema PD (2016) The crystallization of metal soaps and fatty acids in oil paint model systems. Phys Chem 18:10896–10905

Hermans JJ, Keune K, Van Loon A, Iedema PD (2019) Toward a complete molecular model for the formation of metal soaps in oil paints. In: Casadio F, Keune K, Noble P, Van Loon A, Hendriks E, Centeno S, Osmond G (eds) Metal soaps in art: conservation and research. Springer, Cham, pp 47–65

Hess M (1965) Paint film defects, their causes and cure. Barnes & Noble, New York

Higgitt C, Plater MJ (2004) Old masters in the spotlight. Chem World 1(4):40–45

Higgitt C, Spring M, Saunders D (2003) Pigment-medium interactions in oil paint films containing red lead or lead-tin yellow. Natl Gallery Tech Bull 24:75–95

Higgs S, Burnstock A (2019) An investigation into metal ions in varnish coatings. In: Casadio F, Keune K, Noble P, Van Loon A, Hendriks E, Centeno S, Osmond G (eds) Metal soaps in art: conservation and research. Springer, Cham, pp 123–140

Horovitz I (1999) The materials and techniques of European paintings on copper supports. In: Komanecky M (ed) Copper as canvas: two centuries of masterpiece paintings on copper 1575–1775, New York, 1999, pp 63–92

Iddi Z (2015) Een vertroebeld Sprookje. Onderzoek naar een mogelijk origineel vernis op het schilderij Sprookje (ca.1877) van Matthijs Maris (1839–1917). Unpublished master thesis, University of Amsterdam

Jones R, Townsend JH, Stonor K, Duff N (2007) Lead soap aggregates in sixteenth and seventeenth century British paintings. In: AIC paintings specialty group postprints, Providence, 16–19 June 2006, vol 19, pp 24–32

Keune K (2005) Binding medium, pigments and metal soaps characterised and localised in paint cross-sections. Ph.D. thesis, University of Amsterdam, Amsterdam

Keune K, Boéve-Jones G (2014) Its surreal: zinc-oxide degradation and misperceptions in Salvador Dalí's couple with clouds in their heads, 1936. In: Van den Berg KJ, Burnstock A, de Tagle A, de Keijzer M, Heydenreich G, Krueger J, Learner T (eds) Issues in contemporary oil paints. Springer, Cham, pp 283–294

Keune K, Boon JJ (2007) Analytical imaging studies of cross-sections of paintings affected by lead soap aggregate formation. Stud Conserv 52(3):161–176

Klaassen L, Van der Snickt G, Legrand S, Higgitt C, Spring M, Vanmeert F, Rosi F, Brunetti BG, Postec M, Janssens K (2019) Characterization and removal of a disfiguring oxalate crust on a large altarpiece by Hans Memling. In: Casadio F, Keune K, Noble P, Van Loon A, Hendriks E, Centeno S, Osmond G (eds) Metal soaps in art: conservation and research. Springer, Cham, pp 265–282

Laurie AP (1926) The painter's methods & materials. Seeley, Service & Co, London

Mahon D, Centeno S (2005) A technical study of John Singer Sargent's portrait of Madame Pierre Gautreau. Metrop Mus J 40:121–129

Maines CA, Rogala D, Lake S, Mecklenburg M (2011) Deterioration in abstract expressionist paintings: analysis of zinc oxide paint layers in works from the collection of the Hirshhorn Museum and Sculpture Garden, Smithsonian Institution. In: Vandiver P et al (eds) Materials issues in art and archaeology IX, vol 1319. Materials Research Society, Boston

Mérimée JFL (1839) De la peinture à l'huile ou des procédés matériels employés dans ce genre de peinture, depuis Hubert et Jean Van-Eyck Jusqu'a Nos Jours. Mme Huzard, Paris

Middelkoop N, Noble P, Wadum J, Broos B (1998) Rembrandt under the scalpel the anatomy lesson of Dr Nicolaes Tulp dissected. Six Art Production, Amsterdam/The Hague, pp 51–74

Noble P, Boon JJ (2007) Metal soap degradation of oil paintings: aggregates, increased transparency and efflorescence. In: Mar Parkin H (ed) AIC paintings specialty group postprints, Providence, 16–19 June 2006, vol 19, pp 1–15

Noble P, Van Loon A (2007) Rembrandt's Simeon's song of praise, 1631: spatial devices in the service of illusion. Art Matters: Netherlands Tech Stud Art 4:19–36

Noble P, Wadum J, Groen K, Heeren R, Van den Berg KJ (2000) Aspects of 17th century binding medium: inclusions in Rembrandt's anatomy lesson of Dr Nicolaes Tulp. In: Actes du congrès, Art et Chimie, la couleur, Paris, 16–18 Sept 1998, pp 126–129

Noble P, Boon JJ, Wadum J (2002) Dissolution, aggregation and protrusion: lead soap formation in 17th century grounds and paint layers. ArtMatters Neth Tech Stud Art 1:46–62

Noble P, Van Loon A, Boon JJ (2005) Chemical changes in old master paintings II: darkening due to increased transparency as a result of metal soap formation. In: Preprints of the ICOM committee for conservation 14th triennial meeting, The Hague, 12–16 Sept 2005, pp 496–503

Noble P, Meloni S, Pottasch C (2008a) Preserving our heritage: conservation, restoration and technical research in the Mauritshuis, English edition 2009. Waanders Publishers, Zwolle

Noble P, Van Loon A, Boon JJ (2008b) Selective darkening of ground layers associated with the wood grain in 17th-century panel paintings. In: Townsend J, Doherty T, Heydenreich G (eds) Preparation for paintings: the Artist's choice and its consequences. Archetype, London, pp 68–78

Ordonez E, Twilley J (1998) Clarifying the haze: efflorescence on works of art. WAAC Newsletter, January 1998, Vol 20(No 1)

Osmond G (2014) Zinc white and the influence of paint composition for stability in oil based paints. In: Van den Berg KJ, Burnstock A, de Tagle A, de Keijzer M, Heydenreich G, Krueger J, Learner T (eds) Issues in contemporary oil paints. Springer, Cham, pp 263–281

Osmond G (2019) Zinc soaps: an overview of zinc oxide reactivity and consequences of soap formation in oil-based paintings. In: Casadio F, Keune K, Noble P, Van Loon A, Hendriks E, Centeno S, Osmond G (eds) Metal soaps in art: conservation and research. Springer, Cham, pp 25–43

Pavlopoulou LC, Watkinson D (2006) The degradation of oil painted copper surfaces. Rev Conserv 7:55–65

Plahter U (1999) Baburen re-examined. In: Skaug E (ed) Conservare Necesse Est, Festskrift til Leif Einar Plahter på hans 70-årsdag. IIC Nordic Group, Oslo, pp 65–67

Raven LE, Bisschoff M, Leeuwestein M, Geldof M, Hermans JJ, Stols-Witlox M, Keune K (2019) Delamination due to zinc soap formation in an oil painting by Piet Mondrian (1872–1944). Conservation issues and possible implications for treatment. In: Casadio F, Keune K, Noble P, Van Loon A, Hendriks E, Centeno S, Osmond G (eds) Metal soaps in art: conservation and research. Springer, Cham, pp 345–357

Robinet L, Corbeil MC (2003) The characterization of metal soaps. Stud Conserv 48(1):23–40

Rogala DV (2019) Everything old is new again: revisiting a historical symposium on zinc oxide paint films. In: Casadio F, Keune K, Noble P, Van Loon A, Hendriks E, Centeno S, Osmond G (eds) Metal soaps in art: conservation and research. Springer, Cham, pp 317–330

Rogala D, Lake S, Maines C, Mecklenburg M (2010) Condition problems related to zinc oxide Underlayers: examination of selected abstract expressionist paintings from the collection of the Hirshhorn Museum and Sculpture Garden, Smithsonian Institution. JAIC 49: 96–113

Shimadzu Y, Keune K, Van den Berg KJ (2008) The effects of lead and zinc white saponification on surface appearance of paint. In: Bridgland J (ed) Preprints of the ICOM committee for Conservation 15th triennial meeting, New Delhi, 22–26 Sept 2008. Allied Publishers Pvt Ltd, pp 626–632

Skaliks A (1999) Blooming: Auswandern von Bindemittelbestandteilen aus ölhaltigen Farbsystemen: Phänomene, mögliche Ursachen und Überlegungen zur Prävention und Restaurierung. Diplomarbeit [Master thesis], Fachhochschule Köln

Spring M, Higgitt C (2006) Analyses reconsidered: the importance of the pigment content of paint in the reinterpretation of the results of examination of binding media. In: Nadolny J (ed) Medieval painting in Northern Europe: techniques, analysis, art history. Archetype, London, pp 223–229

Spring M, Higgitt C, Saunders D (2005) Investigation of pigment medium interaction processes in oil paint containing degraded smalt. Natl Gallery Tech Bull 26:56–69

Tumosa C (2001) A brief history of aluminum stearate as a component of paint. WAAC Newsletter, Vol 23(No 3), pp 10–11

Van den Berg KJ, Burnstock A, Schilling M (2019) Notes on metal soap extenders in modern oil paints: history, use, degradation and analysis. In: Casadio F, Keune K, Noble P, Van Loon A, Hendriks E, Centeno S, Osmond G (eds) Metal soaps in art: conservation and research. Springer, Cham, pp 331–343

Van der Weerd J, Geldhof M, Van der Loeff LS, Heeren R, Boon JJ (2003) Zinc soap aggregate formation in 'falling leaves (les Alyscamps)' by Vincent van Gogh. Zeitschrift für Kunsttechnologie und Konservierung 17(2):407–416

Van Loon A (2008) Color changes and chemical reactivity in seventeenth-century oil paintings. Ph.D. thesis, University of Amsterdam, Amsterdam

Van Loon A (2009) De korrelige verf in Vermeers Gezicht op Delft: Nieuwe onderzoeksgegevens. kM:10–13

Van Loon A, Noble P, Boon JJ (2011) White hazes and surface crusts in Rembrandt's Homer and related paintings. In: Bridgland J (ed) Preprints of the ICOM committee for conservation 16th triennial meeting, Lisbon, 26–30 Sept 2011. Paper 1322 (CD-ROM)

Van Loon A, Noble P, Boon JJ (2012a) The formation of complex crusts in oil paints containing lead white and smalt: dissolution, depletion, diffusion, deposition. In: Meeks N, Cartwright C, Meek A, Mongiatti A (eds) Historical technology, materials and conservation, SEM and microanalysis. The British Museum/Archetype, London, pp 207–209

Van Loon A, Noble P, Burnstock A (2012b) Degradation phenomena in traditional oil and tempera paints as a result of natural ageing processes. In: Hill Stoner J, Rushfield R (eds) The conservation of easel paintings. Routledge, London, pp 214–241

Van Loon A, Keune K, Hermans J, Sawicka A, Thoury M, Bertrand L, Sandt C, Reguer S, Toskarski C, Corkery R, Iedema PD (2016) Investigation of the dynamic processes underlying lead soap-related degradation phenomena in multi-layer model systems. Poster presented at metal soaps in art conference. Rijksmuseum Amsterdam, pp 14–15

Van Loon A, Hartman LE, Van den Burg J, Haswell R, Pottasch C (2019a) The development of an aqueous gel testing procedure for the removal of lead-rich salt crusts on the surface of paintings by Giovanni Antonio Pellegrini (1675–1741) in the "Golden Room" of the Mauritshuis. In: Casadio F, Keune K, Noble P, Van Loon A, Hendriks E, Centeno S, Osmond G (eds) Metal soaps in art: conservation and research. Springer, Cham, pp 285–297

Van Loon A, Hoppe R, Keune K, Hermans J, Diependaal H, Bisschoff M, Thoury M, Van der Snickt G (2019b) Paint delamination as a result of zinc soap formation in an early Mondrian painting. In: Casadio F, Keune K, Noble P, Van Loon A, Hendriks E, Centeno S, Osmond G (eds) Metal soaps in art: conservation and research. Springer, Cham, pp 361–373

Williams SR (1989) Blooms, blushes, transferred images and mouldy surfaces: what are these distracting accretions on art works? In: Wellheiser JG (ed) Proceedings of the 14th annual conference of the international institute for conservation of historic and artistic works – Canadian Group, Toronto, pp 65–84

Zucker J, Boon JJ (2007) Opaque to transparent: paint film defects in the work of Frederic Church and the Hudson River School. In: Mar Parkin H (ed) AIC paintings specialty group postprints, Providence, 16–19 June 2006, vol 19, pp 33–41

Part I
Formation, Migration and Environmental Influences

Chapter 2
Zinc Soaps: An Overview of Zinc Oxide Reactivity and Consequences of Soap Formation in Oil-Based Paintings

Gillian Osmond

Abstract Zinc white – zinc oxide – is a prevalent industrial age pigment which readily reacts with fatty acids in oil-based paints to form zinc carboxylates (zinc soaps). Pigment properties, oil composition, paint additives and environmental conditions are all significant factors in this process. Release of fatty acids via hydrolysis is particularly implicated. Formation and crystallization of zinc soaps in paintings has serious visual and structural consequences. Fourier transform infrared spectroscopy (FTIR) is a highly sensitive technique for detection of metal carboxylates. Progression of soap formation and distributions are distinguishable in paint cross sections through correlation of microscopic images, elemental analysis and FTIR mapping. Aluminum stearate was found to strongly influence zinc stearate formation during investigations of naturally aged oil-based artists' commercial and custom-produced reference paints containing zinc oxide in various oil and pigment combinations. Zinc soaps are implicated in deterioration of increasing numbers of paintings dating from the nineteenth century onward. They may manifest as disfiguring lumps or surface blooms resistant to removal. In other paintings, paint cleavage is associated with high concentrations of crystalline zinc soaps at unstable interfaces. The condition of paintings affected by zinc soaps is challenging conventional approaches to cleaning and consolidation and requires continuing research.

Keywords Zinc white · Zinc soaps · Stearate · Zinc azelate · Fatty acid · Basic zinc carbonate · FTIR · Oil paint · Paintings conservation

---

G. Osmond (✉)
Queensland Art Gallery/Gallery of Modern Art, Brisbane, Australia
e-mail: gillian.osmond@qagoma.qld.gov.au

F. Casadio et al. (eds.), *Metal Soaps in Art*, Cultural Heritage Science,
https://doi.org/10.1007/978-3-319-90617-1_2

2.1 Introduction

A variety of deterioration phenomena associated with zinc soaps in paintings has come to attention in recent years. Many concern late nineteenth-century and early- to mid-twentieth-century paintings consistent with the emerging availability and popularity of zinc oxide among paint producers during this period; however, impacts for more contemporary paintings are also transpiring.

Initially the most commonly reported manifestation of zinc soap formation was lumps disturbing the image plane in a similar manner to instances of lead soap aggregation (O'Donoghue et al. 2006; Osmond et al. 2005; van der Weerd et al. 2003). Figure 2.1 shows some surface details imaged under magnification which illustrate how zinc soap formation can vary even within a single painting, ranging from white waxy protrusions through to translucent lumps creating a bubbled texture. The likely impacts of increased transparency from zinc soap formation have also been considered (Shimadzu et al. 2008; Shimadzu and van den Berg 2006). Increasingly zinc soaps are being implicated in significant structural failure within paint layers (O'Donoghue et al. 2006; Rogala et al. 2010). Additional examples will be provided here and by other contributors to this volume.

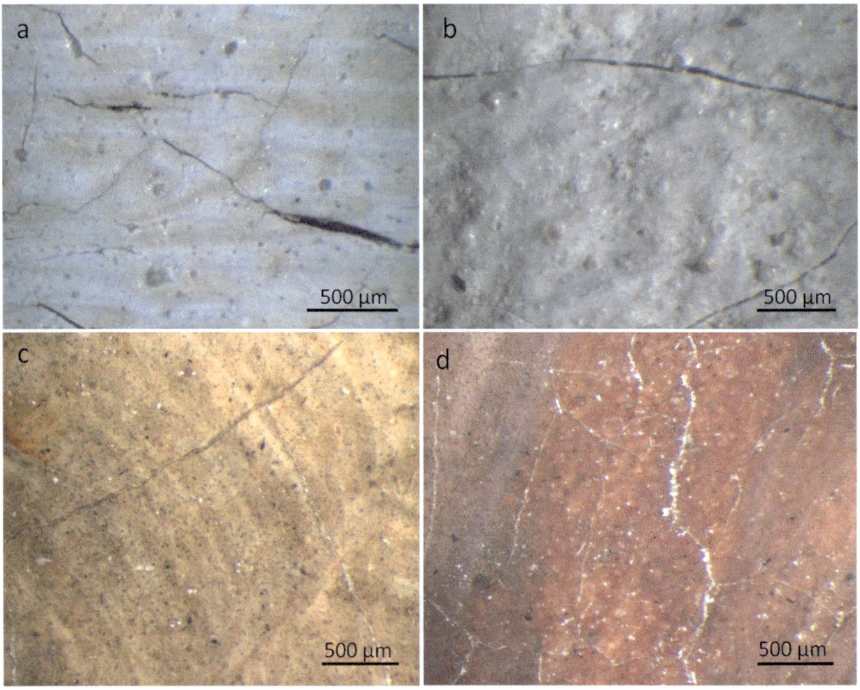

Fig. 2.1 Magnified surface details of zinc soap affected paint from (**a**) RG Rivers, *An alien in Australia* 1904, reworked later (Gift of the Godfrey Rivers Trust 1941) and (**b–d**) RG Rivers, *Woolshed NSW* 1890 (purchased 1903), both oil on canvas, Queensland Art Gallery collection

Zinc soap formation has serious implications for painting collections worldwide. Although there are various potential sources of zinc in modern paint films, this paper focuses specifically on the reactivity of zinc oxide in oil-based media. A detailed review of modern zinc oxide history, pigment properties and industry uptake has been published previously (Osmond 2012). Here, current understanding of factors contributing to the reactivity of zinc white in oil-based paints will be reviewed. Aspects of paint composition and environmental exposure are considered. The application of optical and scanning electron microscopy techniques and FTIR analysis for characterizing the presence and detailed morphology of zinc soaps are applied to give insight into formation mechanisms. Case studies are presented to illustrate a range of possible manifestations. Detailed awareness of the vulnerabilities of paintings containing zinc oxide is necessary to better understand implications for conservation interventions and collection management practices and to guide continuing research.

2.2 Zinc Oxide

Zinc oxide, known in the paint industry as zinc white, was first industrially produced from zinc metal (Indirect/French Process) in early nineteenth-century France as a less toxic alternative to lead white. By the 1850s, a more economical "Direct/American Process" of production was developed in the United States to take advantage of large domestic ore deposits (Brown 1957; Gardner 1917). Australia also had an active zinc oxide production industry from the early twentieth century which additionally supplied zinc concentrates[1] for European markets (Cocks and Walters 1968).

Zinc oxide transformed the early production of ready-mixed decorative paints because of its good suspension properties in oil-based media, typically used in combination with lead white and later titanium white. Paint formulated on a primary base of zinc oxide was less common, but oil-based gloss Ripolin® – much favored by artists such as Picasso and Sidney Nolan – is one notable example (Dredge et al. 2013; Gautier et al. 2009). In artists' oil paints, zinc oxide has inferior hiding power to lead white, but some artists consciously sought zinc white paint for its resistance to yellowing and blackening. Early uptake was notable in France which was the first country to limit production of lead white pigment to protect the health of pigment manufacturers (Bomford et al. 1990; Gardner 1917). Throughout the twentieth century, zinc oxide was unwittingly used by many artists as its prevalence in both white and colored oil paint formulations increased, and there is continuing evidence of this today (Theobald Clark and Osmond 2016).

There are many grades of zinc oxide pigment with a range of particle and crystal properties. Zinc oxide generally is considered to be electrostatically unstable because it has two polar facets at opposing ends of the crystal, one terminating in zinc atoms and one with oxygen (Wöll 2007). Figure 2.2 shows a detail of a

[1]Zinc concentrate is an intermediate product obtained from ore and used as raw material in the production of zinc metal and potentially, in turn, for French Process zinc oxide.

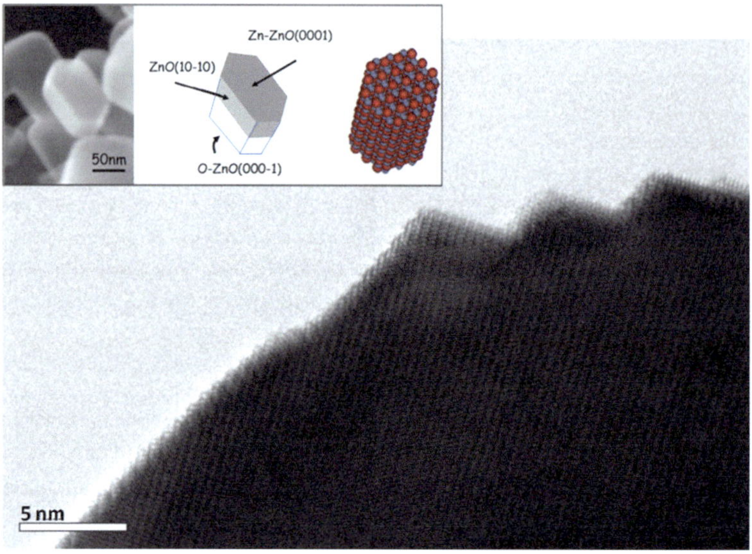

Fig. 2.2 Transmission electron microscopy image detail of a zinc oxide particle captured at 600,000×. The inset image details the facets in a ZnO crystal and illustrates how a more acicular particle increases the surface area of the ZnO(10–10) plane relative to the polar facets at opposing ends. (TEM image: John Drennan. Inset graphic adapted from Wöll 2007)

single zinc oxide crystal imaged at 600,000× where it is possible to resolve rows of zinc atoms terminating in the 0001 plane. Particle size and shape may emphasize the influence of one crystal facet over another, for example, in acicular grades the relatively neutral ZnO plane will be accentuated in comparison to nodular-shaped particles.

Lattice variations also influence pigment properties, particularly interactions with light, affecting fluorescence characteristics and energy dissipation. Interstitial zinc atoms or other lattice irregularities commonly associated with early French Process zinc oxide have been associated with initiating photochemical reactions in adjacent materials (Morley-Smith 1958; Winter and Whittem 1950). More recently, improved analytical capabilities have revealed distinct differences in the photoluminescence properties of individual crystallites in historic zinc white pigment samples at the submicrometric scale (Bertrand et al. 2013). Extension of this research may allow for more detailed understanding of why some zinc oxide-based paints are more susceptible to deterioration than others.

The significance of crystal properties for paint performance can be illustrated with reference to changes in pigment production which occurred in Australia in the post-World War II period. 'Superior' French Process pigment production was introduced for Australian house paints in 1947 but was almost immediately linked to durability issues which had not been apparent with Direct Process zinc oxide used prior to this date (Morley-Smith 1950). The Australian climate and more intense sunlight initiated premature breakdown of exterior paints which did not obviously occur in milder European conditions. The resulting crisis triggered significant research into zinc oxide pigment properties and performance-enhancing modifications. In Australian conditions, the higher purity of French Process pigment did not confer sufficient advantages to offset the increased reactivity – properties compounded by a smaller particle size (and higher reactive surface area) relative to Direct Process zinc oxide (Morley-Smith 1950). Today much tighter control over particle properties is possible, and high-purity French Process zinc oxide dominates international production (personal communication, R Shelton, Umicore Australia Limited, email 28 June 2010).

2.3 Fatty Acid Composition in Oil Mediums

In addition to pigment properties, reactivity of zinc oxide is strongly dependent on the fatty acid composition of a paint. Typical drying oils comprise esters of five principal unsaturated and saturated fatty acids variously combined in triglyceride structures. Acids with 18 carbon atoms in unbranched chains predominate (Mills and White 1987). The proportional representation of constituent fatty acids varies with the source of the oil (Table 2.1). Modern artists' oil paints encompass a substantially wider range of possible oils, preprocessing treatments and additives than those traditionally used (Schilling et al. 2007). Variations are even more pronounced when oil-based house paints used by artists are considered (Standeven 2011). These factors all potentially influence the fatty acid profile and reactivity of a formulated paint.

Table 2.1 Approximate fatty acid compositions of commonly used oil mediums

	Fatty acid weight percentages						
Oil	Palmitic C16:0	Stearic C18:0	Oleic C18:1	Linoleic C18:2	Linolenic C18:3	Others	
Castor	1	1	7	3		Ricinoleic C18:1 hydroxy acid	87
Linseed	6	4	19	24	47		
Oiticica	6	10	14	21		Licanic C18:3 keto acid	48
Poppy	10	2	11	72	5		
Rapeseed	1	1	15	15	8	Erucic C22:1	60
Safflower	7	3	18	69	4		
Soybean	11	4	23	54	8		
Sunflower	7	5	23	65			
Tall oil			46	41	3	Rosin (resin acids)	8
Tung	2	2	5	3		Eleostearic C18:3 conjugated	87

Adapted from Mills and White (1987) and Schilling et al. (2007)

Film formation in a drying oil occurs primarily with oxidative polymerization of the doubly and triply unsaturated fatty acids. Low molecular weight products are produced concurrently within the cross-linked structure via β-scission degradation reactions and oxidative cleavage of double bonds and hydrolysis of glycerol ester bonds (van den Berg et al. 1999; van Gorkum and Bouwman 2005). The liberation of fatty acids via hydrolysis in a paint film is a key process in their availability to react with metal ions to form soaps. Prevailing environmental conditions will influence degradation pathways and the likely representation of various reaction products in the paint. Relative humidity levels above 50% are associated with increased susceptibility of oil paints to photolytically induced cleavage of fatty acid side chains to produce high amounts of diacids (Saunders and Kirby 2004; Schilling et al. 1999). Elevated temperatures have been associated with hydrolysis reactions which release relatively high amounts of saturated fatty acids (Schilling et al. 1999).

Palmitic and stearic acids are preferentially located in triglyceride positions 1 and 3 which are more readily hydrolyzed than second position acids (Erhardt et al. 2005). These two acids, together with azelaic acid – a common oxidative degradation product of the predominant unsaturated C18 fatty acids – comprise the three saturated acids principally found in free form within a typical cured oil paint film (Higgitt et al. 2003; Keune and Boon 2007). While palmitate and stearate ions comprise only a minor proportion of a typical drying oil, they are the components most frequently implicated in defects associated with aggregation of zinc and other metal soaps, and also with crystalline surface deposits, or efflorescence – another condition affecting collections of paintings. Other potential sources of stearic acid include a number of common additives used to modify properties in the formulated paint, with aluminum stearate a notable candidate (Osmond 2014b). Zinc stearate itself may also be present as a manufacturing component. Metal soaps used in paint formulation and their susceptibility to hydrolysis are covered in more detail by van den Berg et al. (2019).

2.4 FTIR Characterization of Zinc Soaps

Fourier transform infrared spectroscopy (FTIR) is a highly sensitive technique for determining the presence of metal carboxylates in paint films. The two spectra shown in Fig. 2.3 illustrate how peak positions change when stearic acid reacts with zinc oxide to form zinc stearate. A dramatic physical change accompanies saponification as illustrated by the inset vial images of zinc oxide and stearic acid combined in toluene (Osmond 2014a). Soap formation sees the C=O and C-O stretching vibrations in the acid replaced by symmetric and antisymmetric vibrations of the carboxylate group in the soap (Robinet and Corbeil 2003). Spectral changes are particularly obvious in the case of zinc soap formation as there is no contributing absorption from zinc oxide itself between 600 and 4000 cm^{-1}.

Different carboxylic acids and their metal soaps can be characterized by subtle variations in the number and position of CH_2 progression bands between 1150 and

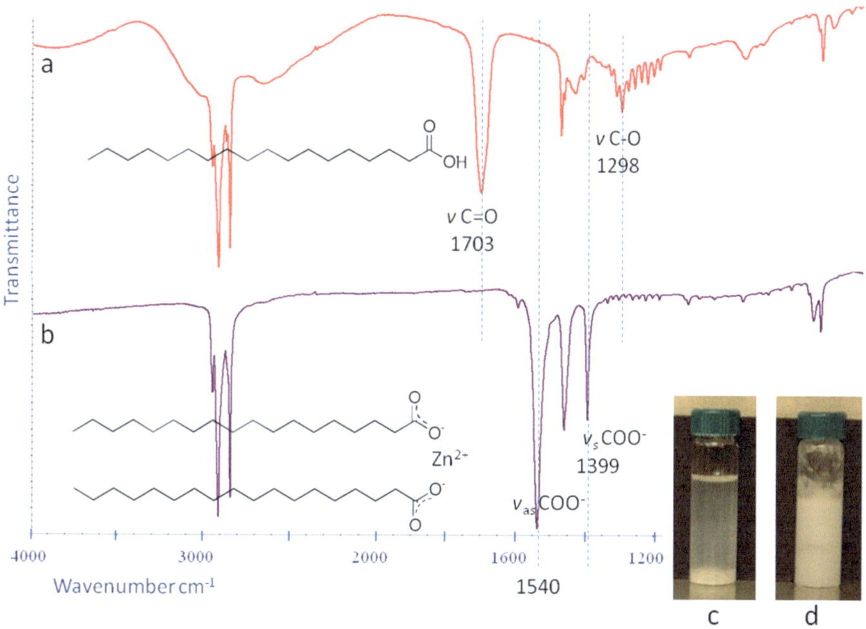

Fig. 2.3 Transmission FTIR spectra for (**a**) stearic acid and (**b**) zinc stearate. Inset vial images show a mixture of 30 nm ZnO and stearic acid 1:10 by weight (approx 1:3 mol) in toluene after (**c**) 30 min and (**d**) 2 months showing a dramatic physical expansion of the solid phase accompanying formation of zinc stearate

1380 cm^{-1} which relate to the length of carbon chain[2]; in zinc soaps, these features are usually only discernible in spectra from relatively pure samples. Zinc oleate (C18:1) can be most readily distinguished from zinc stearate (C18:0) by the presence of a small C=C stretch at 3003 cm^{-1} in combination with split carboxylate bands. The zinc azelate spectrum (C9 dicarboxylate) has additional complexity in the CH_3 and CH_2 stretching region between 2800 and 3000 cm^{-1} and CH_2 rocking bands between 720 and 775 cm^{-1}. Detailed reference spectra for these soaps have been published previously (Helwig et al. 2014). FTIR spectra for soaps of different metals may also be distinguished by the position of the symmetric and antisymmetric carboxylate vibrations which vary according to the coordination with the metal ion. The relevant peaks for various soaps of zinc, lead and copper have been published elsewhere (Robinet and Corbeil 2003). The soaps of primary relevance to this study, zinc palmitate (C16:0) and zinc stearate, are virtually indistinguishable from one another in FTIR spectra from paint. Fortunately, in the context of this research, both soaps are typically present simultaneously in common degradation phenomena based on studies where FTIR analysis has been supplemented by mass spectrometric

[2]Reference spectra are shown in transmittance format where the logarithmic intensity scale assists visual representation of weak CH_2 progression bands.

techniques (Helwig et al. 2014; Keune and Boon 2007). For this reason and for convenience in this paper, the term zinc stearate will be used to signify the presence of zinc stearate and/or zinc palmitate when attribution is based on the major FTIR peaks common to both.

It is also important to note that the defined carboxylate vibrations evident in FTIR spectra of synthesized zinc stearate are not always apparent in FTIR spectra from oil-based paints containing zinc oxide, which more typically exhibit broad absorption between 1530 and 1650 cm^{-1}. Sharp carboxylate peaks indicate a highly ordered crystalline arrangement of the soap molecules, whereas broad absorption has been attributed to paint where zinc carboxylates remain connected to the glycerol backbone and distributed throughout the polymerized oil network (Hermans et al. 2015, 2016). It is in the aggregated crystalline state that zinc soap-related deterioration is typically observed.

2.5 Morphology Change

The conversion of zinc oxide to zinc soap can be discerned in backscattered electron (BSE) images of paint cross sections, most clearly when at an advanced, aggregated stage, due to the atomic contrast between oxide and carboxylate forms (Keune 2005). Paint cross sections also allow correlation of optical and ultraviolet (UV) fluorescence characteristics with elemental analysis and FTIR mapping. Soap formation involves loss of defined pigment particles. BSE images together with supporting analysis of paint layers containing a mix of zinc oxide particle sizes indicate smaller, typically nodular particles react before larger or acicular particles (Osmond 2014a). Zinc oxide particles in mixed pigment paints have been shown to react more readily than lithopone, lead white and a range of colored pigments (Osmond 2014a; Osmond et al. 2005, 2014; Shimadzu et al. 2008). At high magnification, BSE images can reveal early-stage in situ saponification indicated by zinc-based regions with poorly defined boundaries, relatively low pigment density and reduced atomic brightness. These characteristics may assist differentiation from similarly composed metal soaps present as original components which are more commonly discerned as discrete particle-free areas with clear boundaries.

2.6 Zinc White Paints with Added Stearates

The significance for zinc soap formation in paints with added aluminum stearate has been investigated previously using reference paints from the Smithsonian's Museum Conservation Institute containing zinc oxide in various oils and pigment combinations (Osmond 2014b; Osmond et al. 2012). Paints were prepared as drawdowns on Melinex between 1978 and 1999. FTIR spectra obtained from the exposed top surface of each paint using attenuated total reflection (ATR)-FTIR showed differing intensities of carboxylate absorption but remarkably consistent peak shape and position in the band between 1530 and 1650 cm^{-1} (Fig. 2.4). The broad absorption was present irrespective of commercial origin, pigment combination, oil type, film age, or the presence of additives; however, greater differences were apparent in spectra collected from the protected underside of the paints. Most notably, underside spectra from paints formulated with aluminum stearate recorded a strong, defined peak at 1540 cm^{-1}. Significantly, this peak is characteristic of zinc stearate (ν_{as} COO^-), not aluminum stearate,[3] despite none of the paints incorporating zinc stearate as an original component.

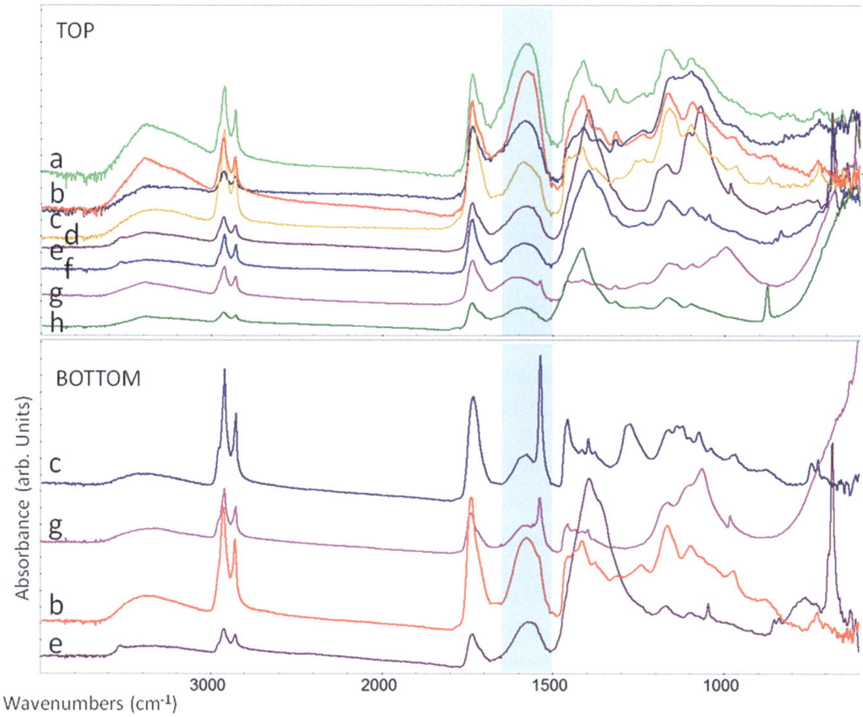

Fig. 2.4 ATR-FTIR spectra obtained from exposed top and protected bottom surfaces of a range of oil-based paints containing zinc oxide from the Smithsonian's Museum Conservation Institute reference collection: (**a**) Winsor & Newton (W&N) zinc white (safflower) cast 1978; (**b**) zinc white control linseed oil + litharge drier 1990; (**c**) Grumbacher zinc white (linseed oil + aluminum stearate) 1978; (**d**) zinc white control boiled linseed oil 1990; (**e**) W&N flake white (safflower oil) 1978; (**f**) W&N Naples yellow 1999; (**g**) titanium white control with zinc oxide, soybean oil + aluminum stearate 1998; (**h**) Gamblin titanium white (with zinc oxide, linseed oil) 1999. The metal carboxylate band is highlighted. A defined peak at 1540 cm^{-1} (ν_{as} COO^- zinc stearate) occurs in the underside spectra of paints formulated with aluminum stearate

[3] Aluminum stearate has its major carboxylate absorption at 1588 cm^{-1}.

The distribution of zinc and aluminum stearate within paint films was further resolved with synchrotron-sourced (SR)-μFTIR mapping of thin cross sections and revealed subtleties not discernible with conventional techniques. Specific aluminum stearate absorptions were rarely detected and were confined to small localized areas of the paint film.

Aluminum stearate in paints increases the availability of stearic acid. Commercial grades typically include some stearic acid in free form, and aluminum stearate is also readily hydrolyzed on the shelf. Model experiments have shown that in the presence of zinc oxide, progressive hydrolysis of aluminum stearate releases stearic acid which then preferentially binds with zinc ions to form crystalline zinc stearate (Osmond 2014b). It is not yet clear whether other metal stearates used in paint formulations behave similarly. The tendency for crystalline zinc stearate to concentrate in the lower margin of studied reference paints may be influenced by the Melinex substrate, although similar distribution patterns have been observed in FTIR mapping of cross sections taken from paintings (Hermans et al. 2015; Raven et al. 2019). The observed separation of zinc stearate from the bulk paint is significant irrespective of where it manifests.

2.7 Formation of Crystalline Zinc Soaps

Zinc soaps have been characterized in numerous paintings in association with both protruding soap masses and at interfaces experiencing poor inter- or intralayer adhesion. Spectra from affected paint may generally include broad carboxylate absorptions, while localized crystalline phases are more specifically implicated in observed deterioration. Higher concentrations of zinc stearate appear to not be readily accommodated within the cross-linked paint network and are more favorably organized in an aggregated crystalline state. Under this scenario, soap formation facilitated by the progressive release of free fatty acids via hydrolysis in paints over time is likely to result in a gradual increase in intensity of the 1540 cm^{-1} peak in spectra from paints. Environmental exposure will influence this progression.

The sharp zinc carboxylate peaks in spectra are sometimes split into doublets but without an accompanying C=C stretch are not attributable to zinc oleate. Investigation of possible explanations for peak splitting by Helwig et al. (2014) concluded a likely cause is structural distortion arising from small amounts of oil and free fatty acids present during gradual soap crystallization in the paint. Examples of carboxylate peak splitting have also been reported in spectra from reference paint films based on soybean and safflower oils which have significantly lower percentages of the triply unsaturated linolenic acid than linseed oil and therefore dry more slowly (Osmond 2014b). Peak splitting in spectra from a large zinc soap aggregate taken from Frederic Leighton, *Winding the skein* 1878 (oil on canvas, Art Gallery of New South Wales collection), occurs in a painting for which the artist recorded using poppy oil, another semi-drying oil (Osmond et al. 2013). In this instance, the fatty acid composition of one aggregate from the painting was determined with gas chromatography-mass spectrometry (GC-MS) which confirmed a predominance of palmitate moieties with lesser amounts of stearic, oleic and azelaic acids – some of which (based on FTIR) remained ester bound (personal communication, K Helwig and J Poulin, Canadian Conservation Institute, email 21, 23 September 2013). This supports the attribution of split carboxylate peaks to crystallization occurring within a complex chemical environment.

2.8 Zinc Soaps and Paint Cleavage

High concentrations of crystalline zinc stearate with both single and split carboxylate FTIR spectral bands are indicated at the interface of serious delamination in a painting by Australian artist, Davida Allen (b. 1951). *Paris painting* (oil on canvas, Queensland Art Gallery collection, purchased 1984) was painted on site for a 1983 exhibition in Paris, commencing with Australian-produced Art Spectrum oil paints and reworked in large areas with French-sourced Sennelier oils. Although paint initially appears to be flaking from the artist-applied chalk and zinc-based priming, delamination is actually occurring within the upper margin of this priming layer. A superficial amount of priming has detached from the bulk of the layer and is visible on the underside of lifting paint (Fig. 2.5). An FTIR spectrum obtained from the detached priming has a single if somewhat broadened peak at 1540 cm^{-1}, whereas more defined peak splitting occurs in the spectrum obtained from the surface of retained priming exposed by the flaking.

The source of fatty acids facilitating soap formation in the upper ground in this instance is almost certainly the oil medium from the dramatically cracked copper phthalocyanine blue paint; a spectrum obtained from the primed but unpainted tacking margin displays only the broad general zinc carboxylate band with no defined crystalline peaks (Fig. 2.5e).

The distinct carbonyl doublet in FTIR spectra at around 1734 and 1717 cm^{-1} in paint layers and exuded medium indicates the presence of fatty acids in both free and ester-bound forms; analysis of weeping drips in other areas of the painting suggests the medium may incorporate a high proportion of semi-drying oil consistent with other paints prone to dripping (personal communication, J Boon, email 24 March 2015; Boon and Hoogland 2014; Bronken and Boon 2014). Penetration of medium into the ground layer is likely to have enhanced saponification, evident as altered morphology in the upper margin of the ground in the BSE image of a cross section from the area (Fig. 2.5d).

Similar flaking of cadmium paints and carbon-based blacks from zinc oxide-based underlayers has been reported in paintings by mid-twentieth-century abstract expressionists although without corresponding published FTIR analysis (Rogala et al. 2010). High levels of oleic acid were detected in affected zinc oxide-based paints with pyrolysis GC-MS. Similarly high levels of oleic acid were measured in reference paints pigmented with a minimum of 25% zinc oxide by weight, in comparison to negligible levels in similarly aged films without zinc white (Maines

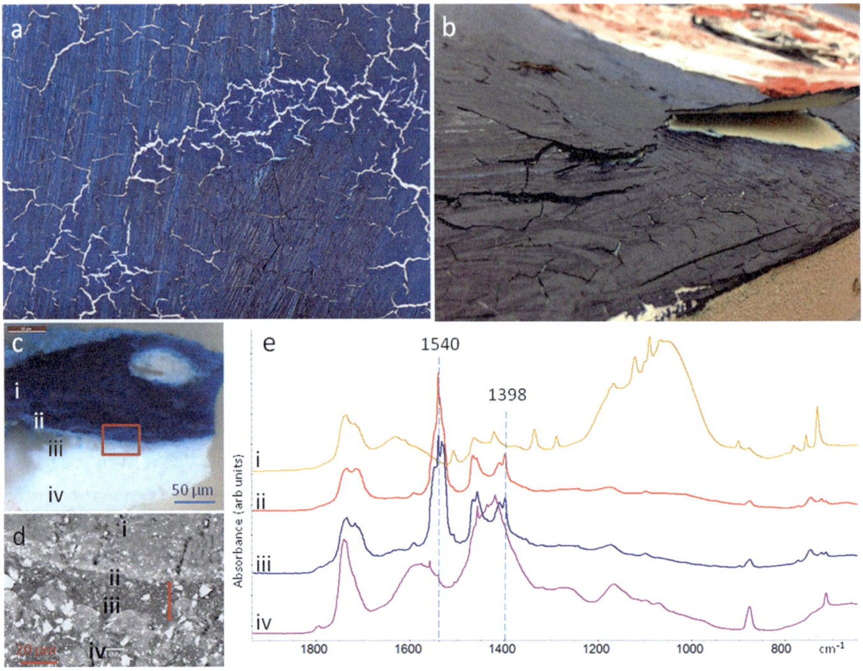

Fig. 2.5 Davida Allen *Paris painting* 1983 details. (**a–b**) Cohesive failure is apparent in the artist-applied chalk and zinc-based priming in association with widespread cracking of the copper phthalocyanine pigmented blue paint (images: Natasha Harth). In cross section (**c**) including higher-magnification BSE image (**d**), altered morphology is apparent in the upper margin of the ground (red arrow). Corresponding FTIR spectra (1850–750 cm^{-1}) from (**i**) cracked blue paint; (**ii**) priming attached to lifting paint; (**iii**) retained priming exposed by lifting paint; (**iv**) unpainted priming show that delamination is occurring in areas of high crystalline zinc stearate concentration

et al. 2011). The authors hypothesize that rapid formation of crystalline zinc oleate has shielded the *cis* double bond from oxidation and prevented the usual conversion of oleic acid to azelaic acid during paint cure. It is not possible to confirm that the high oleic acid levels in these paints are in the form of zinc oleate without FTIR results. FTIR spectra obtained from flaking paint observed in later twentieth-century Australian paintings suggest zinc stearate predominates at regions of failure. Crystalline zinc stearate comprises zinc-coordinated sheets with splayed carboxylate chains in layered structures (Hermans et al. 2014). It is not yet clear if this layered arrangement contributes to observed propensity for delamination.

High concentrations of crystalline zinc stearate have also been found in the upper margin of zinc oxide-containing ground layers in paintings susceptible to loss at this interface during recent research into the paintings of Australian artist William Robinson (b. 1936) (Theobald Clark and Osmond 2016). Robinson's paintings produced between 1987 and 2013 experience a high incidence of isolated small round-edged losses rather than broadscale flaking. Paint is lost cleanly

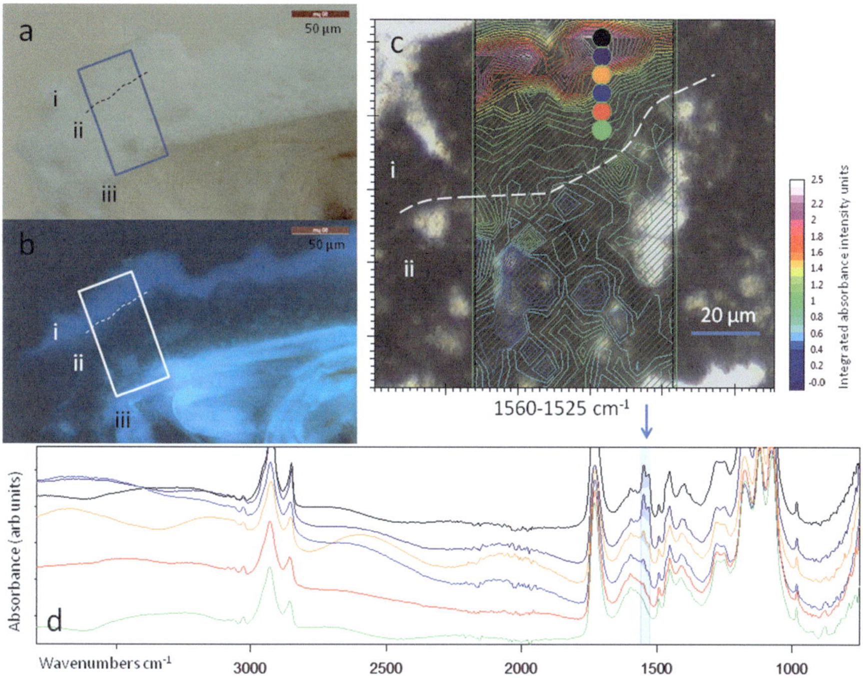

Fig. 2.6 An embedded ground sample from William Robinson *Early Light Coomera Gorge* 1994. (**a**) Incident light, (**b**) UV fluorescence image, (**c**) thin-section detail of boxed area with overlaid SR-µFTIR integrated absorption intensity map for wavenumber range 1525–1560 cm^{-1} and points from which extracted spectra derive as shown in (**d**). Two layers (i and ii) with underlying sizing (iii) are visible. Crystalline zinc stearate occurs toward the top surface of layer (i) with broader zinc carboxylate absorption lower in the same layer. Note some penetration of embedding resin is evident in spectra

from the ground and typically occurs without any preceding cracking to warn of instability. Paintings with a variety of ground compositions including lead white are represented in this group, but all contain zinc oxide; FTIR spectra of surface scrapings sampled from ground layers below losses consistently show distinct sharp peaks at 1540 cm^{-1}.[4]

Figure 2.6 shows an embedded ground sample in cross section from Robinson's *Early Light Coomera Gorge* 1994 (oil on canvas, QUT Collection). Two priming layers with underlying sizing are apparent in the sample. No paint layers are present as the weak bond between paint and ground prevented sampling of complete painting stratigraphy. The lower ground contains titanium white and chalk, while the upper layer is predominantly barium sulfate with zinc white.

[4]There is no corresponding indication of lead stearate formation (characterized by carboxylate peaks at 1513 and 1419 cm^{-1}) in grounds with lead.

An SR-μFTIR map through the cross section was assembled from individual spectra captured from adjacent 5 × 5 μm^2 apertures allowing greater spatial resolution than is possible with conventional instruments. The integrated absorption intensity map for wavenumber range 1525–1560 cm^{-1} indicates that crystalline zinc stearate is concentrated in the top margin of the ground (Fig. 2.6c). Spectra extracted from a line through the upper ground confirm a transition from broad to defined split zinc carboxylate peaks within the layer from bottom to top (Fig. 2.6d).[5]

2.9 Zinc Soap Aggregates

Robinson's paintings from the early 1980s also contain unusual examples of zinc soap formation. *Family Portrait* 1980 (oil on canvas, QUT Collection) is notably deteriorated and has been lined using a thermoplastic acrylic copolymer dispersion adhesive and repeatedly treated for paint cleavage. In addition to widespread cracking and lifting of paint, zinc soaps have manifested as numerous tiny lumps visible on the surface throughout the pale, barium sulfate-rich background paint and in some brown passages. FTIR analysis of various extracted aggregates recorded spectra consistent with crystalline zinc stearate with both single and split carboxylate peaks; however, others appeared more characteristic of zinc oleate (distinguished by a small C=C stretch at 3003 cm^{-1}) and zinc azelate (Theobald Clark and Osmond 2016). Material characterized as zinc azelate derived from lumps containing a relatively yellow glassy substance (Fig. 2.7). The associated spectrum contains carbonyl bands attributable to the oil medium not present in the zinc azelate reference spectrum, and CH_2/CH_3 stretching bands (inset) are less defined for the same reason; however, there is close agreement of carboxylate and more specifically CH_2 progression and rocking bands. Although lead azelate has previously been characterized at the perimeter of lead soap aggregates in a nineteenth-century primed canvas (Keune and Boon 2007), this is thought to be the first reported finding of an aggregate substantially comprising zinc azelate. Its presence suggests oxidative degradation of the paint has preceded soap formation, in this case possibly as a consequence of the painting's early storage in hot, humid conditions (Theobald Clark and Osmond 2016).

[5]The small peak at 3027 cm^{-1} reflects aromatic bonding present in the embedding medium rather than unsaturation in the carboxylate chain. While undesirable, the minor penetration of embedding medium evident in spectra from the periphery of the sample does not interfere with the ν_{as} COO^- absorption bands for zinc stearate.

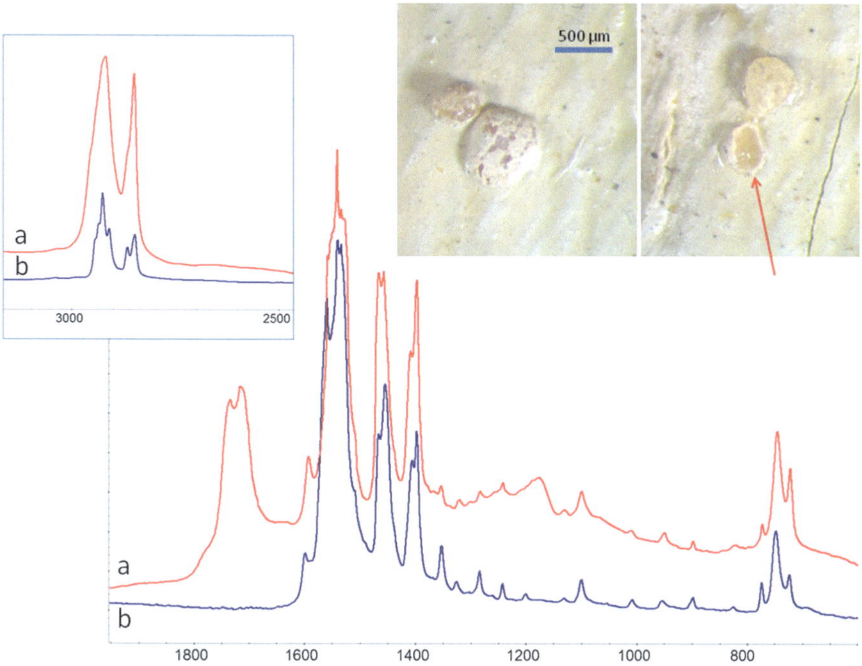

Fig. 2.7 Magnified surface details from William Robinson *Family Portrait* 1980 and transmission FTIR spectrum (**a**) extracted from the lump shown (red arrow) and (**b**) synthesized zinc azelate reference

2.10 Mineralized Zinc Soap Aggregates

Another phenomenon observed with large zinc soap aggregates is the presence of more complex carboxylates incorporating carbonate and sulfate functionality. Carbonate was first assigned in association with zinc carboxylate aggregates in a painting by van Gogh (Keune 2005). More conclusive and detailed characterization of basic zinc carbonate crystallization within aggregates has been acquired using higher-resolution SR-µFTIR mapping in a series of paint samples from paintings by France-based Australian artist Emanuel Phillips Fox (1865–1915) (Osmond 2014a).

One example derives from an undated (early twentieth century) wax-lined oil on canvas, *Untitled* (*Ploughing*) (the University of Melbourne Art Collection, gift of Dr. and Mrs. Eric Stock 1989), with a commercially prepared zinc oxide-based oil ground. A BSE image detail of a cross section from the sky shows unusual morphology within both zinc white and Prussian blue paint layers (Fig. 2.8). The detail from white paint and SEM energy dispersive X-ray spectrum (not shown) from the brightly UV fluorescent region in the optical image (encircled yellow) has lower particle density corresponding with high carbon counts. Small nodular zinc oxide particles in these areas are not visible and appear to have reacted away, leaving

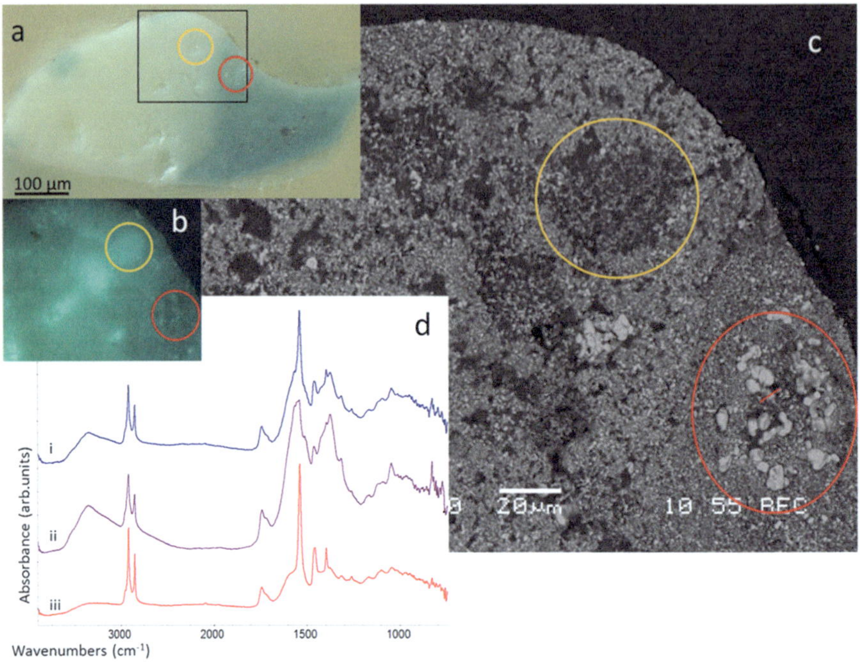

Fig. 2.8 E Phillips Fox *Untitled* (*Ploughing*, not dated) embedded cross-sectional sample. (**a**) Incident light. Details from boxed area (**b**) UV fluorescence and (**c**) back scattered electron image. Circles denote the same features in each image. Unusual morphology is evident in the BSE image reflecting zinc oxide pigment dissolution and soap formation and subsequent crystal growth. SR-µFTIR spectra (**d**) obtained from adjacent points along the red line indicated through an emerging crystal show a transition from zinc stearate (**iii**) to basic zinc carbonate (**ii**). Spectrum (**i**) contains features of both. Carbonate absorptions include CO_3^{2-} ν_3 doubly degenerate antisymmetric stretch (1507, 1987 cm^{-1}), ν_1 symmetric stretch (1045 cm^{-1}), and ν_2 bending mode (836 cm^{-1})

an amorphous zone containing dispersed acicular zinc oxide particles surrounded by zinc carboxylates. Another detail from blue paint (encircled red) reveals whitish zinc-based crystals which appear to emerge from an area of low electron density. In optical images, this area appears translucent and less fluorescent than surrounding paint. Spectra produced with SR-µFTIR mapping and extracted from adjacent 5×5 μm^2 points in a line through an emerging crystal show a transition from zinc stearate to basic zinc carbonate.

Basic zinc carbonate has been characterized in aggregates from other early twentieth-century paintings by Phillips Fox (Osmond 2014a) and also within large aggregates occurring in Leighton's *Winding the skein* 1878 (Osmond et al. 2013). Intended to have a porcelain smooth surface, some areas of the Leighton now display an unfortunate pimpled texture. The phenomenon relates to specific areas of the composition, most notably figures. Zinc-based aggregates originating from a lead, zinc, and chalk-based preparatory layer have expanded to deform the image plane.

SR-μFTIR maps and extracted spectra again show carbonate and carboxylate bands occurring within the same region.

A different form of mineralization has been documented in distinctive zinc-based aggregates in paintings by Vietnamese artist Nguyễn Trọng Kiệm (1933–1991) (Ebert et al. 2011; Osmond et al. 2014). In the *Portrait of My Wife* 1963 (oil on canvas, Witness Collection), elemental analysis combined with FTIR of large white protrusions indicates zinc carbonates and sulfates are represented together with zinc carboxylates. Micro X-ray diffraction (μXRD) indicates the possible presence of layered basic zinc salts with brucite-type zinc hydroxide structure (Osmond et al. 2014).[6] This class of compound can potentially accommodate carboxylate, carbonate and sulfate ions all within hydroxide layers forming stable layered basic zinc salts (Hofmeister and Platen 1992).

As is the case for mineralization documented in instances of lead soap protrusions (Higgitt et al. 2003; Keune and Boon 2007), the source of the carbonate and/or sulfate in the aggregates is not clear. One possibility is diffusion of gases from exposure to air; however, internal sources may also contribute. Sulfur dioxide is produced during the manufacture of Direct Process zinc oxide and may lead to formation of complex sulfites and sulfates on the pigment surface (Brown 1957). The active surface of zinc oxide particles additionally favors conversion to carbonate or sulfide during storage of dry pigment stock (Kühn 1986). Carbonate functionality was detected during routine FTIR analysis of contemporary zinc oxide pigment stocks in the course of recent research but was not present in sufficient amounts to register in corresponding XRD analysis (Osmond 2014a); however, some semicrystalline phases of zinc carbonate hydroxide hydrate were discerned during investigation of historic pigment samples (Clementi et al. 2012). Pigment purity or exposure history prior to paint formulation will influence reactivity to varying degrees, although the significance for later mineralization within soap aggregates is unclear.

2.11 Zinc Salt Formation and Surface Bloom

Other zinc salts characterized in paintings include those of shorter-chain lactate and oxalate ions. Incidences of zinc lactate occur in a number of paintings by Kiệm (Osmond et al. 2014) but also intriguingly in a single sample from a painting by Phillips Fox (Osmond 2014a). Unlike the larger masses of zinc stearate, zinc lactate is not associated with any distinctive morphological features, and characteristic spectra were identified in a variety of paint layers. Although potential sources of lactic acid have not been determined in these examples, it does appear to originate from within the painting. In this sense, these instances differ from the first reported

[6]Brucite group minerals have formula $M^{2+}(OH)_2$ with hexagonal crystal system and a layered lattice structure (www.mindat.org/min-29278 accessed 15 March 2017).

case of zinc lactate assigned to a disfiguring surface accretion affecting a mid-twentieth-century Canadian painting by Andor Weininger (Helwig et al. 2014). This study concluded that atmospheric pollutants were the most likely candidate for a source of lactic acid which reacted with zinc white at the painting surface.

Zinc lactate is water-soluble, which can be advantageous if present as a disfiguring surface accretion, but is less desirable when the salt occurs within paint layers. Also concerning are instances of insoluble metal oxalate crusts forming on paintings. More commonly associated with calcium in the context of both paintings and stone surfaces (Bordignon et al. 2008; Sutherland et al. 2013), oxalic acid will also react with zinc oxide. Poli et al. (2019) have studied terpenic acids present in different natural resins and found zinc white to be one of the pigments most likely to form metal soaps in their presence. Accelerated aging suggests that zinc (and copper) in adjacent pigment will reduce resin stability and simultaneously favor formation of associated oxalates. This has an unwelcome aesthetic consequence and may also significantly reduce the solubility and reversibility of natural resin varnishes. Zinc oxalates on the surface of paintings by Picasso, Mondrian and Munch are discussed by Monico et al. (2013) with evidence presented for both internal (cadmium-based pigment and natural resin) and external (pollutant) sources. Identification of zinc and a range of other metal oxalates using non-invasive reflection mid-FTIR spectroscopy is discussed and expanded upon by Rosi et al. (2019). There seems real potential for disfiguring and insoluble layers to form on the surface of zinc oxide-containing paints following reaction with acids from a range of internal, applied and environmental sources. These will challenge conventional approaches to cleaning.

2.12 Environmental Influences

Zinc soap formation in paintings will be influenced by environmental exposure. Zinc oxide-containing paints have been found to have a relatively high permeability to water vapor (Suryanarayana 1970). In the wet state, the presence of any water in zinc oxide-containing paints was shown to liberate fatty acids via hydrolysis of glycerol ester bonds and facilitate soap formation (Hansen and Klauss 1971). Exposure of a zinc oxide-containing paint reconstruction to high relative humidity and elevated light levels resulted in a higher degree of hydrolysis, oxidation and zinc soap formation relative to the same paint aged under ambient conditions (Keune et al. 2009).

The logical recommendation for paintings with potential for zinc soap formation may seem to be to limit exposure to high humidity and to maintain cool conditions. While links between humidity and oil hydrolysis appear unambiguous, with respect to temperature, however, anecdotal evidence exists that lowering temperatures encourages saturated fatty acids to effloresce at the paint surface – a phenomenon personally witnessed in storage facilities following a 3 °C set-point reduction. Although fatty acid efflorescence can often be brushed away, questions remain about the possible impacts of its removal and typical reoccurrence. In paintings containing zinc oxide, it is also possible that the migration of fatty acids promoted by temperature changes may increase opportunities for soap formation and aggregation. The presence of free fatty acids at the painting surface may additionally encourage formation of intractable metal soap blooms in susceptible paintings.

2.13 Conclusion

The reactivity of zinc oxide in oil-based paints is dependent on a variety of factors including the inherent properties of the zinc oxide particle and the fatty acid composition of the paint. Environmental conditions will influence the release and availability of fatty acids. Zinc oxide, whether used alone or in combination with other pigments, is susceptible to reaction with fatty acids in a variety of oil mediums but is particularly susceptible to high stearic acid concentrations such as may derive from aluminum stearate. Although it is still not clear what might cause zinc soaps to aggregate and crystallize as lumps in the paint rather than migrate to an interface, accumulation between layers or at the surface has its own structural or aesthetic consequences for the painting. The susceptibility of paintings to hydrolysis and saponification during transient exposure to heat or moisture such as may occur during conservation treatments is also not well understood. Continuing research is required to improve our capacity to manage vulnerable collections and to develop innovative treatments for conditions which are challenging conventional approaches to conservation.

Acknowledgments Sophie Theobald Clark, Marion Mecklenburg, Catherine Nunn, Bettina Ebert, Paula Dredge, Jaap Boon, Katrien Keune, Joen Hermans, Kate Helwig, Anne Carter, Amanda Pagliarino, Nicole Tse and John Drennan have all generously shared samples and information or otherwise supported my research over many years. I acknowledge the facilities and the scientific and technical assistance of the Australian Microscopy and Microanalysis Research Facility at the Centre for Microscopy and Microanalysis, University of Queensland, and at the Infrared Microspectroscopy Beamline at the Australian Synchrotron, Victoria, Australia. QUT facilitated the William Robinson Research Project. The continuing support of QAGOMA is gratefully acknowledged.

References

Bertrand L, Refregiers M, Berrie B, Echard J-P, Thoury M (2013) A multiscalar photoluminescence approach to discriminate among semiconducting historical zinc white pigments. Analyst 138:4463–4469. https://doi.org/10.1039/c3an36874b

Bomford D, Kirby J, Leighton J, Roy A (1990) Art in the making: impressionism. National Gallery Publications Limited, London

Boon JJ, Hoogland FG (2014) Investigating fluidizing dripping pink commercial paint on van Hemert's *Seven*-series works from 1990–1995. In: Van den Berg KJ, Burnstock A, de Tagle A, de Keijzer M, Heydenreich G, Krueger J, Learner T (eds) Issues in contemporary oil paints. Springer, Cham, pp 227–246

Bordignon F, Postorino P, Dore P, Tabasso ML (2008) The formation of metal oxalates in the painted layers of a medieval polychrome on stone, as revealed by micro-Raman spectroscopy. Stud Conserv 53(3):158–169

Bronken IAT, Boon JJ (2014) Hard dry paint, softening tacky paint, and exuding drips on *Composition* (1952) by Jean-Paul Riopelle. In: Van den Berg KJ, Burnstock A, de Tagle A, de Keijzer M, Heydenreich G, Krueger J, Learner T (eds) Issues in contemporary oil paints. Springer, Cham, pp 247–262

Brown HE (1957) Zinc oxide rediscovered. New Jersey Zinc Company, New York

Clementi C, Rosi F, Romani A, Vivani R, Brunetti B, Miliani C (2012) Photoluminescence properties of zinc oxide in paints: a study of the effect of self-absorption and passivation. Appl Spectrosc 66(10):1233–1241

Cocks EJ, Walters B (1968) A history of the zinc smelting industry in Britain. George G. Harrap & Co. Ltd, London

Dredge P, Schilling MR, Gautier G, Mazurek J, Learner T, Wuhrer R (2013) Lifting the lids off Ripolin: a collection of paint from Sidney Nolan's studio. J Am Inst Conserv 52(4):213–226

Ebert B, MacMillan Armstrong S, Singer B, Grimaldi N (2011) Analysis and conservation treatment of Vietnamese paintings. In: Bridgland J, Antomarchi C (eds) Preprints of the ICOM committee for conservation 16th triennial meeting. ICOM-CC, Lisbon, p 1305

Erhardt D, Tumosa CS, Mecklenburg MF (2005) Long-term chemical and physical processes in oil paint films. Stud Conserv 50(2):143–150

Gardner HA (1917) Paint researches and their practical application. Press of Judd and Detweiler, Inc, Washington, DC

Gautier G, Bezur A, Muir K, Casadio F, Fiedler I (2009) Chemical fingerprinting of ready-mixed house paints of relevance to artistic production in the first half of the twentieth century. Part I: inorganic and organic pigments. Appl Spectrosc 63(6):597–603

Hansen CM, Klauss HC (1971) Mechanism of paint seeding. Ind Eng Chem Prod Res Dev 10(2):189–192

Helwig K, Poulin J, Corbeil M-C, Moffatt E, Duguay D (2014) Conservation issues in several 20th-century Canadian oil paintings: the role of zinc carboxylate reaction products. In: Van den Berg KJ, Burnstock A, de Tagle A, de Keijzer M, Heydenreich G, Krueger J, Learner T (eds) Issues in contemporary oil paints. Springer, Cham, pp 167–184

Hermans JJ, Keune K, Van Loon A, Corkery RW, Iedema PD (2014) The molecular structure of three types of long-chain zinc(II) alkanoates for the study of oil paint degradation. Poyhedron 81:335–340

Hermans JJ, Keune K, Van Loon A, Iedema PD (2015) An infrared spectroscopic study of the nature of zinc carboxylates in oil paintings. J Anal At Spectrom 30:1600–1608

Hermans JJ, Keune K, Van Loon A, Corkery RW, Iedema PD (2016) Ionomer-like structure in mature oil paint binding media. RSC Adv 6:93363–93369

Higgitt C, Spring M, Saunders D (2003) Pigment-medium interactions in oil paint films containing red lead or lead-tin yellow. Natl Gallery Tech Bull 24:75–95

Hofmeister W, Platen HV (1992) Crystal chemistry and atomic order in Brucite-related double-layer structures. Crystallogr Rev 3(1):3–26

Keune K (2005) Binding medium, pigments and metal soaps characterised and localised in paint cross-sections. Ph.D. thesis, University of Amsterdam, Amsterdam

Keune K, Boon JJ (2007) Analytical imaging studies of cross-sections of paintings affected by lead soap aggregate formation. Stud Conserv 52(3):161–176

Keune K, Hoogland F, Boon JJ, Peggie D, Higgitt C (2009) Evaluation of the added value of SIMS: a mass spectrometric and spectroscopic study of an unusual Naples yellow oil paint reconstruction. Int J Mass Spectrom 284(1–3):22–34

Kühn H (1986) Zinc white. In: Feller RL (ed) Artists' pigments, vol 1. Cambridge University Press, Cambridge/New York, pp 169–186

Maines CA, Rogala D, Lake S, Mecklenburg MF (2011) Deterioration in abstract expressionist paintings: analysis of zinc oxide paint layers in works from the collection of the Hirshhorn Museum and Sculpture Garden, Smithsonian Institution. Mater Res Soc Proc 1319:275–284

Mills JS, White R (1987) The organic chemistry of museum objects. Butterworths, London
Monico L, Rosi F, Miliani C, Daveri A, Brunetti BG (2013) Non-invasive identification of metal-oxalate complexes on polychrome artwork surfaces by reflection mid-infrared spectroscopy. Spectrochim Acta A Mol Biomol Spectrosc 116:270–280
Morley-Smith CT (1950) The development of anti-chalking French process zinc oxides. J Oil Colour Chem Assoc 33:484–501
Morley-Smith CT (1958) Zinc oxide – a reactive pigment. J Oil Colour Chem Assoc 41:85–97
O'Donoghue E, Johnson AM, Mazurek J, Preusser FD, Schilling MR, Walton MS (2006) Dictated by media: conservation and technical analysis of a 1938 Joan Miró canvas painting. In: Saunders D, Townsend J, Woodcock S (eds) The object in context: crossing conservation boundaries: contributions to the Munich congress, 28 Aug–1 Sept 2006. IIC, London, pp 62–68
Osmond G (2012) Zinc white: a review of zinc oxide pigment properties and implications for stability in oil-based paintings. AICCM Bulletin 33:20–29
Osmond G (2014a) Zinc oxide-centred deterioration of modern artists' oil paint and implications for the conservation of twentieth century paintings. Ph.D. thesis, University of Queensland, Brisbane
Osmond G (2014b) Zinc white and the influence of paint composition for stability in oil based paints. In: Van den Berg KJ, Burnstock A, de Tagle A, de Keijzer M, Heydenreich G, Krueger J, Learner T (eds) Issues in contemporary oil paints. Springer, Cham, pp 263–281
Osmond G, Keune K, Boon J (2005) A study of zinc soap aggregates in a late 19th century painting by R.G. Rivers at the Queensland Art Gallery. AICCM Bull 29:37–46
Osmond G, Boon JJ, Puskar L, Drennan J (2012) Metal stearate distributions in modern artists' oil paints: surface and cross-sectional investigation of reference paint films using conventional and synchrotron infrared microspectroscopy. Appl Spectrosc 66(10):1136–1144
Osmond G, Ives S, Dredge P, Drennan J, Puskar L (2013) From porcelain to pimples: a study of synchrotron sourced infrared spectroscopy for understanding the localised aggregation of zinc soaps in a painting by Sir Frederick Leighton. Poster presented at the 7th international workshop on infrared microscopy and spectroscopy with accelerator based sources, Lorne, 10–14 Nov 2013
Osmond G, Ebert B, Drennan J (2014) Zinc oxide-centred deterioration in 20th century Vietnamese paintings by Nguyễn Trọng Kiệm. AICCM Bull 34:4–14
Poli T, Piccirillo A, Nervo M, Chiantore O (2019) Aging of natural resins in presence of pigments: metal soaps and oxalates formation. In: Casadio F, Keune K, Noble P, Van Loon A, Hendriks E, Centeno S, Osmond G (eds) Metal soaps in art: conservation and research. Springer, Cham, pp 143–154
Raven L, Bisschoff M, Leeuwestein M, Geldof M, Hermans JJ, Stols-Witlox M, Keune K (2019) Delamination due to zinc soap formation in an oil painting by Piet Mondrian (1872–1944). In: Casadio F, Keune K, Noble P, Van Loon A, Hendriks E, Centeno S, Osmond G (eds) Metal soaps in art: conservation and research. Springer, Cham, pp 345–357
Robinet L, Corbeil MC (2003) The characterization of metal soaps. Stud Conserv 48(1):23–40
Rogala D, Lake S, Maines C, Mecklenburg M (2010) Condition problems related to zinc oxide underlayers: examination of selected abstract expressionist paintings from the collection of the Hirschhorn Museum and Sculpture Garden, Smithsonian Institution. J Am Inst Conserv 49(2):96–113
Rosi F, Cartechini L, Monico L, Gabrieli F, Vagnini M, Buti D, Doherty B, Anselmi C, Brunetti BG, Miliani C (2019) Tracking metal oxalates and carboxylates on painting surfaces by non-invasive reflection mid-FTIR spectroscopy. In: Casadio F, Keune K, Noble P, Van Loon A, Hendriks E, Centeno S, Osmond G (eds) Metal soaps in art: conservation and research. Springer, Cham, pp 175–192
Saunders D, Kirby J (2004) The effect of relative humidity on artists' pigments. Natl Gallery Tech Bull 25:62–72

Schilling MR, Carson DM, Khanjian HP (1999) Gas chromatographic determination of the fatty acid and glycerol content of lipids. IV. Evaporation of fatty acids and the formation of ghost images by framed oil paintings. In: Bridgland J, Brown J (eds) Preprints of the ICOM committee for conservation12th triennial meeting, Lyon, 1999. James and James, London, pp 242–247

Schilling MR, Mazurek J, Learner TJS (2007) Studies of modern oil-based artists' paint media by gas chromatography/mass spectrometry. In: Learner TJS, Smithen P, Krueger JW, Schilling MR (eds) Modern paints uncovered, Tate Modern, London, 16–19 May 2006. The Getty Conservation Institute, pp 129–139

Shimadzu Y, van den Berg KJ (2006) On metal soap related colour and transparency changes in a 19th century painting by Millais. In: Boon JJ, ESB F (eds) Reporting highlights of the de Mayerne Programme. Netherlands Organisation for Scientific Research (NWO), The Hague, pp 43–52

Shimadzu Y, Keune K, van den Berg KJ (2008) The effects of lead and zinc white saponification on surface appearance of paint. In: Bridgland J (ed) Preprints of the ICOM committee for conservation: preprints of the 15th triennial meeting, New Delhi, 22–26 Sept 2008. Allied Publishers Pvt Ltd, pp 626–632

Standeven H (2011) House paints, 1900–1960: history and use. Getty Conservation Institute, Los Angeles

Suryanarayana NP (1970) Critical pigment volume concentration of some oxide pigments. J Colour Soc 9(2):2–6

Sutherland K, Price BA, Lins B, Passeri I (2013) Oxalate-rich surface layers on paintings: implications for interpretation and cleaning. In: Mecklenburg MF, Charola EA, Koestler RJ (eds) New insights into the cleaning of paintings. Proceedings from the cleaning 2010 international conference Universidad Politecnica de Valencia and Museum Conservation Institute. Smithsonian Institution Scholarly Press, pp 85–87

Theobald Clark S, Osmond G (2016) The materials and techniques of William Robinson. AICCM Bull 37(2):51–65

van den Berg J, van den Berg K, Boon J (1999) Chemical changes in curing and ageing oil paints. In: Bridgland J (ed) Preprints of the ICOM committee for conservation 12th triennial meeting, Lyon, pp 248–253

van den Berg KJ, Burnstock A, Schilling M (2019) Notes on metal soap extenders in modern oil paints: history, use, degradation and analysis. In: Casadio F, Keune K, Noble P, Van Loon A, Hendriks E, Centeno S, Osmond G (eds) Metal soaps in art: conservation and research. Springer, Cham, pp 331–343

van der Weerd J, Gelddof M, van der Loeff LS, Heeren R, Boon J (2003) Zinc soap aggregate formation in 'Falling leaves (les Alyscamps)' by Vincent van Gogh *Zeitschrift für*. Kunsttechnologie und Konservierung 17(2):407–416

Van Gorkum R, Bouwman E (2005) The oxidative drying of alkyd paint catalysed by metal complexes. Coord Chem Rev 249:1709–1728

Winter G, Whittem RN (1950) Fluorescence and photo-chemical activity of zinc oxides. J Oil Colour Chem Assoc 33:477–483

Wöll C (2007) The chemistry and physics of zinc oxide surfaces. Prog Surf Sci 82:55–120

Chapter 3
Toward a Complete Molecular Model for the Formation of Metal Soaps in Oil Paints

Joen J. Hermans, Katrien Keune, Annelies Van Loon, and Piet D. Iedema

Abstract An overview is presented of the current state of understanding of the chemical pathways that lead to the formation of crystalline metal soap phases in oil paints, based on recent experimental work by the authors and supported by relevant literature. Improved (quantitative) interpretation of Fourier-transform infrared (FTIR) spectra has revealed that metal ions are bound to carboxylate functionalities of the oil polymer during oil paint aging, a state similar to ionomeric polymers. Tailored ionomer-like systems based on linseed oil were synthesized to study the structure of the mature oil paint binding medium, and such systems were used as the starting point for studies on metal soap crystallization. Additionally, series of differential scanning calorimetry (DSC) studies shed light on the driving forces and kinetics of metal soap crystallization, and electron microscopy studies have been used to image the initial stages of metal soap crystallization. The results have been used to construct a model of the chemical reactions leading to metal soaps from a mixture of pigment and oil. Additionally, the model provides insight into diffusion mechanisms for metal ions and fatty acids and potential physical transitions in the structure of metal soaps. The mechanisms described are helpful in explaining the different morphologies of lead and zinc soaps observed in actual samples from historic paintings and the locations within paint films where these are typically found.

J. J. Hermans (✉) · P. D. Iedema
Van't Hoff Institute for Molecular Sciences, University of Amsterdam, Amsterdam, The Netherlands
e-mail: j.j.hermans@uva.nl

K. Keune
Conservation Department, Rijksmuseum, Van't Hoff Institute for Molecular Sciences, University of Amsterdam, Amsterdam, The Netherlands
e-mail: K.Keune@rijksmuseum.nl

A. Van Loon
Conservation Department, Rijksmuseum, Amsterdam, The Netherlands

F. Casadio et al. (eds.), *Metal Soaps in Art*, Cultural Heritage Science,
https://doi.org/10.1007/978-3-319-90617-1_3

3.1 Introduction

Since metal soaps were first identified in large protrusions in oil paint layers, considerable efforts have been made to characterize them and the mechanisms by which they form. We now know that lead and zinc are the most common metal ions to form metal soaps, originating from pigments and/or driers like zinc white (ZnO) (Hermans et al., 2015; Keune and Boevé-Jones, 2014; Osmond et al., 2012), lead white ($2\,Pb(CO_3) \cdot Pb(OH)_2$) (Keune et al., 2011), lead-tin yellow (Pb_2SnO_4) (Higgitt et al., 2003), chrome yellow ($PbCrO_4$) (Monico et al., 2016), red lead (Pb_3O_4), litharge (PbO), or lead acetate ($Pb(CH_3COO)_2$), though copper and calcium soaps have also been found. The fatty acids in these metal soap phases derive from the drying oil binder or paint additives (for instance, stearate salts used as dispersing agents in modern paints). They are almost exclusively the saturated fatty acids palmitic, stearic, or azelaic acid (Plater et al., 2003), with the occasional presence of oleic acid (Helwig et al., 2014). In terms of morphology, it has been observed that lead soaps tend to form large metal soap aggregates like in Fig. 3.1. This material seems to be capable of deforming the paint layers under which it grows, and it can undergo remineralization to carbonates, sulfates, chlorides, or oxides (Boon et al., 2006; Keune et al., 2011). In contrast, zinc soaps tend to remain more dispersed throughout a paint layer (Hermans et al., 2015). Though not all metal soap formations are detrimental to paint stability, it is unquestionable that both lead and zinc soaps have been associated with paint flaking, protrusions, transparency, brittleness, and delamination, examples of which are well-documented in some of the other contributions to this volume.

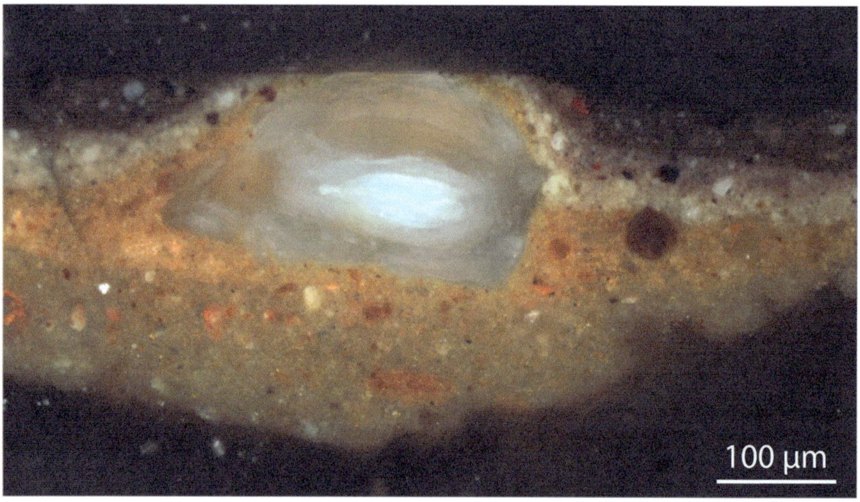

Fig. 3.1 Dark-field cross-polarized microscopy image of a large inclusion formed in the orange ground identified as lead soap in a paint cross-section from the painting *Let the Children Come to Me* by Jan de Bray (1663, Frans Halsmuseum). (Picture by L. Abraham)

For the conservation of paintings that are negatively affected by metal soap formation, a comprehensive understanding of the chemical pathways that lead from a fresh oil paint to a degraded painting should be the ultimate goal. When it is possible to characterize all the stages in the metal soap formation pathway and accurately diagnose the state of a specific paint layer, targeted strategies can be developed to slow down degradation at that particular stage. Besides determining a set of chemical reactions occurring in the drying oil medium and at the interface with pigments, it is important to develop an understanding of the reaction rates and transport processes. These rates not only determine how fast degradation takes place under certain circumstances, but they may also affect the size and location of metal soap phases.

In the past 5 years, we have conducted research on the structure of mature oil paint binding media and the crystallization of metal soaps using a combination of Fourier-transform infrared spectroscopy (FTIR), X-ray diffraction (XRD), differential scanning calorimetry (DSC), and electron microscopy. An adaptable experimental model system that mimics mature oil paint binding media was developed to allow the isolation and control of parameters, like metal concentration, extent of polymerization, and degree of oxidation, and study their effect on metal soap formation. This system also offers an alternative to artificially aged reconstructions of oil paints for the study of cleaning or restoration methods.

In this paper, an overview is given of our current level of understanding of the processes that lead to metal soap formation, which is based on experimental evidence gathered by the authors, supported by relevant literature from the oil paint conservation field as well as the wider chemical literature. To structure the discussion, we classify oil paint aging processes in two sets: processes that mostly contribute to the drying and stabilization of an oil paint (Sect. 3.2) and processes that have a tendency to cause paint deterioration (Sect. 3.3). The two classes of processes are not separated in time; in reality, all processes are likely to co-occur on all timescales to some extent, with paint stabilization being dominant only on relatively short time scales. The discussion of mechanisms presented here mostly concerns metal soap formation in mixtures of pigment and drying oil. The presence of lead driers or additives like aluminum stearate may change the order in which processes take place, but many of the individual reaction or transport steps described here should be relevant to most cases of metal soap formation. It should be noted that the systems described in the chapter are simplified with respect to historic artists' paints (which can have significant formulation differences between lead white and zinc white paints in terms of particle sizes, additives, impurities, and driers), so as to be able to derive widely applicable mechanistic information on paint aging. Finally, we discuss important gaps remaining in our comprehension of metal soap formation and suggest experiments that could reduce their magnitude.

3.2 Oil Paint Drying: The Formation of Ionomer Networks

As soon as a pigment is mixed with a drying oil (typically linseed oil), reactions will start to take place that persist well after the paint has dried to the touch, some of which ultimately lead to the formation of metal soaps. In what follows, it is very important to make the distinction between *metal soaps* (complexes of metal ions and fatty acids, in oil paints usually saturated mono- or dicarboxylic fatty acids) and the much broader class of *metal carboxylates*, which may have any sort of molecular functionality or structure behind the carbon atom in the carboxylic headgroup. The main reaction steps believed to be contributing to the stability of the polymer medium are illustrated in Fig. 3.2 and will be discussed in detail below.

3.2.1 Carboxylate Formation During Oil Paint Drying

The early stages of oil paint drying are described extensively in the literature. Upon exposure to oxygen, drying oils such as linseed oil form a densely cross-linked polymer network mostly through radical polymerization reactions (Mallégol et al., 1999, 2000a,b,c, 2001). In this polymerization process, one reaction pathway leads to scission of fatty acid chains containing a peroxyl radical to yield aldehydes.

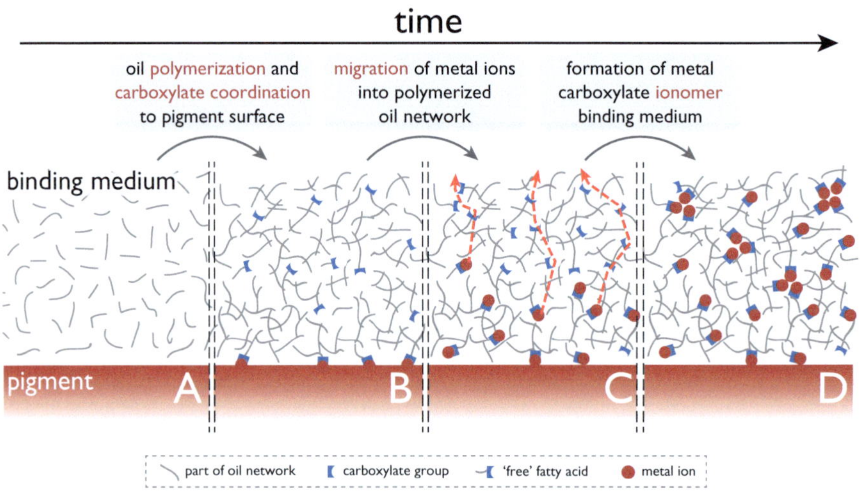

Fig. 3.2 Illustration of the processes near the pigment/medium interface occurring during drying in oil paints containing pigments prone to develop metal soaps. The drying oil binding medium (**a**) acquires carboxylate groups during autoxidative drying (**b**), after which diffusion of metal ions from the pigment surface (**c**) leads to the formation of an ionomeric system (**d**). Dashed red arrows illustrate the diffusion of metal ions from the pigment surface into the polymerized binding medium mediated by negatively charged carboxylate moieties

FTIR spectroscopy has revealed that these aldehydes are at least partially oxidized to carboxylic acid groups (Mallégol et al., 1999). Carboxylic acids can also be generated by hydrolysis of the ester bond in the triacyl glyceride molecules that make up the oil, as discussed in detail in Sect. 3.3.1. Polymers containing carboxylic acid functionalities have been shown to adsorb on metallic zinc, especially when that zinc surface was treated with base to increase the concentration of hydroxyl groups on the oxidized zinc surface (Taheri et al., 2013). Fatty acids are also routinely used to coat ZnO nanoparticles (Badre et al., 2007; Sehmi et al., 2016). The ability of carboxylate groups to adsorb on metal oxide surfaces makes fatty acids and carboxylic polymers common stabilizers for colloidal dispersions (Cesarano and Aksay, 1988; Lin et al., 2005; Shen et al., 1999). These carboxylic-metal surface interactions are expected to be a common phenomenon as well for many pigments in polymerized linseed oil, as the coordination of metal ions on the pigment surface by carboxylate groups could help lower the interfacial energy between pigment and binding medium (Fig. 3.2b).

3.2.2 Migration of Metal Ions Throughout the Polymeric Medium

In cross-sections of oil paintings that contain metal soaps, it has often been observed that the metal soap aggregates are separated from pigment particles by stretches of binding medium. This important observation implies that either metal soaps have formed near the pigment-oil interface and migrated as complex or that metal ions have diffused through the binding medium (assuming that free saturated fatty acids can be found throughout the binding medium). Considering the mass, the bulkiness, and, as we shall discuss in detail later, the very low solubility of metal soap complexes (Hermans et al., 2016b), the latter case is highly likely to be the dominant process. As experimental support of this idea, we observed that characteristic zinc carboxylate vibration bands can be found in equal intensity in transparent areas of a ZnO model paint film and in areas containing pigment, indicating that zinc ions must have migrated through the binding medium, as shown in Fig. 3.3. We propose that metal ions such as zinc and lead are able to diffuse through the binding medium by “hopping” from one carboxylate group to the next, as illustrated in Fig. 3.2c. Such a transport mechanism has been studied both experimentally and through molecular dynamics simulations for several synthetic polymers (Castagna et al., 2011; Hall et al., 2012; Lin and Maranas, 2012; Tudryn et al., 2012). Before the transfer of a metal ion from one carboxylic group to a second can occur, the bound and unbound carboxylic groups need to come in close proximity, which requires mobility of the polymer segments. The rate at which metal ion migration through the medium can occur is therefore strongly increased by conditions that increase polymer dynamics. The swelling of paint films by solvent, for instance, plasticizes the polymeric binding medium and is likely

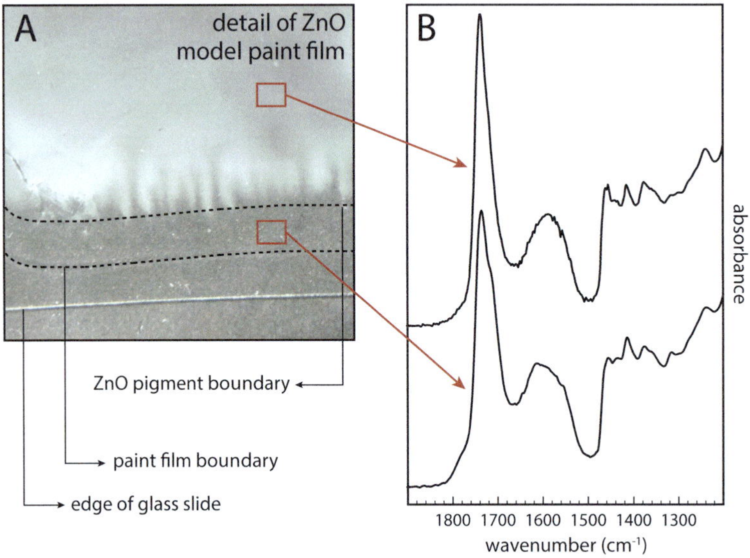

Fig. 3.3 (**a**) Detail of a thin model "paint" film consisting of a low concentration of ZnO in linseed oil, with a band of transparent polymerized medium on the outer perimeter. The film was dried at room temperature for 7 weeks. (**b**) FTIR spectra corresponding to sections of the film with and without ZnO, both showing a broad zinc carboxylate band centered around 1580 cm^{-1}, indicating that zinc ions have migrated through the binding medium

to have a strong effect on metal ion diffusion rates. Like all polymers, oil paint binding media also have a glass transition temperature (T_g), the temperature range at which a polymer changes from a rigid glass-like state to a softer rubbery state. There is still some uncertainty about typical T_g values for oil paint binding media or the influence of pigments on T_g. However, an interesting study by Phenix (2009) reported broad glass transitions between $-2\,°C$ and $45\,°C$ for a variety of fully dried pigmented paint films, with some zinc white paint films only showing gradual thermal softening.

3.2.3 Formation of Ionomer-Like Binding Medium

When dissociation of metal ions from the pigment surface and subsequent metal ion migration through the polymerized binding medium are spontaneous processes, they are likely to continue until many of the free carboxylate groups in the binding medium are bound to metal ions. Such a binding medium filled with metal ions can be classified as an ionomeric polymer network (Fig. 3.2d). We consider this stage in oil paint curing as part of the drying process, because despite the necessary loss of pigment during the formation of an ionomeric system, the metal ions can in fact contribute to the stability of a polymerized oil network. Divalent ions such as Zn^{2+} and Pb^{2+} need to coordinate to two carboxylate groups to balance their charge, so they effectively form additional cross-links between sections of polymer chain. Though these bonds are more easily dissociated than covalent cross-links, at room temperature, these strong metal carboxylate links will increase the stiffness of the binding medium. The formation of ionomer-like structures in oil paint binding media has been suggested before as an interpretation of broad metal carboxylate bands shifted toward higher wavenumbers compared to pure metal soap references often observed in FTIR spectra of mature oil paints. Ionomers have also been offered as an explanation for the somewhat surprising remaining structural stability of oil paints that have suffered extensive breakdown of the polymer network through aging (Boon et al., 2006; Tumosa et al., 2005; van den Berg et al., 2001; van der Weerd et al., 2005).

To investigate the validity of the metal migration processes just described experimentally, we have compared oil paint binding media to a synthetic ionomer based on linseed oil with controllable metal ion concentrations. The most successful method was to synthesize complexes of either lead or zinc with sorbic acid (2,4-hexadienoic acid), mix this insoluble powder with linseed oil, and cure a thin layer of the resulting paste in an air-circulated oven at 150 °C. Transparent homogeneous orange/brown films were obtained that showed no remaining metal sorbate bands in FTIR spectra or crystalline peaks of the metal sorbate complex in X-ray diffractograms (Hermans et al., 2015). Instead, a broad metal carboxylate IR band appeared with a maximum around 1585 and 1530 cm^{-1} for zinc and lead ionomers, respectively (see Fig. 3.4). The position and shape of these bands are very similar to those observed in many samples from naturally aged oil paints

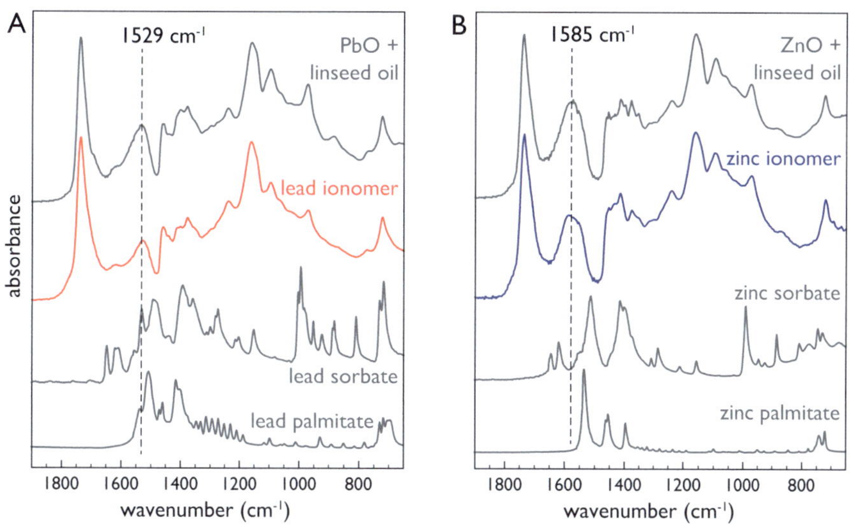

Fig. 3.4 A comparison of FTIR spectra of metal palmitate, metal sorbate salt, linseed oil-based ionomer, and model paint containing the metal oxide for (**a**) lead and (**b**) zinc. The dashed line marks the typical position of the broad metal carboxylate band for both metals

containing ZnO or lead-containing pigments (Higgitt et al., 2003; Osmond et al., 2012; Thoury et al., 2019; van der Weerd et al., 2005). Small-angle X-ray scattering (SAXS) revealed that these linseed oil-based ionomers, similar to commercial less cross-linked ionomer systems, show characteristic peaks at $d = 70–90$ Å caused by the formation of clusters of ionic metal carboxylate groups within the polymerized binding medium (Hermans et al., 2016a). The properties and concentration of these ionic clusters in oil paint binding media could have an important effect on metal ion migration rates and the mechanical properties of the paint.

Several hypotheses have been put forward to account for the broad metal carboxylate band in FTIR spectra of oil paint samples, such as the top spectra in Fig. 3.4a, b. Besides the ionomer hypothesis, it has been suggested that carboxylate groups could be adsorbed on the surface of pigments with a variation of binding geometries and bond lengths, causing the metal carboxylate band to broaden (Clementi et al., 2012; Osmond et al., 2012). Alternatively, the band could be associated with metal soaps that have not yet crystallized. In molten (i.e., amorphous) form, the metal soaps are known to show a shift in the metal carboxylate bands (Ishioka et al., 2000). In order to distinguish between these three hypotheses, we can consider the potential maximum number of metal carboxylate bonds in each of them (Hermans et al., 2016a). The number of noncrystalline metal soap complexes that can be formed in a fully cured binding medium is limited by the fraction of the total fatty acid content (present as esters) in linseed oil that is saturated. These saturated fatty acids are stearic and palmitic acid naturally present in linseed oil, and diacids such as azelaic acid formed through oxidation. The fraction of saturated fatty acids in linseed oil is usually around 7–13% (van den Berg et al., 2004). In the case of adsorption of carboxylates on pigments, on the other hand, the maximum number of metal carboxylate bonds is limited by the available pigment surface area, which in turn depends on the pigment-to-oil weight ratio and the average particle size of the pigments.

Using the linseed oil ionomer model system, we prepared a calibration series varying in metal content that allows calculation of the molar ratio of metal carboxylate groups to ester groups (COOM/COOR) in any real oil paint sample or model paint. A determination of the absolute concentration of metal carboxylates is not (yet) possible, because the concentration of ester bonds is difficult to determine in insoluble solid polymer films. Typical oil paint films showing a broad and shifted metal carboxylate band had a COOM/COOR ratio around 0.5 or higher (Hermans et al., 2016a). This concentration of metal carboxylate groups is too high to be explained completely by either amorphous metal soaps or the adsorption of carboxylates on pigment surfaces or a combination of the two. In each of these cases, the COOM/COOR ratio would never increase beyond 0.15 for a realistic paint formulation (Hermans et al., 2016a). We may conclude, therefore, that the broad metal carboxylate bands observed in FTIR spectra of oil paint films are largely caused by metal ions bound to the linseed oil polymer network. In those cases, the binding medium has become an ionomer-like system.

The classification of the oil paint binding medium as an ionomer-like system has a big impact on our understanding of metal soap-related degradation pathways and paint stability. If the ionomer model for aged oil paints is appropriate, it means that metal ions are distributed throughout the binding medium as carboxylate complexes, where they may act as a catalyst or reactant in subsequent reactions. Therefore, in theory, the formation of metal soaps could occur anywhere where there are free saturated fatty acids available to form a complex with. These free acids can be located either within the paint layer or on the interface with, for instance, a medium-rich layer. In the next section, we will discuss potential molecular mechanisms for forming large crystalline metal soap phases from ionomeric binding media.

3.3 Oil Paint Deterioration: Metal Soap Formation in Ionomer Networks

Ionomer networks are potentially stable for long periods of time. When there are no other significant concentrations of negatively charged ions and the number of carboxylic groups in the system reaches a constant value in a fully dried paint film, there is no net movement of metal ions through the system once all carboxylate functionalities are bound to metal ions. If crystalline metal soaps are to form in an ionomeric system, a significant fraction of the ester bonds in the triacylglycerides that make up the drying oil polymer need to be hydrolyzed (or the system needs to be exposed to an external source of fatty acids, e.g., beeswax), as the concentration of free saturated fatty acids (SFAs) in liquid drying oil is rather low. The processes leading to metal soap formation in ionomeric oil paint media are illustrated in Fig. 3.5.

3.3.1 Free Fatty Acid Generation Through Ester Bond Hydrolysis

There are two distinct ways in which ester hydrolysis can occur, both of which require the presence of water molecules. Firstly, spontaneous hydrolysis, the reaction of an ester group with a water molecule, would lead to the formation of a carboxylic acid group and an alcohol. When the specific fatty acid chain is completely saturated, this reaction would lead to the generation of a carboxylic acid that may diffuse through the polymerized binding medium, a free SFA. However, we need to consider that, in the case of an ionomeric binding medium, there are lead or zinc ions dispersed throughout the system and these ions may act as Lewis acids. The ions are coordinated to carboxylic groups, but the fact that they are capable of diffusing through the binding medium means that it is a rather dynamic type of coordination. The width of the metal carboxylate band in FTIR spectra of ionomers also suggests that metal coordination is not very well-defined in these systems. Given this situation, it is reasonable to suppose that ester carbonyl moieties can

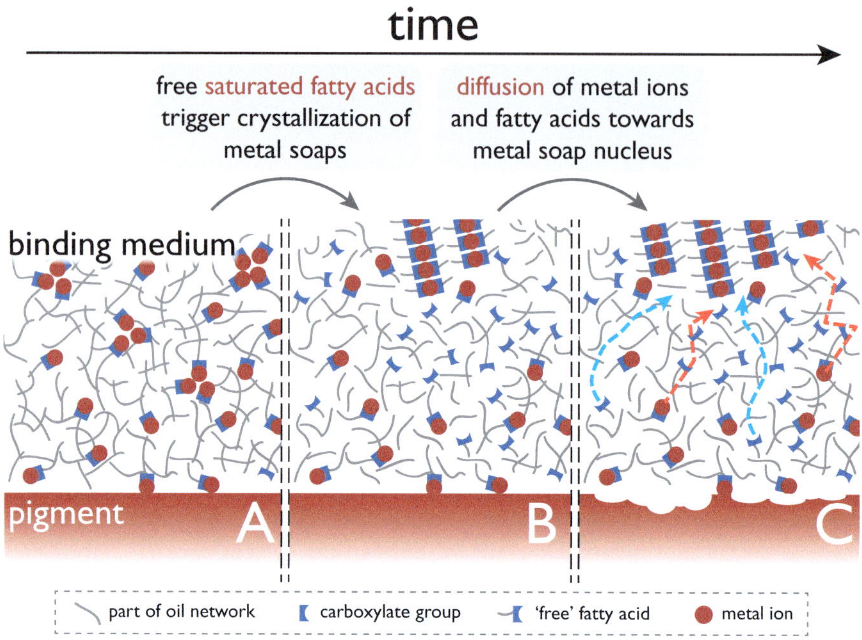

Fig. 3.5 Illustration of metal soap-related degradation processes in ionomeric binding media (**a**), triggered by the presence of free SFAs (**b**) and resulting in the formation of large crystalline metal soap phases (**c**). Dashed arrows illustrate the diffusion of metal ions (red) and free SFAs (blue) toward the growing crystalline metal soap aggregate

transiently bind to the metal ion as well. Lewis acids such as lead and zinc ions draw away electron density from the ester carbonyl bond, destabilizing the C–O bond that links the fatty acid to the glycerol backbone. This destabilization makes the ester more prone toward hydrolysis. Cases have been reported in the literature where a zinc carboxylate complex catalyzes transesterification of a wide variety of methyl esters (Iwasaki et al., 2008). MacDonald and co-workers found that ethyl linoleate is quickly hydrolyzed to form ethanol and metal linoleate complexes in the solutions of either zinc or lead acetate in methanol, while ethyl linoleate seemed to remain intact when potassium acetate was used (MacDonald et al., 2016). This experimental result supports the idea that, given the presence of metal ions in the polymerized binding medium, these ions themselves play a role in liberating fatty acids through hydrolysis, a proportion of which will be saturated.

In light of these findings, the common addition of aluminum stearate (either as $AlSt(OH)_2$ or $AlSt_2OH$, with St = stearate) to commercial titanium white and zinc white oil paint formulations could prove doubly problematic (Boon and Hoogland, 2014; Gabrieli et al., 2017; Osmond et al., 2012). Besides adding a source of saturated fatty acids to the system that could crystallize with zinc ions to form zinc soaps, Al^{3+} is a relatively strong Lewis acid. If the carboxylate or hydroxyl

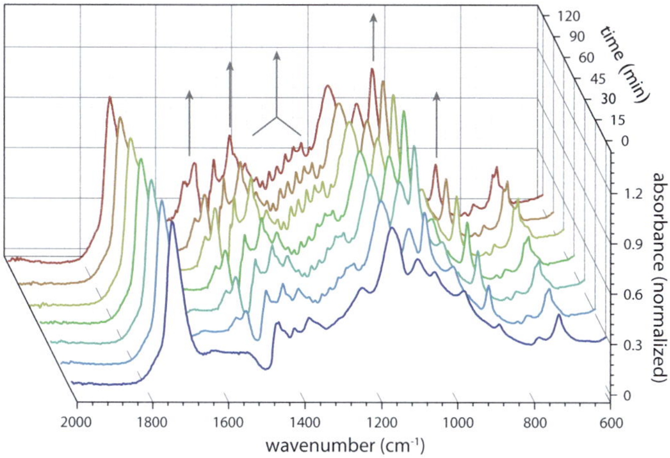

Fig. 3.6 Time series of ATR-FTIR spectra of a lead ionomer of linseed oil (16 mg lead sorbate with 200 mg linseed oil) exposed to a solution of palmitic acid in ethanol (143 mM). All spectra were normalized to the ester carbonyl band at 1740 cm^{-1}. Arrows indicate characteristic crystalline lead palmitate bands that increase in intensity over time

ion coordination to aluminum is temporarily disrupted, the aluminum ion may well have a strong influence on ester hydrolysis reactions.

Even if ester hydrolysis occurs mostly near metal ions, diffusion of metal ions and fatty acids can still happen. In fact, the diffusion rates of metal ions and/or fatty acids may play an important role in determining how fast crystalline metal soap phases grow in oil paint layers, as is discussed below.

3.3.2 The Crystallization of Metal Soaps

When the transition has occurred from an ionomer to a binding medium that contains a concentration of free SFAs as well as metal ions, a state is reached where all the necessary ingredients for metal soap formation are present (Fig. 3.5b). To investigate whether the availability of these components is really a *sufficient* condition for the formation of crystalline metal soaps, i.e., whether free SFAs are causing the formation of metals soaps, we carried out experiments in which lead and zinc ionomers based on linseed oil were exposed to a solution of palmitic acid in ethanol. Figure 3.6 shows a typical time series of ATR-FTIR spectra of a lead ionomer that shows the immediate increase of IR bands associated with crystalline lead palmitate, the lead carboxylate band at 1510 cm^{-1}, and the alkyl chain progression bands in the 1200–1350 cm^{-1} range being the most prominent. XRD analysis confirmed that the metal soaps were indeed crystalline. The same experiment was carried out with zinc ionomers, which showed crystalline zinc

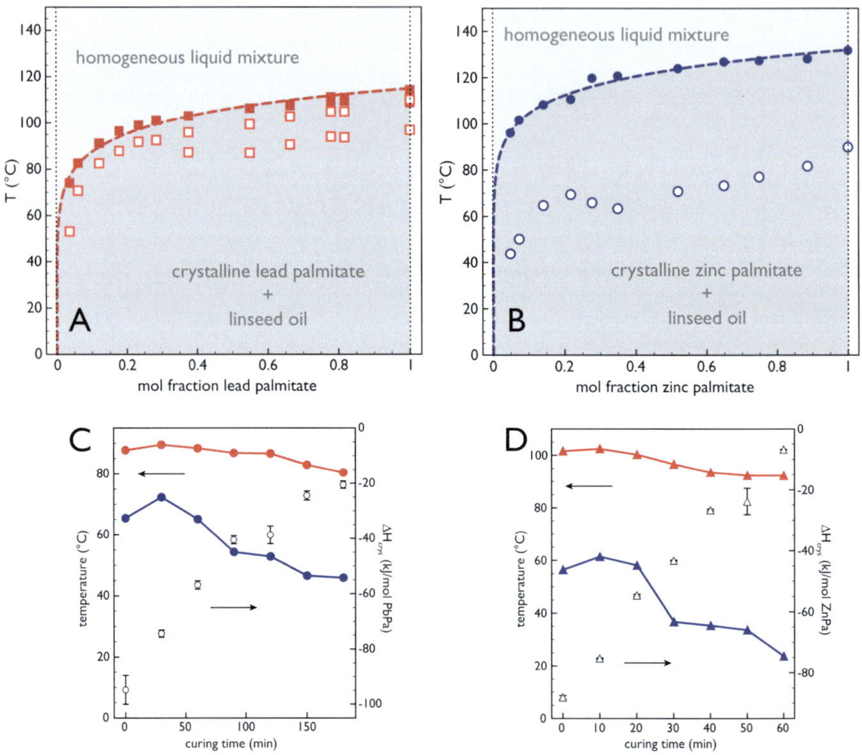

Fig. 3.7 Phase diagrams of (**a**) lead palmitate and (**b**) zinc palmitate in liquid linseed oil. Closed and open symbols represent melting and crystallization temperatures, respectively. Dashed lines indicate the approximate phase boundary as determined by fitting the ideal relation between the saturation concentration of a solute and temperature to the data. Bottom figures show the melting (red) and crystallization (blue) temperature (left axis), as well as the crystallization enthalpies (open symbols, right axis) for (**c**) lead palmitate and (**d**) zinc palmitate in linseed oil as a function of heating time at 150 °C (concentration of metal soaps is at a mol fraction of 0.11). See Hermans et al. (2016b) for further details

palmitate formation on a similar timescale. The metal soap formation reactions are very quick in these experiments, with metal palmitate bands appearing just seconds after starting exposure. This high reaction rate suggests that, in the reaction pathway toward crystalline metal soap phases in oil paint layers, the reaction between metal ions and fatty acids to form crystalline metal soaps is not rate limiting.

To help understand why metal soaps would crystallize and form aggregates in oil paint binding media, we studied the melting and crystallization temperatures of lead and zinc palmitate in both liquid and partially polymerized linseed oil with differential scanning calorimetry (DSC) (Hermans et al., 2016b). The resulting phase diagrams for these two metal soaps are shown in Fig. 3.7a, b. It is immediately clear that the solubility of both lead and zinc soaps is very low in linseed oil at room temperature, in the order of 10^{-3}–10^{-4} mol%, and that this solubility only starts to increase significantly around 80 °C for lead palmitate and around 90 °C for zinc palmitate. It was also found that the melting temperatures of both metal soaps did not change much as the linseed oil started to polymerize due to heat treatment, as shown in Fig. 3.7c, d. This means that the solubility of metal soaps is not greatly affected by oil polymerization (Hermans et al., 2016b). Therefore, soon after a part of the ester bonds are hydrolyzed and the resultant free carboxylates have bound to metal ions (which will not be difficult if hydrolysis is truly catalyzed by metal ions), the binding medium has become supersaturated with metal soaps. In other words, the system has become metastable. Though a metastable state of uncrystallized metal soaps in an ionomer is potentially long-lived under some conditions, the system will tend to show spontaneous crystallization of metal soaps. Whether and at what rate that crystallization does actually occur depends partially on the energy barrier that needs to be overcome for the first metal soap "units" to come together and form a crystalline nucleus. In the case of long-chain metal soap formation in ionomeric oil paint media, the largest driving force for crystallization is probably the alignment of the alkyl chain to maximize Van der Waals interactions, because the metal ions are already coordinated by carboxylate groups (though this coordination might not be ideal).

The degree of supercooling (i.e., the difference between the crystallization and melting temperature) needed for the formation of a critical crystal nucleus and the start of crystallization is a qualitative measure for the crystallization rate. It was found that lead and zinc palmitate are very different in this respect, with zinc palmitate in linseed oil starting crystallization approximately 50 °C below its melting point, while this value was around 6–9 °C for lead palmitate in linseed oil (Fig. 3.7a, b). Furthermore, this difference between melting and crystallization temperatures increased steadily as the linseed oil medium became more polymerized, as shown in Fig. 3.7c, d. These observations could mean that in the case of a fully polymerized binding medium, zinc soaps may be kinetically trapped in an amorphous state, while lead soaps would crystallize relatively fast, which can be a partial explanation for the difference in morphology observed for lead and zinc soaps formed in oil paint layers.

Once the crystallization process has started by the formation of crystallization nuclei, the further growth of crystalline metal soap phases in polymeric binding media is likely to be diffusion-limited. When the local concentrations of metal ions and fatty acids close to crystalline metal soaps drop, a concentration gradient arises which causes a net diffusion of those components toward the crystalline phase (illustrated in Fig. 3.5c). At this point, an oil paint can become subject to dramatic changes, because the accumulation of SFAs and metal ions into the crystalline metal soap phases enhances the equilibrium reactions that liberate the fatty acids from the polymerized medium and metal ions from pigments, causing a breakdown of both binding medium and pigment. However, it is unclear which factors control the rate of growth of metal soap crystals. Ostwald ripening may take place, causing the growth of larger metal soap phases at the expense of smaller ones, as is suggested by the frequent observation of very large crystalline lead soap protrusions in samples from oil paintings, such as the one depicted in Fig. 3.1.

The location of metal soap crystallization is also interesting to consider. Crystalline zinc soaps have often been detected especially in the lower parts of zinc white paint layers (Hermans et al., 2015; Osmond et al., 2012), and lead soap aggregates have been often found close to the interface between paint layers. These preferential locations for crystalline metal soap aggregates could be due to a variety of factors that affect either the nucleation or growth of crystals. For instance, if paint layers do not polymerize homogeneously but have a lower cross-link density in the deeper sections of the paint layer, nucleation of crystal phases could be easier in those areas of the paint. Such an effect is perhaps more relevant for paints based on zinc pigments than for lead-rich paint, as lead pigments may act as a through drier in oil paints. Alternatively, a gradient in polymer properties could affect the rate of diffusion of metal ions and/or free SFAs toward crystal nuclei. In multilayer systems, the growth of metal soap phases could also be dependent on the supply of either metal ions or SFAs from a neighboring paint layer, causing metal soap phases to appear near the interface between two paint layers.

In an attempt to gain some insight into the early stages of the formation of crystalline zinc soaps, a combination of scanning electron microscopy (SEM) and transmission electron microscopy (TEM) was used to study a model paint film after a few weeks of aging at room temperature (Hermans et al. 2018). The film consisted of ZnO in linseed oil pretreated with water to promote the formation of metal carboxylates (Fig. 3.8). In the images, the contrast is reversed from the usual convention, with the regions of higher electron density appearing darker. The concentration of ZnO was deliberately kept very low to ensure an excess of saturated fatty acids and to allow an easy distinction between ZnO particles (black), polymeric binding medium (light gray), and (semi-)crystalline metal soaps (dark gray) in the electron microscopy images.

Figure 3.8a shows that there is a large number of dark gray zinc soap aggregates. Interestingly, there seems to be no correlation between the location of remaining ZnO particles and the zinc soaps; crystallization obviously does not happen near the surface of pigment particles. This observation suggests that the zinc ions are in fact dispersed throughout and that crystallization may take place anywhere in the polymer, which is consistent with the ionomer model discussed previously. A TEM image of one of the zinc soap accumulations in Fig. 3.8c shows the internal structure of the aggregate zinc soap phase in greater detail. The accumulation is composed of many small (semi-)crystalline domains, each clearly showing the layered structure of zinc soaps. The spacing between the layers was calculated by taking the average of many measurements of the interlayer distance within one domain oriented roughly perpendicular to the plane of the image, and was found to be 45 ± 4 Å, which is similar to the interlayer spacings reported for zinc palmitate (38.5 Å) and zinc stearate (42.8 Å) (Hermans et al., 2014). The relatively wide range of measured spacings and the abundance of rough edges, breaks, and curvature in the zinc-containing planes suggest that there is still significant disorder in the packing of alkyl chains and/or metal ion coordination.

An ATR-FTIR spectrum of a bulk sample of this system is shown in Fig. 3.8b. Though the SEM and TEM images suggest that the system contains a mixture of

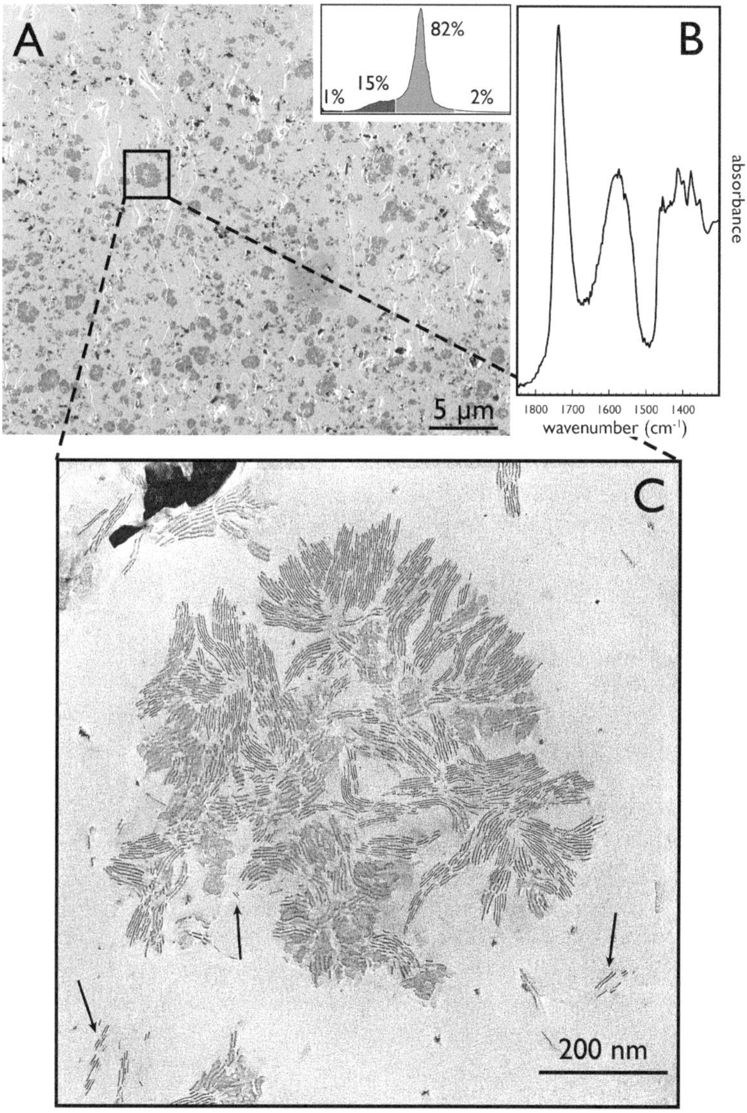

Fig. 3.8 (**a**) Backscatter electron SEM image (Zeiss, 12 keV) of the surface of a model ZnO paint. The surface was prepared by cutting the sample with a microtome blade inside the microscope chamber. Darker colors represent higher electron density. The inset shows a gray value histogram with frequencies of black, dark gray, light gray, and white pixel regions. The black spots are remaining ZnO particles, light gray is the polymer medium, and dark gray areas are zinc soaps. The white lines are artifacts as a result from microtoming. (**b**) ATR-FTIR spectrum of the model ZnO paint, showing a strong broad zinc carboxylate band that is associated with an ionomer-like linseed oil network. (**c**) Detailed TEM image (FEI Tecnai F30 FEG-TEM, 300 kV) of an aggregated zinc soap in a 300 nm thick microtomed slice of a different section of the same model paint film, showing the layer structure of the crystalline zinc soap phase. Arrows indicate very small zinc soap fragments (Hermans et al. 2018)

(semi-)crystalline zinc soaps and ionomer-like polymer medium, the IR spectrum only contains a broad zinc carboxylate band associated with amorphous zinc carboxylates. The lack of any crystalline metal soap features could mean that the zinc soap accumulations are still too disordered to exhibit sharp IR bands and actually contribute to the broad band in Fig. 3.8b. This interpretation is supported by the work of MacDonald and co-workers, who found very similar (semi-)crystalline metal carboxylate domains in ethyl linoleate reacted with lead or zinc acetate (MacDonald et al., 2016). In that system, XRD of an isolated metal carboxylate accumulation showed some crystalline features, while an FTIR spectrum of the same sample exhibited only a broad band associated with amorphous metal carboxylates. Alternatively, it could be the case that, despite the apparent abundance of zinc soaps in Fig. 3.8a, the zinc soap only constitutes a very small fraction of the total population of zinc carboxylate species, meaning that the FTIR spectrum is still dominated by ionomeric zinc carboxylates. Extended research on zinc- and lead-containing systems combining SEM and TEM with high-resolution FTIR microscopy or nanoIR (AFM-IR) would be able to further characterize this early stage of metal soap crystallization in oil paint layers.

3.4 Outlook

The experiments on metal soaps and ionomers described in the literature and those carried out by the authors have contributed to an understanding of the chemical processes occurring in oil paints that lead to the formation of large crystalline metal soap aggregates (Hermans, 2017). There is increasing evidence that oil paint binding media go through an ionomer-like state before metal soap crystallization takes place and that the hydrolysis of ester bonds by various mechanisms is a critical process. However, there are parts in the reaction pathway where we are still mostly limited to speculation. Moreover, the kinetics of nearly the entire pathway to the formation of metal soaps are uncertain. The reaction or diffusion rates can be strongly dependent on environmental conditions or paint composition, and these rates may make the difference between fast degradation and degradation that is too slow to be observed. Even when it is possible to measure directly the rate of a reaction or diffusion in a tightly controlled model system, this rate needs to be “translated” to more accurate systems that resemble mature oil paints. Finally, in order to develop improved conservation strategies, it is crucial that the more fundamental knowledge on the aging processes in oil paint is applied to the context of the practice of cleaning, restoring, and storage of oil paintings.

One of the important gaps in our current knowledge about the behavior of oil paints is the interactions between the liquid or polymerized oil medium and the surface of pigment particles. While it is almost certain that polar molecular groups like carboxylates and alcohols adsorb onto the surface of metal salt pigments, it is unclear exactly how this binding leads to the dissociation of metal ions from the surface into the binding medium. The stability of the pigment structure and its surface properties both play an important role, and they could be one of the reasons why metal soap formation is only regularly observed in paints with a limited selection of pigments. There have been interesting studies on the coordination of carboxylates on the surface of treated zinc surfaces (Taheri et al., 2013), but more detailed experiments are needed that focus on the rate of metal ion release in relation to the concentration of carboxylates or other charged groups in the surrounding medium. The role of water in these processes should be considered as well, since humidity is consistently found to influence the degree of metal soap-related degradation and we have observed that pretreating model oil paints with a little water leads to a fast generation of broad metal carboxylate bands in FTIR spectra. Water could have an effect on the pigment surface charge or polarity, or it could directly affect the concentration of carboxylate groups or other polar groups in the polymer medium that make the polymer medium a more hospitable environment for metal ions.

While we have shown that the appearance of intense broad metal carboxylate bands in FTIR spectra of oil paints signifies a linseed oil polymer network that contains metal ions in its structure (Hermans et al., 2016a), it is at present not certain whether a full ionomer-like state is absolutely necessary to form crystalline metal soap phases. If the release of metal ions from certain pigments is much slower than the generation of free saturated fatty acids (either within the paint layer or in an adjacent one), it could be the case that metal ions do migrate through the binding medium by an ion-hopping mechanism but that they react with free SFAs quick enough to prevent the buildup of a concentration of metal carboxylates bound to the polymeric binding medium high enough to be detected with FTIR spectroscopy. In this scenario, especially likely for the more stable lead-containing pigments, one would expect only to detect crystalline metal soaps.

An interesting set of experiments to start investigating the questions raised above would be to prepare mixtures of zinc white or various lead pigments with linseed oil, with and without water pretreatment, and to track the appearance of both broad and sharp types of metal carboxylate band with FTIR as the mixtures polymerize to dry paints. Using calibration methods similar to those in Hermans et al. (2016a), it is possible to quantify the concentration of reaction products and find potential mechanistic differences for the formation of metal soaps with different pigments.

Another important point to be investigated is the hydrolysis of esters in drying oils. If hydrolysis is really the key process that triggers the crystallization of metal soaps, we need to verify whether metal carboxylates or other metal species present in paints have an active role in catalyzing this reaction, how strongly humidity affects the reaction, and what measures can be taken to minimize the rate of hydrolysis. A good starting point would be to test the potential catalytic effect of insoluble lead and zinc carboxylates on the hydrolysis of esters with model compounds in various solvents with different concentrations of water.

The rate at which an ionomer state is reached in oil paint binding media will be affected by the diffusion rate of metal ions through the system. The mobility of metal ions in ionomers can be studied by measuring the conductivity of the polymer system (Fragiadakis et al., 2008; Kutsumizu et al., 1994; Tudryn et al., 2012). In a pilot experiment, we have found that the ionomer model system based on linseed oil has a low conductivity that seems to be a function of metal ion concentration. It will be very interesting to see how the mobility of metal ions is affected by an increase in temperature or a humid environment.

The same environmental conditions could also have a strong effect on the probability of forming a critical nucleus for metal soap crystallization and the rate with which this nucleus can grow into a large crystalline metal soap phase. The relation between all these rates, diffusion rates on the one hand and reaction rates on the other, will ultimately determine the location and size of crystalline metal soap phases in oil paint layers. For example, when diffusion is slow and crystallization fast in an ionomeric system, one expects a large number of small metal soap crystals to form that will only gradually increase in size through Ostwald ripening. However, when the situation is reversed (fast diffusion and slow crystallization) while keeping the concentration of reactants the same, one expects to find only few metal soap crystals that grow to a large size once they are formed. If conditions like humidity, temperature, or exposure to solvents have a strong effect on the relative rates of diffusion and crystallization in oil paint binding media, one can even imagine that mechanisms of metal soap formation are altered due to external influences. Detailed time-resolved FTIR studies on reaction-diffusion systems of linseed oil-based ionomers exposed to palmitic acid solutions are currently underway.

3.5 Conclusions

The processes in the pathway by which metal soaps are formed in oil paint systems have been reviewed, based on experiments on synthetic linseed oil-based ionomer systems conducted by the authors and supported by literature. Improved interpretation of metal carboxylate bands in FTIR spectra has shown that many oil paints containing pigments prone to metal soap formation go through a state in which the metal ions are bound to carboxylate groups attached to the polymer network. This characterization of oil paint media as an ionomer-like system is very helpful, because it allows us to make use of the existing literature on the structure and diffusion mechanisms in ionomers to understand molecular and macroscopic degradation phenomena in oil paintings.

When oil binding media are in an ionomer-like state, the presence of free fatty acids like palmitic and stearic acid (generated through ester hydrolysis) may cause a fast crystallization of metal soaps, as these complexes are practically insoluble in linseed oil. The location, size, and number of crystalline metal soap phases are determined in part by the relative rates of metal ion and fatty acid diffusion and the rate of metal soap crystallization. Therefore, kinetic studies on these processes are very important to improve our understanding of oil paint degradation phenomena related to metal soaps, and they are a logical starting point to investigate the effect of cleaning, restoration, and storage on the condition of oil paints.

Acknowledgements The authors are indebted to Dr. John Drennan and his co-workers at the Australian Microscopy and Microanalysis Research Facility (AMMRF) for the collaborative effort to visualize the early stages of zinc soap crystallization. Lambert Baij, Robert Corkery, Ties Korstanje, and Silvia A. Centeno are thanked for sharing their knowledge and inspiring discussions.

References

Badre C, Dubot P, Lincot D, Pauporte T, Turmine M (2007) Effects of nanorod structure and conformation of fatty acid self-assembled layers on superhydrophobicity of zinc oxide surface. J Colloid Interface Sci 316(2):233–237

Boon JJ, Hoogland FG (2014) Investigating fluidizing dripping pink commercial paint on van Hemert's seven-series works from 1990–1995. In: Van den Berg KJ, Burnstock A, de Tagle A, de Keijzer M, Heydenreich G, Krueger J, Learner T (eds) Issues in contemporary oil paints. Springer, Cham, pp 227–246

Boon JJ, Hoogland F, Keune K (2006) Chemical processes in aged oil paints affecting metal soap migration and aggregation. In: AIC annual meeting, Providence, pp 18–25

Castagna AM, Wang W, Winey KI, Runt J (2011) Structure and dynamics of zinc-neutralized sulfonated polystyrene ionomers. Macromolecules 44:2791–2798

Cesarano J, Aksay IA (1988) Processing of highly concentrated aqueous α-alumina suspensions stabilized with polyelectrolytes. J Am Ceram Soc 71(12):1062–1067

Clementi C, Rosi F, Romani A, Vivani R, Brunetti BG, Miliani C (2012) Photoluminescence properties of zinc oxide in paints: a study of the effect of self-absorption and passivation. Appl Spectrosc 66(10):1233–1241

Fragiadakis D, Dou S, Colby RH, Runt J (2008) Molecular mobility, ion mobility, and mobile ion concentration in poly(ethylene oxide)-based polyurethane ionomers. Macromolecules 41:5723–5728

Gabrieli F, Rosi F, Vichi A, Cartechini L, Pensabene Buemi L, Kazarian SG, Miliani C (2017) Revealing the nature and distribution of metal carboxylates in Jackson Pollock's Alchemy (1947) by micro-attenuated total reflection FT-IR spectroscopic imaging. Anal Chem 89(2):1283–1289

Hall LM, Stevens MJ, Frischknecht AL (2012) Dynamics of model ionomer melts of various architectures. Macromolecules 45:8097–8108

Helwig K, Poulin J, Corbeil MC, Moffatt E, Duguay D (2014) Conservation issues in several twentieth-century Canadian oil paintings: the role of zinc carboxylate reaction products. In: Van den Berg KJ, Burnstock A, de Tagle A, de Keijzer M, Heydenreich G, Krueger J, Learner T (eds) Issues in contemporary oil paints. Springer, Cham, pp 167–184

Hermans JJ (2017) Metal soaps in oil paint: structure, mechanisms and dynamics. Ph.D. thesis, University of Amsterdam

Hermans JJ, Keune K, Van Loon A, Corkery RW, Iedema PD (2014) The molecular structure of three types of long-chain zinc(II) alkanoates for the study of oil paint degradation. Polyhedron 81:335–340

Hermans JJ, Keune K, Van Loon A, Iedema PD (2015) An infrared spectroscopic study of the nature of zinc carboxylates in oil paintings. J Anal At Spectrom 30:1600–1608

Hermans JJ, Keune K, Van Loon A, Iedema PD (2016a) Ionomer-like structure in mature oil paint binding media. RSC Adv 6:93,363–93,369

Hermans JJ, Keune K, Van Loon A, Iedema PD (2016b) The crystallization of metal soaps and fatty acids in oil paint model systems. Phys Chem Chem Phys 18:10896–10905

Hermans JJ, Osmond G, Van Loon A, Iedema PD, Chapman R, Drennan J, Jack K, Rasch R, Morgan G, Zhang Z, Monteiro M, Keune K (2018) Electron microscopy imaging of zinc soaps nucleation in oil paint. Microsc Microanal. 24(3):1–5. https://doi.org/10.1017/S1431927618000387

Higgitt CL, Spring M, Saunders DR (2003) Pigment-medium interactions in oil paint films containing red lead or lead-tin yellow. Nat Gallery Tech Bull 24:75–95
Ishioka T, Maeda K, Watanabe I, Kawauchi S, Harada M (2000) Infrared and XAFS study on structure and transition behavior of zinc stearate. Spectrochim Acta A 56:1731–1737
Iwasaki T, Maegawa Y, Hayashi Y, Ohshima T, Mashima K (2008) Transesterification of various methyl esters under mild conditions catalyzed by tetranuclear zinc cluster. J Org Chem 73:5147–5150
Keune K, Boevé-Jones G (2014) It's surreal: zinc oxide degradation and misperceptions in Salvador Dalí's couple with clouds in their heads, 1936. In: Van den Berg KJ, Burnstock A, de Tagle A, de Keijzer M, Heydenreich G, Krueger J, Learner T (eds) Issues in contemporary oil paints. Springer, Cham, pp 283–294
Keune K, Van Loon A, Boon JJ (2011) SEM backscattered-electron images of paint cross sections as information source for the presence of the lead white pigment and lead-related degradation and migration phenomena in oil paintings. Microsc Microanal 17:696–701
Kutsumizu S, Hashimoto Y, Hirasawal E, Yanot S (1994) dc conduction properties of a model ethylene-methacrylic acid ionomer. Macromolecules 27:1781–1787
Lin KJ, Maranas JK (2012) Cation coordination and motion in a poly(ethylene oxide)-based single ion conductor. Macromolecules 45:6230–6240
Lin CL, Lee CF, Chiu WY (2005) Preparation and properties of poly(acrylic acid) oligomer stabilized superparamagnetic ferrofluid. J Colloid Interface Sci 291(2):411–420
MacDonald MG, Palmer MR, Suchomel MR, Berrie BH (2016) Reaction of Pb(II) and Zn(II) with ethyl linoleate to form structured hybrid inorganic-organic complexes: a model for degradation in historic paint films. ACS Omega 1(3):344–350
Mallégol J, Gardette JL, Lemaire J (1999) Long-term behavior of oil-based varnishes and paints I. Spectroscopic analysis of curing drying oils. J Am Oil Chem Soc 76(8):967–976
Mallégol J, Gardette JL, Lemaire J (2000a) Long-term behavior of oil-based varnishes and paints. Fate of hydroperoxides in drying oils. J Am Oil Chem Soc 77(3):249–255
Mallégol J, Gardette JL, Lemaire J (2000b) Long-term behavior of oil-based varnishes and paints. Photo- and thermooxidation of cured linseed oil. J Am Oil Chem Soc 77:257–263
Mallégol J, Lemaire J, Gardette JL (2000c) Drier influence on the curing of linseed oil. Prog Org Coat 39(2–4):107–113
Mallégol J, Gonon L, Lemaire J, Gardette JL (2001) Long-term behaviour of oil-based varnishes and paints 4. Influence of film thickness on the photooxidation. Polym Degrad Stab 72:191–197
Monico L, Janssens K, Cotte M, Sorace L, Vanmeert F, Brunetti BG, Miliani C (2016) Chromium speciation methods and infrared spectroscopy for studying the chemical reactivity of lead chromate-based pigments in oil medium. Microchem J 124:272–282
Osmond G, Boon JJ, Puskar L, Drennan J (2012) Metal stearate disributions in modern artists' oil paints: surface and cross-sectional investigation of reference paint films using conventional and synchrotron infrared microspectroscopy. Appl Spectrosc 66(10):1136–1144
Phenix A (2009) Thermal mechanical transitions in artists' oil paints and selected conservation materials: a study by dynamic mechanical analysis (DMA). In: AIC paintings specialty group postprints, vol 22, pp 72–89
Plater M, De Silva B, Gelbrich T, Hursthouse MB, Higgitt CL, Saunders DR (2003) The characterisation of lead fatty acid soaps in 'protrusions' in aged traditional oil paint. Polyhedron 22:3171–3179
Sehmi SK, Noimark S, Pike SD, Bear JC, Peveler WJ, Williams CK, Shaffer MSP, Allan E, Parkin IP, MacRobert AJ (2016) Enhancing the antibacterial activity of light-activated surfaces containing crystal violet and ZnO nanoparticles: investigation of nanoparticle size, capping ligand, and dopants. ACS Omega 1(3):334–343
Shen L, Laibinis PE, Hatton TA (1999) Bilayer surfactant stabilized magnetic fluids: synthesis and interactions at interfaces. Langmuir 15(2):447–453
Taheri P, Ghaffari M, Flores JR, Hannour F, De Wit JHW, Mol JMC, Terryn H (2013) Bonding mechanisms at buried interfaces between carboxylic polymers and treated zinc surfaces. J Phys Chem C 117:2780–2792

Thoury M, Van Loon A, Keune K, Hermans JJ, Réfrégiers M, Berrie BH (2019) Photoluminescence micro-imaging sheds new light on the development of metal soaps in oil paintings. In: Casadio F, Keune K, Noble P, Van Loon A, Hendriks E, Centeno S, Osmond G (eds) Metal soaps in art: conservation and research. Springer, Cham, pp 213–226

Tudryn GJ, O'Reilly MV, Dou S, King DR, Winey KI, Runt J, Colby RH (2012) Molecular mobility and cation conduction in polyether-ester-sulfonate copolymer ionomers. Macromolecules 45:3962–3973

Tumosa CS, Erhardt D, Mecklenburg MF, Su X (2005) Linseed oil paint as ionomer: synthesis and characterization. In: Materials research society symposium proceedings, vol 852, pp 25–31

van den Berg JDJ, van den Berg KJ, Boon JJ (2001) Determination of the degree of hydrolysis of oil paint samples using a two-step derivatisation method and on-column GC/MS. Prog Org Coat 41:143–155

van den Berg JDJ, Vermist ND, Carlyle L, Holčapek M, Boon JJ (2004) Effects of traditional processing methods of linseed oil on the composition of its triacylglycerols. J Sep Sci 27:181–199

van der Weerd J, Van Loon A, Boon JJ (2005) FTIR studies of the effects of pigments on the aging of oil. Stud Conserv 50(1):3–22

Chapter 4
Understanding the Dynamics and Structure of Lead Soaps in Oil Paintings Using Multinuclear NMR

Jaclyn Catalano, Anna Murphy, Yao Yao, Nicholas Zumbulyadis, Silvia A. Centeno, and Cecil Dybowski

Abstract To understand the mechanisms and factors that trigger soap formation and the dynamics of the reactive compounds in paints, advanced nuclear magnetic resonance (NMR) and X-ray analyses, complemented by analysis with FTIR spectroscopy, were performed on a series of lead carboxylates and model paint samples. Similar spectroscopy and lead coordination were observed for lead carboxylates of carbon chain length of 9 to 11 and 16 (palmitic acid) and 18 (stearic acid). Experiments as a function of temperature and humidity provided insight into the factors that increase soap formation. The local dynamics of palmitic acid and lead palmitate, in a linseed oil matrix at different temperatures (T) were measured by ^{2}H NMR spectroscopy. The results show the extent of mobility of palmitic acid and lead palmitate in the paint matrix, how they differ, and how they depend on T. The kinetics of soap formation in model paint films subjected to different relative humidities was monitored by ^{13}C NMR spectroscopy; the rate of soap formation increases with relative humidity. The results are discussed in the context of their implications for the conservation and preservation of works of art affected by lead soap formation.

J. Catalano
Montclair State University, Montclair, NJ, USA

A. Murphy · Y. Yao · C. Dybowski
University of Delaware, Newark, DE, USA

N. Zumbulyadis
Independent Researcher, Rochester, NY, USA

S. A. Centeno (✉)
Department of Scientific Research, The Metropolitan Museum of Art, New York, NY, USA
e-mail: Silvia.Centeno@metmuseum.org

F. Casadio et al. (eds.), *Metal Soaps in Art*, Cultural Heritage Science,
https://doi.org/10.1007/978-3-319-90617-1_4

4.1 Introduction

The application of nuclear magnetic resonance (NMR) spectroscopy in cultural heritage has been discussed in several reviews (Capitani et al. 2012; Capitani and Proietti 2015; Lambert et al. 2000), and it is particularly useful in providing chemical and structural information. Specifically, solid-state NMR (ssNMR) spectra of lead soaps and precursors to them have been extremely valuable because ssNMR methods provide structural information about the geometry around the lead site(s). This information is not accessible by single-crystal X-ray diffraction due to the poor solubility of lead soaps that makes it difficult to obtain sufficiently large single crystals (Catalano et al. 2014a, b, 2015). The relationship between ^{207}Pb ssNMR parameters and solid-state structure has been reviewed by Fayon et al. (1997), by Neue et al. (1996), and more recently by Dmitrenko et al. (2008). There is a strong dependence of the NMR chemical-shift parameters on local structure, particularly on the coordination geometry around the lead atom. Hence, principal elements of the ^{207}Pb chemical-shift tensor and the isotropic chemical shift are indicative of the chemical identity of a lead center, to the extent that different fatty acids may have different coordination geometries. However, acquiring ^{207}Pb NMR spectra can be challenging, due to the large chemical-shift anisotropy that causes the signal of a powder sample to be spread over thousands of ppm, decreasing the signal-to-noise ratio. In addition, extra care has to be taken to ensure uniform excitation over the range. One recent advancement in the use of NMR pulse sequences, WURST-CPMG NMR spectroscopy (Macgregor et al. 2011), has provided quick and accurate depiction of lead NMR tensors because it allows more uniform excitation over a broad band, and the effects of the generated echoes provide enhancement of the signal through the development of the so-called spikelet spectrum. In addition, since ^{13}C chemical shifts are sensitive to crystal symmetry, the conformation, and the dynamics of the acyl chains, lead carboxylates have been studied in detail using ^{13}C ssNMR (Catalano et al. 2014a, b, 2015; Feio et al. 1991, 1993).

Knowledge of the local dynamics of fatty acids and carboxylates is important to understand the factors that trigger soap formation; however, little is known about the dynamic state of these molecules in the oil binder and in oil paint films. ssNMR spectroscopy is quite informative of the dynamics of soaps, as well as the reactivity of soap formation. For example, a combination of ^{13}C, ^{207}Pb, and ^{119}Sn ssNMR gave in-depth information on the reactivity of lead-tin yellow type I with palmitic acid (Catalano et al. 2014a). ^{207}Pb and ^{13}C NMR spectroscopies have also been used to study the mixture and reactivity of lead white oil paint with palmitic acid (Catalano et al. 2014a).

In this study, we compare the structure and spectroscopy of long-chain and short-chain lead carboxylates to lead palmitate, lead stearate, and lead azelate, which are commonly found in soaps formed in oil paintings (Higgitt et al. 2003; Keune 2005). To provide insight into the local dynamics of the compounds in an oil matrix and in an oil paint film, the temperature dependence of the magnetic resonance parameters of deuterons incorporated into the methyl groups of palmitic acid and lead palmitate was analyzed by ^{2}H ssNMR (Catalano et al. 2018). In addition, the effects of humidity on the reactivity of free palmitic acid with a lead white paint film were monitored by ^{13}C NMR spectroscopy.

4.2 Materials and Methods

4.2.1 Synthesis of Materials

Lead soaps were synthesized by methods adopted from previously published protocols (Robinet and Corbeil 2003; Catalano et al. 2014b). Specifically, equimolar amounts (3.25 mmoles) of lead nitrate dissolved in 20 ml of water and the respective fatty acid dissolved in 50 ml of ethanol were mixed with an equimolar amount of potassium hydroxide (5 M solution) and reacted for 20 min at 80 °C. The reaction mixture was cooled to room temperature, filtered, and washed with water, methanol, ethanol, and acetone. In each case, the resulting material was dried and recrystallized from ethanol. For the deuterated samples, palmitic acid-16,16,16-D_3 was purchased from Cambridge Isotope Laboratories, Inc. and used without further purification. The deuterated palmitic acid was also used in the procedure described above to synthesize lead palmitate-16,16,16-D_3. Lead azelate was synthesized (Plater et al. 2003) from equimolar amounts (12.8 mmoles) of lead nitrate dissolved in a mixture of 50 ml of water, 50 ml of ethanol, and 20 ml of methanol and azelaic acid dissolved in 50 ml of ethanol. Both solutions were sonicated at 40 °C and subsequently mixed. The reaction mixture was put in a water bath at 40 °C and allowed to react overnight. A pipette was used to remove the solvent above the crystals, after which the crystals were filtered and washed four times with a mixture of 8 ml ethanol and 2 ml water. The purity of samples was verified by FTIR and XRD.

For the preparation of deuterated co-melted samples, about 50 milligrams of palmitic acid-16,16,16-D_3 or lead palmitate-16,16,16-D_3 was co-melted in about 450 milligrams of linseed oil and then cured at 40 °C in an oven until dry.

4.2.2 NMR Methods

^{207}Pb, ^{2}H, and ^{13}C ssNMR spectra were recorded at 11.75 tesla (104.63 MHz ^{207}Pb frequency, 76.77 MHz ^{2}H frequency, and 125.76 MHz ^{13}C frequency) with a standard Bruker 4 mm probe. Approximately 100 mg of sample was packed in a 4 mm rotor. Solid lead nitrate was used as a secondary external reference for the ^{207}Pb spectra, the isotropic chemical-shift being −3491 ppm relative to tetramethyllead (TML) at 298 K (Neue et al. 1996). A plot of spinning speed versus isotropic chemical shift of lead nitrate was obtained to compensate for the temperature increase due to spinning. Glycine was used as a secondary external reference for the ^{13}C spectra, with an isotropic chemical shift of the carbonyl peak of 176 ppm relative to tetramethylsilane.

^{207}Pb WURST-CPMG spectra of the samples were recorded using the parameters of MacGregor et al. WURST pulse widths were 50 μs, with pulse shapes created via the shape tool in Topspin 3.1. Seventy-five Meiboom-Gill loops were acquired for the WURST-CPMG experiments, with a 200-μs echo time and a sweep range of 0.5 MHz in all cases. The recycle delay was 7 s. High-power proton decoupling (CW at 100 kHz) was used to suppress dipolar couplings. Multiple WURST-CPMG spectra were collected at different carrier frequencies by shifting the carrier frequency a multiple of the spikelet separation (981.934 ppm) from spectrum to spectrum. Each spectrum was 10,240 scans. The collected spectra were superimposed to form the final spectrum.

^{13}C spectra were acquired using cross polarization with magic-angle spinning (MAS) at 12 kHz. Acquisition parameters for these experiments included a contact time of 1 ms and a recycle delay of 5 s. During acquisition, proton decoupling (SPINAL-64 at 100 kHz) was used to suppress dipolar couplings.

^{2}H spectra were acquired using the quadrupole-echo pulse sequence (Davis et al. 1976) with a $\pi/2$ pulse width of 7 μs and a delay between pulses of 40 μs. The recycle delay was 5 s. The temperature of the experiments was controlled by the Bruker Topspin 3.1 program from 235.0 to 335.0 K at approximately 10 K intervals. The number of scans was 512 for all static experiments.

4.2.3 Fitting Methods

The analysis of the ^{207}Pb chemical-shift tensors was performed by fitting the WURST-CPMG envelope or fitting the MAS sideband pattern from the spectrum using HBA 3.1 (Eichele 2013). The analyses of deuterium spectra were performed by fitting the powder pattern or spinning sideband pattern with a sum of a broad quadrupolar-broadened component and a narrow central component. Fits were aided by simulation of the patterns with the program WSOLIDS (Eichele 2013). The "solid-like" fraction was determined from the areas of the broad component and the sharp central peak.

4.2.4 Reaction of Samples at Different Relative Humidities

Palmitic acid (1-^{13}C, 99%) was purchased from Cambridge Isotope Laboratories, Inc. and was dried in an oven at 100 °C for about an hour to dehydrate before use. Lead white paint was made by mixing basic lead white with 22% by mass linseed oil to form films that were aged under standard laboratory conditions for 3 years. The palmitic acid and the lead white paint samples were conditioned for 24 h before mixing and made ready for NMR experiments by placing them in a plastic chamber (Thermo Scientific Nalgene autoclavable plastic chamber) with a specific relative humidity at room temperature (controlled by glycerol-water solutions) (Forney and Brandl 1992). The humidity was monitored with a compact humidity gauge, and the error in the humidity was ±5%. The samples were mixed and ground under liquid nitrogen with a mortar and pestle. The resultant material was packed in a 4 mm rotor for analysis. Inert SiO_2 was used to fill excess rotor space to lighten the sample for spinning. The weight of palmitic acid was approximately 10% of the weight of paint for every sample. Immediately after packing the rotor, the sample was examined with ^{13}C NMR spectroscopy using cross polarization and MAS at 12 kHz every 30 min for 3 days. The number of scans for each spectrum was 128.

4.3 Results

4.3.1 Structure of Lead Soaps

^{207}Pb ssNMR studies show that the lead coordination in lead palmitate is similar to that of lead stearate, but the structures of both are different from that of lead azelate (Catalano et al. 2014b). Further NMR studies of lead monocarboxylates of chain lengths from C_6 to C_{11} have shown that compounds containing nine carbons or more have similar ^{207}Pb NMR spectra to lead palmitate and lead stearate and that the spectra of lead carboxylates of eight or fewer carbons are quite different. Thus, lead nonanoate (C_9) serves as a structural model for lead palmitate and lead stearate. Consistent with this notion, an X-ray diffraction study of lead nonanoate indicates that the local structure around the lead site is quite different from that around lead in lead azelate and lead heptanoate (C_7) (Catalano et al. 2015).

The crystal structure of lead azelate, a dicarboxylic acid having a nine-carbon chain, has been published (Plater et al. 2003) and shows that the compound has a different coordination environment from lead nonanoate, which is mirrored in the differences in the ^{207}Pb WURST-CPMG spectra. Figure 4.1 shows spectra of lead palmitate, lead azelate, and lead heptanoate, as well as schematic depictions of the local structure surrounding the lead atom (Catalano et al. 2015). The ^{207}Pb NMR spans of lead azelate and lead heptanoate are rather large (~2600–2700 ppm), and coordination environments around the lead center are similar. The structure around the lead site in the azelate and the heptanoate is called hemidirecting (Shimoni-Livny et al. 1998). In contradistinction, for the nonanoate, the ^{207}Pb NMR span is observed to be relatively small (~730–750 ppm), and X-ray diffraction shows that the local structure about the lead site is more nearly spherical, a situation referred to as holodirecting (Shimoni-Livny et al. 1998).

Carbon ssNMR spectra (Fig. 4.2) show resonance doubling of the carbons closest to the lead site (C1) for all lead carboxylates studied, indicating two carboxylate conformations in the asymmetric unit. This result is in agreement with previous

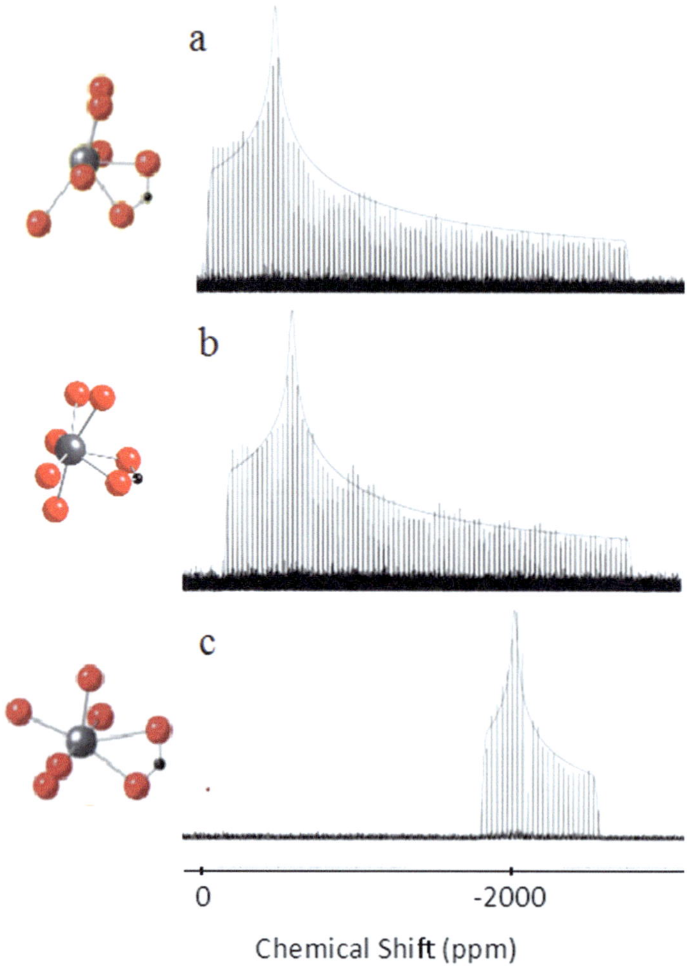

Fig. 4.1 The local lead coordination environment and the ^{207}Pb WURST-CPMG spectra for (**a**) a short-chain carboxylate, lead heptanoate, (**b**) lead azelate, and (**c**) a long-chain lead carboxylate, lead palmitate. Structures taken from the crystal structure of lead heptanoate (Lacouture et al. 2001), lead azelate (Plater et al. 2003), and lead nonanoate, the structural model for lead palmitate (Catalano et al. 2015)

NMR reports for lead pentanoate (Martínez Casado et al. 2008), lead decanoate, and octadecanoate (Feio et al. 1991), electron diffraction measurements of Langmuir-Blodgett films of lead octadecanoate (Stephens and Tuck-Lee 1969), and published crystal structures of lead heptanoate (Lacouture et al. 2001), lead azelate (Plater et al. 2003), and lead nonanoate (Catalano et al. 2015). ^{13}C ssNMR spectra of the lead soaps also indicate a difference between the short-chain and the long-chain lead carboxylates, in the magnitude of the separation of the two carboxylate resonances.

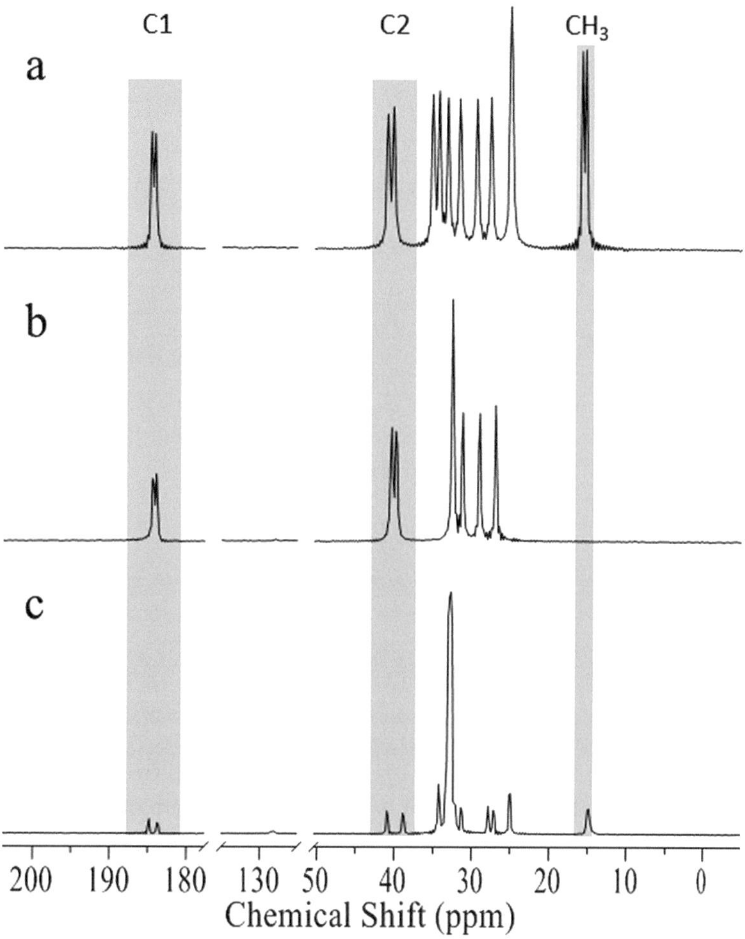

Fig. 4.2 ^{13}C ssNMR spectra representative of (**a**) lead heptanoate, (**b**) lead azelate, and (**c**) lead palmitate. The carbon resonances close to lead are doubled

For the long-chain carboxylates (C_9, C_{10}, C_{11}, C_{16}, and C_{18}), the separation of the two carboxylate peaks is in the range of 1.13–1.25 ppm. For the short-chain carboxylates (C_6, C_7, and C_8) and lead azelate, the separation of the two carboxylate peaks is in the range of 0.50–0.69 ppm.

A similar trend is seen for the α-carbon, where the resonance is also doubled, with chemical-shift separations in the range of 2.0–2.1 ppm for C_9, C_{10}, C_{11}, C_{16}, and C_{18} and in the range of 0.50–0.69 ppm for C_6, C_7, and C_8. The separation of the peaks for lead azelate is similar to that of the short-chain lead carboxylates.

Another spectroscopic distinction between the two groups of lead monocarboxylates is seen in the methyl region of the ^{13}C spectrum. For short-chain soaps, the

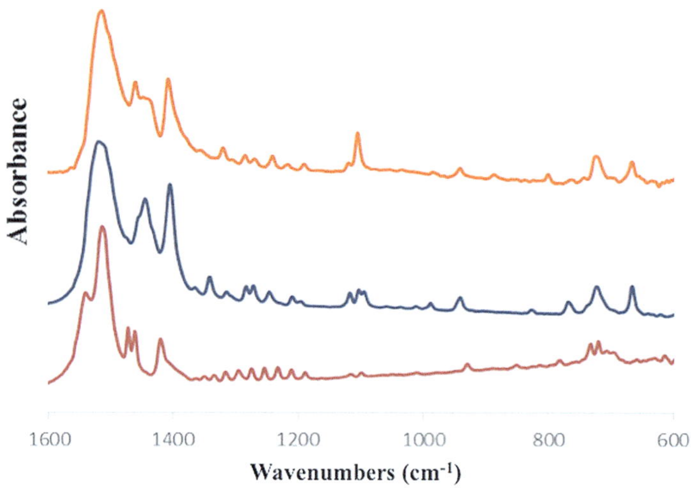

Fig. 4.3 FTIR spectra of lead heptanoate (orange), lead azelate (blue), and lead palmitate (red)

methyl resonance is doubled. For the long-chain group, the resonance is a singlet. These observations suggest that there are two conformations of the chain end for the short-chain soaps, whereas for the long-chain soaps, there is only a single conformation.

The effects of structure and conformation are also seen in similarities and differences in the IR spectra of the soaps, examples of which are shown in Fig. 4.3. For example, the absorption assigned to the carboxylate antisymmetric stretching mode consists of a distinct doublet for long-chain lead soaps, with peaks at ~1515 and 1540 cm^{-1}. For the short-chain soaps and lead azelate, there is only a broad peak centered around 1525 cm^{-1}. Additionally, there are differences in the low-frequency region, where band positions are quite sensitive to the alkyl chain packing (Ellis et al. 2005). For the short-chain soaps such as lead octanoate, a peak assigned to a CH_2 rocking mode appears around 720 cm^{-1} (Mesubi 1982). In lead palmitate, this region contains multiple overlapping features, over which one detects a doublet (Fig. 4.3). These observations are consistent with reports of doublets in this region for C_{10}, C_{12}, C_{14}, and C_{16} soaps (Mesubi 1982). In the range between 1150 and 1350 cm^{-1}, long-chain lead soaps like lead palmitate (Fig. 4.3) show the regular band progression associated with the wagging and twisting of the hydrocarbon chains (Mesubi 1982), whereas these progressions are not so well-defined for short-chain carboxylates such as lead octanoate (Fig. 4.3).

The carbon NMR spectra, the lead NMR spectra, and the IR spectra, as well as the limited X-ray diffraction results for the lead soaps, all suggest that there are two structural motifs for these materials in the solid state, one adopted by the short-chain soaps (and the azelate) and one adopted by the long-chain soaps. The local structure near the lead center is either asymmetric (for the short-chain soaps and the azelate) or more nearly spherical (for the long-chain soaps), with two different carboxylate

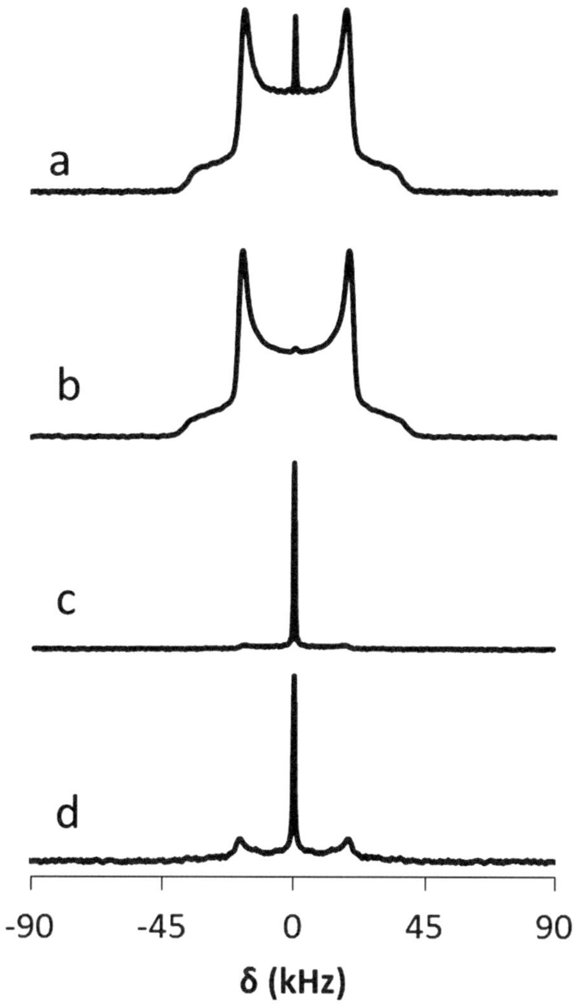

Fig. 4.4 Deuterium spectra of methyl-labeled palmitic acid and lead palmitate at 295 K, (**a**) pure palmitic acid, (**b**) pure lead palmitate, (**c**) palmitic acid in linseed oil film, (**d**) lead palmitate in linseed oil. There are two components to the line shape, as described in the text. The relative amounts of the two components depend on the sample, on the environment, and on the temperature

conformations. Interestingly, the close similarity of the two carboxylate resonances for the short-chain soaps suggests that they are more nearly similar to each other than the two carboxylates on the long-chain soaps. The long-chain soaps appear to have a single chain-end conformation, but two conformations in the carboxylate region, and a regular array in the middle of the chain, whereas the short-chain soaps appear to have two different chain-end conformations, and the lack of a well-defined chain progression for the short-chain soaps suggests there may be multiple structures present in this region.

4.3.2 Local Molecular Dynamics of Soaps

The structure of a lead soap depends on the length, as discussed above, and also on the saturation of the fatty acid chain (Catalano et al. 2014b). In addition, the dynamic state of the solid may be influenced by other factors that affect the local environment. In particular, the presence of other materials such as the binding medium may have a measurable effect on the dynamic state. For example, the IR spectra of the various materials show that chain vibrations strongly depend on the chain length (Fig. 4.3), as discussed above.

One may address dynamics of alkyl chains by a variety of spectroscopic methods. We have chosen to study the dynamics by their effects on the NMR spectroscopy of deuterium incorporated in the chain of the fatty acid in a variety of samples. The NMR spectroscopy of deuterium is well-known to be sensitive to rotational motions (Spiess 1978). Briefly, the deuterium NMR response is determined primarily by the coupling of the nuclear quadrupole moment to an electric field gradient at the nuclear site. For a sample that consists of a random powder of the material under study, the spectrum consists of a band, the shape of which is determined by the time-averaged coupling of the nucleus to the electric field gradient (which is determined primarily by the electrons in the bonds to the deuterium nucleus). For example, rapid isotropically random motion of the C-D bond in an organic material results in a sharp resonance, whereas a static solid or an anisotropically moving molecule results in a broad band (Spiess 1983, 1986).

Deuterium spectra were measured for samples with deuterons incorporated at the methyl group of the palmitic moiety. Both palmitic acid and lead palmitate were examined in the pure state and as mixed with linseed oil to approximate the situation in a paint sample (Fig. 4.4). Palmitic acid and lead palmitate are also models for stearic acid and lead stearate since they have a similar structure and spectroscopy. The deuterium spectra are comprised of a broad band, which reflects a solid-like phase in the sample, and a narrow resonance that indicates a relatively isotropically mobile phase. Figure 4.4 shows that, at 295 K, palmitic acid and lead palmitate mixed with linseed oil have larger amounts of the mobile phase than do the pure materials. The increased number of molecularly mobile palmitate moieties in the film suggests the presence of the linseed oil increases the propensity for palmitic and lead palmitate molecules to exhibit local dynamics.

At 295 K, palmitic acid behaves differently from lead palmitate in both the pure materials and in the oil films. The fraction of palmitic acid molecules that show isotropic mobility (i.e., contribute to the narrow component) is larger than the mobile-component fraction of lead palmitate. In fact, in the pure lead palmitate, virtually all of the molecules are static at 295 K.

We investigated the temperature dependence of the deuterium NMR spectra of the four samples between 250 K and 350 K, as shown in Fig. 4.5. At every temperature, the spectrum of a sample could be parsed into a broad component and a narrow component, the fractions of which depend on the temperature and the sample. At sufficiently low temperatures, the spectra of pure palmitic acid and

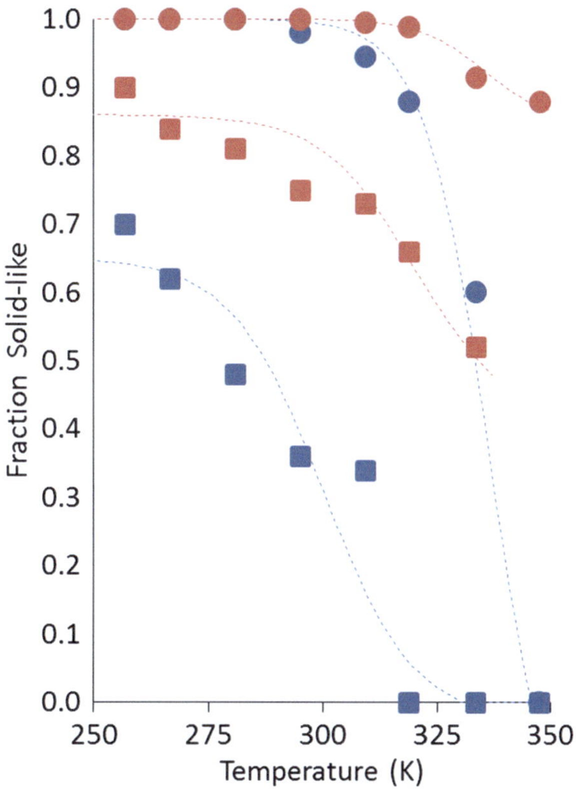

Fig. 4.5 The solid-like fraction of palmitic acid and lead palmitate from 250 to 350 K, as determined by deuterium NMR spectroscopy of the methyl group. ● Pure palmitic acid, ● lead palmitate, ■ palmitic acid in a linseed oil matrix, ■ lead palmitate in a linseed oil matrix. The dotted lines indicate trends, but are not fits to the data

pure lead palmitate indicate that all molecules in the samples are in a solid-like phase. On the other hand, mixing either palmitic acid or lead palmitate with linseed oil produces samples that have some isotropically mobile molecules, even at the lowest temperature investigated, 257 K. At all temperatures studied, the mixtures have a smaller fraction of solid-like molecules than the analogous pure material, suggesting that mixing the materials with linseed oil creates an environment that results in larger molecular mobility.

As temperature is increased, the solid-like fraction decreases, particularly for the materials mixed with linseed oil. For the pure materials, there is also a decrease in the solid-like fraction as temperature increases that seems to be associated with the melting phenomenon or a rotator phase prior to the formation of a liquid crystalline phase. The melting point for palmitic acid is 336 K, but measurable amounts of mobile material are detected at temperatures of 300 K and above. Similarly, pure lead palmitate also shows some mobility at temperatures above 335 K. Its reported melting point is 385 K.

These experiments demonstrate that, even though some parts of the palmitic acid and lead palmitate mixed with linseed oil are similar to the pure materials in some ways, there is a portion of the sample that is dynamically different in the mixture.

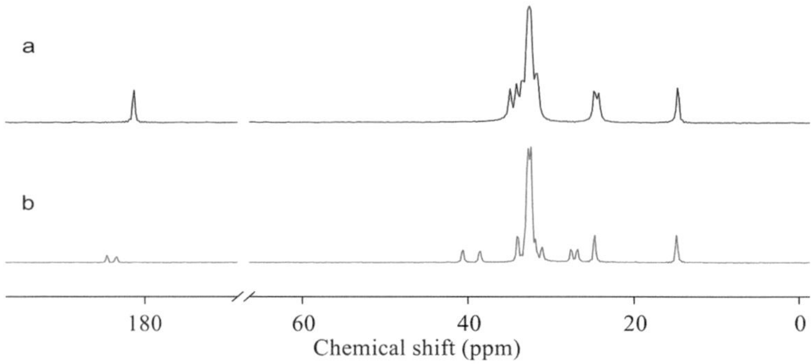

Fig. 4.6 ^{13}C CP-MAS spectra of (**a**) palmitic acid and (**b**) lead palmitate. The spectra indicate there are distinct resonances to distinguish the two species from each other

In the mixtures, palmitic acid is in a different sample matrix than the pure material, which causes the melting point to decrease. In particular, it is tempting to suggest that the existence of molecular mobility in part of the sample provides evidence that fatty acids may migrate through a linseed oil matrix to participate in reaction with pigment species such as lead ions. This fact suggests that reactions may be occurring at temperatures at which materials are stored and displayed.

4.3.3 Kinetics of Soap Formation

Previous studies of soap formation have shown that water can increase reactivity (Cotte et al. 2006). The reaction can readily be monitored with ssNMR ^{13}C spectroscopy because the carboxyl resonances of free palmitic acid and lead palmitate can be distinguished easily (Fig. 4.6).

Lead white paint samples were prepared and aged in the laboratory for 3 years as described in the materials and methods section. Importantly, in separate containers, lead white paint and palmitic acid were allowed to equilibrate at a specific relative humidity before being mixed. The two materials were mixed to create a sample that was ~10% palmitic acid by weight and immediately placed in a sealed NMR sample rotor. Spectra were taken at regular intervals over a period of 3 days. The spectra were analyzed to determine the fraction of material that remained as palmitic acid by fitting the resonances to Lorentzian line shapes (Fig. 4.7).

The progress of the reaction for samples conditioned at different relative humidities is shown by the disappearance of the signal of palmitic acid in the ^{13}C spectrum in Fig. 4.7. Humidity clearly affects the reaction rate, with the palmitic acid disappearing more quickly for samples exposed to higher relative humidity.

The disappearance of palmitic acid is a complex function of time. Rather than proposing a model for the reaction, we characterized the rate by defining the parameter T_{50}, the time at which 50% of the total signal is due to palmitic acid.

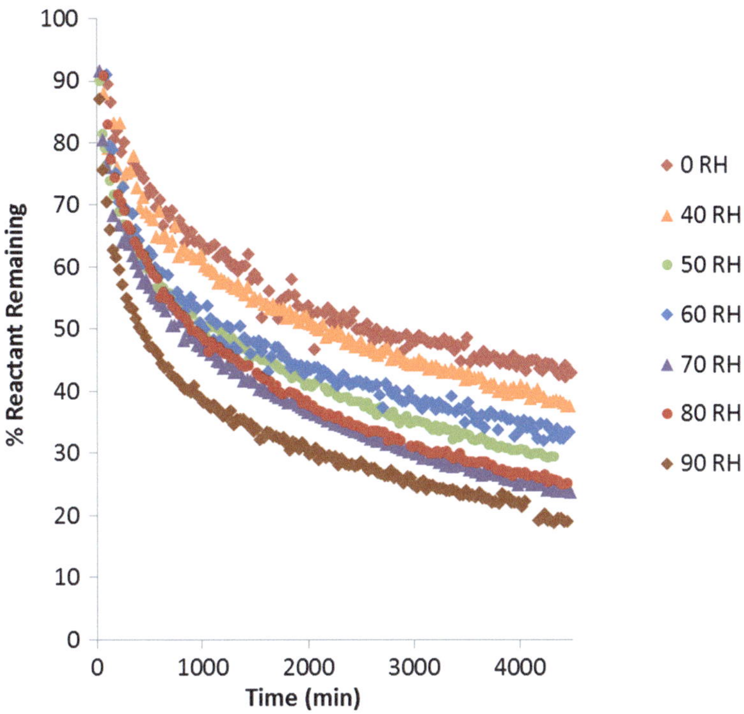

Fig. 4.7 Percentage of reactant remaining as a function of time at different relative humidities

In Fig. 4.8, a plot of T_{50} versus relative humidity for the samples examined in this investigation is shown. Although there is quite a bit of scatter, the figure shows that the reaction depends on the exposure to humidity.

Although it is not possible at the moment to model the reaction explicitly, Fig. 4.7 indicates that there must be at least two time regimes in the process. As Fig. 4.8 demonstrates, the time T_{50} is a monotonically decreasing function of relative humidity.

4.4 Conclusions

NMR spectroscopy is a sensitive technique to probe structure and dynamics of molecules in paint films. For example, packing of the chains of palmitic acid and lead palmitate produces spectroscopic effects that allow the discrimination and quantification of these two materials in a mixture. Additionally, electronic effects at the site of the lead ion allow one to infer geometric structure from the manner in which distribution of oxygen centers, for example, affect the lead NMR spectroscopy. The NMR results obtained in the model paint samples directly

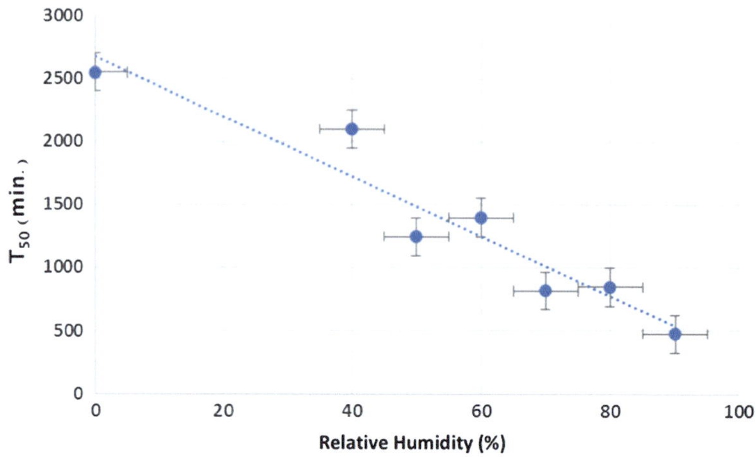

Fig. 4.8 T_{50} versus relative humidity for the formation of lead palmitate from palmitic acid in a lead white paint film

correlate with features of the IR data, so they are useful to interpret FTIR spectra acquired in micro-samples removed from works of art. Also, these ssNMR data set the basis for minimally invasive NMR studies in micro-samples with the development of advanced techniques such as dynamic nuclear polarization (DNP) techniques.

Local rotational dynamics in palmitic acid and lead palmitate indicate that when these materials are present in a linseed oil matrix, such as is found in paints, the environment causes the molecules to be substantially more rotationally free than they would be in the pure solid. From studies of reaction with added free palmitic acid, we observe the kinetics of reaction with NMR spectroscopy. T_{50}, the time for half of the free palmitic acid to disappear, depends on the relative humidity at which the sample is conditioned.

Acknowledgments CD and SC acknowledge the support of the National Science Foundation under Grants CHE-1139192, DMR-1608366, CHE-1139180, and DMR-1608594. We acknowledge helpful discussions with Shi Bai and Guangjin Hou about the implementation of ^{207}Pb and ^{2}H spectroscopy.

References

Capitani D, Proietti N (2015) NMR in cultural heritage. Magn Reson Chem 53:1

Capitani D, Di Tullio V, Proietti N (2012) Nuclear magnetic resonance to characterize and monitor cultural heritage. Prog Nucl Mag Res Sp 64:29–69

Catalano J, Murphy A, Yao Y, Alkan F, Zumbulyadis N, Centeno SA, Dybowski C (2014a) 207Pb and 119Sn solid-state NMR and relativistic DFT studies of the historic pigment Lead-Tin yellow type I and its reactivity in oil paintings. J Phys Chem A 18:7952–7958

Catalano J, Yao Y, Murphy A, Zumbulyadis N, Centeno SA, Dybowski C (2014b) NMR spectra and 207Pb chemical-shift tensors of lead carboxylates relevant to soap formation in oil paintings. Appl Spectrosc 68:280–286
Catalano J, Murphy A, Yao Y, Yap GP, Zumbulyadis N, Centeno SA, Dybowski C (2015) Coordination geometry of lead carboxylates – spectroscopic and crystallographic evidence. Dalton Trans 44:2340–2347
Catalano J, Murphy A, Yao Y, Zumbulyadis N, Centeno S, Dybowski C (2018) Molecular dynamics of palmitic acid and lead palmitate in cross-linked linseed oil films: Implications from deuterium magnetic resonance for lead soap formation in traditional oil paintings. Sol State Nucl Magn Reson 89:21–26
Cotte M, Checroun E, Susini J, Dumas P, Tchoreloff P, Besnard M, Walter P (2006) Kinetics of oil saponification by lead salts in ancient preparations of pharmaceutical lead plasters and painting lead mediums. Talanta 70:1136–1142
Davis JH, Jeffrey KR, Bloom M, Valic MI, Higgs TP (1976) Quadrupolar echo deuteron magnetic resonance spectroscopy in ordered hydrocarbon chains. Chem Phys Lett 42:390–394
Dmitrenko O, Bai S, Beckmann PA, Van Bramer S, Vega AJ, Dybowski C (2008) The relationship between Pb-207 NMR chemical shift and solid-state structure in Pb(II) compounds. J Phys Chem A 112:3046–3052
Eichele K (2013) HBA 3.1 and WSOLIDS. University of Tubingen, Tubingen
Ellis HA, White NAS, Taylor RA, Maragh PT (2005) Infrared, X-ray and microscopic studies on the room temperature structure of anhydrous lead(II) n-Alkanoates. J Mol Struct 738:205–210
Fayon F, Farnan I, Bessada C, Coutures J, Massiot D, Coutures JP (1997) Empirical correlations between Pb-207 NMR chemical shifts and structure in solids. J Am Chem Soc 119:6837–6843
Feio G, Burrows HD, Geraldes CFGC, Pinheiro TJT (1991) Multinuclear NMR studies of lead (II) soaps. 13C and 1H studies of solid and liquid-crystalline phases. Liq Cryst 9:417–432
Feio G, Geraldes CFGC, Pinheiro TJT (1993) Multinuclear nuclear magnetic resonance, studies of lead(ll) soaps. Studies on the liquid phase with reference to the behaviour of the corresponding acids. J Chem Soc Faraday Trans 89:3117–3122
Forney CF, Brandl DG (1992) Control of humidity in small controlled-environment chambers using glycerol-water solutions. HortTechnology 2:52–54
Higgitt C, Spring M, Saunders D (2003) Pigment-medium interactions in oil paint films containing red lead or lead tin yellow. Natl Gallery Tech Bull 24:75–91
Keune K (2005) Binding medium, pigments and metal soaps characterised and localised in paint cross-sections. Ph.D. thesis, University of Amsterdam
Lacouture F, François M, Didierjean C, Rivera J-P, Rocca E, Steinmetz J (2001) Anhydrous lead(II) Heptanoate. Acta Crystallogr C 57:530–531
Lambert JB, Shawl CE, Stearns JA (2000) Nuclear magnetic resonance in archaeology. Chem Soc Rev 29:175–182
Macgregor AW, O'dell LA, Schurko RW (2011) New methods for the acquisition of ultra-wideline solid-state NMR spectra of spin-1/2 nuclides. J Magn Reson 208:103–113
Martínez Casado FJ, Ramos Riesco M, Sánchez Arenas A, García Pérez MV, Redondo MI, López-Andrés S, Garrido L, Cheda JAR (2008) A novel rotator glass in lead(II) pentanoate: calorimetric and spectroscopic study. J Phys Chem B 112:16601–16609
Mesubi MA (1982) An infrared study of zinc, cadmium, and lead salts of some fatty-acids. J Mol Struct 81:61–71
Neue G, Dybowski C, Smith ML, Hepp MA, Perry DL (1996) Determination of Pb-207(2+) chemical shift tensors from precise powder lineshape analysis. Solid State Nucl Mag 6:241–250
Plater MJ, De Silva B, Gelbrich T, Hursthouse MB, Higgitt CL, Saunders DR (2003) The characterisation of lead fatty acid soaps in 'protrusions' in aged traditional oil paint. Polyhedron 22:3171–3179
Robinet L, Corbeil MC (2003) The characterization of metal soaps. Stud Conserv 48:23–40
Shimoni-Livny L, Glusker JP, Bock CW (1998) Lone pair functionality in divalent lead compounds. Inorg Chem 37:1853–1867

Spiess HW (1978) Rotation of molecules and nuclear spin relaxation. In: Diehl P, Fluck E, Kosfeld R (eds) NMR basic principles and progress. Springer, Berlin

Spiess HW (1983) Pulsed deuteron NMR investigations of structure and dynamics of solid polymers. J Mol Struct 111:119–133

Spiess HW (1986) Deuteron NMR investigations of structure and dynamics in solid polymers, liquid crystalline polymers and polymer model membranes. Makromol Chem-M Symp 4:227–230

Stephens JF, Tuck-Lee C (1969) Structure of a multilayer of lead stearate. J Appl Crystallogr 2:1–10

Chapter 5
Historical Evolutions of Lead-Fat/Oil Formula from Antiquity to Modern Times in a Set of European Pharmaceutical and Painting Treatises

Marine Cotte, Laurence De Viguerie, Emilie Checroun, Jean Susini, and Philippe Walter

Abstract Recipes describing the controlled mixture and reaction of fat/oil with lead-based compounds have been known and employed since antiquity until modern times. Two major fields have developed such practices: pharmacy, for the preparation of lead plasters (i.e., dressings made of lead-based compounds and fat/oil) and related unguents, and painting, for the preparation of thickened oils. Here, we review and analyze a set of almost 500 recipes, from historical texts from the Mediterranean area, dating from Pharaonic Egypt to the nineteenth century. Recipes show similarities (e.g., type of lead ingredients, advice regarding the grinding level of powders, risks associated with overheating the mixtures) and differences (e.g., type of fat/oil, introduction of additional compounds, use of sunlight) which can be explained by the different purposes and applications of the formed pastes: plasters must be soft, with a thick consistency, and adhere to the skin for a long time without drying; conversely, "thickening" oil by reaction with lead-based compounds is intended mainly to speed up drying reactions in paintings. Also, some historical evolutions may be noticed, e.g., the choice of lead ingredients and the progressive

M. Cotte (✉)
European Synchrotron Radiation Facility, Grenoble, France

Laboratoire d'archéologie moléculaire et structurale (LAMS), Sorbonne Université, CNRS, UMR 8220, Paris, France
e-mail: marine.cotte@esrf.fr

L. De Viguerie · P. Walter
Laboratoire d'archéologie moléculaire et structurale (LAMS), Sorbonne Université, CNRS, UMR 8220, Paris, France
e-mail: marine.cotte@esrf.fr

E. Checroun
Checroun RCPM, Magny-le-Hongre, France

Epitopos, Strasbourg, France

J. Susini
European Synchrotron Radiation Facility, Grenoble, France

F. Casadio et al. (eds.), *Metal Soaps in Art*, Cultural Heritage Science,
https://doi.org/10.1007/978-3-319-90617-1_5

replacement of lead-based compounds by zinc oxide for toxicological reasons, the interpretation in chemical terms (in particular the identification of the formed lead soaps and glycerol), and, for lead plasters, the quasi-systematic introduction of water from the eighteenth century and the standardization of recipes.

Keywords Lead plasters · Pharmacy · Cosmetics · Lead-oil paint media · Saponification · Treatises

5.1 Introduction

Recipes advising to mix fat or oil with metallic compounds to produce metallic soaps have been known and employed since antiquity. As an example, calcium soaps were used as durable and waterproof lubricants in the wheels of fourteenth-century B.C. Egyptian wheel bearings (Corkery 1998). The study of historical texts shows a continuous and frequent use of such hybrid metal-fat preparations. Nowadays, metal soaps are essential ingredients in various industries. They are used as antifoaming agents in aircraft lubricating oils; detergents in dry cleaning; waterproofing agents on fabric, paper, masonry, and metals; and heat and light stabilizers in plastics. Also, they find applications in fungicide treatments of wood (Corbeil and Robinet 2002; Kastens and Hansen 1949). They are also employed as precursors of nanofilms, nanocomposites, and nanoparticles in homogeneous catalysis (Mishra et al. 2007) or as chemical thickeners in greases, anticaking agents for powdered substances, and solid lubricants for drawing wire or metal tubing (Corkery 1998). In addition to these uses, metal soaps are fundamental ingredients in the industry of paints and in pharmacology. More particularly, lead soaps have been present in these two domains over centuries, until the twentieth century.

5.1.1 Pharmaceutical Domain

While today, they are excluded from modern pharmacopeias, lead plasters (in French, "emplâtre," not to be confused with "plâtre," i.e., plaster of Paris) and unguents have been used extensively since antiquity. As an example, lead soaps were identified in a thirteenth-century B.C. Egyptian cosmetic preserved in a reed (Cotte et al. 2005). Ancient Greeks were naming plasters *emplasta,* meaning "to form a mass, to coat and to fill in." The main purpose of these products was indeed *to be applied on the body and to stay in place for a long time, allowing agents contained in the plaster to act efficiently and at length* (Lemery 1764, p. 1146). They had many pharmaceutical functions: to treat burns, abscesses, and ulcers, to help skin to heal, etc. It has to be noted that not all plasters were made with lead/metal ingredients. As an example, of the 124 plaster recipes given by Lemery, only 45 include lead-based compounds (Fig. 5.1 shows an extract from the table of contents of Lemery's pharmacopeia, Lemery 1764, p. 1265). Mixing fat with lead-based compounds ends, under certain conditions, in the formation of some lead soaps, solid at room

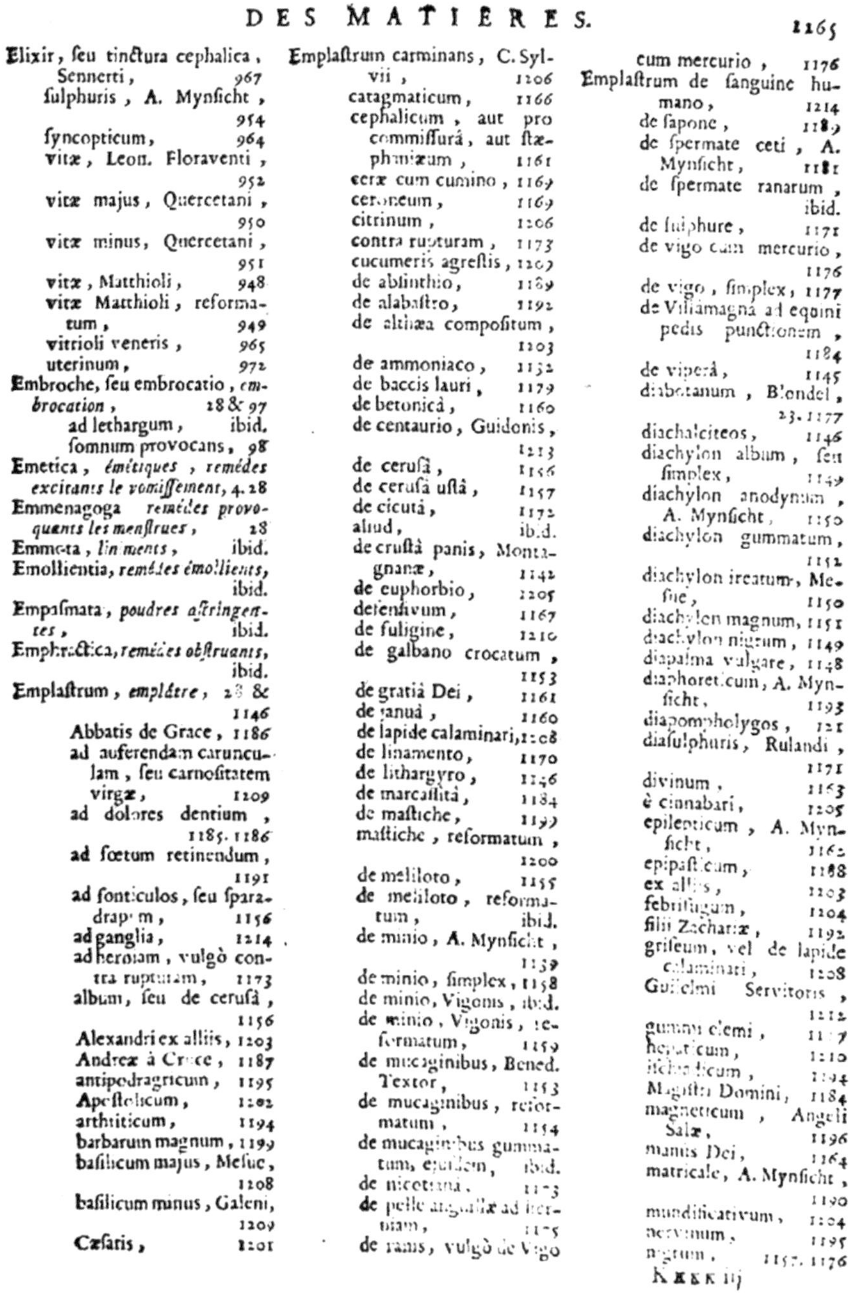

DES MATIÈRES. 1165

Elixir, seu tinctura cephalica, Sennerti, 967
sulphuris, A. Mynsicht, 954
syncopticum, 964
vitæ, Leon. Floraventi, 952
vitæ majus, Quercetani, 950
vitæ minus, Quercetani, 951
vitæ, Matthioli, 948
vitæ Matthioli, reformatum, 949
vitrioli veneris, 965
uterinum, 972
Embroche, seu embrocatio, *embrocation*, 18 & 97
ad lethargum, ibid.
somnum provocans, 98
Emetica, *émétiques, remédes excitants le vomissement*, 4. 28
Emmenagoga *remédes provoquants les menstrues*, 28
Emmeta, *liniments*, ibid.
Emollientia, *remédes émolliens*, ibid.
Empasmata, *poudres astringentes*, ibid.
Emphractica, *remédes obstruants*, ibid.
Emplastrum, *emplâtre*, 28 & 1146
Abbatis de Grace, 1186
ad auferendam carunculam, seu carnositatem virgæ, 1209
ad dolores dentium, 1185. 1186
ad fœtum retinendum, 1191
ad fonticulos, seu sparadrapum, 1156
ad ganglia, 1214
ad herniam, vulgò contra rupturam, 1173
album, seu de cerusâ, 1156
Alexandri ex alliis, 1203
Andreæ à Cruce, 1187
antipodagricum, 1195
Apostolicum, 1202
arthriticum, 1194
barbarum magnum, 1199
basilicum majus, Mesue, 1208
basilicum minus, Galeni, 1209
Cæsaris, 1201
Emplastrum carminans, C. Sylvii, 1206
catagmaticum, 1166
cephalicum, aut pro commissurâ, aut staphisagrium, 1161
ceræ cum cumino, 1169
ceroneum, 1169
citrinum, 1206
contra rupturam, 1173
cucumeris agrestis, 1207
de absinthio, 1189
de alabastro, 1192
de althæa compositum, 1203
de ammoniaco, 1132
de baccis lauri, 1179
de betonicâ, 1160
de centaurio, Guidonis, 1213
de cerusâ, 1156
de cerusâ ustâ, 1157
de cicutâ, 1172
aliud, ibid.
de crustâ panis, Montagnanæ, 1142
de euphorbio, 1205
defensivum, 1167
de fuligine, 1210
de galbano crocatum, 1153
de gratiâ Dei, 1161
de januâ, 1160
de lapide calaminari, 1208
de linamento, 1170
de lithargyro, 1146
de marcasitâ, 1184
de mastiche, 1199
mastiche, reformatum, 1200
de meliloto, 1155
de meliloto, reformatum, ibid.
de minio, A. Mynsicht, 1159
de minio, simplex, 1158
de minio, Vigonis, ibid.
de minio, Vigonis, reformatum, 1159
de mucaginibus, Bened. Textor, 1153
de mucaginibus, reformatum, 1154
de mucaginibus gummatum, ejusdem, ibid.
de nicotianâ, 1173
de pelle anguillæ ad herniam, 1175
de ranis, vulgò de Vigo cum mercurio, 1176
Emplastrum de sanguine humano, 1214
de sapone, 1189
de spermate ceti, A. Mynsicht, 1181
de spermate ranarum, ibid.
de sulphure, 1171
de vigo cum mercurio, 1176
de vigo, simplex, 1177
de Villamagna ad equini pedis punctionem, 1184
de viperâ, 1145
diabotanum, Blondel, 23. 1177
diachalciteos, 1146
diachylon album, seu simplex, 1149
diachylon anodynum, A. Mynsicht, 1150
diachylon gummatum, 1152
diachylon ireatum, Mesue, 1150
diachylon magnum, 1151
diachylon nigrum, 1149
diapalma vulgare, 1148
diaphoreticum, A. Mynsicht, 1193
diapompholygos, 1201
diasulphuris, Rulandi, 1171
divinum, 1153
è cinnabari, 1205
epilepticum, A. Mynsicht, 1162
epipasticum, 1188
ex alliis, 1203
febrifugum, 1204
filii Zachariæ, 1192
griseum, vel de lapide calaminari, 1208
Guilelmi Servitoris, 1212
gummi elemi, 1197
hepaticum, 1210
ischiadicum, 1194
Magistri Domini, 1184
magneticum, Angeli Salæ, 1196
manus Dei, 1164
matricale, A. Mynsicht, 1190
mundificativum, 1204
nervinum, 1195
nigrum, 1157. 1176
Kkkk iij

Fig. 5.1 A page of Lemery's pharmacopeia that includes a table of contents showing the variety of names and recipes for making plasters (Lemery 1764) © BNF

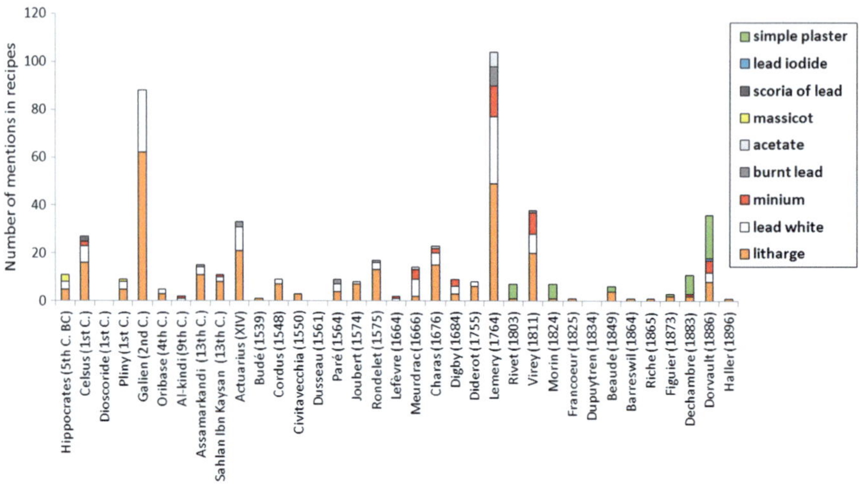

Fig. 5.2 Number of occurrences of the different lead-based compounds in the studied pharmaceutical recipes. "Simple plaster" refers to recipes employing "simple plaster" as an ingredient (produced by using litharge)

temperature (Cotte et al. 2006). The product can have various consistencies, from a cream to a paste. Lead-based compounds act both in the control of the physical *plastic* and adhesive properties of the plaster in particular through the formation of lead soaps and in their pharmaceutical properties. Pharmaceutical properties are also tailored by the set of additional ingredients included in the various recipes, which could be of bioorganic and/or mineral origin (see an example below).

Since, to our knowledge, there are few scientific reviews of pharmaceutical recipes of lead plasters and ointments in European history, it seems valuable to complete this introduction with one particular and representative example of lead plaster recipe, showing its basic principles and historical evolutions, called *simple plaster*. As shown in Fig. 5.1, the diversity of plaster recipes is huge. While in 1676, Charas states that *the variety of plasters' composition and that of medical products which are introduced in these plasters are such that it would be impossible to set general rules for the preparation of these kinds of ingredients* (Charas 1676, p. 704), from the eighteenth century onward, most of the recipes are based on a basic formula: the so-called simple plaster, which can be considered as the basis of the majority of recipes (see Figs. 5.2 and 5.3). It basically consists of mixing and heating oil with finely crushed litharge (PbO) and water (usually in a mass ratio 2-1-2). This *simple plaster* was generally not used as such but as the excipient in most of the composed plasters (Dorvault 1886, p. 832). For this purpose, besides fat- and lead-containing compounds, recipes include various additional agents (organic substances, such as plant juice, decoction of leaves, mucilage, gum, wax, and resin and/or minerals) making the plaster composition and pharmaceutical properties more specific. The recipe for *simple plaster* appears rather late in the pharmacopeia, only from the

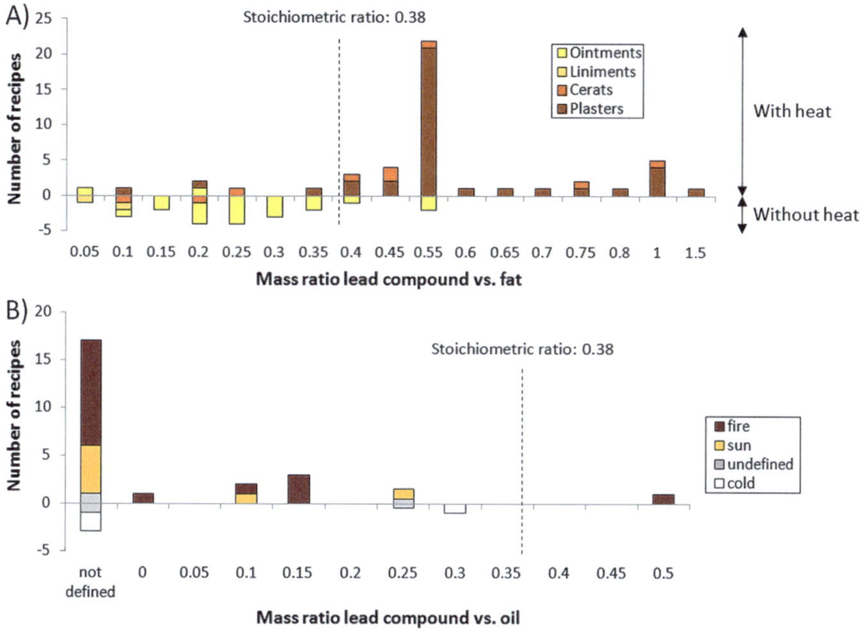

Fig. 5.3 Comparison of fat/lead-based compound ratio and working conditions in some pharmaceutical and paint recipes. (**a**) Statistical analyses of 75 recipes of lead/fat-based ointments, liniments, cerats, and plasters, as quoted in (Lemery 1764). (**b**) Statistical analyses of 29 recipes from de Mayerne's manuscript, mentioning the use of lead-based compounds in oil preparations (Mayerne 1620–1646) (De Viguerie et al. 2016). For each recipe, the mass ratio of lead-based compound over fat is calculated. A theoretical stoichiometric ratio of 0.38 was calculated in the case of a mixture PbO/ triolein. Cumulated frequencies were calculated over bin ranges of 0.05 width (values with no frequency are not displayed). In the vertical axes, a distinction is made for (**a**) between recipes made at room temperature or with fire and for (**b**) between recipes made at room temperature or with sun light or fire

nineteenth century onward; it is absent in (Lemery 1764) treatise (see Fig. 5.2). It is a simplified version of more ancient plasters such as *diapalm plaster* or *diachalciteos plaster*, which were known and used since antiquity. *Diachalciteos plaster* differs from *diapalm plaster* in the addition of "chalcitis." The meaning attributed to this term differs from one author to another. Joubert mentions *calchitidis aut vitrioli albi*, i.e., white vitriol (Joubert 1574, p. 331). Charas indicates that *chalcitis or calcinated vitriol ("vitriol rubifié") will be used to prepare red plaster while white vitriol will be used if one wants to obtain a white plaster* (Charas 1676, p. 525).

As early as the second century A.D., Galen (131–210), in the book I of his *Composition of Drugs*, Chapter IV (Galien et al. 1552), writes that *diapalm plaster* is obtained by cooking together axungia (animal fat), aged olive oil, litharge, and chalcitis on a small fire and by constantly stirring with a freshly cut palm stick. The name *diapalm* comes from the use of a palm stick.

Almost identical recipes for *diapalm plaster* can be found in Arabic recipes from 1290 (edited in 1953 by Sbath and Avierinos 1953): *Pulverize the colcothar [translated by Sbath and Avierinos into "calcinated iron sulfate"] and litharge to the degree of dust; place everything in a pot and stir with the stem of a palm branch. Progressively, cut the part of the stem which is drying up and continue until complete extraction of the juice. Then discard the spent rod to take another one and so on until drug concentration. It has a more energetic effect to strengthen the organ and to stop the matter that it attracts if more palm juice is added.*

In 1574, Joubert quotes Mesué (also known as Yuhanna Ibn Masawaih, 776-c.885), *advising that those who cannot used fresh & new palm can do the same with cane root (Joubert* 1574, *p. 331).* He also proposes oakwood as an alternative.

In 1676, Charas recommends to use a decoction of young palm shoots, instead of palm sticks (Charas 1676, p. 526): *Any learner too, could be blamed for arguing that some palm branches bound together are suitable instruments to agitate & unify litharge with oil & fat when cooked, because they cannot prevent the litharge from remaining at the bottom of the pan, and the plaster from burning.*

Similarly, Lemery mentions: *Those who work know that we cannot successfully prepare this plaster following exact descriptions of the Ancients, who do not ask any other aqueous humidity to cook litharge with oil and fat, than that which can come out from the palm spatula because the plaster would darken and it would never acquire good consistency, but usually we add water to make it boil: indeed palm decoction would be more suitable than pure water, if we want to follow the intent of the author, who plans to provide the plaster with the quality of this tree* (Lemery 1764, p. 1147).

In the nineteenth century, most of the plaster recipes are based on the *simple plaster*, to which different ingredients are added. Accordingly, the *diapalm plaster* recipe is simplified as:

Simple plaster, ...	2 kilograms
Zinc sulphate dissolved in required amount of water, ...	8 decagrams
White wax, ...	16 decagrams

Cut the wax in small pieces; put it with plaster and the zinc sulphate solution, in a basin; make it liquefy and evaporate moisture at a gentle heat; then let cool and form rolls (Rivet 1803, p. 337).

This recipe, although called *diapalm plaster,* no longer includes any palm-based ingredient, as in its earlier versions.

In this recipe, *zinc sulfate whitens the mixture, through the formation, by double decomposition, of zinc soap and lead sulfate, both being white* (Figuier 1873, p. 450).

This example shows that, despite the heterogeneity and spread over time and space of our corpus (see details below), recipes sharing the same denominations have been used over almost two millennia, from the Mediterranean basin to the Middle East. The historical texts reveal a constant basic protocol and some evolutions. It seems therefore interesting to see if/which general trends can be observed along these historical evolutions.

5.1.2 *Paint Domain*

On a more reduced timescale (starting mainly in the fifteenth century, for what concerns Europe, with a few earlier exceptions), similar recipes for preparing mixtures of lead-based compounds and oil have been developed in the field of oil paintings, in particular to prepare "lead media," i.e., oil binding media with lead-containing compounds. These hybrid compositions are tightly linked to the development of oil painting during the Renaissance in Europe (Maroger 1948). These preparations were called "fat oil" ("huile grasse"), "olio cotto," "olio coto," "boiled oil" (names found in Merrifield 1999), or "litharge oil" ("huile de litharge") (name found in Mayerne 1620–1646) and are nowadays usually called "thickened oil." The reaction of lead-based compounds (e.g., litharge) with oil, and particularly with the action of heat, modifies the chemical composition of the oil and consequently its physical properties, in particular it "thickens" it (De Viguerie et al. 2008) and increases the drying speed (De Viguerie et al. 2016). Such thickened oils can then be used as such or mixed with other organic binders (e.g., wax, resin, untreated oil) in the preparation of ground layers, paint layers, glazes, and varnishes and in gilding.

Even if the intended uses of the paint media and pharmaceutical plasters/ointments were completely different, the recipes share common aspects: both domains require mastering the plastic properties of a fat-based medium, which can serve as the basic binder of various compounds (organic substances and/or minerals), and both fields use(d) lead-based compounds in reaction with fat/oil for this purpose. However the types of compositions differ in their respective functions and, therefore, in their long-term behaviors: plasters must be soft with a thick consistency and adhere to the skin for a long time without drying, while in paint recipes, the "thickening" of the oil by the reaction with lead-based compounds is mainly intended to speed up the drying, and the oil usually remains liquid. These different objectives result in the use of different fat/oil ingredients and different operational conditions, which are presented below. In particular, the present work focuses on the analysis of the type of lead ingredients and of fat/oil ingredients, their respective proportions, and on the use of fire, sunlight, and water for the preparation. The effects of these different parameters are discussed in the context of the intended final properties of the resulting lead-fat/oil media.

5.2 Text Corpus

This work aims to highlight some historical trends in both the pharmacy and paint domains, but it does not intend to be exhaustive at all. Sources have been selected primarily from the Mediterranean basin, with very few additional references from the Middle East and Asia. Specific reviews of texts/practices produced by these cultures would be required to further extend the present work and are not targeted here. Our discussion is based on a total of 37 books and manuscripts for pharmaceutical recipes (representing a total of 389 recipes of lead plasters/unguents) and 30 publications for paint recipes (representing a total of 100 recipes of lead-containing media). Even if the corpus is important, the unequally divided datasets, the influence of the geographical origin and time period on a source, and the influence of the nature and purpose of a source on the character of the recipes must be taken into account. All the translations of Latin and French texts quoted in this paper were done by the authors.

5.2.1 *Pharmaceutical Treatises*

For practical reasons, research in this field has been mainly focused on documents available online, in French, Latin, and English. More particularly, the corpus contains 25 French references dating from the sixteenth to the nineteenth centuries, comprising 18 treatises on medicine or pharmacy (Beaude 1849; Budé 1539; Charas 1676; Dechambre 1883; Di Civitavecchia 1550; Digby 1684; Dorvault 1886; Dupuytren 1834; Dusseau 1561; Joubert 1574; Lemery 1764; Morin 1824; Paré 1564; Rivet 1803; Rondelet 1575; Virey 1811), 5 treatises on chemistry (Barreswil 1864; Haller and Muller 1896; Lefevre 1751; Meurdrac 1999; Riche 1865), and 3 more generic books or encyclopedias (Diderot and D'alembert 1755; Figuier 1873; Francoeur and Lenormand 1825). The corpus also includes the pharmacopeia of a sixteenth-century German physician and botanist (Cordus 1548). There are few texts dating before the sixteenth century. To cover earlier practices, the corpus has been extended to the Mediterranean basin, including ancient Egypt. These sources include the Ebers Papyrus dating back to 1500 B.C. that provides unique medical recipes (Bardinet 1995); texts from Greek authors with recipes from Hippocrates (460 BC–390 BC) (Littré 1839–1861), Dioscorides (c. 40 AD; 90 AD) (Wellman 1914), Galen (129 AD–c. 216 AD) (Galien et al. 1552), and Oribase (c. 320 AD–c. 395 AD) (Oribase 1851–1876); texts by Roman authors, Pliny the Elder (23–79) (Plinel'ancien 1983) and Aulus Cornelius Celsus (c. 25 BC–c. 50 AD) (Celse 1876); and a few references from the Middle East (Al-Kindi (801–873)) (Al-Kindi 1966), Nejibeddine Assamarkandi (thirteenth century) (Assamarkandi 1994), Sahlān Ibn Kaysān and Rašīd al-Dīn Abū Ḫulayka (thirteenth century) (Sbath and Avierinos 1953), and Joannes Actuarius (c. 1275–c.1330) (Actuarius 1546).

5.2.2 *Painting Treatises*

The corpus of paint recipes considered here comprises mainly European sources, dating approximately from the ninth century A.D. to the nineteenth century. It is based on the reviews previously made by Nadolny, covering European texts dating before c. 1550 (Nadolny 2008); Merrifield, who focuses on Medieval and Renaissance treatises, namely, the manuscript of Eraclius, the Bolognese manuscript, the

Brussels manuscript, the Paduan manuscript, and the Volpato manuscript, and on texts by Cennini, Pacheco, Baldinucci, De Piles, Fra Fortunato, Palomino, and Chilone (Merrifield 1999); and Maroger, on the practices supposedly developed by Antonello da Messina, Leonardo da Vinci, Giorgione, Titian, and Tintoretto (Maroger 1948) and also on the Tegernsee manuscript (Kneepkens 2012) and on recipes from Vasari (Stols-Witlox et al. 2008), J. Smith (1687) (White and Kirby 1994), and Willem Beurs (1692) (Noble et al. 2008). Carlyle also reviewed more modern (eighteenth to nineteenth century) recipes (Carlyle 1999). In total, the set of all these documents provides a corpus of 39 recipes for preparing mixtures of lead-based compounds with oil. A reproduction of these recipes can be found in the supporting information of our recent publication (Cotte et al. 2017). Besides, our corpus contains 29 lead/oil media recipes given by T. de Mayerne (1620) (Mayerne 1620–1646) and detailed in (De Viguerie et al. 2016), 17 recipes given by Watin (Watin 1773), and 6 recipes given by Mérimée (Mérimée 1830). Some statistical analyses of these 100 recipes have recently been published (Cotte et al. 2017) and will not be discussed in this manuscript. Only the main trends are reported here, as a support to the discussion.

5.3 Choice of Ingredients and Proportions

5.3.1 Lead-Based Ingredients

Different lead-based compounds recommended for making lead plasters and lead-based media are listed below. Their identification in chemical terms is not always easy and reliable, in particular in the most ancient treatises (see below the example of Egyptian texts). For example, the same term could refer to different minerals depending on the time and author. The different lead-based compounds have different colors; chemical properties, in particular solubility and basicity that lead to different saponification efficiencies (Cotte et al. 2006); and drying and pharmaceutical properties.

5.3.1.1 Pharmaceutical Domain

Figure 5.2 presents an overview of the frequency with which lead-based ingredients are mentioned in the recipes studied. From the Greek and Latin medical texts considered in this work until more modern pharmacopeias, the lead-based compounds mentioned more frequently are lead oxides – in particular litharge (α-PbO, usually named *spuma argenti* or *lithargyrum* in many Latin texts), minium (Pb_3O_4), and lead white. As a reminder, simple plaster recipes, shown in green in Fig. 5.2, call for the use of litharge, demonstrating the predominance of this compound in lead plaster recipes. The earliest texts also mention massicot (β-PbO, $\mu o\lambda \upsilon \beta \delta \alpha \iota \nu \eta \varsigma$, molibdena), such as Hippocrates' texts (Littré 1839–1861), and *plumbi recrimenti (crasse de plomb, scoria of lead* in two recipes by Celsus), which according to Sallé would also correspond to a generic lead oxide (Sallé 1817, p. 46).

These different lead-based compounds could be used alone or in combination. As an example, Pliny the Elder (23–79) mentions in Book XXXIII § 34 and 35 the use of *spumae lapis* (also called *stimmi*, *stibi*, *alabastrum*, or *larbasom*), mixed with fat, litharge, lead white, and wax, ("cum adipe, ac spuma argenti, cerussaque, etc."), to treat burns (Plinel'ancien 1983, p. 74). In his recipe of *white or cerussa plaster* (*emplâtre blanc ou de ceruse*), Charas indicates that *the virtue of all these limes [lead white, litharge, minium] is much the same, and one can without fear replace one with the other one* (Charas 1676, p. 705). He adds that for this specific recipe, lead white should not be replaced by minium; otherwise it would result in a red plaster; conversely, lead white can be substituted by litharge, which, if well used, also gives a white plaster.

Occasionally, some recipes mention lead acetate, named as such, e.g., in Dorvault (1886, p. 835), but more often as "salt or sugar of Saturn." Earliest examples go back to the seventeenth century (e.g., Meurdrac 1999, p. 183; Charas 1676, p. 751).

In our corpus, the earliest pharmaceutical texts mentioning the use of lead-based compounds mixed with fat/oil are the Ebers papyri, dated to ca. 1500 B.C. (Bardinet 1995), in particular the chapter dealing with eye diseases (Eb. 336 to 431). The translations of the ingredients names in these texts, in particular of the lead-based compounds, are sometimes unclear (hence not indicated in Fig. 5.2). Bardinet translates some of the hieroglyphs (phonetically corresponding to *mesdemet*) by *galena*, with additional terms sometimes. As an example, Eb. 359 (Bardinet 1995): *Other (remedy) for eye care: galena: 1; red ochre (tjerou): 1; djaret-plant: 1; gesefen-galena: 1; male part of galena: 1. (This) will be prepared as a homogeneous mass and applied to the eyes.*

In an earlier translation, the term *mesdemet* has been translated in a more nuanced way as *black makeup for eyes (Grapow* 1958*)*, creating the possibility that the term could correspond to a mixture of different ingredients and not to a single one. The X-ray analysis of Egyptian cosmetics dating from 2000 to 1200 BC revealed indeed the presence of galena (PbS) but also of other lead-based compounds such as cerussite ($PbCO_3$), laurionite (PbOHCl), and phosgenite ($Pb_2Cl_2CO_3$) (Walter et al. 1999). The translation of the lead-based ingredients named in the most ancient texts is indeed still a matter of debate. In later European pharmacopeia considered here, the term "galena" is found only occasionally in lead plaster recipes (e.g., *Pompholyx ointment, by Nicolas of Alexandria*, (Virey 1811, p. 261)). A few recipes mention the use of the so-called burnt lead (*plomb brûlé*, *plumbum combusti*, *plumbum ustum*). According to some sources, this could refer to lead sulfides obtained by mixing and calcinating lead with sulfur (Lemery 1764, p. 433; De Lens and Mérat 1833, p. 377). Conversely, recent reconstructions based on heating lead white highlight the formation of different lead oxides (Aubin et al. 2017).

In very few cases, the use of metallic lead is mentioned, for example, Liébault proposes to treat condyloma with an ointment made of linseed oil and egg yolk, lengthily ground in a lead mortar (Liébault 1651, p. 254).

One anecdotic mention of lead iodide is found in (Dorvault 1886, p. 462) to prepare a *plaster of hemlock and lead iodide*.

With advances in chemistry, the compounds resulting from the reaction of oil with lead-based compounds were identified as metallic soaps, first by Pierre Joseph Macquer (1718–1784) in 1758 and, later around 1768, by Carl Wilhelm Scheele (1742–1786). The saponification process itself was first described in 1823 by Michel Eugene Chevreul (1786–1889) (Corkery 1998). From this date onward, pharmaceutical recipes for making *simple plaster* are usually additionally named as *lead soap* or "stearate." In 1824, Morin notes that the plasters can be divided into four classes, the first one being *metallic plasters that comprise those which own their consistency, their properties, to the combination of a metallic oxide with oil and grease, and these are kinds of metal soaps* (Morin 1824, p. 289). Since the chemical reaction involved in the preparation of lead plasters had by then been identified, an alternative protocol was proposed by M. Gélis to prepare the *simple plaster* by the so-called double decomposition, i.e., by mixing a solution of ordinary soap with a solution of lead acetate (Béral 1837). However, as noticed by Riche, *lead soap prepared [like this] cannot be used as a plaster, because it is too dry. This is due to the fact that all fatty matter has disappeared from soap, while when lead plaster is prepared directly, there always remains a certain amount of unsaponified olein, which binds the matter* (Riche 1865, p. 424).

Since the nineteenth century, the toxicity of lead-based compounds was perceived as a critical factor, motivating the replacement of lead by other metals. In pharmaceutical treatises, zinc oxide was proposed as a good alternative to litharge (Dorvault 1886), and today the mixture oil/ZnO is a component of many dermatological creams.

5.3.1.2 Paint Domain

A detailed analysis of the frequency with which different lead-based compounds are mentioned in historical recipes involving oil binding media and lead-containing compounds has been recently published (Cotte et al. 2017). Results are extremely similar: the most frequently mentioned lead-based compound is litharge, followed by lead white and minium. As for pharmaceutical recipes, the use of metallic lead and massicot is mentioned only occasionally. Lead acetate is also mentioned but in a relatively later period (from the nineteenth century, cf. Carlyle 1999). Galena is not mentioned, which can be explained by its low reactivity with oil (Cotte et al. 2006) and its black color.

Lead-tin yellow is recommended in some paint recipes and is absent from medical recipes studied here. This can be explained by the specific use of this compound as a pigment in paintings. Of course, many other lead-based pigments were used in painting (e.g., chrome yellows) but for their coloring properties, not to control the oil drying.

In contrast to the pharmaceutical domain, the chemical identification of the saponification reaction and of the properties of metallic soaps led to their industrial production, particularly of lead soaps, to be used as additives in paints (Kastens and Hansen 1949).

Similar to the field of pharmacy, lead toxicity led to concerns in the nineteenth century. In 1850, Michel-Eugène Chevreul was asked by the French Academy of Science to determine if lead white could be efficiently substituted by zinc oxide, a less toxic drier (Chevreul and France 1850). The frequency of the use of lead-based driers progressively decreased from then.

5.3.2 *Fat Ingredients*

5.3.2.1 Pharmaceutical Domain

In pharmaceutical texts, many types of fat ingredients are mentioned: usually the recipes call for *axungia*, *tallow*, and *common oil*. Sometimes, it is specified that the oil be old. Many other organic ingredients, from vegetal and animal sources, could be included, leading to a variety of preparations. The oils employed in pharmaceutical recipes were usually non-drying oils. As noticed by Dechambre (Dechambre 1883, p. 70), *the choice of fat [for lead plasters] is not trivial: the soap given by poppy oil is often too soft, drying out in the long term and covered with a brittle crust; castor oil would be preferable, if the plaster obtained was not coloured; then preference should be given to olive oil; it is advantageous to add pork fat to it which, used alone, would give too soft a product, but mixed with olive oil provides a plaster with suitable consistency.*

5.3.2.2 Paint Domain

For the preparation of paint binders, the set of fat ingredients reacting with lead-based compounds is much more restricted, consisting only of siccative oils, almost systematically linseed oil or walnut oil, and with very few mentions of poppy seed oil (e.g., Watin 1773). In contrast to lead plasters, olive oil would be completely inappropriate for painting, as it is not a drying oil.

5.3.3 *Proportions of Lead and Fat/Oil Compounds*

While plasters and paint media are roughly based on the same types of ingredients, the proportions of these ingredients clearly differ. This can be related to their different purposes.

5.3.3.1 Pharmaceutical Domain

In pharmaceutical treatises, the proportion of lead-based compounds and fat strongly depends on the desired plastic properties of the product. As shown in Fig. 5.3a, the statistical analysis of 75 recipes of fat/lead-based compounds mixtures given by Lemery in his *Pharmacopée Universelle (Lemery* 1764*)* shows that *ointments* and *liniments* are usually synthesized with an excess of fat over lead-based compounds and at room temperature, while in *plasters*, the mixture usually contains an excess of lead-based compounds over fat and it is heated. For *cerat*, classification is not obvious. As shown experimentally (Cotte et al. 2006), the saponification rate, and consequently the plasticity of the product, increases with the lead-based compound concentration and with the preparation temperature. Charas details: *plasters are compositions that are applied externally & that are used as ointments & cerats, but their consistency has to be much stronger, such that they can be prepared as rolls or "magdaléons" when cooked and cooled down; and can simply be wrapped in paper, when we want to keep them, while we ordinarily put ointments and cerats in pots, because of their softness and the difficulty of keeping them otherwise* (Charas 1676, p. 703).

5.3.3.2 Paint Domain

Figure 5.3b presents a similar statistical analysis of oil paint binding media prepared with lead-containing compounds of 29 recipes given by T. de Mayerne (Mayerne 1620–1646; De Viguerie et al. 2016). When provided in the recipe, the mass ratio lead-based compound/oil is indicated. As shown here, and also discussed in (De Viguerie et al. 2008), in comparison with pharmaceutical recipes, these lead-based media usually require a lower amount of lead-containing compounds, with a mass ratio litharge/oil varying from 1–4 to 1–16. The lower amount of lead-containing compounds results in a lower saponification rate as observed experimentally, hence a more fluid product, in agreement with the desired final properties.

5.4 Controlling the Operating Conditions: Use of Fire, Water, and Sunlight and the Grinding Level of Lead-Based Compounds

5.4.1 Use of Fire and Control of the Temperature by the Use of Water

As shown in Fig. 5.3, a distinction can be made between preparations made at room temperature and those requiring the use of fire. Increasing the temperature will speed up the saponification process. However, as stated in the quotations of Charas and

Lemery about the preparation of diapalm plaster in the Introduction, the control of the firing conditions, in particular of the temperature, is a major concern if one wants to control the final product and obtain a white paste.

5.4.1.1 Pharmaceutical Domain

Many recommendations are given to finely crush the powders and to control the temperature of the mixture. The following are examples in early Arabic texts:

> 16.20 Suppository that narrows the vagina,
>
> Put one waq of litharge reduced to powder in a marmite. Pour on it 2.5 waq of oil and let it stand. Set on small fire until complete dissolution of litharge (Pharmacopeia by Nejibeddine Assamarkandi (1222), Assamarkandi 1994).

> 128. Description of a red poultice for boils and abscesses, torn body, some abscesses and other ailments
>
> [. . .] all ingredients are well crushed. The fat is melted on a low fire, powders are added and well mixed [...]. (Medical formulary by Al-Kindi Aqrabadhin, 870 (Al-Kindi 1966))

And even earlier:

> Other makeup to be prepared during summer, winter and during flood season: galena. (It) will be milled in the terep-goose fat, in the morning, without allowing it to fall (?) in the fire. Make up with (it), at night. (Ebers Papyri, Ebers 389 (61, 6–8) (Bardinet 1995))

Other authors insist on the fact that the mixture may overflow if it is not carefully checked and shaken and if it is heated too much; as an example, Dorvault advises to use a *basin at least three times larger than needed to contain ingredients* (Dorvault 1886).

As shown in Fig. 5.4a, a fundamental evolution in the lead plaster recipes is in the addition of water to the fat/lead mixture. In this figure, 280 lead plaster recipes, all requiring to heat the lead/fat mixture, were classified into five groups.

- Two groups in which no aqueous ingredient is included. In these, there is the group of the so-called burnt plasters, in which the lead-fat mixture is willingly overheated to obtain a brown paste (typical French names are *Emplâtre brun*, *onguent de la mère Thécle*, *emplâtre brûlé*, and *emplâtre de céruse brûlé*).
- Three groups in which at least one aqueous ingredient is included. Among these three groups, we distinguish among those prepared with water, those prepared with decoctions or mucilages, and those based on the so-called *simple plaster*. As mentioned in the Introduction, *simple plaster* recipes include water.

Although limited by the low number and geographical heterogeneity of the sources (in particular before the sixteenth century), this graph shows a quasi-systematic introduction of water in pharmaceutical recipes from the seventeenth century onward, while most of the recipes before that time were based on dry processes. As proved experimentally, the presence of water has a big impact on the chemical properties, particularly on the saponification rate (Cotte et al. 2006) and on the physical properties, mainly on the color and rheological properties (De Viguerie

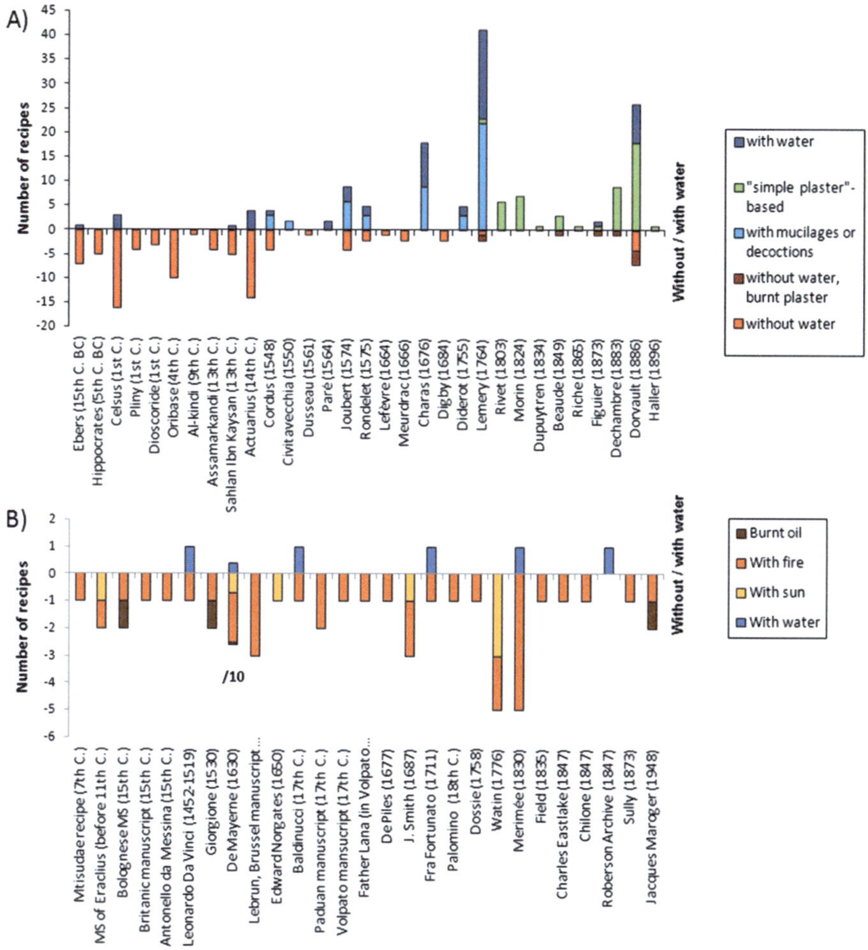

Fig. 5.4 (**a**) Classification of 280 historic recipes of lead plasters in five groups: not involving water, *burnt plasters*, with mucilages or decoctions, based on *simple plaster*, and with water, respectively. (**b**) Classification of 49 historic recipes of lead-based paint media in four groups: burnt oil, with fire, with sun, and with water, respectively. For de Mayerne's recipes, the numbers of recipes are displayed after division by ten to have a final scale similar to that of the other sources

et al. 2008) (see Fig. 5.5). It allows to well control the temperature and the mixture of the components. Diderot et al. perfectly detail the advantages of the use of water in the preparation of lead plasters: *The wonderful aspect, or better the simplicity of this process consists in this: oil and litharge are treated as in a water-bath, and this while the water which makes the bath is present in the same vessel as the materials it heats; it is indeed useless to put it in a separate vessel, because it does not have any chemical actions on these materials. Yet it is important to expose these materials just at the level of heat, because part of the oil could be burnt if the temperature*

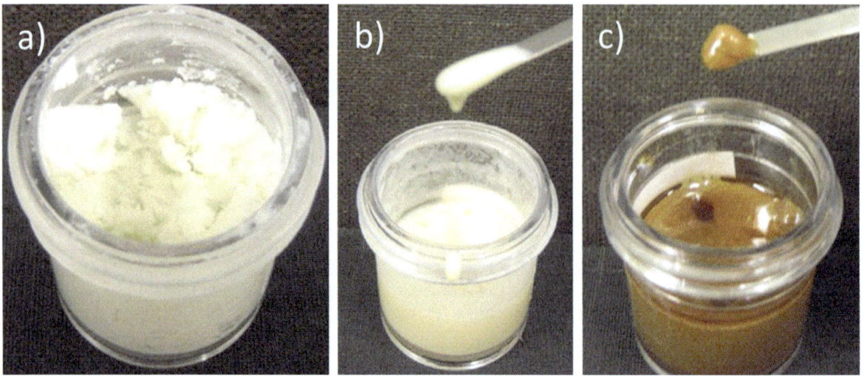

Fig. 5.5 Pictures of lead media/plaster prepared according to the following recipe: oil (4 g) + PbO (1 g) heated at 100 °C during 2 h under continuous stirring, (**a**) olive oil with added water (same amount as oil, the water evaporates during the heating); (**b**) walnut oil with added water; (**c**) walnut oil without water

was higher, leading to formation of coal, and calcined white lead [litharge] could be reduced, or at least blackened: both effects would affect the plaster's elegance, assuming that elegance would not depend on blackness [...] If water would miss before [the combination of ingredients] is reached, which can be detected as soon as the plaster mass swells & rises more than before, and that it falls & then collapses almost suddenly, boiling water is added that must be ready, or that must be heated removing the pan from heat during this time. Cold water cannot be used, because this liquid introduced in the mass of the plaster, which is currently hot to the degree of boiling water, as we will observe, and being suddenly put into expansion, would suddenly make the plaster expand, spread and might hurt the artist, and even cause a fire (Diderot and D'alembert 1755, p. 804).

5.4.1.2 Paint Domain

Figure 5.4b shows a similar classification for paint media in historical recipes. Sorting these was relatively more difficult since painting recipes are usually less detailed than pharmaceutical recipes. Four classes were made. One group includes the recipes that call for the use of water, all of which require the mixture be heated, and the other three groups comprise recipes that do not require water. The three groups recommending the dry processes include recipes in which the medium is exposed to sunlight, those in which the medium is heated on a fire, and those in which the medium is overheated to burn the oil during. The distinction between the last two classes is indeed not always easy due to the lack of description of the properties (in particular color) of the final mixtures. Probably many of the recipes sorted in the "with fire" category end as a brown-black medium (cf. Fig. 5.5).

In contrast with the field of lead plasters (Fig. 5.4a), no clear trend appears over centuries regarding the preferred use of dry or wet processes, and most of the recipes exclude the introduction of water. According to Maroger's personal interpretations, Leonardo da Vinci presumably would have been the first to introduce water in the oil/litharge mixture, thus controlling the firing temperature and the color of the final product (Maroger 1948): *The formula for the second lead medium, the probable technique of Leonardo da Vinci* (*Maroger* 1948, p. 157): *Litharge I part; Raw Linseed or Walnut Oil. 3 to 4 parts; Water 3 to 4 parts. After weighing, the litharge and oil are mixed thoroughly by grinding the litharge first with a little of the oil, then adding the rest. This mixture is placed over a low fire and allowed to simmer. (Asbestos mats should be used under the vessel to distribute the heat evenly and prevent burning.) It is necessary to stir the mixture often with a spatula. As soon as the water begins to boil, a strong yellowish froth will appear on the surface. It will increase rapidly in volume and after a certain time – about half an hour according to the quantity with which one is working – this froth will become whitish. The operation can be considered terminated when the bubbles no longer have small particles of litharge attached to their periphery. It is advisable to replace the water as fast as it evaporates, but this must be done very cautiously because the introduction of too much water at one time retards the operation* [...].

To illustrate this process, Fig. 5.5 shows pictures of paint media prepared by mixing oil (4 g) and PbO (1 g), heating the mixture at 100 °C for 2 h while stirring continuously and using different oils (olive and walnut) with and without water. The color and consistency of the resulting pastes are clearly affected by the operating conditions.

Figure 5.4 shows that even though most recipes for making lead plaster include water from the seventeenth century onward, only few recipes (nine) for paint media do so. This difference may be related to the wish to obtain lead media free of water (that dries faster; and it is better suited for use with hydrophobic pigments). Indeed, some recipes for making lead media recommend the use of a water bath to prevent the oil carbonization (e.g., Mayerne 1620–1646, p. 364; Eastlake 1847, quoting Dreme, p. 354; Merrifield 1999, quoting Palomino, p. ccxli). On the contrary, in pharmaceutical preparations, the presence of residual water in the final paste is not critical since this product is not required to dry.

The use of well crushed lead-based compounds is also regularly recommended in lead-media recipes (e.g., cf. Baldinucci Merrifield 1999. p. ccxxxxix). Alternatively, lead-containing compounds can be placed in a rag during cooking to avoid residual particles in the products (e.g., Pierre Lebrun, *Brussels manuscript, Recueil des essays des merveilles de la peinture*, 1635 (Merrifield 1999, p. ccxxxviii and p. 816–818)). This aspect is particularly important since remaining coarse litharge particles may cause "litharge age" effect (Bergeon and Faillant-Dumas 1980), affecting the final paint through the formation of lead soaps aggregates (Cotte et al. 2017).

5.4.2 *Use of Sunlight*

Figure 5.4b also highlights a class of recipes in which paint media are exposed to sunlight. Sunlight is not only a source of moderate heat but also causes photooxidation, initiating the formation of radicals and accelerating drying reaction processes (Mallégol et al. 2000). In opposition, very few occurrences of sunlight were found in the selected corpus of pharmaceutical texts (e.g., see Nicodeme oil in Lemery 1764, p. 983). Again, this difference is explained by the different purposes of the products: lead media should dry fast, and lead plasters should not dry. Besides, sunlight can be used to bleach oil. Again, this is important in painting, but not so much for pharmaceutical applications.

5.5 Conclusions

Eastlake noticed in his *Materials for a History of Oil Painting* the many connections between the early history of painting and that of medicine. He highlighted in particular the fundamental role played by religious confraternities in preserving written materials and their interest in both Medicine and Decorative Arts (Eastlake 1847). Here, we have shown that mixtures of fat and lead-based compounds have been extensively used since antiquity until modern times, both in pharmacy and painting domains. Recipes show similarities and differences that can be related to the desired functions of the final products.

Lead-based ingredients are globally the same, throughout Western history, with a strong preference for litharge, lead white, and minium. This can possibly be explained by the fact that these compounds, litharge in particular, will dissolve well in fat/oil, giving a homogeneous distribution of lead (as a pharmaceutical agent or as a drier). Litharge also has the highest saponification rate, required to render the liquid oil more viscous or "plastic" in reference to the etymology of "plasters." Litharge and lead white are the preferred compounds to produce white pastes, while minium is recommended to obtain a red product. Lead sulfides are mentioned only in pharmaceutical treatises, mainly in Egyptian texts, and in agreement with the chemical composition of Egyptian cosmetics, particularly galena which gives a black color of special esthetical value to blacken the eyes. Conversely, lead-tin yellow is mentioned only in (few) paint media recipes, offering the painter the possibility to obtain a colored ground layer (e.g., Vasari, 1550 (Stols-Witlox et al. 2008, p. 79)). Lead acetate is found occasionally in both fields, in more recent treatises. The use of metallic lead (as recipient) is mentioned in few recipes, in both fields. During the nineteenth century, concerns about lead toxicity motivate both pharmacists and painters to find alternative ingredients, and in both fields zinc oxide is identified as a good candidate.

The choice of fat/oil ingredients completely differs between both fields. In lead plaster recipes, fatty ingredients are various and can be used together. Olive oil is regularly employed. An important aspect is that the plasters should stay in place without drying. On the contrary, paint media recipes usually advise the use of only one type of oil, generally linseed oil or walnut oil, these too being excellent drying oils.

The lead/fat ratios are usually higher for lead plasters, in comparison to lead ointments and liniments and to paint media. This is directly linked to the targeted consistency: strong and almost solid for plasters (more soaps) and soft and almost liquid for the others (less soaps).

Regarding the operational conditions, in both fields, emphasis is placed on finely crushing and thoroughly mixing of the ingredients as well as controlling the temperature. In pharmaceutical recipes, the use of water as an extra ingredient (in particular since the eighteenth century onward) helps to fulfil these two last requirements. In contrast, water is rarely used as an ingredient for the preparation of paint media (probably to prevent residual humidity in the medium). Alternatively, a water bath can be used to control the temperature. Besides, paint media can be exposed to sunlight to speed up the drying reaction (not required for lead plasters).

In conclusion, by controlling the saponification of fat with lead-based compounds, a large set of products has been invented over history, with different consistencies, viscosities, colors, pharmaceutical properties, and drying properties, offering the physicians and the painters the appropriate materials to develop their respective arts.

Acknowledgments The authors thank Marlène Aubin for the discussion. They are extremely grateful for the major work of reviewers and editors, thanks to whom this manuscript has been deeply revised and improved.

References

Actuarius J (1546) Actuarius de medicamentorum compositione, Ed. de Parisiis,apud Iacobum Bogardum

Al-Kindi (1966) The medical formulary or aqrabadhin of Al-Kindi, translated with a study of its materia medica by Martin Level. The University of Wisconsin Press, Madison

Assamarkandi NE (1994) Pharmacopée sur la classification des causes, traduction et commentaire par Dr G Tohmé, Librairie du Liban Editeurs

Aubin M, Labonnelie M, Walter P, Bellot-Gurlet L (2017) Studying the thermal transformation of lead carbonates involved in Antique Roman eye care medicine. Technart, Bilbao

Bardinet T (1995) Les papyrus médicaux de l'Egypte pharaonique, Paris, Fayard

Barreswil LC (1864) Dictionnaire de chimie industrielle Paris, F. Tandou

Beaude (1849) Dictionnaire de médecine usuelle, éd. de Paris,Didier

Béral M (1837) Note sur la préparation de l'emplâtre simple. Journal de chimie médicale, de pharmacie, de toxicologie, et revue des nouvelles scientifiques nationales et étrangères 3:303–306

Bergeon S, Faillant-Dumas L (1980) Restauration des peintures. Catalogue de l'exposition, paris, 1980, Éditions de la Réunion des musées nationaux

Beurs W (1692) De groote Waereld in het Kleen Geschildert. Johannes en Gillis Ansonius van Waesberge, Amsterdam

Budé G (1539) De curandis articularibus morbis commentarius, Parisiis, Petrum Regnault
Carlyle L (1999) Paint driers discussed in 19th-century British oil painting manuals. J Am Inst Conserv 38:69–82
Celse (1876) Traité de médecine de A.C. Celse: traduction nouvelle, Masson, Paris, Edition du Dr. Védrènes
Charas M (1676). Pharmacopée royale galénique et chymique, Ed. de Paris, chez l'Auteur
Chevreul ME., France IND (1850). Recherches expérimentales sur la peinture à l'huile, F. Didot frères
Corbeil M-C, Robinet L (2002) X-ray powder diffraction data for selected metal soaps. Powder Diffract 17:52–60
Cordus V (1548) Novum Valerii Cordi dispensatorium hoc est. Pharmacorum conficiendorum ratio, Ed. de Parisiis,apud Jacobum Gazellum
Corkery R (1998) Artificial biomineralisation and metallic soaps
Cotte M, Dumas P, Richard G, Breniaux R, Walter P (2005) New insight on ancient cosmetic preparation by synchrotron-based infrared microscopy. Anal Chim Acta 553:105–110
Cotte M, Checroun E, Susini J, Dumas P, Tchoreloff P, Besnard M, Walter P (2006) Kinetics of oil saponification by lead salts in ancient preparations of pharmaceutical lead plasters and painting lead mediums. Talanta 70:1136–1142
Cotte M, Checroun E, De Nolf W, Taniguchi Y, De Viguerie L, Burghammer M, Walter P, Rivard C, Salomé M, Janssens K, Susini J (2017) Lead soaps in paintings: Friends or foes? Stud Conserv 62:2–23
De Lens AJ, Mérat FV (1833) Dictionnaire universel de matière médicale et de thérapeutique générale: contenant l'indication, la description et l'emploi de tous les médicaments connus dans les diverses parties du globe, J.-B. Baillière, Méquignon-Marvis
De Viguerie L, Ducouret G, Cotte M, Lequeux F, Walter P (2008) New insights on the glaze technique through reconstruction of old glaze medium formulations. Colloids Surf A Physicochem Eng Asp 331:119–125
De Viguerie L, Payard PA, Portero E, Walter P, Cotte M (2016) The drying of linseed oil investigated by Fourier transform infrared spectroscopy: Historical recipes and influence of lead compounds. Prog Org Coat 93:46–60
Dechambre A (1883) Dictionnaire encyclopédique des sciences médicales, Ed. de Paris, G. Masson, P. Asselin
Di Civitavecchia B (1550) In antidotarium Joannis filii Mesuae censura: cum declaratione simplicium medicinarum, & solutione multorum dubiorum ac difficilium terminorum: ad haec, receptarium quam castigatissimum cum suo repertorio, in calce apposuimus, Ed. de Lugduni, apud Joannem Frellonium
Diderot, D'alembert (1755) Encyclopédie ou Dictionnaire raisonné des sciences, des arts et des métiers, Ed. de Paris: chez Briasson, David l'aîné, le Breton, Durand
Digby K (1684) Remèdes souverains et secrets expérimentez, Ed. de Paris, G. Cavelier
Dorvault F (1886) Officine ou Répertoire général de pharmacie pratique, Paris, Ed. Asselin et Houzeau
Dupuytren G (1834) Traité théorique et pratique des blessures par armes de guerre, Ed. de Paris, J.-B. Baillière
Dusseau M (1561) Enchirid, ou Manipul des miropoles sommairement traduit & commenté suivant le texte latin pour les inerudits & tyroncles dudit estat, en forme de théorique, Ed. de A Lion, par Ian de Tournes
Eastlake CL (1847) Materials for a history of oil painting
Figuier L (1873) Les merveilles de l'industrie ou Description des principales industries modernes, Edition de Paris, Jouvet, Furne
Francoeur L-B, Lenormand LS (1825) Dictionnaire technologique ou nouveau dictionnaire universel des arts et métiers, et de l'économie industrielle et commerciale, Paris, Thomine, Fortic
Galien C, Winther J, Rouillé G (1552) Claudii Galeni, de Compositione medicamentorum per genera, Joanne Andernaco, interprete, Ed. de Lugduni, apud Gulielmum Rouillium

Grapow H (1958) Die medizinischen Texte in hieroglyphischer Umschreibung autographiert, Akademie-Verlag
Haller A, Muller PT (1896) Chimie organique, Ed. de Paris, G. Carré
Joubert L (1574) Traitté des arcbusades, contenant la vraye essence du mal, et la propre curation, par certaines et méthodiques indications, avec l'explication de divers problèmes touchant cette matière, Ed. de Lyon, Jean de Tournes
Kastens ML, Hansen FR (1949) Drier soap manufacture. Ind Eng Chem 41:2080–2090
Kneepkens I (2012) Understanding historical recipes for the modification of linseed oil. Master of Arts, University of Amsterdam
Lefevre N (1751) Cours de chymie, pour servir d'introduction à cette science, Ed. de Paris, J.-N. Leloup
Lemery N (1764) Pharmacopée universelle: contenant toutes les compositions de pharmacie qui sont en usage dans la médecine, tant en France que par toute l'Europe: leurs vertus, leurs doses, les manières d'opérer les plus simples et les meilleures: avec un lexicon pharmaceutique, plusieurs remarques, et des raisonnemens sur chaque opération, Ed. de Paris: De Saint et Saillant, J.-T. Herissant, Nyon
Liébault J (1651) Thresor universel des pauvres et des riches ou Recueil de remèdes faciles, pour toute sorte de maladies qui surviennent au corps humain, depuis la plante des pieds, jusqu'au sommet de la teste, tant intérieures qu'extérieures... ([Reprod.]) / par M. Jean Liébault, médecin de Paris, Paris, Gervais Clousier
Littré É (1839–1861) Oeuvres complètes d'Hippocrate, Paris, Ballière
Mallégol J, Gardette J-L, Lemaire J (2000) Long-term behavior of oil-based varnishes and paints. Photo- and thermooxidation of cured linseed oil. J Am Oil Chem Soc 77:257–263
Maroger J (1948) The secret formulas and techniques of the masters. Studio Publications, New York
Mayerne TTD (1620–1646) Pictoria, Sculptoria, Tinctoria, et quae subalternarum artium (manuscript). British Museum London, Ms. Sloane 2052
Mérimée JFL (1830) De la peinture à l'huile ou des procédés matériels employés dans ce genre de peinture, depuis Hubert et Jean Van-Eyck Jusqu'a Nos Jours, Paris, Mme Huzard
Merrifield MMP (1999) Medieval and renaissance treatises on the arts of painting: Original texts with English translations. Courier Dover Publications, Mineola, New York
Meurdrac M (1999) La chymie charitable et facile en faveur des dames (1666) CNRS Editions
Mishra S, Daniele S, Hubert-Pfalzgraf LG (2007) Metal 2-ethylhexanoates and related compounds as useful precursors in materials science. Chem Soc Rev 36:1770–1787
Morin J (1824) Manuel théorique et pratique des gardes-malades et des personnes qui veulent se soigner elles-mêmes, ou L'ami de la santé, Ed. de Paris: Roret
Nadolny J (2008) European documentary sources before c. 1550 relating to painting grounds applied to wooden supports: translation and terminology. Preparation for painting: the artist's choice and its consequences (London, 2008), pp 1–13
Noble P, Van Loon A, Boon J (2008) Selective darkening of the ground and paint layers associated with the wood structure in seventeenth-century panel paintings. Preparation for paintings: the artist's choice and its consequences, pp 68–78
Oribase (1851–1876) Oeuvres complètes avec texte grec et traduction française établie par les docteurs Daremberg et Bussemaker, Tome 5, livre II, Paris, Imprimerie Nationale
Paré A (1564) Dix livres de la chirurgie: avec le magasin des instrumens necessaires à icelle, Ed. de Paris: de l'impr. de Jean Le Royer
Plinel'ancien (1983) Histoire Naturelle, traduction H. Zehnacker, Paris, Les Belles Lettres
Riche A (1865) Leçons de chimie: professées à l'école supérieure de pharmacie de Paris et à Sainte-Barbe, Ed. de Paris: F. Didot frères, fils
Rivet J-B (1803) Dictionnaire raisonné de pharmacie-chimique, théorique et pratique, Ed. de Lyon: Reymann; Paris: Brunot
Rondelet G (1575) Gulielmi Rondeletii,... Methodus curandorum omnium morborum corporis humani: in tres libros distincta; De dignoscendis morbis; De febribus, Ed. de Lugduni: apud Guliel. Rouillium

Sallé L (1817) Cours élémentaire d'histoire naturelle des médicaments, Paris, Imprimerie de Feugueray
Sbath P, Avierinos C (1953) Sahlān Ibn Kaysān et Rašīd al-Dīn Abū Ḫulayka: Deux Traités Medicaux. Bullétin de l'Insitut d'Egypte 25:43–75
Smith J. C. M. (1687) The art of painting in oyl, 2nd edition, London
Stols-Witlox M, Doherty T, Schoonhoven B (2008) Reconstructing Seventeenth-century Streaky Imprimatura Layers Used on Panel Paintings. Preparation for Painting: The artist's choice and its consequences, pp 79–91
Virey JJ (1811) Traité de pharmacie théorique et pratique, Paris, Rémont, Ferra aîné
Walter P, Martinetto P, Tsoucaris G, Brniaux R, Lefebvre MA, Richard G, Talabot J, Dooryhee E (1999) Making make-up in Ancient Egypt. Nature 397:483–484
Watin J-F (1773) L'art du peintre, doreur, vernisseur: ouvrage utile aux artistes et aux amateurs, Paris, Editions de Paris: Grangé
Wellman M (1914) Dioscoridis Pedanii, De Materia Medica, libri quinque, Vienna, Ed. Weidmannsche verlagsbuchhandlung
White R, Kirby J (1994) Rembrandt and his circle: seventeenth-century Dutch paint media re-examined. Natl Gallery Tech Bull 15:64–77

Chapter 6
Impact of Lead Soaps on the Formation of Age Craquelure

Katrien Keune, Rick P. Kramer, Suzanne Stangier, and Margriet H. van Eikema Hommes

Abstract The painted wall hanging (Andries Warmoes, signed and dated 1778) in the rear salon of the eighteenth-century Hofkeshuis in Almelo, Netherlands, offers a unique opportunity to study the relationship between lead soap formation and age craquelure in the oil paint on these canvasses. The proportion of intact lead white pigments to lead soap-rich areas (LW/LS ratio) was deduced from backscattered electron images of paint cross sections. The LW/LS ratio was determined with a newly developed computational imaging analysis method.

This paper presents insights regarding two craquelure phenomena related to lead soap formation. The first concerns areas of paint with pronounced age craquelure alternating with areas with minimal age craquelure. These areas are only a few centimeters apart. The paint with extreme craquelure shows considerable lead soap formation, while the lead white in the paint film with minimal craquelure is much better preserved. This difference can be linked to the preparation of the canvasses. The second phenomenon occurs in paint that is partially covered by oak panelling that was installed in 1956 to encase the original marble mantelpiece. The paint behind the oak panelling shows less cracking and more lead soap formation, in contrast to the uncovered paint. The saponification is explained by the local humid environment created by the panelling. The low degree of craquelure is ascribed to small fluctuations in relative humidity and temperature behind the wood panelling.

K. Keune (✉)
Conservation Department, Rijksmuseum, Van't Hoff Institute for Molecular Sciences, University of Amsterdam, Amsterdam, The Netherlands
e-mail: K.Keune@rijksmuseum.nl

R. P. Kramer
Eindhoven University of Technology, Eindhoven, Netherlands

S. Stangier
Restauratie Atelier Stangier, Guttecoven, Netherlands

M. H. van Eikema Hommes
Delft University of Technology, Delft, Netherlands

Cultural Heritage Agency of Netherlands, Amsterdam, Netherlands

F. Casadio et al. (eds.), *Metal Soaps in Art*, Cultural Heritage Science,
https://doi.org/10.1007/978-3-319-90617-1_6

Keywords Lead soaps · Oil paint degradation · SEM/EDX · Indoor climate · Chemical marker · Age craquelure

6.1 Introduction

Lead soaps have been used for the first time to understand factors determining the formation of age craquelure in oil paintings. Close examination of a large painted wall hanging (1778) by Andries Warmoes (1748–1793) offered a unique opportunity to study this relationship. The wall hanging (2.5 m high and nearly 20 m wide) decorates three walls of the rear salon that measures approximately 7.5 × 6 m on the first floor of the Hofkeshuis, a private neoclassical-style house in Almelo built in 1775 for the textile entrepreneur Egbert Hofkes (1738–1822).

The wall hanging, situated above wainscoting, covers all the walls apart from the wall with windows on the east. It is interrupted only by the fireplace on the south wall and the door to the adjoining room in the northern corner of the west wall and is thus divided into four canvasses. It depicts a Roman triumphal procession painted in imitation relief surrounded by a trompe l'oeil-painted frame for which only shades of brown were used, what is referred to as a *brunaille* (Van Eikema Hommes and Bakker 2016; Van Eikema Hommes et al. 2016). The wall hanging has a double ground: a silicate-containing layer followed by a chalk-containing layer with traces of lead white. The ground layers were applied before the canvasses were mounted on the wooden framework in the room. Unpainted ground layers were found on the canvas where the edges were folded over for mounting. Thus Warmoes painted his representation after the canvasses had been installed in the room. Both glue and oil were found in the ground layers (sample taken from the canvas edge and analyzed with FTIR and GC-MS). It is hypothesized that the ground layers consist of a glue binder, while the oil is assumed to come from a thin layer of oil applied on top of the ground layers to reduce oil paint absorption (Van Eikema Hommes and Bakker 2016). The artist used very few pigments: lead white, chalk, red and brown earth pigments, and some fine black. Warmoes used these pigments for the entire representation but varied their proportions to render the imitation relief convincingly. FTIR and GC-MS analyses confirmed oil as the binder of the paints (Van Eikema Hommes et al. 2016). The paint layer structure is very simple consisting of only one or two thin layers. The wall hanging has never been lined nor has it been removed from the wall and is still mounted as it was in 1778. It has had very few cleaning interventions and remained unvarnished until 1956, when it received its current thin varnish layer.

In a previous study, lead white degradation in 18 paint samples from the trompe l'oeil-painted frame was studied with scanning electron microscopy (SEM) (Keune et al. 2016). Backscattered electron (BSE) images of paint cross sections visualize the pigmentation and paint morphology (Fig. 6.1). In these contrast images, the lead soap-rich areas appear darker in the BSE image than the intact and semi-intact lead white pigment particles (brightest particles), due to their higher organic content, but not as dark as the organic binding medium and the chalk and earth

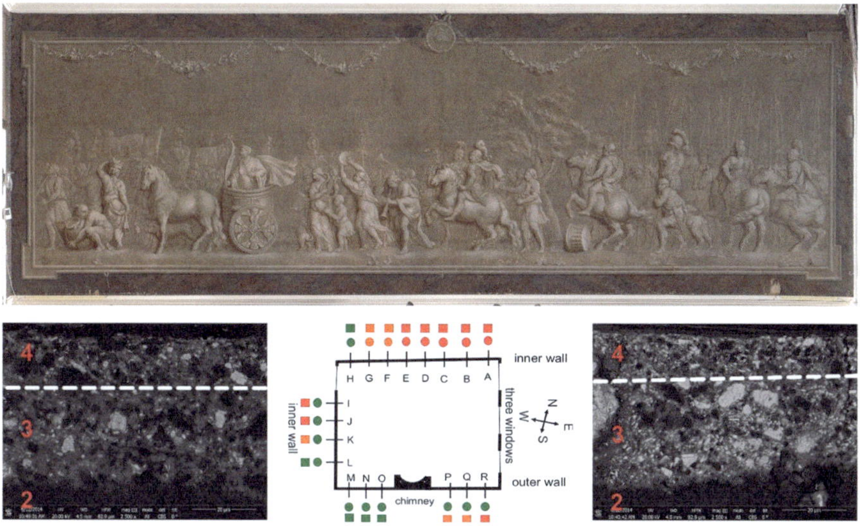

Fig. 6.1 Above: Wall hanging, north wall, *The Triumph of the Roman General Quintus Fabius Maximus* (1778) by Andries Warmoes, Hofkeshuis, Almelo. (photograph: Rik Klein Gotink, 2013). Below left: BSE image of paint sample taken from trompe l'oeil-painted frame close to the door where the lead white is well preserved. Below right: BSE image of paint sample taken from trompe l'oeil-painted frame close to window where lead white is highly saponified. Below middle: Floor plan of the room with locations of samples of A-R. The degree of lead white degradation of layers 3 and 4 is divided into three classes, high (red), medium (orange), and low (green), and is indicated in the BSE images and at the sample locations. The measurements obtained from layer 3 are indicated by squares and those from layer 4 by circles

pigment particles. The lead soap-rich areas are clearly visible around the lead white pigment particles. The BSE images were analyzed using a newly developed computational imaging method to quantify the ratio of intact lead white particles to lead soap-rich areas (referred to below as LW/LS ratio). This method proved to be quantifiable and is consistent with previous visual observations (Keune et al. 2016). Moreover it eliminates the subjective character of classification based on visual inspection.

The pigmentation and paint layer structure of the 18 paint samples from the wall hanging are identical; only the thickness of the paint layers varies somewhat. The locations of the samples are spaced regularly over the wall hangings and are representative for different local environmental conditions in the room. BSE images from paint samples reveal strong variations in amount of lead soap formation, which are linked to different local climatological conditions within the room (Huijbregts 2017).

The formation of age craquelure in paintings is caused by differences in mechanical behavior of the various layers due to shrinkage or expansion. The causes behind different age craquelure patterns in oil paintings are diverse and not always fully understood (Hess 1951; Bucklow 2012). This case study shows the relationship of two interesting age craquelure phenomena observed in the painted wall hanging related to differences in lead white degradation:

1. Variations in the age craquelure pattern on the north wall
2. The effect of the oak panelling placed around the mantelpiece in 1956 that resulted in variations in the age craquelure pattern on the south wall

The LW/LS ratio in BSE images of 5 sets of paint cross sections, subjected to the computational imaging analysis method, was used to study both phenomena.

6.2 Results and Discussion

6.2.1 Inconsistencies in Craquelure Formation on the North Wall

6.2.1.1 Craquelure Patterns in the Hofkeshuis

In total three types of age craquelure were observed in the Hofkeshuis paintings (Stangier 2013). The first, the main craquelure pattern in all four paintings, is a fine network of irregular islands with angular contours. The coarseness of the crack network, dimensions of the islands, and degree of cupping vary strongly. These variations can be linked to some extent to differing local environmental conditions at different positions in the room. For instance, cupping is more pronounced in areas where the paint is exposed to direct sunlight. However, the environment alone does not explain the variations observed, and given the close proximity between the locations of different age craquelure patterns, other factors must play a role as well.

The two other types of age craquelure patterns are only found on the paintings hanging on the south wall (external wall). The painting to the right (west) of the fireplace has long parallel vertical cracks and shows no network of cracks or cupping. What appears to be mold can be seen inside these cracks (not identified), which can be directly linked to previous water damage. The third age craquelure type is present in the upper part of both paintings, left and right of the fireplace, although it is more pronounced in the painting to the right. This craquelure pattern has a very broad crack network of square islands. There is hardly any cupping and here too the cracks appear to contain mold.

6.2.1.2 Inconsistencies in the Craquelure Patterns

An area of the wall painting on the north wall shows remarkable variations in craquelure formation that differs from the general age craquelure pattern described

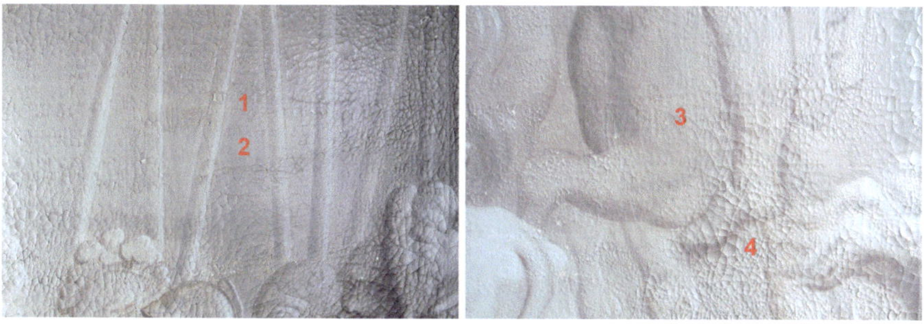

Fig. 6.2 Details of the wall hanging on the north wall depicting the lances of the group of soldiers in the background (left, photo, S. Stangier) and part of cavalier and horse (right, photo, M.H. van Eikema Hommes). The numbers correspond to the sample areas: 1 = Lances_craq; 2 = Lances_nocraq; 3 = Horse_nocraq; 4 = Horse_craq

above.[1] This comes from an area depicting the lances of a group of soldiers in the background of the painting (Fig. 6.2 left). Here, at distances of only a few centimeters apart, areas with extreme craquelure and cupping alternate with areas of minimal craquelure and no cupping. It would seem that horizontal zones of affected and more intact paint alternate (Fig. 6.2 left). There is a comparable area 30 cm below this region[2]: the cavalier's calf and his horse's chest where no extreme craquelure and cupping can be observed, while the cavalier's feet and knee and the horse's leg, tail, flank, and hind quarter have significant craquelure and cupping (Fig. 6.2 right). In both cases differences in the age craquelure are independent of the painted depiction and modelling and therefore cannot be explained by differences in paint composition and structure. We can eliminate differences in local environmental conditions because the affected and more intact areas are so close to each other. Paint cross sections were taken from adjacent areas showing extreme and minimal craquelure formation in order to investigate the differences.

6.2.1.3 Lances and Horse Paint Samples

Figure 6.2 gives the locations of the paint cross sections, Lances_craq, Lances_nocraq, Horse_craq, and Horse_nocraq. The Lances cross sections have the following paint structure: first a silicate glue ground (20–50 μm, layer 1); a chalk glue ground (100–200 μm, layer 2); two light beige-grey layers consisting of lead white, chalk, fine earth pigment, and very fine black pigment (layers 3, 4); a lead white layer (approx. 5 μm, layer 5); and a thin varnish layer (1–2 μm) applied

[1]This area is 40×20 cm and some 150 cm from the window.

[2]This area is 10×20 cm and some 150 cm from the window.

in 1956 (layer 6). The Horse cross sections have the same layer buildup as the samples from the Lances, but without layer 5.

6.2.1.4 Lances and Horse Samples: BSE Images and Contrast Values

BSE images reveal a noticeable difference in the paint morphology of layers 3–5 (Fig. 6.3). Samples from paint areas with minimal age craquelure (*Lances_nocraq* and *Horse_nocraq*) have relatively large amounts of lead white particles as compared to samples from the highly cracked areas (*Lances_craq* and *Horse_craq*) (Fig. 6.3a, b). The contrast value determined using the computational imaging analysis method (Keune et al. 2016)[3] expresses the LW/LS ratio in the sample. The higher this value, the better preserved the lead white pigment. Table 6.1 gives the contrast values for layers 3, 4, and 5 for the Lances and Horse samples. There is a clear trend: paint from heavily cracked areas has far fewer intact lead white particles. For instance, the lead white in layer 3 of *Lances_craq* contains 67%

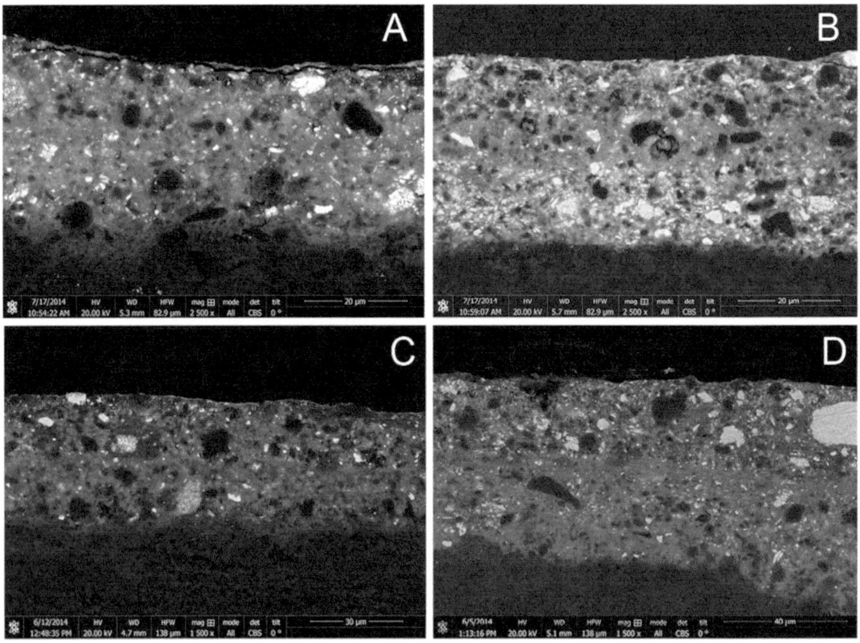

Fig. 6.3 BSE image of samples Lances_craq (**a**), Lances_nocraq (**b**), Horse_craq (**c**), and Horse_nocraq (**d**)

[3]Note that these values cannot be compared directly to those published for the painted frame because the paint composition of layers 3 and 4 is different (Keune et al. 2016).

Table 6.1 Contrast values for layers 3 and 4 + 5 of the Lances and Horse samples

	Lances_nocraq	Lances_craq	Horse_nocraq	Horse_craq
Layer 3	33	10	21	17
Layer 4/5	26	12	30	22

less intact lead white particles and thus a higher conversion to lead soaps than sample *Lances_nocraq*. The Horse samples show a smaller difference in intact lead white as compared to the Lances samples, but the same trend is seen here as well. In conclusion, paint with a relative high LW/LS ratio corresponds with minimal cracking and no cupping, while paint with a relative low LW/LS ratio is associated with heavy cracking and cupping.

6.2.1.5 *Lead Soaps* Versus *Mechanical Strength*

These results indicate that the mechanical strength of the paint layer changes due to lead soap formation and loss of intact lead white particles. Lead soaps are homogenously distributed in the paint. No aggregation or obvious crust formation was found in the dozens of Hofkeshuis samples investigated. The relationship between mechanical properties and the degree of lead soap formation in oil paints has so far not been studied in detail. Studies by Erhard and Mecklenburg show that the older the lead white oil paint, the more brittle it becomes (Erhardt et al. 2005; Mecklenburg and Tumosa 2001). Erhart et al. postulate that this increase in stiffness is due to the increase in oil hydrolysis and formation of lead salts over time. Studies on ionomeric systems mixed with metal soaps show that crystallization of metal soap affects stiffness (Wakabayashi and Register 2006). Ionomers with crystalline magnesium soaps have a higher modulus than ionomers containing noncrystalline soaps (Wakabayashi and Register 2006). The sharp asymmetric lead carboxylate band at 1510 cm^{-1} in an FTIR spectrum published in our previous paper indicates that the Hofkeshuis sample contains crystalline lead soaps (Keune et al. 2016). The brittleness of synthesized ionomeric oil paint systems with lead as the metal ion increases with the concentration of lead ions and thus lead carboxylate groups (Baij and Hermans, University of Amsterdam, personal communication, 2016).

6.2.1.6 Canvas Preparation

The substantial differences in saponification of areas close to each other cannot be explained by differences in specific environmental conditions related to temperature and/or relative humidity, as was concluded in our previous paper (Keune et al. 2016). Instead, in this case, the availability of fatty acids seems to play an important role in the degree of saponification. Intact paint areas are visible as stripy horizontal zones (Fig. 6.2 left), suggesting that the difference in saponification is related to the painting technique. As mentioned, the modelling of the paint does not play a role; instead the most likely source of the fatty acids is the preparation layers of the canvas, in particular the thin oil layer on top of the ground. Presumably this oil layer was applied swiftly with a broad brush, resulting in a streaky uneven application. We assume this oil layer is very thin or almost absent in the relatively intact areas. Here the availability of fatty acids would be far less, resulting in less saponification and thus less craquelure.

6.2.2 Impact of Oak Panelling on Lead Soap Formation

6.2.2.1 Historical Use of the Room

The Hofkeshuis remained in the family for four generations before it was sold in 1928. A 1927 account indicates that the first-floor rear salon with the painted wall hanging was then the most important reception room in the house. The iconography of the wall hanging even suggests it may have been intended originally as a semiofficial meeting room and may not have been used on daily basis (Van Eikema Hommes and Bakker 2016). The 1927 account also reveals that the room had been used for some time as an official dining room, also in keeping with the suggestion that it was not intended for daily use. We can therefore assume that during the first 150 years of the room's existence, its shutters were mostly closed and that it was only heated occasionally. Originally an open fireplace would have provided heat, and later a coal stove may have been used. In 1928 the Hofkeshuis was turned into a shoe shop, and the rear salon was used for storage. The first floor of the Hofkeshuis became a family residence in 1956, and the rear salon with the wall hanging was turned into a living room. From then on the room was used intensively and was heated on a daily basis. A central heating system, with radiators under the windows, was installed in 1956. Since then the shutters have always been open without curtains or other protection from direct sunlight.

6.2.2.2 Inconsistencies in Age Craquelure Caused by the Oak Panelling

In 1956, when the salon with the wall hanging became a living room, the original neoclassical marble mantelpiece was covered with oak panelling, which also covered a strip of some 3 centimeters of the paintings on either sides of the fireplace (Fig. 6.4).

There is a distinct visible difference in appearance between the paint in areas covered by the panelling and exposed areas of paint. The paint surface behind the wooden panelling looks hazy and matt. Remarkably, it also exhibits much less craquelure than exposed areas of the same canvas, which display both minimal and severe age craquelure (Fig. 6.4 above right). This indicates this particular type of age craquelure must have formed after 1956. The paint covered by the panelling on the right (west) side of the mantelpiece has relatively less craquelure and shows no large differences in the age craquelure of the covered and exposed areas of paint.

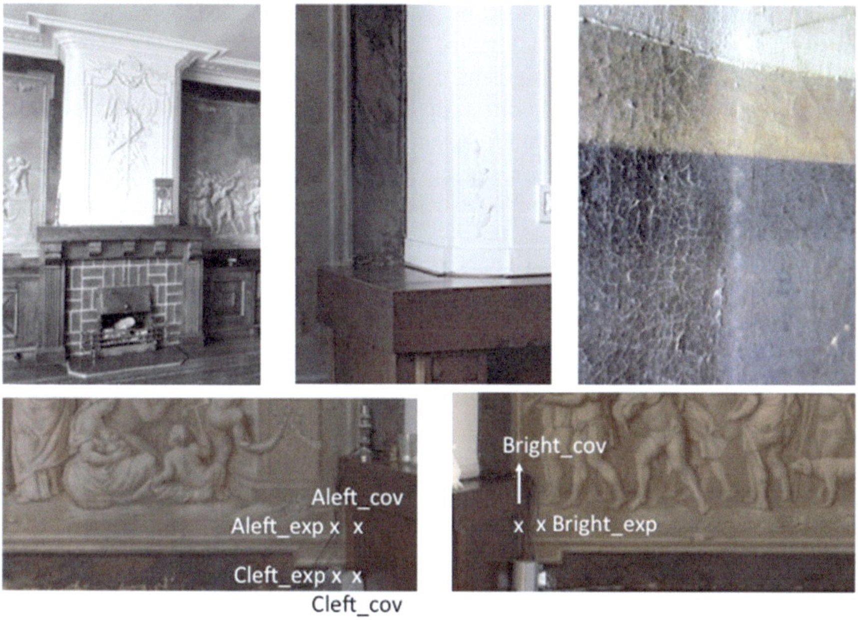

Fig. 6.4 Photograph taken in 1974 of the fireplace with oak panelling on the south wall (photo: RCE) (above left), detail of left side of the panelling covering part of the painting (photo: M.H. van Eikema Hommes, 2013) (above middle) and detail of painting left (east side) to the fireplace, left part is exposed to the room and right part is covered by the panelling (above right) (photo: Keune, 2013). Lower left and right, detail of the paintings left and right to the fireplace with indication of sample areas of A-left_cov, A-left_exp, B-right_cov, B-right_exp, C-left_cov, and C-left_exp

6.2.2.3 Paint Sample Sets: A-Left, B-Right, and C-Left

In order to investigate this, three sample sets (each with one sample of covered and one of exposed paint) were taken:

1. Set A-left from a brown painted area to the left of the fireplace. The distance between the sample from the exposed area (A-left_exp) and the sample from covered area (A-left_cov) was 4.8 cm.
2. Set C-left from a dark brown painted border 7 cm below the sample set A-left. Sample C-left_exp came from an exposed area and C-left_cov from a covered area. The distance between the samples was 5.8 cm.
3. Set B-right was taken from a brown painted area comparable to that of sample set A-left, but from an area of canvas to the right of the fireplace. Sample B-right_exp came from an exposed area and B-right_cov from a covered area. The distance between the samples was 4 cm.

All paint samples have a double ground consisting of first the silicate-containing and then the chalk-rich layer (light microscopic images not shown). The third layer in sets A-left and B-right has a light beige-grey paint layer (lead white, chalk, fine earth pigment, and very fine black pigment), while the fourth layer is a brown color (lead white, chalk, fine earth pigment, and very fine black pigment). All samples have a thin layer of varnish and dirt at the top. The interface between layers 3 and 4 is barely visible in sample A-left_cov. Layer 3 in set C-left contains more fine black pigment and is grey in color, while layer 4 is dark brown and contains less lead white as compared to layer 3.

6.2.2.4 A-Left, B-Right, and C-Left: BSE Images and Contrast Values

Interestingly, the BSE images of all three sets of samples reveal distinct differences in the morphology of the pigment particles in the paint covered by the oak panelling as compared to the exposed paint. In the paint to the left (east) of the fireplace, the lead white pigments covered by the panelling, A-left and C-left, are severely degraded (low LW/LS ratio), while the exposed paint 4–5 cm to the left is much better preserved (high LW/LS ratio) (Fig. 6.5). Table 6.2 shows the contrasting values of the LW/LS ratios in layers 3 and 4, as analyzed by the computational imaging method. The absolute values of sets A-left and C-left cannot be compared because they are different paints; nevertheless, the absolute difference between exposed and covered paints dropped, and the amount of lead white pigment particles decreased by 37–57%. The paint in set B-right is generally much better preserved. The brown painted area in A-left and B-right is assumed to be identical or almost the same paint. The lead white pigment in set B-right is more intact, consistent with the trend found with the painted frame of the work described by Keune et al. 2016. In that study, the best-preserved lead white paint was in the samples to the right of the fireplace where the local temperature is a few degrees lower than that on the left of the fireplace (Huijbregts 2017). Although the lead white in B-right is generally better preserved than in A-left and C-left, the same trend is found here too in the passages of paint covered by the panelling where it was found that the amount of lead white pigment particles had decreased by about 27–38%.

6.2.2.5 Role of Oak Panelling

The explanation for increased lead white degradation, in other words increased saponification of the paint covered by the oak panelling, can be found in the microclimate created behind the panelling due to trapped humidity released from the exterior wall. Here there is therefore constant high relative humidity, while the relative humidity in the uncovered paint fluctuates much more because it is exposed to the room air. Since we know that water plays a crucial role in the saponification of lead white with fatty acids (Cotte et al. 2006, 2007; Catalano et al. 2014), the constant presence of water in the paint behind the panelling leads to more saponification of lead white.

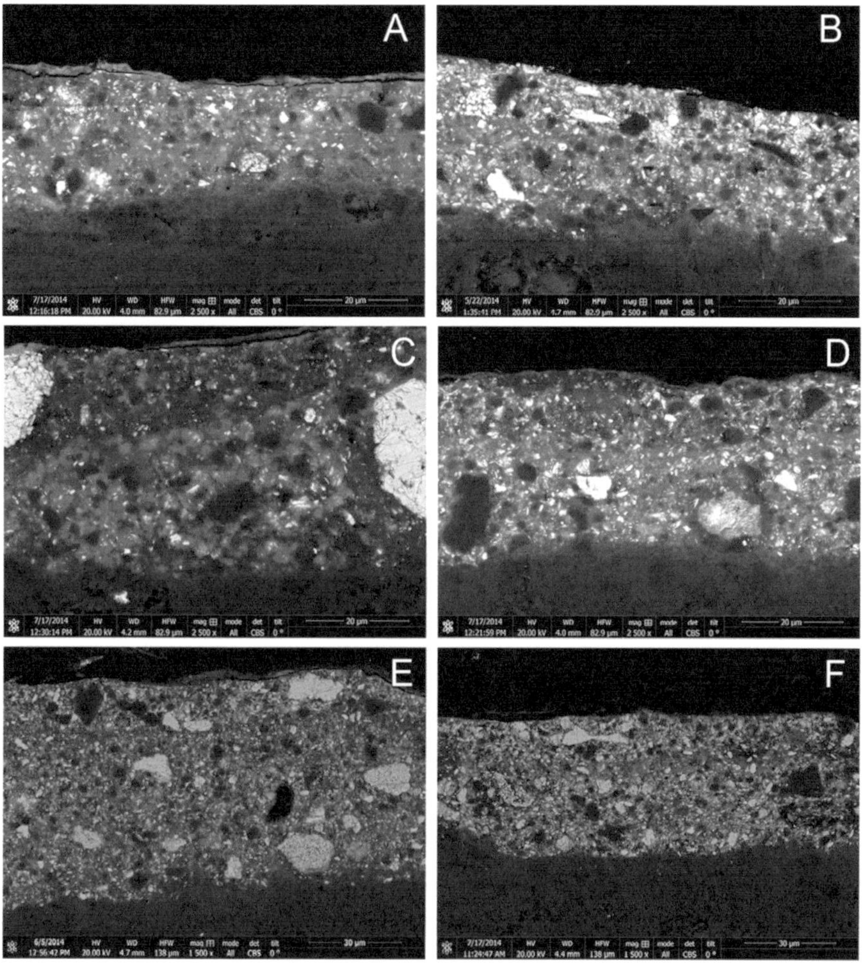

Fig. 6.5 BSE image of samples A-left_cov (**a**), A-left_exp (**b**), C-left_cov (**c**), C-left_exp (**d**), B-right_cov (**e**), and B-right_exp (**f**)

Table 6.2 Contrast values for layers 3 and 4 of the A-left, B-right, and C-left sample sets

	A-left_cov	A-left_exp	B-right_cov	B-right_exp	C-left_cov	C-left_exp
Layer 3	15	24	39	63	10	21
Layer 4	18	35	47	64	11	26

The role of acids released from oak in saponification, however, cannot be excluded since oak is known to emit volatile acetic or formic acids (Chiavari et al. 2008; Gibson 2010), which have been identified in zinc soap formation (Keune and Boéve-Jones 2014).

6.2.2.6 Comparison of A-Left, B-Right, and C-Left with Lances/Horse

Unlike the extreme age craquelure found in the saponified paint of the Lances and Horse paint areas on the wall painting on the north wall, the opposite is found in samples A-left, B-right, and C-left from the canvasses on the south wall. Here, the saponified paint shows relatively less craquelure compared to the more intact paint. This can be explained by the local environmental conditions behind the panelling, which have remained rather constant. Thanks to minimal fluctuations in relative humidity and/or temperature in the saponified paint layer, stresses have been low resulting in elastic deformation and no perceivable age craquelure. Conversely, in the canvasses on the south wall, the exposed areas that are better preserved with more intact lead white particles are exposed to fluctuations in relative humidity and temperature and therefore show greater cracking. The difference in age craquelure between the covered and exposed paint is more extreme in the wall hanging at left of the fireplace as compared with the wall hanging on the right. This is consistent with results of indoor climate monitoring over a 2-year period (Huijbregts 2017). The paint surface to the right of the fireplace was found to be cooler with smaller temperature fluctuations as compared with the paint surface to the left of the fireplace. These smaller fluctuations cause less stress in the paint resulting in less craquelure.

Assuming the installation of the oak panelling in 1956 more or less preserved the craquelure pattern that already existed, it is notable that the degree of age craquelure in the paint next to the area covered by the panelling has increased dramatically since 1956. This shows that intensive use of the room has a significant impact on crack formation in the paintings. Until 1956 the room was only used on special occasions. We can assume that most of the time, the room was unheated with the shutters closed. After 1956, the shutters were always open resulting in large fluctuations in temperature and relative humidity. Moreover, the room was maintained at comfortable temperatures by central heating.

6.3 Conclusion

The proportion of intact lead white pigment to lead soap-rich areas in the paint is used to understand inconsistencies in the age craquelure in the painted wall hanging in the Hofkeshuis. On the north wall, areas of paint with heavy age craquelure alternate with areas with minimal age craquelure. These areas are only a few centimeters apart. The paint film with heavy craquelure shows more lead soap formation, while the lead white in the paint with minimal craquelure is much better preserved. This phenomenon appears linked to the preparation of the canvasses. It is postulated that a thin oil layer applied on top of the glue-based ground layers was unevenly applied in a streaky fashion, resulting in varying amounts of paint saponification.

On the two wall hangings on the south wall, the paint surface is partially covered by oak panelling that was installed in 1956. The paint film behind the oak panelling was found to be highly saponified compared with nearby areas of exposed paint. The increased saponification can be explained by the local humid environment created by the panelling. Interestingly, the saponified paint behind the oak panelling shows less cracking than the exposed paint. It is postulated that fluctuations in relative humidity and temperature behind the panelling are far less than elsewhere in the room, leading to less cracking. The increase in craquelure due to intensified use of the room and installation of central heating after 1956 is notable.

It can be concluded that differences in the extent of saponification of lead white in the paintings were induced by local differences in the relative humidity and temperature and from the availability of fatty acids from an oil preparation layer on the canvas. Thus, the extent and distribution of age craquelure is not only dependent on fluctuations in relative humidity and temperature but also on the degree of saponification.

Acknowledgments This research is part of the 5-year (2012–2017) research project From Isolation to Coherence: An Integrated Technical, Visual and Historical Study of Seventeenth- and Eighteenth-Century Dutch Painting Ensembles (supervised by Dr. Margriet H. van Eikema Hommes), supported by the Netherlands Organisation for Scientific Research (NWO: Innovational Research Incentives Schemes Vidi Grant), and part of the PAinT project, supported by the Science4Arts program of the Dutch Organization for Scientific Research (NWO).

We wish to thank Zara Huijbregts (TU/e), Henk L. Schellen (TU/e), Joen Hermans (UvA) and Lambert Baij (UvA), the Stichting Hofkeshuis in Almelo, and Hans Niers and André Hovink, owner and tenant of the Hofkeshuis, respectively.

Appendix

Experimental

Samples

The description and locations of the ten samples taken can be found in the results. The samples were embedded in a polyester resin (Polypol) and dry polished with Micro-mesh® polishing cloths (final step 12,000 mesh).

Light Microscopy

The paint cross sections were examined under a Zeiss Axioplan 2 microscope with both incident polarized light and incident UV light (from a xenon lamp and a mercury short arc photo optic lamp HBO, respectively). The filter set UV H365 used for examination in UV light consists of the following filters: excitation BP 365/12, beam splitter FT 395, and emission LP 397.

Scanning Electron Microscopy

Scanning electron microscopy in combination with energy dispersive X-ray analysis (SEM-EDX) studies of the Hofkeshuis samples was performed on a Verios 460 high vacuum electron microscope (FEI, Eindhoven, Netherlands) equipped with an Energy Dispersive X-ray system (X-Max 80 mm^2 SDD detector, Oxford) with spot analysis and elemental mapping facilities. BSE images of the cross sections were taken at a 20 kV accelerating voltage, at a 5 mm eucentric working distance and with current density of approximately 130pA. Prior to SEM-EDX analysis, samples were gold coated (3 nm thickness) in a SC7640 gold sputter coater (Quorum Technologies, Newhaven, East Sussex, UK) to improve surface conductivity.

Computational Imaging Analysis

The texture contrast of the BSE images was calculated to determine the relative degree of degraded lead white pigment. To quantify the texture contrast, an image analysis procedure was developed using MATLAB release 2013b (technical computing software of the MathWorks, Inc.). A gray-scale image, containing n by m pixels, is represented in MATLAB by an n by m matrix. In this matrix a numerical value is assigned to every pixel ranging from 0 (black) to 255 (white). Detailed descriptions of the image analyses procedure can be found elsewhere (Keune et al. 2016).

References

Bucklow S (2012) Classification of craquelure, chapter 16. In: Hill Stoner J, Rushfield R (eds) Conservation of easel paintings: principles and practice. Butterworth-Heinemann, London/New York, pp 285–290

Catalano J, Yao Y, Murphy A, Zumbulyadis N, Centeno SA, Dybowski C (2014) Nuclear magnetic resonance spectra and 207Pb chemical-shift tensors of lead carboxylates relevant to soap formation in oil paintings. Appl Spectrosc 68:280–286

Chiavari C, Martini C, Prandstraller D, Niklasson A, Johansson L-G, Svensson J-E, Aslund A Bergsten CJ (2008) Atmospheric corrosion of historical organ pipes: the influence of environment and materials. Corros Sci 50(9):2444–2455

Cotte M, Checroun E, Susini J, Dumas P, Tchoreloff P, Besnard M, Walter P (2006) Kinetics of oil saponification by lead salts in ancient preparations of pharmaceutical lead plasters and painting lead mediums. Talanta 70:1136–1142

Cotte M, Checroun E, Susini J, Walter P (2007) Micro-analytical study of interactions between oil and lead compounds in paintings. Appl Phys A Mater Sci Process 89:841–848

Erhardt D, Tumosa C, Mecklenburg MF (2005) Long-term chemical and physical processes in oil paint films. Stud Conserv 50(2):143–150

Gibson LT (2010) Acetic and formic acids emitted from wood samples and their effect on selected materials in museum environments. Corros Sci 52(1):172–178

Hess M (1951) Hess's paint film defects: their causes and cure. Chapman and Hall Ltd, London Hamburg HR, Morgans WM (eds), 3rd edn., 1979

Huijbregts, Z. (2017). Experimental and numerical analysis of climate change induced risks to historic buildings and collections. Eindhoven: Technische Universiteit Eindhoven

Keune K, Boéve-Jones G (2014) Its surreal: zinc-oxide degradation and misperceptions in Salvador Dalí's couple with clouds in their heads, 1936. In: Van den Berg KJ, Burnstock A, de Tagle A, de Keijzer M, Heydenreich G, Krueger J, Learner T (eds) Issues in contemporary oil paints (ICOP), symposium proceedings, March 2013. Springer, Heidelberg/Amersfoort, pp 283–294

Keune K, Kramer RP, Huijbregts Z, Schellen HL, Stappers MHL, van Eikema Hommes MH (2016) Pigment degradation in oil paint induced by indoor climate: comparison of visual and computational backscattered electron images. Microsc Microanal 22(2):448–457

Mecklenburg MF, Tumosa CS (2001) Traditional oil paints: the effects of long-term chemical and mechanical properties on restoration efforts. MRS Bull 26(1):51–54

Stangier S (2013) Craquelé onderzoek aan de behangselschilderingen in het Hofkeshuis, Almelo Overzicht. Unpublished report

Van Eikema Hommes MH, Bakker P (2016) A triumph with no battle: the significance of a painted wall hanging (1778) in the Hofkeshuis in Almelo. Oud Holland 129(2):47–118

Van Eikema Hommes MH, Keune K, Bakker P, Verslype I (2016) The triumphant procession (1778) of Andries Warmoes in the Hofkeshuis. In: Wallert A (ed) Proceedings of painting techniques. Rijksmuseum, Amsterdam, pp 182–191

Wakabayashi K, Register RA (2006) Ethylene/(meth)acrylic acid ionomers plasticized and reinforced by metal soaps. Polymer 47(8):2874–2883

Chapter 7
An Investigation into Metal Ions in Varnish Coatings

Sally Higgs and Aviva Burnstock

Abstract This study investigated the hypothesis that metal ions may migrate into varnish coatings applied to paintings. Cross-sectional samples from selected paintings in oil media were examined using scanning electron microscopy with energy dispersive X-ray spectroscopy (SEM-EDX) and Fourier transform infrared (FTIR) imaging to map elemental distribution and characterize metal carboxylates. Results showed that in some samples elemental lead, zinc, calcium, and potassium were present in oil-resin varnish. The presence of metal ions in a varnish layer may influence its solubility using conventional solvents by conservators during cleaning.

Keywords Metal ions · Oil-resin varnish · Solubility · Cleaning · Metal carboxylate · Aggregates · SEM-EDX · FTIR

7.1 Introduction

Conservators treating paintings regularly face the challenge of removing degraded and discolored varnish from the surface of paintings. The removability of varnish is influenced by a number of factors, including the composition of varnish and paint and the difference in solubility between the paint and coating. The physical and chemical similarities between paint and coating may be particularly problematic for removal where varnishes contain an oil component. In addition, the solubility of a coating is often observed to vary across different areas of a painting. A number of conservators have noted the difficulty in removing residues of oil-resin varnish from paintings, in particular in passages of sky paint that contain lead white pigments, or separating varnish from degraded copper-containing paints.

S. Higgs (✉)
Conservator in Private Practice, London, UK
e-mail: sallyrhiggs@gmail.com

A. Burnstock
Department of Conservation and Technology, Courtauld Institute of Art, London, UK
e-mail: aviva.burnstock@courtauld.ac.uk

F. Casadio et al. (eds.), *Metal Soaps in Art*, Cultural Heritage Science,
https://doi.org/10.1007/978-3-319-90617-1_7

A significant number of studies have established that metal ions in pigments or driers may migrate within the paint layers in oil paintings (e.g., Keune et al. 2011, 698). This study investigates the hypothesis that metal ions may also migrate into the varnish coating. The presence of metal ions and insoluble metal carboxylates in the varnish could change the solubility of the layer and may explain some of the challenges conservators face in removing coatings using conventional organic solvents.

7.1.1 Metal Ion Migration

The transport mechanism of metal ions through semipermeable oil paint films has been the focus of many recent studies.[1] It is now established that metal ions present in the inorganic materials used for painting react with fatty acids in the oil medium, to form metal carboxylates. Multivalent metals, such as lead and zinc, can form coordination bonds with carboxylic acids, which create linkages between different parts of the cross-linked oil network. This system is described as an ionomer and is the present working model for explaining the remarkably tough and durable properties of many aged oil paint films (Boon et al. 2007).

If the paint is attacked by acids in its chemical environment, or anions, the relatively weak bond at the carboxylic acid can be broken. A carboxylic acid could lose its connection with the network and migrate, either in isolation or as a metal carboxylate. Thus migration through paint layers of fatty acids and metal carboxylates may occur. Further, metal ions from the surface of selected pigments or driers have been proposed to move through the polymeric binding medium by hopping from one carboxylic acid to the next (Hermans 2017).

The migration of metal carboxylates causes a range of physical phenomena in paintings, including protrusions and inclusions that are visible on the surface of paintings. Metal carboxylates are also observed to migrate without aggregation, moving toward the surface of the painting, where they react with atmospheric gasses and remineralize to form a crystalline crust (Van Loon, Noble and Boon 2011, 2). The possibility that metal ions may react with reactive species including saturated fatty acids or other acidic species in degraded varnish to form insoluble products is also feasible.

7.1.2 The Effect of Metal Ions on the Aging and Solubility of Oil-Resin Varnishes

It has been shown that the presence of metal ions affects the curing and aging of drying oils, increasing the rate of oxidation and stabilizing the film through the formation of an ionomeric network (Tumosa and Mecklenburg 2005, 40).

[1] Research is currently led by the PAinT (Paint Alterations in Time) research group at the University of Amsterdam. The mission statement, current research topics and lists of publications may be found at their webpage http://www.s4a-paint.uva.nl (date accessed 20th October 2014).

Studies of the aging of megilps, typically composed of oil and mastic resin, have shown that metal ions may also affect the aging of the resin fraction in oil and resin mixtures. Megilps made from drying oils and mastic varnish heated with lead carbonate were found to have more highly oxidized resin components than mastic varnish alone (Carlyle 2001, 104–104). Similar mixtures of oil and resins have been used for traditional picture varnishes from the seventeenth century into the twentieth century (Phenix and Townsend 2012, 252–263).

Therefore, migration of metal ions into oil-resin varnish may potentially change the solubility of the layer, making it more difficult for conservators to remove using conventional solvent methods.

7.2 Experimental

Samples previously prepared as embedded cross sections from paintings of different historic periods and with varnish layers of different compositions were selected from the archive of paint samples in the Department of Conservation and Technology, Courtauld Institute of Art. Paintings with an oil-resin varnish were the focus for the study; however, samples from works with resin varnishes were included for comparison. One sample from a painting in egg tempera was also included. Samples examined in this study are given in Table 7.1.

Samples from paintings with an oil-resin varnish:

Seurat_5b, from Georges Seurat, *Le Chahut* (1889–1890), oil on canvas
Dobson_7, from William Charles Thomas Dobson, *Christ Raiseth from Death the Widow's Son of Nain* (1868), oil on canvas
Goya_1 and **Goya_2,** both from Francisco de Goya, *Don Francisco de Saavedra* (1798), oil on canvas
Olver_9, from Kate Elizabeth Olver, *Return from the Front* (c. 1916), oil on canvas
Laszlo_9, from Philip de Laszlo, *Wolmer Woods* (c.1930), oil on canvas

Samples from paintings with resin varnish:

Manet_2, from Edouard Manet, *Le Déjeuner sur l'herbe* (1863–1868), oil on canvas
Gainsborough_1, from Thomas Gainsborough, *Elizabeth and Mary Linley* (1772), oil on canvas
Gerino_16A, from Gerino d'Antonio Gerini, *Virgin and Child Enthroned with Saint Lawrence, Saint John the Baptist, Saint Monica and Saint Augustine* (1510), oil on panel

Sample from egg tempera painting:

Daddi_L, Bernardo Daddi, *Triptych: The Virgin and Child Enthroned with Saints* (1338), egg tempera on panel

The samples, embedded in polyester clear casting resin (from Tiranti, www.tiranti.co.uk product code 405–210), were repolished using a handheld polisher supplied by Prof. JJ Boon, Boon Enterprises, Amsterdam. This improved the quality of imaging using light microscopy.

Table 7.1 Information about cross-sectional samples analyzed. Pigments were identified using light microscopy and EDX analysis (elements identified with SEM-EDX are shown in parenthesis)

Sample	Artist/title	Paint and ground (from ground upward)	Varnish coating(s)	Metal ion(s) detected in varnish	Further information
Seurat_5b	Georges Seurat, *Le Chahut* (1889–1890) Kröller-Müller Museum, Otterlo Oil on canvas	1. Lead white (Pb) 2. Blue: cobalt blue, lead white (Co, Pb, Al) 3. Orange: vermilion and chrome yellow (Hg, Cr, Pb) 4. Blue: cobalt blue, lead white (Co, Pb, Al) 5. Dark blue: cobalt blue, lead white, chalk, emerald green (Co, Pb, Cu, As, Ca, K)	6. Linseed oil and diterpenoid resin (TMAH-Py-GC-MS)	Calcium, potassium, lead	Condition report at Kröller-Müller Museum, Otterlo
Goya_1	Francisco de Goya *Don Francisco de Saavedra* (1798) Courtauld Gallery, London Oil on canvas	1. Chalk, lead white, and iron oxide pigments (Ca, Pb, Fe) 2. Blue: Prussian blue (Pb) 3. Red: iron oxide pigments, lead white, red lake (Fe, Pb, Al, K)	4. Oil layer (organic staining tests) 5. Dammar resin varnish (documented as applied 1959)	Lead	Courtauld Gallery accession number P.1947.LF.180. Condition report: Anon (Courtauld Institute of Art, Department of Conservation and Technology, London, 1959)
Goya_2	Francisco de Goya *Don Francisco de Saavedra* (1798) Courtauld Gallery, London Oil on canvas	1. Chalk, lead white, and iron oxide pigments (Ca, Pb, Fe, K) 2. Brown: iron oxide pigments, lead white (Fe, Pb, K)	3. Oil layer (organic staining tests)	Lead, potassium	As above

Olver_9	Kate Elizabeth Olver, *Return from the Front* (c. 1916) Private Collection Oil on canvas	1. Chalk and lead white (Ca, Pb) 2. Lead white (Pb) 3. Grey: lead white, zinc white, titanium white, iron oxide pigments, carbon black, cobalt blue (Pb, Zn, Ti, Fe, Mn, Co) 4. Brown: lead white, zinc white, titanium white, iron oxide pigments, carbon black (Pb, Zn, Ti, Fe, Mn)	5. Alkyd varnish (GC-MS)	Calcium, titanium	Courtauld Institute accession number CIA 2169. Condition report: C. Shepherd (Courtauld Institute of Art, Department of Conservation and Technology, London, 2012)
Dobson_7	William Charles Thomas Dobson, *Christ Raiseth from Death the Widow's Son of Nain* (1868) Private collection Oil on canvas	1. Lead white and chalk (Pb, Ca) 2. Pigmented oil layer 3. Pale blue: lead white, cobalt blue (Pb, Co, Al)	4. Thin oil –resin varnish layer (organic staining tests) 5. Thicker oil-resin varnish layer (organic staining tests)	Lead	Courtauld Institute accession number CIA 2264. Condition report: S. Bayliss, (Courtauld Institute of Art, Department of Conservation and Technology, London, 2014)
Laszlo_9	Philip de Laszlo *Wolmer Woods* (c.1930) Private collection Oil on canvas	1. Titanium white and chalk (Ti, Ca) 2. Green paint: emerald green, zinc white, lead white, scattered viridian particles, chalk (As, Cu, Zn, Cr, Ca, Pb) 3. Green paint: emerald green, zinc white, scattered viridian particles, chalk (As, Cu, Zn, Cr, Ca)	4. Oil varnish (organic staining tests)	Zinc, calcium, lead	Courtauld Institute accession number CIA 2264. Condition report: L. Ackerlund (Courtauld Institute of Art, Department of Conservation and Technology, London, 2012)
Manet_2	Edouard Manet, *Le Déjeuner sur l'herbe* (1863–1868) Courtauld Gallery, London Oil on canvas	1. Lead white and barium sulfate (Pb, Ba) 2. Green: viridian, iron oxide pigments, bone black, French ultramarine, chalk and barium sulfate, emerald green (Cr, Fe, P, Ca, As, Cu, Al, Na, Si, Ba)	3. Ketone resin varnish AW2 (untested: documented as applied 1976)	None	Courtauld Gallery accession number P.1932.SC.232. Condition report: R.B. Gardener, (Courtauld Institute of Art, Department of Conservation and Technology, London, 1976)

(continued)

Table 7.1 (continued)

Sample	Artist/title	Paint and ground (from ground upward)	Varnish coating(s)	Metal ion(s) detected in varnish	Further information
Gainsborough_1	Thomas Gainsborough, *Elizabeth and Mary Linley* (1772) Dulwich Picture Gallery, London Oil on canvas	1. Lead white, red iron oxide, chalk (Pb, Fe, Ca) 2. Lead white, red iron oxide, chalk (Pb, Fe, Ca) 3. Yellow: lead white, yellow ochre, brown iron oxide, green earth (Pb, Fe, Al, P, Si)	4. Residue of aged varnish (unknown composition) 5. Natural resin varnish (untested)	Lead in residue of aged varnish layer	Technical report: A. Burnstock, (Courtauld Institute of Art, Department of Conservation and Technology, London, 1999)
Gerino_16A	Gerino d'Antonio Gerini, *Virgin and Child Enthroned with Saint Lawrence, Saint John the Baptist, Saint Monica and Saint Augustine* (1510) Courtauld Gallery, London Oil on panel	1. Calcium sulfate (Ca, S) 2. Calcium sulfate (Ca, S) 3. White imprimatura: lead white, lead tin yellow, and glass (Pb, Sn, Si, Mg, Al) 4. Black: carbon black, chalk and glass (Ca, K, S, Mg, Al, Si) 5. Blue: lead white, azurite (Cu, Pb) 7. Greyish retouching: lead white, carbon black (Pb)	6. Varnish below retouching (unknown composition)	None	Courtauld Institute accession number CIA 1841.Condition report: S. Wohler (Courtauld Institute of Art, Department of Conservation and Technology, London, 2014)
Daddi_L	Bernardo Daddi, *Triptych: The Virgin and Child Enthroned with Saints* (1338) Courtauld Gallery, London Egg tempera on panel	1. Calcium sulfate (Ca, S) 2. Bole: red iron oxide (Fe) 3. Gilding: gold (Au) 4. Grey paint: lead white and carbon black 5. Red paint: vermilion (Hg, S) 6. Varnish 7. Red retouching: red lake, iron oxide pigments, lead white, chalk extender (Al, K, Fe, Pb, Ca)	6. Varnish below retouching (unknown composition)	None	Courtauld Gallery accession number P.1978.PG.81: Technical Report: A. Tate-Harte, A. Miller and R. Morrison Condition report: J.M. Brealey, (Department of Conservation and Technology, London, 1957)

7.2.1 *Analytical Methods*

The cross-sectional samples were examined using a Zeiss Axioplan 2 light microscope (LM) with incident polarized light and incident UV light (halogen optic lamp and a mercury short-arc photo optic lamp HBO). The filter set "UV 01" used for examination in UV light consists of the following filters: excitation band pass 365 nm and barrier filter long pass 397 nm so that wavelengths observed were > 397 nm. The samples were photographed using a Zeiss Axiocam color camera at Rijksdienst voor het Cultureel Erfgoed (RCE), Amsterdam.

Scanning electron microscopy with energy dispersive X-ray spectroscopy (SEM-EDX) was performed on a FEI Quanta 650 FEG microscope (Natural History Museum, London). Backscattered electron (BSE) images of uncoated cross sections were taken at 12 kV acceleration voltage at a spot size of 3.0. The chamber pressure was 50 Pa and the aperture 30 μm, and the detector was CBS (concentric backscattered). Elemental mapping was performed using a Bruker Flash Quad 5060F detector at 12 kV acceleration voltage at a spot size of 3.5. The chamber pressure was 30 Pa.

Selected samples were analyzed using Fourier transform infrared (FTIR) imaging to characterize the presence of metal carboxylates, using absorption data from previous studies (Keune and Boon 2007, 162). Five samples were analyzed using FTIR imaging (Seurat_5b, Goya_1, Goya_2, Olver_9, and Laszlo_9). FTIR spectral data was collected at RCE using a Perkin Elmer Spectrum 100 FTIR spectrometer combined with a Spectrum Spotlight 400 FTIR microscope equipped with a 16x1 pixel linear mercury cadmium telluride (MCT) array detector. A Perkin Elmer attenuated total reflection (ATR) imaging accessory consisting of a germanium crystal was used.

7.3 Results and Discussion

Figures 7.1, 7.2, 7.3, 7.4, and 7.5 show imaging results for the five samples with an oil-resin varnish: Seurat_5b, Goya_1, Goya_2, Dobson_7, and Laszlo_9. Figure 7.6 shows imaging results for a sample with a resin varnish for comparison: Manet_2. Results from the analysis of other samples are shown in column 5 of Table 7.1.

7.3.1 *Lead*

Elemental lead was detected within the varnish of six of the samples (Dobson_7, Goya_1 and 2, Seurat_5b and Laszlo_9, and Gainsborough_1).

The BSE image of sample Dobson_7 shows highly scattering material at the surface of the lower varnish layer which forms crystalline clusters similar to that

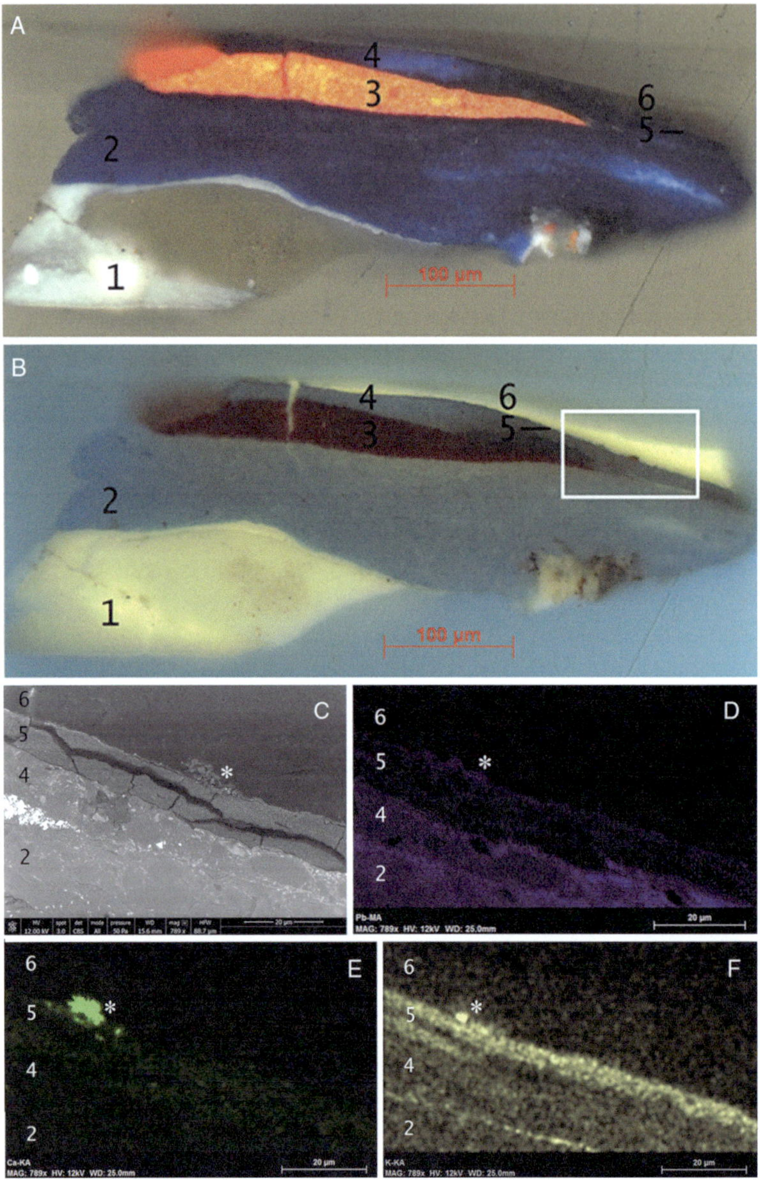

Fig. 7.1 Seurat 5b. *Imaging analysis of sample from Georges Seurat, Le Chahut (1889–1890), oil on canvas.* (**a**): LM image showing paint layer structure. 1: ground, 2: lower blue paint, 3: orange paint, 4: middle blue paint, 5: upper dark blue paint, 6: varnish. (**b**): UV LM image. Fluorescing layer is linseed oil and diterpenoid resin varnish. Box indicates position of BSE image and elemental mapping. (**c**): BSE image. There is an amorphous formation indicative of metal carboxylate aggregation at the surface of the paint layer protruding into the varnish layer (position indicated by *). (**d**): Elemental lead map. Lead is identified in the aggregate at the surface. (**e**): Elemental calcium map. Calcium is identified in the aggregate at the surface and in the varnish layer. (**f**): Elemental potassium map. Potassium is identified in the aggregate at the surface and in the varnish layer

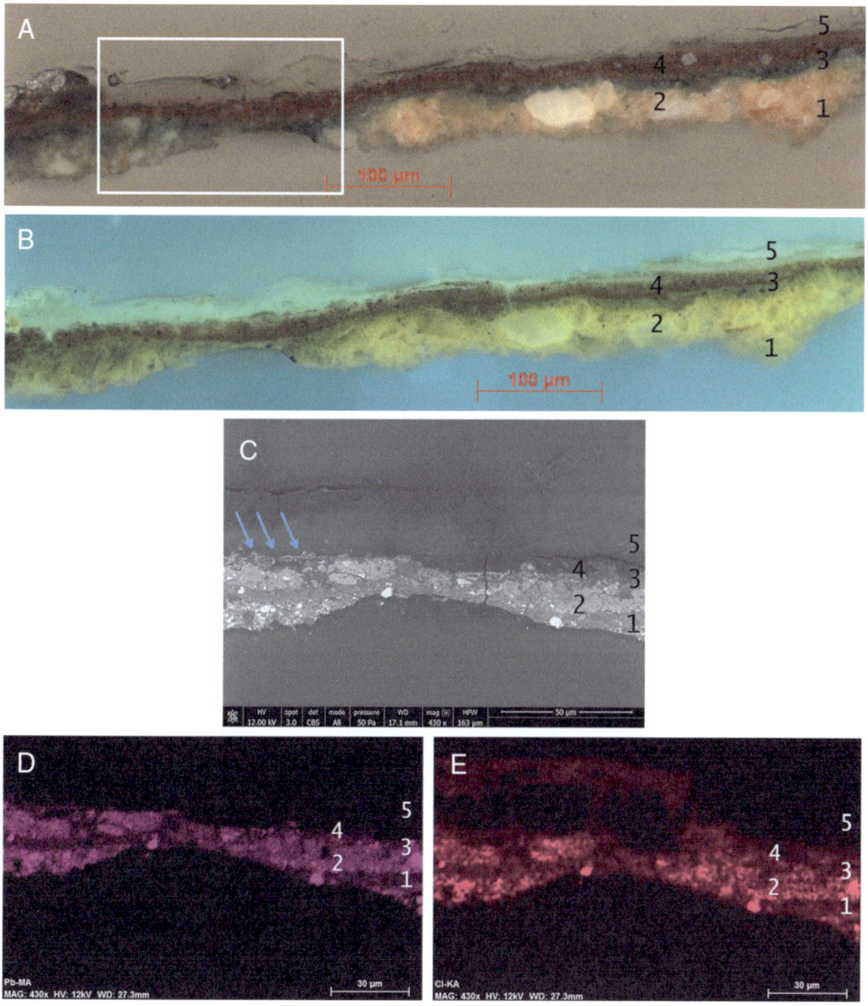

Fig. 7.2 Goya_1. *Imaging analysis of sample from Francisco de Goya, Don Francisco de Saavedra, (1798), oil on canvas.* (**a**): LM image showing paint layer structure. 1: ground, 2: blue paint, 3: red paint, 4: oil varnish, 5; natural resin varnish. Box indicates area examined using BSE imaging and elemental mapping. (**b**): LM UV image. Green fluorescing upper layer is natural resin varnish. Thinner lower layer is an oil-containing varnish layer. (**c**): BSE image. Arrows point to amorphous formations indicative of metal carboxylate aggregates at the surface of the paint layer. (**d**): Elemental lead map. Lead is identified in the aggregates at the surface. (**e**): Elemental chlorine map. Chlorine is identified in the aggregates at the surface and in the upper varnish layer

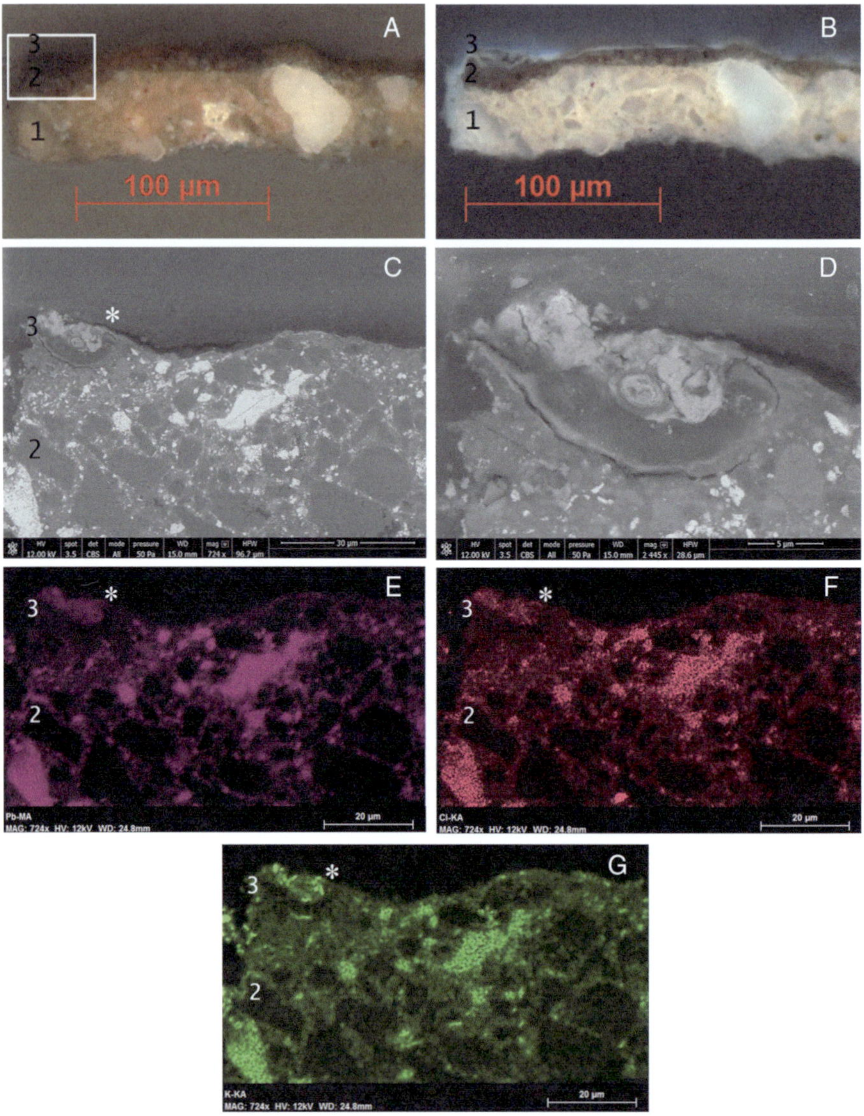

Fig. 7.3 Goya_2. *Imaging analysis of sample from Francisco de Goya, Don Francisco de Saavedra, (1798), oil on canvas.* (**a**): LM image showing paint layer structure. 1: ground, 2: brown paint, 3: oil layer. Box indicates area examined using BSE imaging and elemental mapping. (**b**): UV LM image. Thin fluorescing layer is an oil-containing varnish layer. (**c**): BSE image. There is an amorphous formation indicative of metal carboxylate aggregation positioned at the surface of the paint layer (position indicated by *). (**d**): Close-up BSE image showing the concentric structure of formation. (**e**): Lead elemental map. Lead is identified in the aggregate at the surface. (**f**): Chlorine elemental map. Chlorine is identified in the aggregate at the surface. (**g**): Potassium elemental map. Potassium is identified in the aggregate at the surface

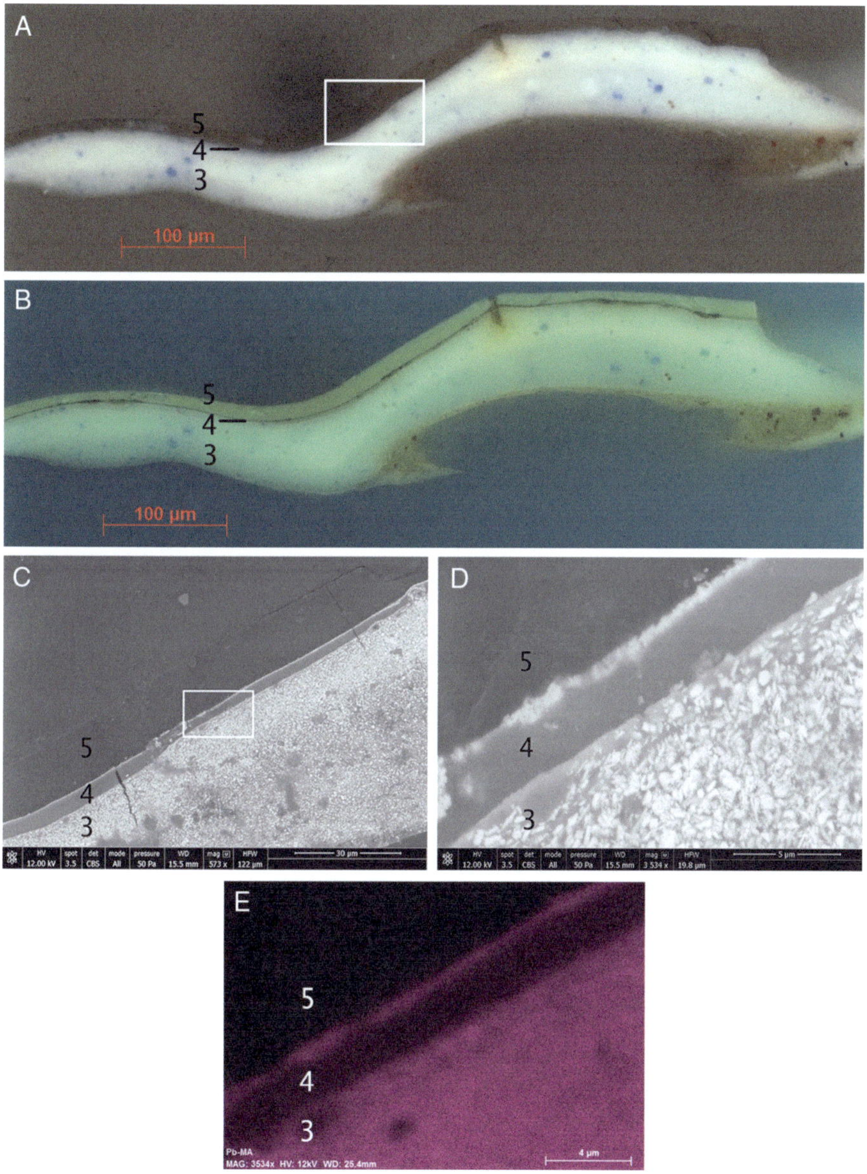

Fig. 7.4 Dobson_7: *Imaging analysis of sample from William Charles Thomas Dobson, Christ Raiseth from Death the Widow's son of Nain (1868), oil on canvas.* (**a**): LM image showing paint layer structure. 3: blue paint, 4: lower varnish, 5: upper varnish. Box indicates area examined using BSE imaging (**c**). (**b**): UV LM image. Thick upper varnish layer with yellowish fluorescence and thin lower layer below dark material. Both oil-containing varnishes. (**c**): BSE image. Highly scattering material indicated at surface of lower varnish layer. Box indicates area examined using BSE imaging and elemental mapping. (**d**): BSE image close-up. Shows the crystalline structure of material. (**e**): Elemental lead map. Lead is identified in the material at the surface of the lower varnish layer and in the lower varnish layer in a lower concentration

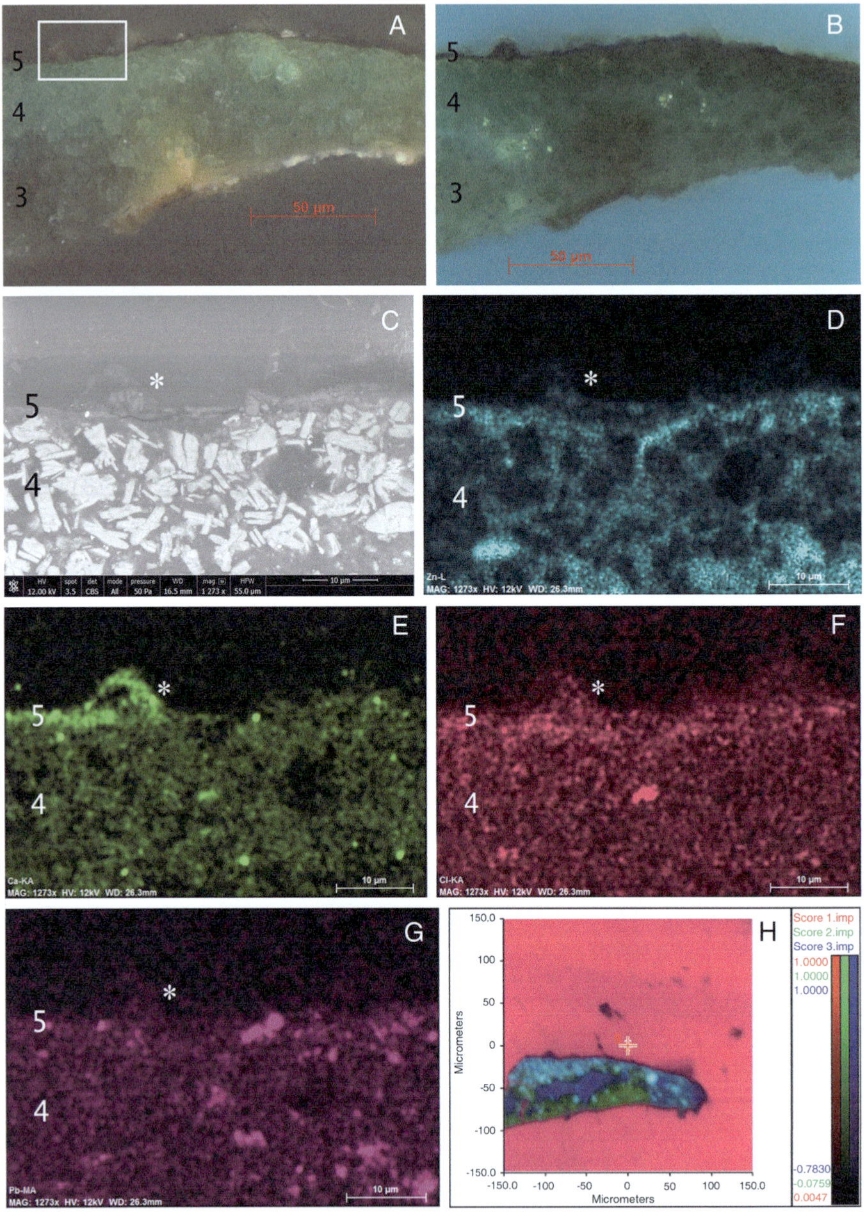

Fig. 7.5 Laszlo_9: *Imaging analysis of sample from Philip de Laszlo, Wolmer Woods (c.1930), oil on canvas.* (**a**): LM image showing paint layer structure. 3: lower green paint, 4: upper green paint, 5: varnish and surface dirt. Box indicates position of BSE image and elemental mapping; (**b**): UV LM image. Dark discolored oil-resin coating with some fluorescence at the surface of the sample. (**c**): BSE image. Grey amorphous material forms bands indicative of metal carboxylates. (**d**): Elemental zinc map. Zinc is identified in the material at the surface. (**e**): Elemental calcium map. Calcium is identified in the material at the surface. (**f**): Elemental chlorine map. Chlorine is identified in the material at the surface. (**g**): Elemental lead map; lead is identified in the material at the surface. (**h**): FTIR image – green indicates zinc carboxylates

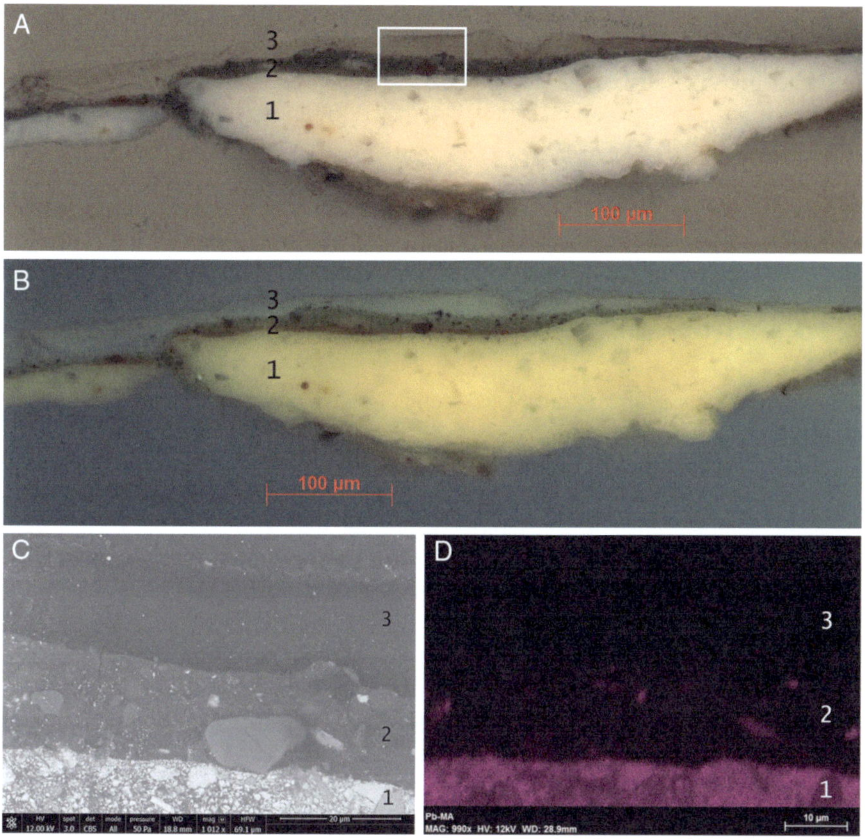

Fig. 7.6 Manet_2. *Imaging analysis of sample from Eduard Manet, Le Déjeuner sur l'herbe (1863–1868), oil on canvas.* (**a**): LM image showing paint layer structure. 1: ground layer, 2: green paint, 3: varnish. Box indicates position of BSE image and elemental mapping. (**b**): UV LM image. Fluorescing layer is ketone resin varnish. (**c**): BSE image. There is no highly scattering material in the varnish layer, or indications of metal soap aggregates. (**d**): Elemental lead map. Lead is identified in the paint layers, but there is no sign of migration into varnish

seen in metal carboxylate efflorescence identified in other studies (Keune et al. 2006, 146–150) (Fig. 7.4c, d). Elemental mapping confirmed the presence of lead (Fig. 7.4e). The BSE image and elemental mapping showed particles characteristic of lead white pigment in the paint layer. It is possible that in this sample lead white pigment has reacted with fatty acids in the oil medium, either in the paint or in the oil-resin varnish, to form lead carboxylates, and the recrystallized salts are now present at the surface of the lower varnish layer. A lead-containing crystalline crust is 2 μm thick and was not visible by eye or using low-powered surface microscopy (Bayliss 2014). Elemental mapping also identified lead within the lower varnish layer, in a lower concentration.

BSE images of samples Goya_1 and Goya_2 show aggregates at the surface of the paint sample (Fig. 7.2) that are similar in appearance to lead soaps identified in other studies (Keune et al. 2011, 700). The images show light grey amorphous concentric structures that contrast with a darker surrounding organic material (Figs. 7.2c and 7.3c, d). Elemental mapping confirmed the presence of lead in the light grey areas (Figs. 7.2d and 7.3e). The aggregates are positioned within the oil-resin varnish layer. It is probable that lead white pigment in the paint layer has reacted with fatty acids, either in the oil medium or in the oil-resin varnish itself, to form lead carboxylates, which have aggregated at the surface and partly recrystallized as soaps.

Elemental mapping identified chlorine together with lead in the surface aggregates in Goya_1 (Fig. 7.2e). Previous studies have found lead chloride mineral phases formed inside lead carboxylate aggregates and in the lead carboxylate surface crusts of paintings, and it has been suggested that chlorides play a part in metal carboxylate formation (Van Loon 2008, 150). Elemental chlorine was found in all the paint layers in samples Goya_1 and Goya_2, with relatively higher concentration in lead-containing areas of the sample. Chlorides may have been introduced into the lead white pigment during the processing (Gettens et al. 1967, 126). Elemental mapping also identified chlorine in the upper portion of the natural resin varnish in sample Goya_1 suggesting that the chloride derives from environmental sources, rather than inorganic materials in this paint film (Keene et al. 1999, 8438).

The BSE image of sample Seurat_5b shows a light grey aggregate with amorphous edges at the paint surface protruding into the varnish (indicated by * in Fig. 7.1c). Elemental mapping confirmed the presence of lead and possibly sulfur in the aggregate (Fig. 7.1d).

Elemental lead was identified in sample Laszlo_9 in low concentration in a protrusion at the surface of the sample (Fig. 7.5g). However, the lead map is not well resolved, and the quantity may be below the threshold for accurate detection.

The BSE image of sample Gainsborough_1 shows surface formations comprising grey and white bands within darker areas indicating organic medium. This is similar in appearance to lead carboxylate formations seen in other studies, and elemental mapping confirmed the presence of lead in the lighter material. It is likely that lead carboxylates have migrated to the surface of the paint layer and accumulated within a layer of residual varnish.

Of the six samples identified as having lead migration into varnish, five had oil or oil-resin varnish layers (see Table 7.1). Conservation reports for the paintings note that the varnishes were difficult or impossible to remove using conventional solvents. Although sample Gainsborough_1 was noted as having a natural resin varnish, the lead identified was within a lower residual varnish layer, which may have an oil content; however, the presence of oil could not be positively confirmed as the thin layer was beyond the resolution of the analytical methods used in this study.

The presence of lead in the oil-resin varnish layers of these paintings may have changed the solubility, making it difficult to remove in the organic solvents commonly used by conservators. Lead ions may have a stabilizing effect on the oil portion of the varnish through the creation of a metal-coordinated ionomer and accelerate oxidation of the natural resin portion. Additionally, in samples where lead carboxylate aggregations were identified within or as crusts at the surface of the varnish, the insoluble lead carboxylates would inhibit the action of the solvent on the varnish, preventing dissolution.

7.3.2 *Zinc*

Zinc was detected in the oil-resin varnish in sample Laszlo_9. The BSE image shows an accumulation of material at the surface of the sample, which appears grey and amorphous and forms bands similar in appearance to material identified as metal carboxylates in other studies (Fig. 7.5c, protrusion indicated with *). Elemental mapping identified zinc in the material (Fig. 7.5d).

FTIR imaging of this sample identified zinc carboxylates in the lower paint layer (Fig. 7.5h). The high proportion of oil medium in this paint layer is likely to have provided free fatty acids for reaction with zinc ions deriving from zinc oxide white pigment, to form zinc carboxylates. It is probable that the surface protrusions formed through migration of zinc carboxylates from the lower paint layer. FTIR imaging did not confirm the presence of zinc carboxylates in the protrusions; however, it may be that they are present in too low a concentration to be identified using this technique (Keune 2005, 4–5).

The presence of zinc within the oil-resin varnish may cause similar changes to the physical properties of the varnish as lead. Zinc ions may have a stabilizing effect on the oil portion of the varnish through the creation of a metal-coordinated ionomer and accelerate oxidation of the natural resin portion. Zinc carboxylate aggregates within the varnish would inhibit the action of the solvent on the varnish, preventing dissolution.

7.3.3 *Calcium*

Elemental mapping identified calcium within the oil-resin varnish layers of samples Seurat_5b (Fig. 7.1e), Laszlo_9 (Fig. 7.5e), and Olver_9. In Seurat_5b elemental calcium was identified within a surface aggregate and dispersed throughout the varnish layer. In Laszlo_9 calcium was identified within surface protrusions. In the sample from Olver_9, calcium was present throughout the varnish layer, with a higher concentration in some areas.

In each of these samples, elemental calcium was also present within the paint layers as calcium carbonate, used either as part of the ground or as an extender for pigments. It is possible that calcium has reacted with fatty acids in the oil medium to form mobile calcium carboxylates, which have migrated into the varnish layer or accumulated at the surface (Ferreira et al. 2011, 1–8).

The presence of calcium carboxylates in an oil-resin varnish layer may change the solubility of the coating. Calcium salts such as calcium oxalate may form at the surface, catalyzed by the action of microorganisms and oxidation of organic material (Cariati et al. 2000, 180–188). Calcium oxalate is insoluble in organic solvents and would inhibit the action of the solvent on the varnish.

7.3.4 Potassium

Potassium was identified within the oil-resin varnish layer of sample Seurat_5b and Goya_2 in surface protrusions within the oil-resin varnish layer (Figs. 7.1f and 7.3g). Potassium carboxylates have been identified in surface crusts of paintings in previous studies and have usually been associated with degradation of smalt pigment (Spring 2004, 77–86). In sample Seurat_5b, potassium is detected in the paint layers, in highest concentration in layer 5. It is possible potassium is a trace element associated with the chalk identified in this layer. In sample Goya_2, potassium is detected throughout the paint and ground layers and may be a trace element in natural iron oxide pigments. Migration of potassium carboxylates from other sources has been found in other studies. In these cases it was thought to originate from lakes (precipitated from potash alum) and earth pigments and in another work from potash alum used as an adhesive in a lining treatment (Noble and Van Loon 2007 19–36; (Boon and Oberthaler 2010), 235–253, 328–335). However, the mechanisms for migration in these cases have not been investigated. The detection of potassium in the varnish layers in this study may suggest that migration of potassium from other sources is more common than previously thought.

7.4 Conclusion

Chemical mapping of paint cross sections from old master paintings in this study highlighted the presence of elements including lead, zinc, calcium, and potassium within the oil-resin varnish coatings of some paintings. Conservation reports for the paintings where metals were identified described difficulties removing the layers using conventional solvent methods. These findings suggest that the presence of metal ions within oil-resin varnish layers changes the solubility of the layer, making it more challenging to remove.

Metals identified within varnish layers in this study were characterized as metal carboxylates by their appearance in BSE images. Aggregates of migrated carboxylates are present as inhomogeneous amorphous masses that sometimes contain striation patterns of lighter grey enriched in lead and darker areas indicating organic material. Migration of lead carboxylates to form a semicrystalline surface crust was identified as highly scattering white material at the surface in sample Dobson_7. The limited resolution of the method of FTIR imaging used in the present study was insufficient to identify metal carboxylates in thin layers (typically less than a few micrometers), in inclusions, and at interfaces where the concentration of metal carboxylates was low.

Metals were not identified in natural or synthetic resin varnish layers in any of the samples. This could indicate that metal ions do not readily react with acidic degradation products from natural resin varnishes. It is also possible that, as these varnishes were applied more recently, migration of metal ions into these layers had not yet occurred. The timescale for migration of lead carboxylates is the focus of current investigation.[2]

The samples investigated were taken from historical paintings with complex material compositions and varied physical histories. Recipes for historical varnishes describe the addition of driers including compounds containing lead, zinc, cobalt, and other metal ions to drying oil and resin. Alternative sources of metal ions may come from environmental agents and surface dirt. The solubility or removability of varnish coatings on paintings may be influenced by the presence of metal carboxylates, soaps, and salts in the coating and the paint, together with a range of environmental factors related to the physical history of each work.

More precise analysis of the composition of the varnish using complimentary analytical techniques (such as SIMS or GC-MS) would provide more detailed context for interpretation of the chemical relationship between paint and varnish coatings in paintings. Understanding the chemistry together with the physical and optical properties of varnish and their application and removal is the key concern for conservation practitioners.

Acknowledgments Klass Jan Van den Berg, Suzan De Groot, and Henk Van Keulen, Rijksdienst voor het Cultureel Erfgoed, Amsterdam
Tomasz Góral, Natural History Museum, London
Katrien Keune, University of Amsterdam
Margje Leeuwestein, Kröller-Müller Museum, Otterlo
Jaap J. Boon, JAAP enterprise for Art Scientific Research, Amsterdam
Sarah Bayliss, Hamilton Kerr Institute, Cambridge

References

Bayliss B (2014) CIA 2264 William Charles Thomas Dobson, Christ Raiseth from Death the Widow's Son of Nain. Conservation documentation, Courtauld Institute of Art, Department of Conservation and Technology, London

Boon JJ, Oberthaler E (2010) Mechanical weakness and chemical reactivity observed in the paint structure and surface of The Art of Painting by Vermeer. In: Vermeer's The Art of Painting, Scrutiny of a Picture, Exhibition catalogue. Kunsthistorisches Museum, Vienna, pp 235–253 and 328–335

[2]Research was led by the PAinT (Paint Alterations in Time) research group at the University of Amsterdam.

Boon JJ, Hoogland F, Keune K (2007) Chemical Processes in aged oil paints effecting metal soap aggregation. In: AIC Paintings Specialty Group Postprints papers presented at the thirty fourth annual meeting of The American Institute for Conservation of Historical and Artistic Works, June 16–19, 2006, vol 19. Washington, DC, pp 145–149

Cariati F, Rampazzi L, Toniolo L, Pozzi A (2000) Calcium oxalate films on stone surfaces: experimental assessment of the chemical formation. Stud Conserv 45(3):180–188

Carlyle L (2001) The artist's assistant: oil painting instruction manuals and handbooks in Britain, 1800–1900, with reference to selected eighteenth-century Sources Archetype Publications, London

Ferreira ESB, Boon JJ, Stampanoni M, Marone F September (2011) Study of the mechanism of formation of calcium soaps in an early 20th century easel painting with correlative 2D and 3D microscopy in ICOM 16th Triennial Conference Lisbon, pp 19–23

Gettens RJ, Kühn H, Chase WT (1967) 3 lead white. Studies Conserv 12(4):126

Hermans JJ (2017) Metal Soaps in oil paint: structure, mechanisms and dynamics, University of Amsterdam, p 115. ISBN:978–94–6295578-3

Keene WC, Aslam M, Khalil K, Erickson DJIII, McCulloch A, Graedel TE, Loberr JM, Aucott ML, Ling Gong S, Harper DB, Kleiman G, Midgley P, Moore RM, Seuzaret C, Sturges WT, Benkovitz CM, Valentin Koropalov TM, Barrie LA, Fan Li Y (1999) Composite global emissions of reactive chlorine from anthropogenic and natural sources: reactive chlorine emissions inventory. J Geophys Res 104(D7):8429–8440

Keune K (2005) Binding medium pigments and metal soaps characterised and localised in paint cross-sections. Ph.D. thesis, MOLART Report 11, AMOLF, University of Amsterdam, Amsterdam

Keune K, Boon J (2007) Analytical imaging studies of paint cross-sections illustrate the oil paint defect of lead soap aggregate formation. Studies in Conservation 5:161–176

Keune K, Kirsch K, Jaap Boon (2006) Lead soap efflorescence in a 19th C painting: appearance, nature and sources of materials. In: AIC Paintings Specialty Group Postprints, Rhode Island (ed. H. Mar Parkin), AIC, Washington. pp 145–149

Keune K, Van Loon A, Boon J (2011) SEM backscattered-electron images of paint cross-sections as information source for the presence of the lead white pigment and lead-related degradation and migration phenomena in oil paintings. Microsc Microanal 17(5):696–701

Noble P, Van Loon A (2007) Rembrandt's Simeon's Song of Praise, 1631: pictorial devices in the service of spatial illusion. ArtMatters Netherlands Techn Stud Art 4:19–36

Phenix A, Townsend J (2012) A brief survey of historical varnishes. In: Conservation of Easel Paintings edited by Joyce Hill Stoner and Rebecca Rushfield, 252–263. Oxon, Routledge

Spring M (2004) Raphael's materials: some new discoveries and their context within early sixteenth-century painting'. In: Raphael's painting technique: working practices before Rome. Proceedings of the Eu-ARTEX+ CH workshop, National Gallery, London, vol 11, pp 77–86

Tumosa CS, Mecklenburg MF (2005) The influence of lead ions on the drying of oils. Rev Conserv (6):39–47

Van Loon A (2008) Color changes and chemical reactivity in seventeenth-century oil paintings Ph.D. thesis, University of Amsterdam

Van Loon A, Noble P, Boon JJ (2011) White hazes and surface crusts in Rembrandt's Homer and related paintings. In: Bridgeland J (ed) ICOM Committee for Conservation, 16th Triennial Meeting, Lisbon, pp 26–30

Chapter 8
Aging of Natural Resins in Presence of Pigments: Metal Soap and Oxalate Formation

Tommaso Poli, Anna Piccirillo, Marco Nervo, and Oscar Chiantore

Abstract The degradation process involving the formation of metal soaps in drying oils is a well-known problem due to the reaction of cations coming from pigments reacting with free fatty acids from the oil. A similar behavior has been observed during an experiment concerning the aging of different natural resins (shellac, dammar, and colophony) in the presence of common historic pigments (smalt, ochre, umber, azurite, lead white, zinc white, and titanium white). In the case of natural resins, the terpenic acids react with the cations in the pigments. Each resin shows a different reactivity and affinity to different cations related to the complex mixture of terpenic structures and acids contained in natural resins. So far it has not been possible to identify one or more specific acids preferentially involved in this "saponification" process. Colophony seems to be the most reactive resin, while zinc white and smalt appear to be the most reactive pigments. The explanation for the different reactivity of the acid components of the three resins is unclear. Studies on acidity of different resins, free acid mobility, and pigment reactivity are ongoing. Generally, the presence of a pigment has been proven to modify the stability of resins that have undergone accelerated aging. Mixtures of resins and pigments have been exposed to natural and different artificial aging conditions (photooxidative, thermal, thermo-oxidative, and wet cycles).

Moreover, it has been observed that some pigments, particularly smalt and zinc white, favor the simultaneous formation of oxalates. The formation of oxalates and different degradation products from natural resins in the presence of pigments is particularly important, as it deeply affects the removability of varnishes and, more generally, the cleaning processes. Furthermore, it permanently modifies the interface between painting and varnish layers as well as the aesthetic properties of the paint surfaces.

T. Poli (✉) · O. Chiantore
Dipartimento di Chimica, Università degli Studi di Torino, Torino, Italy
e-mail: tommaso.poli@unito.it

A. Piccirillo · M. Nervo
Centro Conservazione e Restauro "La Venaria Reale", Piazza Della Repubblica, Venaria Reale (TO), Italy

F. Casadio et al. (eds.), *Metal Soaps in Art*, Cultural Heritage Science,
https://doi.org/10.1007/978-3-319-90617-1_8

8.1 Introduction

In painting layers, the formation of carboxylic salts that are the product of reactions between fatty acids and the cations of pigments is a well-known issue (Meilunas et al. 1990; Boon et al. 1997; Lazzari and Chiantore 1999; Mallégol et al. 1999; Van den Berg et al. 1999; Erhardt et al. 2000; Colombini et al. 2002; Van den Berg 2002; Arbizzani et al. 2004; Dietemann et al. 2009; Van den Brink et al. 2009). In particular, several studies have been published on the interaction between binding media, drying oils in particular, and metallic compounds (Boon et al. 2002; Noble et al. 2002; Higgitt et al. 2003; Plater et al. 2003; Keune 2005; Van der Weerd et al. 2005; Boon 2006; Cotte et al. 2006, 2007; Boon et al. 2007; Mazzeo et al. 2008; Manzano et al. 2009). The effects of these transformation processes of the pictorial layer are different and not completely understood. Metal soaps may work as plasticizer of the film (Antony and De 1999; Wakabayashi and Register 2006; Cotte et al. 2017) slowing the cross-linking (reticulation) processes and the embrittlement but can also deeply affect the appearance and the state of conservation. Actually, fatty acids and metal ions migrate, and metal soaps aggregate in painting layers with mechanisms that are not yet completely understood, creating protrusions or crusts at the surface of paintings. These new formations can determine the weakening and sometimes failures on the pictorial film (Boon et al. 2002; Boon 2006). The causes of the metal soap formation reported in literature are many, but humidity seems to play a key role. This study on metal soap formation with natural resins started from the findings in three case studies: Antonio Molinari (seventeenth century), Giacomo Balla (twentieth century) and Giovanni Ambrogio della Torre (sixteenth century). Based on the finding of these case studies, a series of mock-up films have been designed. The mock-up films are composed of natural resin (shellac, dammar, and colophony) or stand oil mixed with some common pigments (smalt, ochre, umber, azurite, lead white, zinc white, and titanium white). The metal soap formation in the mock-ups with the natural resin binder is compared with the results obtained using stand oil as binder.

Natural resins are historically and widely employed as finishing layers or varnishes, for their appreciated optical and preservative properties. In many cases, they have been used and mixed with drying oils in order to obtain oleoresinous binders with particular aesthetic properties, or else they are used purely as binders in retouching paintings (Gunn et al. 2002). In all these conditions, the resins, and in particular the terpenic acid fraction, come in contact with pigments by direct mixing or by interaction with the underlying painting layer, often under-bound due to the degradation of the natural binder. Moreover, past and present conservation treatments required the removal of old varnish and application of fresh natural resin, which may come in contact with an under-bound painting layer as consequence of previous solvent-based cleanings. Nevertheless, very few studies (Carlyle 2001; Arbizzani et al. 2004; Doménech-Carbó et al. 2006) focus on the reactivity of natural resins with pigment, and most of those available in the literature are related to verdigris (copper acetate) and copper resinate recipes. This paper reports and summarizes the results of a systematic study on the formation of metal soaps in presence of natural resins (Poli et al. 2017). Shellac, even if less frequently used in easel paintings, has been chosen in order to evaluate the influence of the mobility of terpenic acids on soap formation. Shellac contains a large reticulate fraction due to the presence of aleuritic acid that is supposed to reduce the mobility of terpenic acids. The mock-ups were then artificially aged (thermal and photooxidative aging) and analyzed by means of Fourier transform infrared (FTIR) spectroscopy in transmission and imaging mode.

8.2 Materials and Methods

A mixture of pigment and binder (2:1 w/w) has been applied on glass microscope slides and on silicon wafers for the imaging measurements. Azurite, lead white, titanium white, zinc white, red ochre, and raw umber pigments were purchased by Kremer Pigmente, Germany. Smalt was purchased from Zecchi, Italy. Stand oil (linseed oil heated in absence of oxygen), colophony, dammar, and shellac were purchased from Kremer Pigmente, Germany. Five sets of samples have been prepared and differently aged in five different ways:

- 3 years at room condition with no direct sunlight exposure (22 °C and 55% RH)
- A total of 1000 h (with an intermediate measurement after 600 h) in oven at 50 °C (20% RH)
- 1000 h of simulated solar irradiation in an UV solar box Heraeus Suntest CPS equipped with a filtered (coated quartz glass simulating a 3 mm window glass, cutting $\lambda < 300$ nm) xenon lamp and with an average irradiation of 750 W/m^2 and an internal temperature of about 50 °C
- Wet (3 days, 22 °C and 95% RH) and dry (7 days in oven at 35 °C, 20% RH) alternated cycles up to 30 days
- 20 s at 250 °C

All aging started after 1 month of film curing.

The FTIR measurements on paints fragments (from paintings and from mock-ups) have been carried out by sampling the painting layers (at different time intervals in the case of aged mock-ups) and using a diamond anvil cell (High Pressure Diamond Optics, Inc.) in transmission modality. FTIR transmission spectra (64 scans) recorded using a diamond anvil cell were obtained on a Bruker Vertex 70 spectrophotometer coupled with a Bruker Hyperion 3000 IR microscope equipped with an MCT detector (Infrared Associates Inc.), working in the spectral range from 4000 to 600 cm^{-1} with an average spectral resolution of 4 cm^{-1}.

The silicon wafer paints were analyzed in the optical bench in transmission modality in the range 4000–400 cm^{-1}.

FTIR imaging/mapping measurements were recorded on the same instrument equipped, in this setup, with a photovoltaic mercury cadmium telluride (MCT) FPA (focal plane array) detector, Santa Barbara Infrared Inc., working in a spectral range from 4000 to 900 cm^{-1} (64 scans) with an average resolution of 4 cm^{-1}. The detector consists in a quadratic (128 × 128) raster of detector elements for an effective operative matrix of 64 × 64 pixels and an investigated area of 150 μm × 150 μm operating with a 15× objective. The measurements have been carried out on thin films applied on the silicon wafers.

8.3 Results and Discussion

This study originated from some evidence collected during the conservation treatment of three easel paintings. The first case study is a 201 × 214 cm oil on canvas from the seventeenth century (Birth of Mary, Antonio Molinari, Church of Ospedaletto, in Venice) exposed to overheating because of a fire. The huge amount of metal soaps found on the blackened surface suggested a possible fast, even if extreme, way to induce carboxylates formation on mock-ups (Fig. 8.1). The second case study is a Giacomo Balla's masterpiece "Grido dimostrazione in piazza del Quirinale" (private collection, 1915, 100 × 70 cm) on which the presence of metal soaps and zinc oxalates has been pointed out in presence of natural resins and oleoresinous binder (Rava et al. 2013). The third one is a painting on wood panel from the sixteenth century (Pietà with St. Ambrogio and Girolamo, Giovanni Ambrogio Della Torre, Church of Santa Maria Assunta, Paderno d'Adda, 205 × 148 cm). In this painting, UV fluorescence image (Fig. 8.2) and FTIR spectra pointed out the presence in particular areas of different signals not only attributable to the presence of drying oil and its decay products.

Figure 8.2b shows a detail of the different types of fluorescence of the glove (the lower part of the glove has a particular orange fluorescence) still present even after the removal of the varnish. In visible light and under the microscope observation, there is no trace of retouching or discontinuity in the red painting layer of the glove presenting the two different fluorescence. Nevertheless, FTIR analysis of two superficial samples coming from the two areas, after varnish removal, showed two completely different situations. The upper part of the glove (Fig. 8.2, violet spectrum) shows the typical spectrum of an aged oil painting: a layer very poor in binder with strong signals coming from the carbonate stretching vibration of lead white and the asymmetric vibration band of lead carboxylates. The lower part of the glove (Fig. 8.2, red spectrum) presents a different superficial composition rich in natural resin (shoulder at 1712 cm^{-1}), oxalates (sharp band at 1320 cm^{-1}), and probably resin soaps (broad band at 1540 cm^{-1}). This case study shows that the oxalate most likely is related to the presence of the resin, as they are found only in combination with the resin and are absent in the oil paint. It is very probable that something happened between the original varnish and the paint layer in the area that contains oxalates and resin soaps, since it is very unlikely that the artist changed the composition of the binder only for the lower part of the gloves. Two hypotheses for the differences in surface composition in the glove can be formulated: a resin

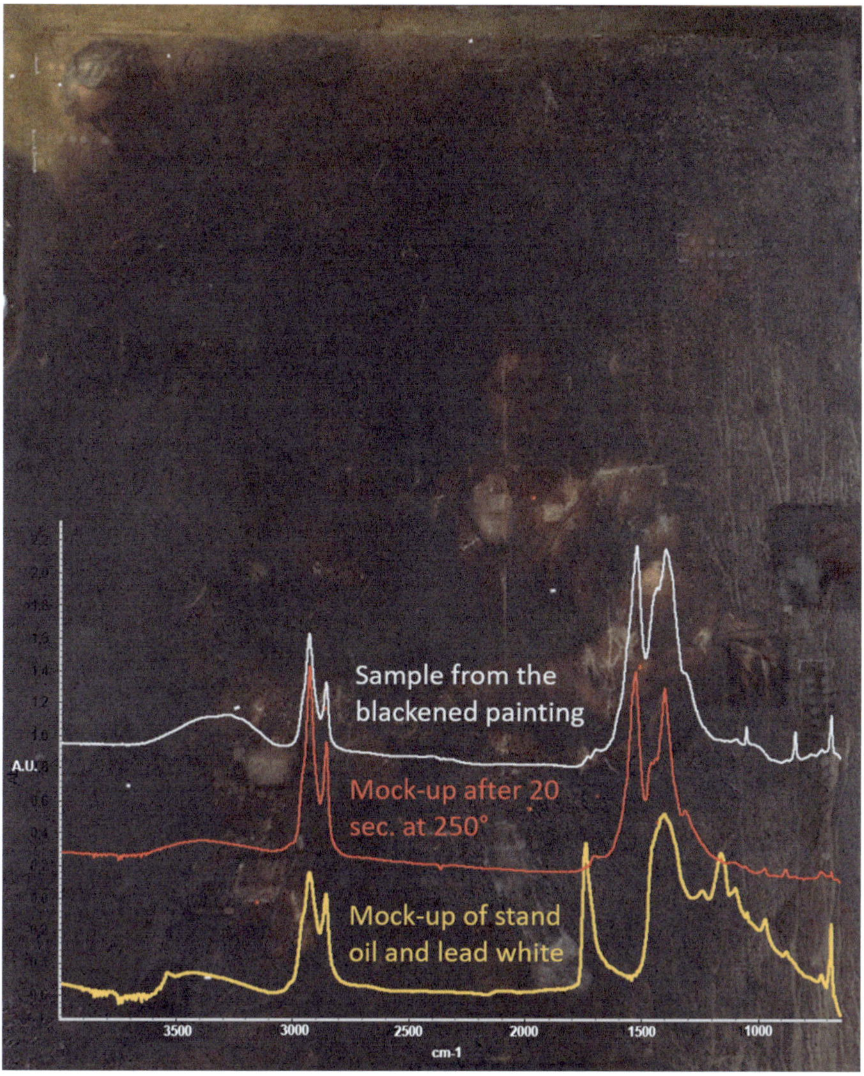

Fig. 8.1 FTIR transmission spectra reported show the correspondence between the mock-up of stand oil and lead white heated for 20 s at 250 °C (red line) with a sample coming from the painting on background (Birth of Mary, Antonio Molinari, Church of Ospedaletto, in Venice, seventeenth century) exposed to a fire (white line). The yellow spectrum belongs to the mock-up before the heating

"glaze," applied in order to obtain a particular shadow effect, is no longer detectable. A second hypothesis is that different cleaning approaches were taken in the glove area, causing two different states of conservation in the paints: an area locally cleaned many times involving the saint face and chest casually including the upper

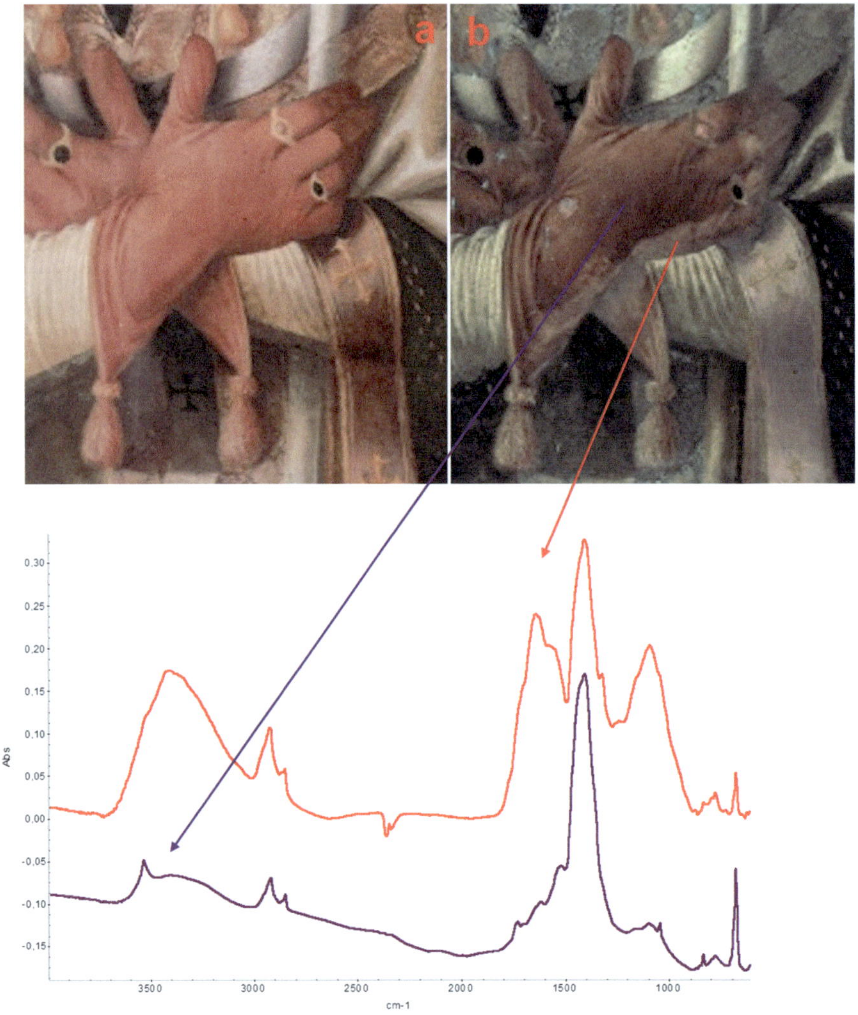

Fig. 8.2 Detail of a painting on wood panel (Pietà with St. Ambrogio and Girolamo, Giovanni Ambrogio Della Torre, Church of Santa Maria Assunta, Paderno d'Adda, 205 × 148, sixteenth century) after the varnish removal (**a**, visible light; **b**, UV fluorescence). The violet spectrum is related to the upper part of the glove, while the red one to the lower and UV-fluorescent part

part of the glove and an uncleaned area in the lower part of the glove, where the resin has been allowed to be in contact with the pigment for a longer period of time. This suggests, in both cases, that natural resins are actively involved in the soap and oxalate formation.

So far, this study aimed at demonstrating that terpenic acids coming from the resins can form soaps with cations from pigments in a similar way as fatty acids coming from oil. Results on aged mock-ups confirmed that natural resins can easily

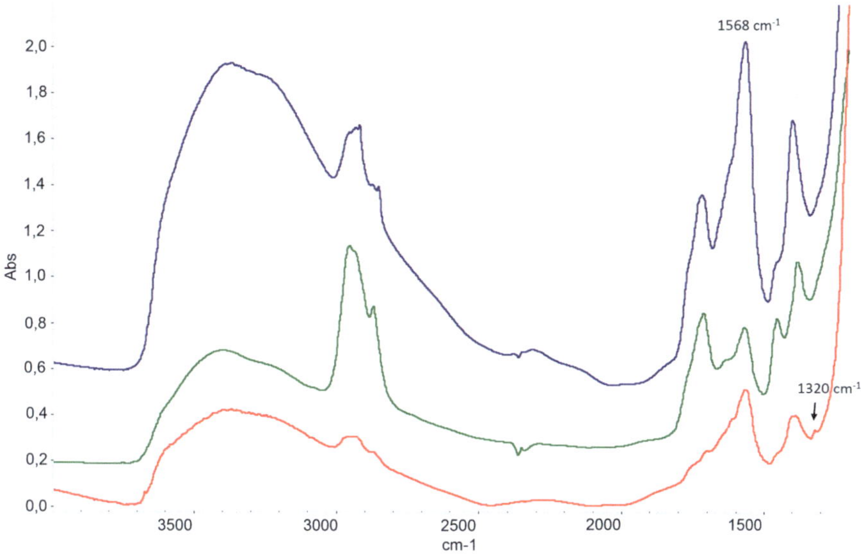

Fig. 8.3 Spectra of shellac (blue line), dammar (green line), and colophony (red line) mixed with smalt after 900 h of photooxidative aging

react with some pigment-forming metal soaps and, in case of photooxidative aging, oxalates (Fig. 8.3). The specific carbonyl absorption MIR bands of these resin soaps are at very similar wave numbers of the carboxylates coming from drying oil (1590 cm^{-1} with zinc white, 1568 cm^{-1} with smalt, and 1580 cm^{-1} with azurite) (Poli et al. 2017). In Fig. 8.3, spectra of the three considered natural resins mixed with smalt show in all cases the formation of a band (1568 cm^{-1}) attributable to metal soap formation after 900 h of simulated solar irradiation. Colophony (red line in Fig. 8.3) seems to be the most decayed resin with the almost complete disappearance of methylenic (2920 and 2850 cm^{-1}) and carbonyl (1708 cm^{-1}) stretching bands and the appearance of the C-O stretching band related to oxalate formation at 1320 cm^{-1}. It has not been possible to identify the nature of oxalates since only the signal at 1320 cm^{-1} was detectable. This behavior has also been found in presence of zinc white pigment, another very reactive pigment where the formation of zinc oxalates and zinc resin soaps has been assessed (Poli et al. 2014), as in the case of the Giacomo Balla's painting (Rava et al. 2013). In the mock-ups, the interaction of the natural resin, in particular shellac and colophony and after photooxidation, with zinc oxide leads to a significant formation of zinc oxalates confirmed by the characteristic carbonyl stretching band at 1363 cm^{-1} together with the C-O stretching band at 1320 cm^{-1} indicating the presence of bridging oxalates, with all four oxygen ions coordinated to the metal ions (Gabal et al. 2003). Generally speaking, a minimal formation of oxalates has been observed in all the photooxidized samples containing resins (particular shellac and colophony) and the pigments smalt, zinc white, lead white, and azurite. On the contrary, the oxalate formation has not been observed in the mock-up samples with stand oil.

It is important to point out that in the presence of smalt and zinc oxide, the formation of metal soaps is unmistakable within all the resins just a few days after the film casting even without any accelerated aging. The fast kinetic of reaction of smalt with natural resins means that potassium ions become rapidly available in sufficient amount. Moreover, this means that terpenic acids, seizing cations from pigment grains to form metal soap, can work together with fatty acids from oil favoring the discoloration of smalt in the cases of varnished paintings. Probably resins are more reactive than oil because most of terpenic acids are immediately available, while in the case of oil, most of the fatty acids became available after the hydrolysis of the triglycerides.

Red earth and umber earth show a minimal amount of soap formation (metal carboxylate band at 1548 cm^{-1}). Differences between the two iron-based pigments, showing higher reactivity of umber earth, are probably related to the latter's manganese content, since pure iron oxides (used as reference) did not show any metal carboxylate formation in the aging conditions used for this study. On the contrary, infrared spectra of mock-ups containing these earth pigments provide evidence of slight protection of the terpenic structures (in all the spectra, the carbonyl band shows a minimal broadening, and the fingerprint region appears almost unaltered) from photooxidation on the part of iron-containing pigments. Titanium white, as predictable, is pretty stable and does not induce the formation of any soap but seems to favor oxidative processes resulting in a significant broadening of the carbonyl band, but no soaps nor oxalates have been observed.

The behavior of lead white, one of the most reactive pigments with drying oil, was particularly interesting as it showed minimal interaction with the resins. Only after the thermo-oxidative aging (20 s at 250 °C), a reactivity similar to the drying oil has been showed, i.e., significant transformation of the organic binder in metal soaps (Fig. 8.1) with the appearance of the carboxylate band at 1528 cm^{-1}.

Azurite proved to be stable even if, in a few samples, grains of discolored pigment that were rich in resin soaps have been found (Fig. 8.4) by means of micro-FTIR localized measurements. Only on the discolored green grain, a well-defined band of carboxylates (at 1594 cm^{-1}) is clearly detectable (Fig. 8.4, green spectrum). In all the other areas of the chemical image and in the diamond anvil cell measurements, the obtained spectra do not show a detectable carboxylates band (Fig. 8.4, blue spectrum) or present only a shoulder probably referable to soaps (Fig. 8.4, red spectrum).

Considering the low average reactivity showed by azurite with resins in the mock-ups, it is not clear if the observed color loss (the green grain in Fig. 8.4) is a consequence of the saponification process or if the soap formation occurred with a more reactive grain present in the azurite pigment powder. It is important to point out that the complex carbonate bands of azurite can mask weak carboxylate stretching bands for newly formed products that are present in low concentration.

All the aging conditions used for this study induced similar effects with the exception of the thermo-oxidative one (25 s at 250 °**C**), which forced, as predictable,

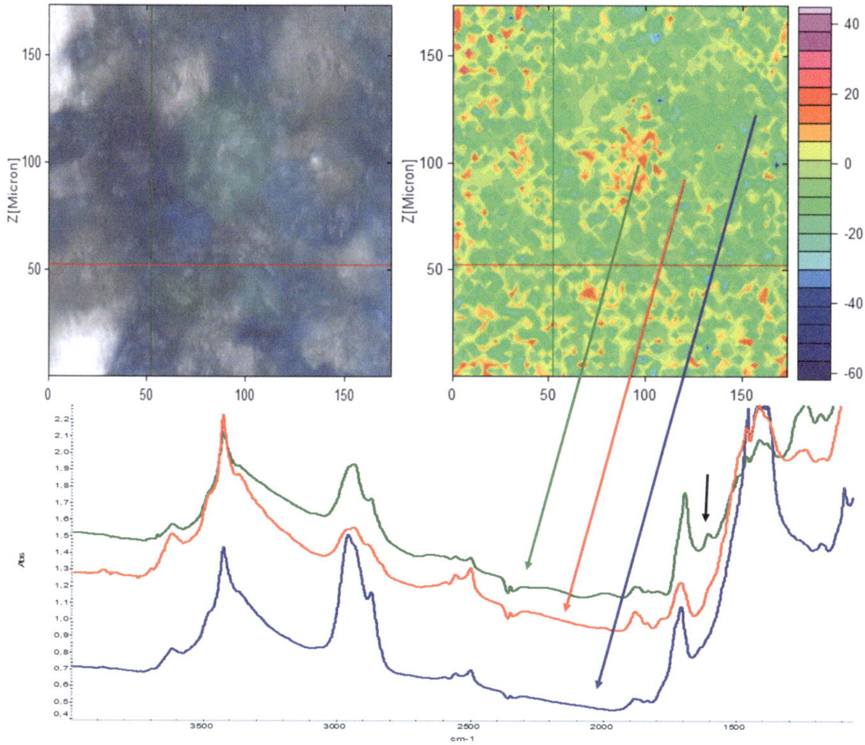

Fig. 8.4 Light microscopic image (left) and chemical image of azurite grains in colophony medium (right). Chemical image has been integrated in the spectral range from 1550 to 1600 cm^{-1}. Only the green grain appears rich in metal soaps in the chemical image. FTIR transmission spectra (obtained averaging signals coming from similar areas) of the discolored green grain (green spectrum) and the intact blue particles and colophony medium (blue and red spectrum)

an extended decay of the organic component and a relatively large degree formation of metal soaps. In Fig. 8.1 the spectrum obtained from a sample of the blackened surface of the historical painting is comparable to the one obtained with a mock-up heated at 250 °**C** for 25 s. This extreme accelerated aging has been included to confirm that overheating can be a significant cause of metal soap formation. No oxalate formation has been observed.

The aging with the least perceptible effects on the artificial deterioration of the mock-ups has been the thermal one (50 °C, 1100 h), which actually induced less soap formation than the room conditions. This is, probably, related to the very low humidity content in the fan-assisted oven and to a faster drying (and to the evaporation of low molecular weight fractions) with the consequent reduction of mobility of the system.

8.4 Conclusions

The case studies presented illustrate that natural resins play a role in the formation of metal soaps and metal oxalates. In this paper, the possibility of metal soap formation when natural resin and pigments are present has been demonstrated in mock-up films. Terpenic acids quickly react with cations coming from some pigments in order to form carboxylic salts as resin soaps. Smalt and zinc white were the most reactive pigments immediately forming metal carboxylates, even without any artificial aging. Azurite and natural earths also proved to be somehow reactive. The rapid formation of resin soaps in the presence of smalt points to the fact that terpenic acids can favor the leaching of potassium ions contributing, together with fatty acid from drying oil, to the discoloration of the pigment in varnished easel paintings. Lead white showed a lower affinity to the natural resins compared to its reactivity in stand oil. This will be further studied by testing lead white with different contents of cerussite and hydrocerussite and other lead pigments such as lead oxides (minium).

The ability of natural resins to form metal soaps means that at the resin-paint interface, unpredicted competitive reactions among fatty acids coming from the oil painting layer and terpenic acids coming from the varnish could occur. Oil-derived metal soaps coming from the painted layer could, for example, exchange cations with terpenic acids when they reach the varnish layer, playing a role in the superficial crust formation. Generally, the presence of one or more layers of natural resins could affect aggregation and the ionomeric structure of the resulting soaps.

Particularly interesting is the oxalate formation observed in the majority of mock-up samples containing active pigments (such as Zn- and Co-containing pigments) and natural resins aged in photooxidative conditions. Studying reaction products with infrared spectroscopy alone has made it possible to define the specific oxalate cation only in the case of samples containing zinc white. Oxalate formation during the resin aging in the presence of a reactive pigment could be an explanation of the strange superficial behavior for a painting where the oxalate formation has been observed only in the area where natural resin is still present. Results allow to hypothesize that the hard-to-clean thin oxalate patinas frequently found under varnishes in easel paintings can be related to resin degradation more than to drying oil.

A new experimentation is ongoing in order to expand the number of pigments and natural resins included in the study and to replicate and validate results described here, in particular the ones related to oxalate formation. This is very important since the presence of oxalates deeply affects the cleaning process of paintings. The new ongoing experiments are focused on:

- Understanding why natural resins have different behaviors (acidity, free acid amounts, and mobility)
- Testing oleoresinous binders and possibly studying the results of competition between fatty and terpenic acids
- Understanding if the different reactivities are related to a particular affinity of specific terpenic acids toward certain cations or to the ability of one of the resin components to modify a pigment's chemistry, increasing the cation availability

References

Antony P, De SK (1999) The effect of zinc stearate on melt-processable ionomeric blends based on zinc salts of maleated high-density polyethylene and maleated EPDM rubber. Polymer 40:1487–1493

Arbizzani R, Casellato U, Fiorin E, Nodari L, Russo U, Vigato PA (2004) Decay markers for the preventative conservation and maintenance of paintings. J Cult Herit 5:167–182

Boon JJ (2006) Processes inside paintings that affect the picture: chemical changes at, near and underneath the paint surface. In: Boon JJ, Ferreria E (eds) Reporting highlights of the De Mayerne Programme. NWO, The Hague, pp 21–32

Boon JJ, Peulve SL, Van den Brink OF, Duursma MC, Rainford D (1997) Molecular aspects of mobile and stationary phases in ageing tempera and oil paint films. In: Bakkenist T et al (eds) Early Italian paintings: techniques and analysis, symposium, Maastricht, 9–10 Oct 1996, Limburg Conservation Institute, Maastricht, pp 35–56

Boon JJ, van der Weerd J, Keune K, Noble P, Wadum J (2002) Mechanical and chemical changes in old master paintings: dissolution, metal soap formation and remineralization processes in lead pigmented ground/intermediate paint layers of 17th century paintings. In: Vontobel R (ed) 13th triennial meeting of the ICOM Committee for Conservation in Rio De Janeiro Preprints. James & James, London, p 401

Boon JJ, Hoogland F, Keune K (2007) Chemical processes in aged oil paints affecting metal soap migration and aggregation. In: AIC paintings specialty group Postprints, American Institute for Conservation of Historic & Artistic Works, Providence, Rhode Island, 16–19 June 2006, pp 16–23

Carlyle LA (2001) The artist's assistant: oil painting instruction manuals and handbooks in Britain 1800–1900, with reference to selected eighteenth-century sources. Archetype Publications, London

Colombini MP, Modugno F, Fuoco R, Tognazzi A (2002) A GC-MS study on the deterioration of lipidic paint binders. Microchem J 73:175–185

Cotte M, Checroun E, Susini J, Dumas P, Tchoreloff P, Besnard M, Walter P (2006) Kinetics of oil saponification by lead salts in ancient preparations of pharmaceutical lead plasters and painting lead mediums. Talanta 70:1136–1142

Cotte M, Checroun E, Susini J, Walter P (2007) Micro-analytical study of interactions between oil and lead compounds in paintings. Appl Phys A-Mater 89:841–848

Cotte M, Checroun E, De Nolf W, Taniguchi Y, De Viguerie L, Burghammer M, Walter P, Rivard C, Salomé M, Janssens K et al (2017) Lead soaps in paintings: friends or foes? Stud Conserv 62:2–23

Dietemann P, Higgitt C, Kälin M, Edelmann MJ, Knochenmuss R, Zenobi R (2009) Aging and yellowing of triterpenoid resin varnishes–influence of aging conditions and resin composition. J Cult Herit 10:30–40

Doménech-Carbó MT, Kuckova S, de la Cruz-Cañizares J, Osete-Cortina L (2006) Study of the influencing effect of pigments on the photoageing of terpenoid resins used as pictorial media. J Chromatogr A 1121:248–258

Erhardt D, Tumosa CS, Mecklenburg MF (2000) Natural and accelerated thermal aging of oil paint films. Stud Conserv 45:65–69

Gabal MA, El-Bellihi AA, El-Bahnasawy HH (2003) Non-isothermal decomposition of zinc oxalate–iron (II) oxalate mixture. Mater Chem Phys 81:174–182

Gunn M, Chottard G, Rivière E, Girerd J-J, Chottard J-C (2002) Chemical reactions between copper pigments and oleoresinous media. Stud Conserv 47:12–23

Higgitt C, Spring M, Saunders D (2003) Pigment-medium interactions in oil paint films containing red lead or lead-tin yellow. National Gallery Technical Bulletin 24:75–95

Keune K (2005) Binding medium, pigments and metal soaps characterised and localised in paint cross-sections. Dissertation, University of Amsterdam

Lazzari M, Chiantore O (1999) Drying and oxidative degradation of linseed oil. Polym Degrad stabil 65:303–313

Mallégol J, Gardette J-L, Lemaire J (1999) Long-term behavior of oil-based varnishes and paints I. Spectroscopic analysis of curing drying oils. J Am Oil Chem Soc 76:967–976

Manzano E, Navas N, Checa-Moreno R, Rodriguez-Simón L, Capitán-Vallvey LF (2009) Preliminary study of UV ageing process of proteinaceous paint binder by FT-IR and principal component analysis. Talanta 77:1724–1731

Mazzeo R, Prati S, Quaranta M, Joseph E, Kendix E, Galeotti M (2008) Attenuated total reflection micro FTIR characterisation of pigment–binder interaction in reconstructed paint films. Anal Bioanal Chem 392:65–76

Meilunas RJ, Bentsen JG, Steinberg A (1990) Analysis of aged paint binders by FTIR spectroscopy. Stud Conserv 35:33–51

Noble P, Boon JJ, Wadum J (2002) Dissolution, aggregation and protrusion, lead soap formation in 17th century grounds and paint layers. ARTMATTERS–Netherlands technical studies in art I, pp 46–61

Plater MJ, De Silva B, Gelbrich T, Hursthouse MB, Higgitt CL, Saunders DR (2003) The characterisation of lead fatty acid soaps in "protrusions" in aged traditional oil paint. Polyhedron 22:3171–3179

Poli T, Piccirillo A, Zoccali A, Conti C, Nervo M, Chiantore O (2014) The role of zinc white pigment on the degradation of shellac resin in artworks. Poly Degrad Stabil 102:138–144

Poli T, Piccirillo A, Nervo M, Chiantore O (2017) Interaction of natural resins and pigments in works of art. J Colloids Interface Sci 503:1–9

Rava A, Radelet T, Giovagnoli A, Poli T, Chiantore O, Piccirillo A (2013) Short communication: the painting materials in a work of the futurist artist Giacomo Balla. J Am Inst Conserv 52: 227–235

Van den Berg JDJ (2002) Analytical chemical studies on traditional linseed oil paints. MolArt Dissertation, University of Amsterdam

Van den Berg JD, Van Den Berg KJ, Boon JJ (1999) Chemical changes in curing and ageing oil paints. In: Triennial meeting (12th), Lyon, 29 Aug-3 Sept 1999: preprints, vol 1. James & James, pp 248–253

Van der Weerd J, Van Loon A, Boon JJ (2005) FTIR studies of the effects of pigments on the aging of oil. Stud Conserv 50:3–22

Van den Brink OF, Ferreira ES, van der Horst J, Boon JJ (2009) A direct temperature-resolved tandem mass spectrometry study of cholesterol oxidation products in light-aged egg tempera paints with examples from works of art. Int J Mass Spectrom 284:12–21

Wakabayashi K, Register RA (2006) Ethylene/(meth) acrylic acid ionomers plasticized and reinforced by metal soaps. Polymer 47:2874–2883

Chapter 9
Factors Affecting the Reactivity of Zinc Oxide with Different Drying Oils: A Vibrational Spectroscopy Study

Francesca Casadio, Ludovic Bellot-Gurlet, and Céline Paris

Abstract This study focuses on the short-term reactivity of zinc oxide with different types of drying oils. Chemical grade zinc oxide (with diameter <100 nm) was mixed with oils with different proportions and types of fatty acids. Specifically, reactivity of zinc oxide with alkali-refined linseed oil, raw linseed oil, boiled linseed oil, polymerized linseed oil, stand oil, tung oil, and alkyd was studied between 4 days and 10 months. Effects of short-term exposure to temperature, humidity, and pH were also evaluated. The type of metal soaps formed and the kinetics of formation were monitored with attenuated total reflection Fourier transform infrared (ATR- FTIR) spectroscopy. The results confirm that humidity and heat combined favor the formation of both network-coordinated zinc carboxylates (POL-Zn) and zinc complexes of free fatty acids (FA-Zn) irrespective of the type of oil; however, the phenomenon is particularly severe for alkali-refined, raw, and boiled linseed oils, while stand oil and tung oil show the lowest amounts of zinc soap (FA-Zn) formation. The findings advance our knowledge of the reactivity of zinc oxide with an expanded range of drying oils that are commonly encountered in many nineteenth- and twentieth-century paints and reinforce the notion that even short-term exposure to heat with humidity can have irreversible effects on zinc carboxylate formation.

Keywords Zinc carboxylates · Drying oils · ATR-FTIR · Artificial aging · Humidity · Linseed oil · Tung oil · Stand oil · Alkyd

---

F. Casadio (✉)
The Art Institute of Chicago, Chicago, IL, USA
e-mail: fcasadio@artic.edu

L. Bellot-Gurlet · C. Paris
Sorbonne Universités, UPMC Université Pierre et Marie Curie – Paris 6, MONARIS "de la Molécule aux Nano-objets: Réactivité, Interactions et Spectroscopies", UMR 8233, UPMC/CNRS, Paris, France
e-mail: ludovic.bellot-gurlet@upmc.fr; celine.paris@upmc.fr

F. Casadio et al. (eds.), *Metal Soaps in Art*, Cultural Heritage Science,
https://doi.org/10.1007/978-3-319-90617-1_9

9.1 Introduction

9.1.1 Historical Perspectives from Paint Technology Manuals

Zinc oxide is a common white pigment in paints that were formulated in the nineteenth and twentieth centuries. In-depth studies of technical manuals on the early twentieth-century paint-making technology uncovered frequent passages seeming to imply that promoting the reaction of zinc oxide with the paint medium would lead to superior quality enamels (Kokkori et al. 2014a, 2015b). Recommendations for "exceptionally high-grade" enamels included preparing a paste with an excess of zinc white (obtained, e.g., by grinding 65 pounds of condensed zinc white in 28 pounds of heavy oil and 7 pounds of pure turpentine) and letting the paste stand for at least a week, before adding to it 60 pounds of varnish made from pale kauri gum and linseed oil (Mattiello 1943; Kokkori et al. 2014b). Other authors recommended setting away the paste of zinc oxide with medium – often a thermally treated, bodied oil – "to ripen" in closed containers for at least 2 weeks or even longer, before reducing this "white enamel base" with varnish or other diluents to the desired consistency (Uebele 1913). Among those early oil-based enamels are the ones produced by the French brand Ripolin, which had an important role among the materials used by the global artistic avant-gardes and was notably used by artists such as Pablo Picasso (1881–1973), Francis Picabia (1879–1953), and others (Gautier et al. 2009; Muir et al. 2013; Dredge et al. 2013; McMillan et al. 2013; Kokkori et al. 2015a). Interestingly, when studied under the scanning electron microscope, these Ripolin paints show very even dispersion of the pigment particles, much more so than zinc oxide-containing fine artist paints from tubes (Muir et al. 2011). This observation suggests that paint makers might have included those recommendations because of empirical knowledge that the early formation of complexes of zinc ions with fatty acids from the oil would act as an internally formed dispersant, avoiding the clumping of pigment particles. The historical literature thus provided the inspiration for the present study, which focused on the short-term reactivity of zinc oxide with an expanded range of oils, beyond linseed oil alone, which are commonly encountered in industrial and artist's paint manufacture.

9.1.2 Experimental Studies

Previous studies of zinc-based carboxylates (Fauve and Vandenmaele 1964; Robinet and Corbeil 2003; van der Weerd et al. 2005; Barman and Vasudevan 2007; Rogala et al. 2010; Osmond et al. 2012; Hermans et al. 2015) have shed light on the organic/inorganic compounds that form by reaction of the zinc oxide pigment with the lipidic binding medium, identifying mostly zinc palmitates ($Zn(C_{16})_2$), stearates ($Zn(C_{18})_2$), and oleates ($Zn(C_{18:1})_2$) in crystalline and amorphous states (for brevity, the abbreviation FA-Zn is going to be used to identify this class of zinc complexes of free fatty acids). Most recently, it has been convincingly proposed that soaps whose FTIR spectra cannot be correlated to that of a specific type of zinc carboxylate (Robinet and Corbeil 2003; Otero et al. 2014) may consist of network-coordinated zinc soaps, i.e., zinc carboxylate moieties that are covalently bonded to the cross-linked and cured oil (Hermans et al. 2015, 2019; Hermans 2017) (these will hereafter be called POL-Zn). Fourier transform infrared analysis in transmittance of many Ripolin paints has demonstrated that these oil-based enamels have spectra with characteristic broad antisymmetric vibrations of the carboxylate group, centered at approximately 1585 cm^{-1}, which have been attributed to such network-coordinated zinc soaps (Gautier et al. 2009; Muir et al. 2013; McMillan et al. 2013) (POL-Zn). It is interesting to highlight that many paintings believed to be made with Ripolin (some nearing 100 years of age) have been proven to be stable and have not shown characteristic zinc soap protrusions (Casadio et al. 2013; Muir et al. 2013; McMillan et al. 2013) as have been otherwise observed in zinc oxide-containing paints by, for example, Vincent van Gogh (van der Weerd et al. 2003), or delamination of paint as noted for paintings by Piet Mondrian (Raven et al. 2019), abstract expressionist painters (Rogala et al. 2010), and others. It should be noted that it is the crystalline zinc soaps (FA-Zn), which are found in protrusions and at the interface of paints that are delaminating.

In recent years, the deterioration phenomena associated with the reaction between the metal ions of the pigment and the organic matrix have become a major concern for art conservators and the scientists interested in unraveling the chemical processes, crystallization, and kinetics of metal soap formation, as the many chapters in this volume attest. Surprisingly, while our knowledge of metal soaps has significantly advanced in the past 20 years overall, comparably little work has been done on the early stages of formation of zinc soaps. In 2012 Clementi et al. (Clementi et al. 2012) studied Gamblin alkyd medium and linseed oil mixed with zinc oxide (1:2 wt) aged in the dark at 40 °C and 95% RH. FTIR spectra taken at 15, 28, 50, 69, and 88 days demonstrated how, when plotting the ratio of the intensity of the FTIR $\nu C{=}O$ band of the lipidic component of the oil at 1730 cm^{-1} versus the antisymmetric COO^- stretching of newly formed metal carboxylates at 1600 cm^{-1}, the ratio became stable after 15 days of aging. No data was offered before day 15 though, while the historical technical manuals discussed previously clearly seem to indicate reactivity after a few days. Poli and coworkers performed artificial aging of mixtures of Kremer zinc oxide and shellac 1:1, aged for 3 minutes at 300 °C on a heating plate, which led to the formation of disordered or network-coordinated zinc carboxylates with resinous acids (as indicated by the observation of the broad FTIR band at 1585 cm^{-1}) (Poli et al. 2014). As described in this volume (Poli et al. 2019) 1000 h of UV aging in a solar box (cutting $\lambda < 300$ nm) of the same mixture led to the formation of zinc oxalates. Finally, Hermans et al. (Hermans et al. 2015; Hermans 2017) explored the reactivity of mixtures of cold-pressed linseed oil with zinc oxide with and without the addition of demineralized water (to promote soap formation) stirred in a sealed vial at room temperature (RT) for 3 days, then pipetted on a glass slide, and left to dry for 7 weeks, leading to the observation of POL-Zn after a week of curing.

Despite these previous efforts, a systematic study is lacking, focusing on the early-stage reactivity of zinc oxide with an expanded range of drying oils commonly encountered in many nineteenth- and twentieth-century paints. This work aims at filling that gap, by concentrating on the short-term reactivity of zinc oxide and evaluating the effect of the type of oil used. Variables examined included the drying oil's viscosity, degree of prepolymerization, availability of free fatty acids in the initial oil composition (as expressed by their acid value, when available from the manufacturer), and type of free fatty acids contained in the TAGs (triacylglycerides). Specifically, four variously processed linseed oils were used, whose free fatty acid values were provided by the manufacturer, ranging from 0.3 max for alkali-refined linseed oil to 19 max for polymerized linseed; these were chosen to explore whether the initial availability of different amounts of free fatty acids is a factor in the short-term formation of zinc carboxylates and their evolution over time. Bodied oils (polymerized linseed, boiled linseed, and stand oil) were also examined, to evaluate the effect of higher viscosity (when compared to unbodied oils) and prepolymerization on the formation of zinc carboxylates. Finally an oil-based alkyd medium containing 10–20% of safflower oil (which has higher proportions of linoleic but lower proportions of linolenic acid than linseed oil) as well as tung oil (which is very rich in the triply unsaturated conjugated α-eleostearic C18:3 oil) (see Table 2.1 in Osmond 2019) was included to explore the effects of different oil chemistries. Additionally, the effect of short-term exposure to high levels of humidity coupled with elevated temperature as it could be experienced by painted surfaces during treatments such as paint consolidation, flattening, and relining were explored. Lastly, the effect of low pH was also evaluated. The type of metal soaps formed and the kinetics of formation were monitored with ATR-FTIR.

9.2 Materials and Methods

9.2.1 Paint Mixtures

Paints were prepared by hand-mixing with a spatula on a glass slide zinc oxide particles with oil in proportions of 2:1 (v:v) pigment-binder until a viscous, stringy consistency was obtained, which mimicked paint formulations as they would be used by artists. Chemical grade zinc oxide (Sigma-Aldrich) of diameter smaller than 100 nm (as certified by the manufacturer) was used. Because chemical grade zinc oxide is of significantly smaller particle size, and larger surface area than pigment grade, it has been shown to be substantially more reactive than artists' pigment in previous models (Osmond 2014). Therefore it should be expected that reaction products will be formed in a shorter timescale than if experiments were conducted using artist grade zinc oxide.

Several different commercial oils were used; where available, their industrially determined acid value (a measure of the free fatty acids available in the oil (Bailey 1951)) is indicated in parenthesis. These were alkali-refined linseed oil (Welch, Holme and Clark, Inc. (WHC), acid value 0.3 max), raw linseed oil (Alnor, acid value 4 max), boiled linseed oil (WHC, acid value 7.5 max), polymerized linseed oil (Alnor, K-6 grade, acid value 19 max), stand oil (Winsor and Newton), tung oil (WHC, acid value 5 max), and alkyd (Gamblin Galkyd slow dry, certified by the manufacturer to contain 40–60% of alkyd resin and 10–20% safflower oil). Immediately after mixing, a small quantity of each different paint was deposited on a microscope cover glass in the form of a 6-mm-diameter drop of paint of approx. 0.8–1 mm thickness in preparation for artificial aging, while a set of all paint mixtures was left to rest in the dark, at laboratory ambient conditions ($T = 20\ °C \pm 2$ and $RH\% = 50 \pm 5$).

9.2.2 Conditions of Artificial Aging

After mixing, samples were left for 2 days at room temperature before being subjected to the artificial aging regimen described below for 2 days. The short, 48-h aging time was chosen to simulate the effect of a conservation treatment such as lining, consolidation or flattening of paint, which would have a short duration, though it should be noted that at such early stages of curing, some paints were not even touch-dry yet.

Paint mixtures were aged (1) at room temperature (RT in the following), (2) in a dry oven at 80°C (labeled 80 in the following), and (3) at 80°C in a sealed glass jar containing a small beaker where the samples were positioned, surrounded by a saturated solution of KNO_3 (Sigma-Aldrich) to ensure approximately 90% RH (called 80+RH). In a subset of these glass jars, (4) a small vial containing a saturated solution of oxalic acid dihydrate (Sigma-Aldrich) was added to the chamber to simulate an acidic and humid environment (termed 80+RH+H+ in the following). In both of the latter cases, a Gore-Tex membrane was affixed to the top of the beaker with the paint samples to avoid direct condensation of water onto the paint surfaces (Fig. 9.1).

After the 2 days at RT and 2 days of heating in the conditions described above, all samples were removed from the glass jars and left in the dark at room temperature.

9.2.3 Instrumentation

A Bruker Alpha small footprint portable FTIR spectrometer with ATR module with diamond crystal was used to monitor the samples. A small amount of paint (in the form of a chip measuring approximately 2 × 2 mm and 0.5–0.8 mm in thickness) was placed on the diamond crystal face, and, upon applying a consistent

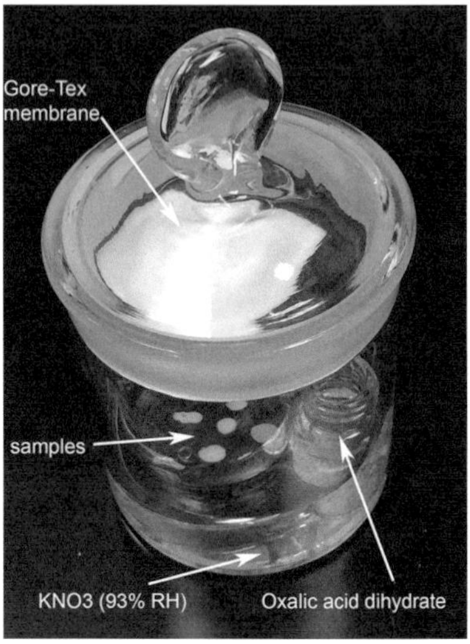

Fig. 9.1 The sealed glass jar used for aging paint mixtures with hot, humid, and acidic air (80+RH+H+). The jar diameter is 8 cm

pressure with the instrument's anvil, 64 scans were acquired at a resolution of 4 cm^{-1} between 4000 and 400 cm^{-1}. Care was taken in making sure that it was always the uppermost portion of the sample (the one in contact with the atmosphere during aging) to be in direct contact with the ATR crystal. The depth of penetration (optical path length) inside the surface of the sample in contact with a diamond crystal is estimated by the manufacturer to be 1.66 μm at 1000 cm^{-1} and with a 45° incidence. The spectra illustrated here represent an average of three measurements for each sample.

The first ATR-FTIR measurements were conducted after 4 days, inclusive of 2 days at room temperature immediately after mixing, and 2 days under the respective regime of artificial aging. Subsequently, all samples were left at room temperature and measured with the ATR-FTIR after 10 months. A subset of samples was also measured at 21 days (Table 9.1).

To account for the different degrees of coverage and contact with the diamond crystal by the samples (especially at 4 and 21 days, as some samples were still jelly-like, whereas others were completely hardened) and in order to achieve a consistent comparison of the formation of different metal carboxylate species, the following procedure was used.

All spectra were normalized by the intensity of the CH stretch at around 2922 cm^{-1}, which can be considered to remain fairly constant under the conditions used. Then maximum band intensity was calculated with the OPUS 7 software after subtraction of the baseline for the νC=O ester band of the oil at 1741–38 cm^{-1} (labeled I_{1740} in the following), as well as the antisymmetric COO^{-} stretching of newly formed network-coordinated metal carboxylates (POL-Zn) at 1585 cm^{-1} (I_{1585}) and of the sharp band at 1540–50 cm^{-1} (I_{1540}) of crystalline metal soaps (FA-Zn) (Table 9.1). This method has limitations, because even after subtraction of the baseline and normalization, the maximum peak intensity of convoluted or

Table 9.1 Normalized band intensities for FTIR peaks characteristic of the νCO of the oil and the $\nu_{as}COO^-$ of newly formed carboxylates (FA-Zn and POL-ZN) for the oil and zinc oxide mixtures studied under the different aging regimes

			10 months				21 days				4 days			
Oil	Acid value	Condition	I(1740)	I(1585)	I(1540)	I(1540)/ I(1585)	I(1740)	I(1585)	I(1540)	I(1540)/ I(1585)	I(1740)	I(1585)	I(1540)	I(1540)/ I(1585)
Alkali-refined linseed	0.3 max	RT	1.25	0.64	N/A	N/A					1.36	N/A	N/A	N/A
		80	1.46	0.23	N/A	N/A	1.52	0.41	N/A	N/A	1.21	0.13	N/A	
		80+RH	0.48	0.31	0.95	3.08					0.77	0.50	0.58	1.16
		80+RH+H+	0.49	0.41	0.92	2.23	0.52	0.33	0.79	2.42	0.88	0.64	0.76	1.20
Raw linseed	4 max	RT	1.30	0.50	N/A	N/A					1.37	N/A	N/A	N/A
		80	1.56	0.20	N/A	N/A	1.52	0.13	N/A	N/A	1.57	0.09	N/A	N/A
		80+RH	0.59	0.52	0.86	1.67					0.77	0.50	0.58	1.16
		80+RH+H+	0.52	0.40	0.87	2.20	0.51	0.39	0.90	2.28	0.88	0.64	0.76	1.20
Boiled linseed	7.5 max	RT	1.47	0.53	N/A	N/A					1.21	N/A	N/A	N/A
		80	0.91	0.17	N/A	N/A	1.41	0.22	N/A	N/A	1.57	0.21	N/A	N/A
		80+RH	0.60	0.44	0.75	1.70					0.79	0.53	0.67	1.26
		80+RH+H+	0.75	0.53	0.70	1.33	0.47	0.25	0.91	3.69	0.77	0.48	0.53	1.09
Polymerized linseed	19 max	RT	1.24	0.27	N/A	N/A					1.17	N/A	N/A	N/A
		80	1.11	0.10	0.11	1.16	1.10	0.08	N/A	N/A	1.13	0.08	N/A	N/A
		80+RH	0.55	0.21	0.64	3.07					0.76	0.34	N/A	N/A
		80+RH+H+	0.48	0.20	0.61	3.00	0.58	0.26	0.55	2.15	0.77	0.34	N/A	N/A

(continued)

Table 9.1 (continued)

			10 months				21 days				4 days			
Oil	**Acid value**	**Condition**	**I(1740)**	**I(1585)**	**I(1540)**	**I(1540)/ I(1585)**	**I(1740)**	**I(1585)**	**I(1540)**	**I(1540)/ I(1585)**	**I(1740)**	**I(1585)**	**I(1540)**	**I(1540)/ I(1585)**
Stand oil		RT	1.24	0.14	N/A	N/A					1.20	N/A	N/A	N/A
		80	1.20	0.09	N/A	N/A	1.20	0.07	N/A	N/A	1.23	0.06	N/A	N/A
		80+RH	0.89	0.30	0.38	1.26					0.97	0.19	N/A	N/A
		80+RH+H+	0.94	0.23	0.26	1.13	0.99	0.26	0.28	1.09	0.98	0.18	N/A	N/A
Tung oil	5 max	RT	1.92	N/A	N/A	N/A					1.39	N/A	N/A	N/A
		80	1.86	N/A	N/A	N/A	1.71	N/A	N/A	N/A	1.70	N/A	N/A	N/A
		80+RH	1.25	0.36	0.34	0.95					1.21	0.41	N/A	N/A
		80+RH+H+	1.25	0.34	N/A	N/A	1.11	0.53	N/A	N/A	1.17	0.48	N/A	N/A
Galkyd		RT	1.98	0.48	N/A	N/A					1.89	0.17	N/A	N/A
		80	1.91	0.48	N/A	N/A	2.25	0.38	N/A	N/A	2.19	0.30	N/A	N/A
		80+RH	1.75	0.80	0.97	1.21					1.72	0.47	0.48	1.02
		80+RH+H+	1.66	0.82	1.11	1.35	1.66	0.82	0.98	1.19	1.68	0.74	0.91	1.22

N/A indicates that a band was not observed; when a space is left blank, it means that the measurement was not carried out

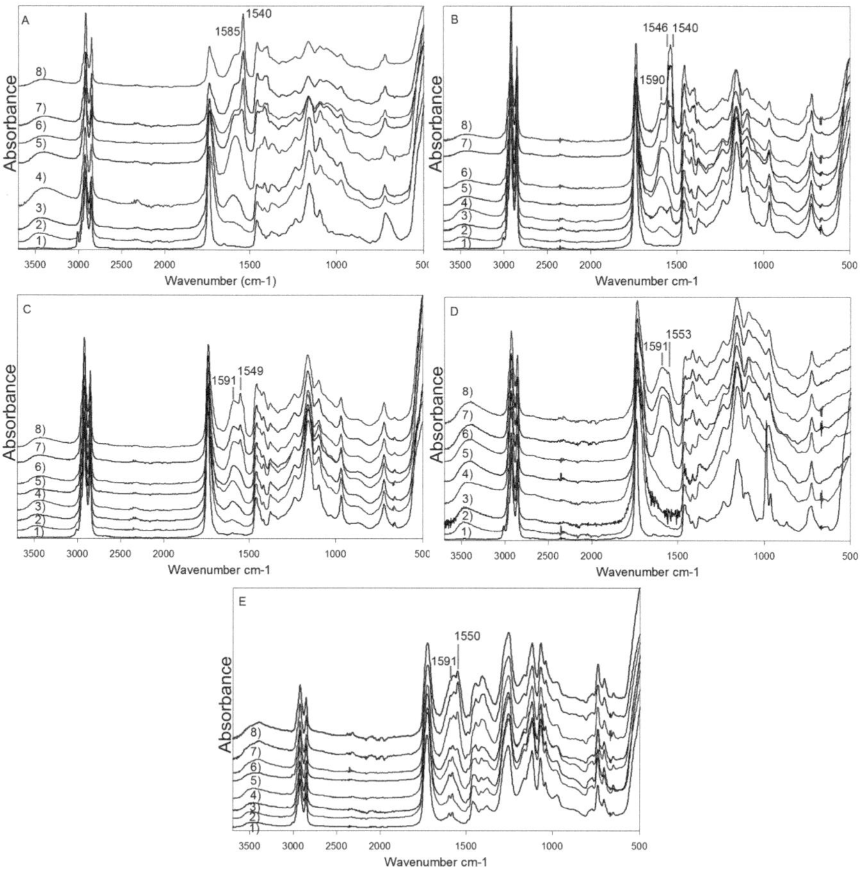

Fig. 9.2 ATR-FTIR spectra of paint mixtures of zinc oxide with (**a**) alkali-refined linseed oil; (**b**) polymerized linseed oil; (**c**) stand oil; (**d**) tung oil; (**e**) Galkyd, at various stages of aging, specifically (1) RT 4 days; (2) 80 4 days; (3) 80 10 months; (4) RT 10 months, (5) 80+RH 4 days; (6) 80+RH+H+ 21 days; (7) 80+RH+H+ 10 months; (8) 80+RH 10 months

overlapping bands (such as the sharp 1540 cm^{-1} band, which overlaps with the broad band at 1585 cm^{-1}) is difficult to calculate precisely due to the reciprocal contributions of overlapping bands. As a result, in this work the intensity of the peak at 1540 cm^{-1} has been marked only when a sharp peak was discernible. Though in such cases complete spectral decomposition and calculation of the area below each individual peak is preferable, comparison of band intensity is a rapid method of evaluation that has been employed by other authors (Clementi et al. 2012) to provide a reliable qualitative visualization of trends for chemical change as reflected in infrared spectra.

All spectra were also corrected to remove any residual water vapor and CO_2 absorptions (atmospheric compensation of the OPUS software). In Fig. 9.2, all spectra are normalized by the intensity of the CH stretch at around 2922 cm^{-1}, for ease of comparison.

9.3 Results and Discussion

Plotting the intensity of the characteristic band for POL-Zn (I_{1585}) and comparing it with that of FA-Zn (I_{1540}), as illustrated in Fig. 9.3, demonstrate that POL-Zn were not yet detectable with ATR-FTIR after 4 days of aging at RT, with the exception of the Galkyd medium, but that POL-Zn had already formed after 4 days for all the samples exposed to heat and humidity. Furthermore, alkali-refined, raw linseed, and boiled linseed oil showed formation of FA-Zn upon exposure to hot and humid conditions (80+RH). The Galkyd showed a similar trend while the heavily bodied oils such as polymerized linseed and stand oil, and the quick-drying tung oil showed no formation of FA-Zn after 4 days both at RT and 80+RH.

Figure 9.4 shows consistent trends over time, i.e., the development of FA-Zn after 10 months for all samples exposed to heat and humidity, a phenomenon that was not apparent immediately after 4 days but started being observed at 21 days and continued to evolve. Over time (Fig. 9.4 and Table 9.1), no oils showed formation of FA-Zn unless they were exposed to the high humidity combined with high-temperature aging regimen (80+RH or 80+RH+H+).

Lowering the pH of the hot and humid air in some cases caused higher overall formation of zinc carboxylates, but in general, as evident from Table 9.1, the effects of the 80+RH+H+ regime were comparable (within the uncertainty of the measurement) to the 80+RH samples.

The trends illustrated in Figs. 9.3 and 9.4 also demonstrated that the amount of free fatty acids initially available in the oil used, as determined through their reported acid value, was not a good predictor of the propensity of a certain oil to form POL-Zn or FA-Zn when mixed with the pigment. To the contrary, the three oils with the lowest reported acid value (alkali-refined, raw, and boiled linseed) appeared

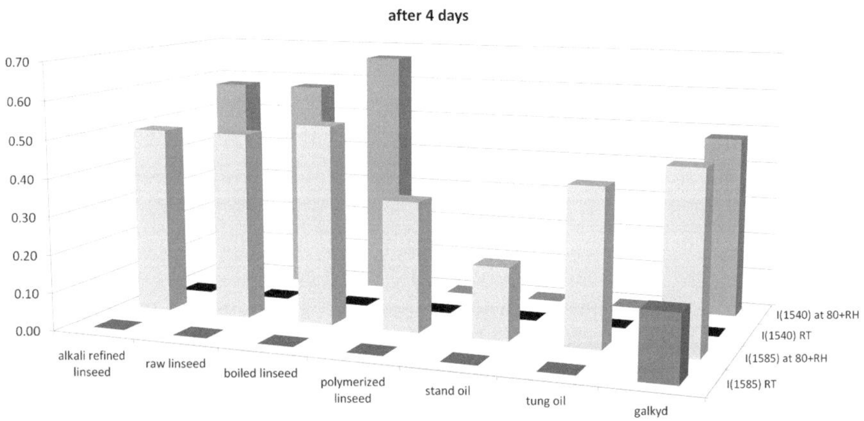

Fig. 9.3 Visualization of the values of I_{1585} and I_{1540} after 4 days of aging in two conditions: RT and 80+RH (for a full list of measured values, see Table 9.1; where a band intensity value is not plotted, it means that the band was not observed)

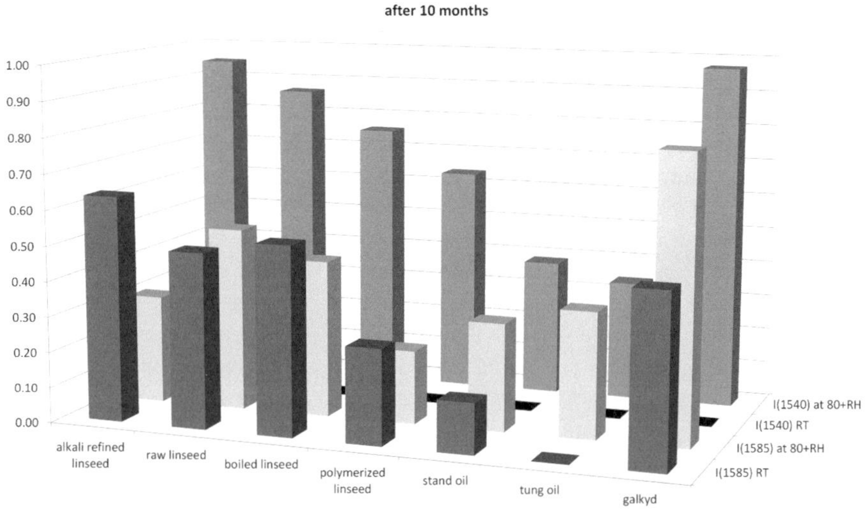

Fig. 9.4 Visualization of the values of I_{1585} and I_{1540} after 10 months of aging in two conditions: RT and 80+RH (for a full list of measured values, see Table 9.1; where a band intensity value is not plotted, it means that the band was not observed)

to form the highest amounts of FA-Zn both at 4 days and 10 months when exposed to 80+RH or 80+RH+H+ conditions. The same applied to the Galkyd (though its acid value was not known). Could it be that the highest amount of free fatty acids originally present in oils, such as the polymerized linseed oil, causes the zinc oxide particles to be "passivated" by a monolayer of free fatty acids, which, though hardly detectable by FTIR (Hermans et al. 2015), could render the Zn ions less available for reaction with the oil at later stages of curing? This hypothesis would nevertheless need to be experimentally demonstrated. The accuracy of the oil acid values provided by the manufacturers also requires verification.

After 4 days at 80+RH or 80+RH+H+, values of I_{1585} are similar for alkali-refined, raw linseed, and boiled linseed oils; however the polymerized linseed oil, one of the four linseed oils with the highest acid value reported, was the one that formed the lowest amount of POL-Zn and no FA-Zn after 4 days.

Alkali-refined linseed and raw linseed behaved in very similar ways, even though they had 0.3 max and 4 max free acid values reported. Together with the Galkyd, they are the media that developed the highest amount of POL-Zn at RT after 10 months of aging (Fig. 9.4).

Looking at these trends over time, it thus seems that viscous (bodied) oils such as polymerized linseed and stand oil have a lower tendency to form zinc carboxylates, perhaps due to issues of lower mobility of the carboxylic groups that are part of the polymer network and other transport phenomena in paints affecting both zinc ions and free fatty acids. The prepolymerization of these oils is also expected to affect their reactivity with the Zn particles. Further mechanistic studies should be conducted to explore this hypothesis further.

Overall, tung oil and stand oil were the least prone to forming zinc carboxylates (POL-Zn and FA-Zn) in the conditions used for this study. Because tung oil is very rich in the triply unsaturated conjugated α-eleostearic C18:3 oil and has much lower proportions of saturated and monounsaturated fatty acids with respect to linseed oil, its fast autoxidation and cross-linking, quickly leading to a hard and cured film, could be one of the reasons for its low tendency to form FA-Zn even at 80+RH.

Examining more specifically the spectra recorded for the different types of oils, several interesting observations can be made (Fig. 9.2). So as to keep the focus of this contribution on zinc carboxylate formation only, it should be noted that, in the discussion of the spectra, the phenomena of carbonyl band broadening and variations in the CH stretching bands that are related to the known processes of aging of the oil binding medium are not addressed, as they are already widely discussed in the literature (Meilunas et al. 1990; Lazzari and Chiantore 1999).

One interesting observation that is consistent for all samples studied (Fig. 9.2 and Table 9.1) is that 10 months after mixing, the amount of POL-Zn detected was actually higher for the samples aged at RT than the ones aged in dry oven at 80 °C. Though the exact reasons for this phenomenon are still under study, this finding further corroborates the limitations of dry heating as an artificial aging regime for oil paints that has also been demonstrated by other researchers (Erhardt et al. 2000). Dry heating alone, even at the relatively mild artificial aging regime of 80 °C and with exposures of only 2 days, initially seemed to accelerate aging. This was demonstrated by the formation of a small amount of POL-Zn after 4 days at 80 °C, which was not observed for the samples aged at RT. However, in the longer term, this study provided evidence that an aging regimen involving only heating at 80 °C will not cause the same type of chemical processes that are observed with natural aging.

For alkali-refined linseed oil (acid value 0.3 max), 4 days after mixing, at RT, there was no evidence of soap formation, while with heating and humid air, two main soaps bands are formed: a broad band approximately centered at 1585 cm^{-1} and a sharp one at 1550–51 cm^{-1}. After 21 days, the 80 and 80+RH+H+ samples showed no substantial increase in the POL-Zn formed (indicated by the broad band at 1585 cm^{-1}). Rather, in the 80+RH+H+ samples at 21 days and 10 months, there was a great increase in the intensity of the 1540 cm^{-1} band (shifting to lower wavenumbers from the initially observed value of 1550 cm^{-1} after 4 days). These trends were rendered evident when observing the variation of intensity of the three bands (I_{1540}, I_{1585}, and I_{1740}) over time for samples of the different oils aged in the same conditions (80+RH+H+) and measured at different times (Fig. 9.5). The hydrolysis of the ester bonds of the oil over time was signaled by the decrease in I_{1740}, especially evident for alkali-refined and polymerized linseed, while the normalized I_{1740} remained fairly constant over time for Galkyd, tung, and stand oil. A decrease or fairly constant value of the POL-Zn species was also observed; specifically, I_{1585} decreased over time for alkali-refined, polymerized, and tung oil, while it showed a mild increase for Galkyd and stand oil. After initial formation at 4 days, the formation of FA-Zn species increases for all oils, except stand oil, for which a peak at 1540 cm^{-1} was detected after 21 days and 10 months (Fig. 9.2).

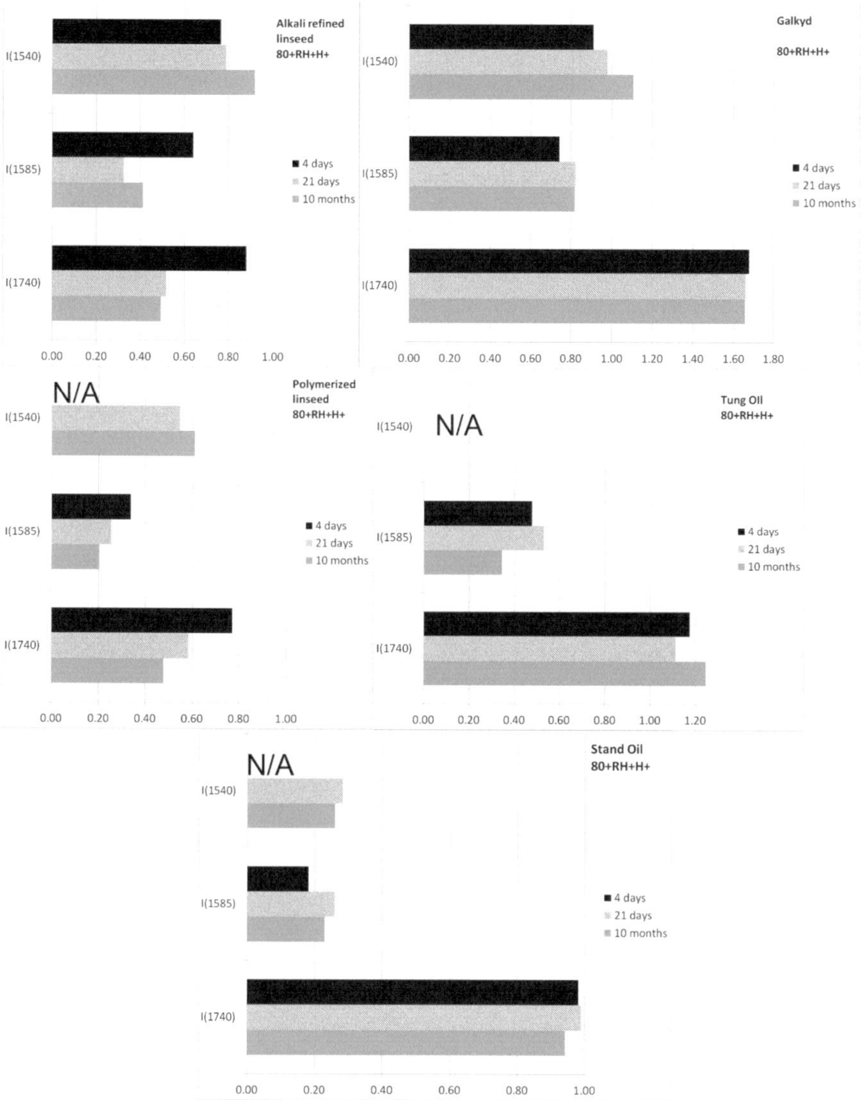

Fig. 9.5 Evolution over time of the normalized ATR-FTIR band intensities I_{1740}, I_{1585}, and $I_{1540\text{–}50}$ for the various oils studied at 80+RH+H+ (for a full list of measured values, see Table 9.1)

The tung oil did not show a distinctive peak at 1540 cm^{-1} in this aging regime, only a shoulder on the broad band centered at 1585 cm^{-1}, as illustrated in Fig. 9.2.

The spectra for polymerized linseed oil (Fig. 9.2) were unique in this group, because at 21 days, the 80+RH+H+ sample showed peaks at 1547 and 1532 cm^{-1}, with possibly a convoluted 1540 cm^{-1} contribution, becoming significantly more intense than the broad absorption at 1585 cm^{-1}. After 10 months, the 80+RH and 80+RH+H+ samples show intense signal for the broad band of zinc carboxylate (POL-Zn) and extremely intense absorption for the crystalline zinc soaps (FA-Zn), with the split peak at 1537–1546 cm^{-1} displaying different ratios of relative intensity for the two samples. It is interesting to note that the mixtures prepared for this work, even after a short time, developed the type of peak splitting of the $\nu_{as}COO^{-}$ absorption in two main bands (centered at 1550/1530 cm^{-1}) that has been observed by other researchers in older paints (Helwig et al. 2014; Osmond 2014, 2019). Though zinc oleate shows the $\nu_{as}COO^{-}$ bands at 1547 and 1527 cm^{-1} (Robinet and Corbeil 2003; Otero et al. 2014) and zinc azelate at 1556 and 1535 cm^{-1}, it's been suggested that in fact the doublet at 1550 and 1530 cm^{-1} may be related to a distortion of the symmetry around the zinc ion (Helwig et al. 2014). Further research on pure, synthetic compounds is needed to understand this aspect and completely exclude a convolution of bands of zinc palmitate with zinc azelate or oleate to explain peak splitting.

Tung oil showed less of a tendency to form FA-Zn, which were visible only as a shoulder (centered at around 1553 cm^{-1}) to the broad POL-Zn band, rather than as a distinct sharp peak, as observed with all other media, even after 10 months in hot and humid conditions (80+RH or 80+RH+H+) (Fig. 9.2).

Similarly, stand oil only developed a small amount of FA-Zn, as evident by a sharp peak at 1549 cm^{-1}, with a shoulder at 1532 cm^{-1} for the 80+RH+H+ sample after 21 days and the 80+RH and 80+RH+H+ after 10 months.

After 10 months the samples of Galkyd aged at RT and those that were briefly heated at 80 °C show approximately the same level of POL-Zn formation and no FA-Zn. Similarly, the 80+RH and 80+RH+H+ samples display roughly equivalent high levels of POL-Zn, with the 80+RH+H+ sample showing slightly higher contribution of the sharp peak at 1550 cm^{-1} attributed to FA-Zn. Overall, these findings demonstrate that alkyd paints are to be considered at risk for the formation of FA-Zn and other zinc carboxylates in the same way as older oil paints, given that they may contain significant proportions of other bio-based oils (such as safflower oil in this case).

9.4 Conclusions

The results of this study advance our knowledge of the reactivity of zinc oxide with an expanded range of drying oils that are commonly encountered in many nineteenth- and twentieth-century paints and shed initial light on some of the parameters that may favor the formation of POL-Zn over FA-Zn.

The experiments and results described in this study showed that 4 days after mixing zinc oxide with various drying oils, no evidence of zinc carboxylate formation is detected with ATR-FTIR at RT. Heating at 80 °C did not have a significant impact on the rate of formation of zinc soaps; however, the detrimental effect of humidity combined with heat was confirmed, even after a short-term exposure of 48 h. It should be noted, though, that the transient exposure to artificial aging conditions took place at the early stages of cure.

Samples subjected to heat and humidity tended to quickly develop higher amounts of POL-Zn even after just 4 days. Over time (as demonstrated by the measurements taken after 10 months) the hydrolysis of the oil medium caused the formation of crystalline zinc soaps such as zinc stearate/palmitate (FA-Zn), at higher rates for the samples that were briefly exposed to hot and humid conditions. FA-Zn have been demonstrated to be potentially detrimental to paintings, given that the crystalline zinc soaps are found in protrusions and at the interface of paints that are delaminating.

Lowering the pH of the hot and humid air in some cases aggravated the process of POL-Zn and FA-Zn formation; in others the effect on zinc carboxylate formation was comparable to heat with high humidity without acidity, so further testing and longer-term study would be needed to fully understand the role of acidity in the formation of zinc soaps.

The viscosity (and initial cross-linking caused by pre-processing) of the oil was shown to matter more than the initial availability of free fatty acids (as indicated by the acid value) to the tendency of forming zinc carboxylates. Viscosity affects the mobility of the polymeric chains, thus slowing down the transport phenomena of metal ions and free fatty acids that have been demonstrated to be at the basis of the formation of detrimental FA-Zn. Obviously, other aspects such as degree of prepolymerization and chemical profiles of constituent fatty acids associated with specific oils also play a role, which should be elucidated further.

From a practical conservation standpoint, this study confirmed the important role of humidity with heat in the formation and evolution of zinc carboxylates. Thus, the impact of lining and consolidation (which may introduce heat and moisture) as well as aqueous cleaning procedures, whereby pH can be modified, should be carefully considered by conservators in their decision-making process, also in terms of the risk factors involving soap formation. Further studies should be conducted to confirm that the phenomena observed for exposures at early stages of cure are also replicable for older and fully cured paints.

Lastly, these qualitative experiments seemed to indicate that the type of industrial processing of the oil plays an important role in regard to soap formation. Of the three heat-processed oils evaluated (polymerized linseed, boiled linseed, and stand oil), the Winsor and Newton stand oil showed the lowest tendency to form both POL-Zn and FA-Zn. Considering that Ripolin paints were formulated with heat-bodied (stand) oils with addition of hard resins, these trends might provide some clues as to the stability of Ripolin paints and may indicate that, below a certain threshold, network-coordinated zinc soaps can, in fact, have a stabilizing effect on historical paints.

Acknowledgments The LabEx MiChem is thanked for offering a visiting professorship to F.C. in 2015 (LabEx MiChem is a part of French state funds managed by the ANR within the Investissements d'Avenir program under reference ANR-11-IDEX-0004-02). At the Art Institute of Chicago, Maria Kokkori is thanked for sharing her knowledge of the early twentieth-century manufacture of paint and Ken Sutherland for providing the oils and for useful discussions on soap formations and oil chemistry. Scientific research at the Art Institute of Chicago is generously supported by grants from the Andrew W. Mellon and Grainger Foundations. Welch, Holme and Clark, Inc., and Alnor are thanked for graciously supplying test samples of their oils.

References

Bailey AE (1951) Industrial oil and fat products. Interscience Publisher, New York/London

Barman S, Vasudevan S (2007) Mixed saturated−unsaturated alkyl-chain assemblies: solid solutions of zinc stearate and zinc oleate. J Phys Chem B 111:5212–5217. https://doi.org/10.1021/jp068675x

Casadio F, Miliani C, Rosi F, Romani A, Anselmi C, Brunetti B, Sgamellotti A, Andral J-L, Gautier G (2013) Scientific investigation of an important corpus of Picasso paintings in Antibes: new insights into technique, condition, and chronological sequence. J Am Inst Conserv 52:184–204. https://doi.org/10.1179/1945233013Y.0000000013

Clementi C, Rosi F, Romani A, Vivani R, Brunetti BG, Miliani C (2012) Photoluminescence properties of zinc oxide in paints: a study of the effect of self-absorption and passivation. Appl Spectrosc 66:1233–1241

Dredge P, Schilling MR, Gautier G, Mazurek J, Learner T, Wuhrer R (2013) Lifting the lids off Ripolin: a collection of paint from Sidney Nolan's studio. J Am Inst Conserv 52:213–226. https://doi.org/10.1179/1945233013Y.0000000011

Erhardt D, Tumosa CS, Mecklenburg MF (2000) Natural and accelerated thermal aging of oil paint films. In: Tradition and innovation, Advances in conservation. IIC, Melbourne, pp 65–69

Fauve M, Vandenmaele J (1964) Interactions between the fatty acids of the oil and zinc oxide in the preparation of white pastes. In: Contributions to the 7th FATIPEC congress in Vichy, Paris, pp 234–244

Gautier G, Bezur A, Muir K, Casadio F, Fiedler I (2009) Chemical fingerprinting of ready-mixed house paints of relevance to artistic production in the first half of the twentieth century. Part I: inorganic and organic pigments. Appl Spectrosc 63:597–603

Helwig K, Poulin J, Corbeil M-C, Moffatt E, Duguay D (2014) Conservation issues in several twentieth-century Canadian oil paintings: the role of zinc carboxylate reaction products. In: Van den Berg KJ, Burnstock A, de Tagle A, de Keijzer M, Heydenreich G, Krueger J, Learner T (eds) Issues in contemporary oil paints. Springer, Cham, pp 167–184

Hermans J (2017) Metal soaps in oil paint: structure, mechanisms and dynamics. University of Amsterdam

Hermans J, Keune K, Van Loon A, Iedema PD (2015) An infrared spectroscopic study of the nature of zinc carboxylates in oil paintings. J Anal At Spectrom 30:1600–1608. https://doi.org/10.1039/C5JA00120J

Hermans J, Keune K, Van Loon A, Iedema PD (2019) Toward a complete molecular model for the formation of metal soaps in oil paints. In: Casadio F, Keune K, Noble P, Van Loon A, Hendriks E, Centeno S, Osmond G (eds) Metal soaps in art: conservation and research. Springer, Cham, pp 47–65

Kokkori M, Casadio F, Sutherland K, Vermeulen M (2014a) Charting the development of oil-based enamel paints through the correlation of historical paint technology manuals with scientific analysis. In: Van den Berg KJ, Burnstock A, de Tagle A, de Keijzer M, Heydenreich G, Krueger J, Learner T (eds) Issues in contemporary oil paints. Springer, Cham, pp 117–125

Kokkori M, Casadio F, Boon J (2014b) A complete study of early 20th-century oil-based enamel paints: integrating industrial technical literature and analytical data. In: ICM-CC preprints of the 17th triennial conference, Melbourne

Kokkori M, Hubert M-O, Balcar N, Barabant G, Sutherland K, Casadio F (2015a) Gloss paints in late paintings by Francis Picabia: a multi-analytical study. Appl Phys A Mater Sci Process 122:1–11. https://doi.org/10.1007/s00339-015-9532-2

Kokkori M, Sutherland K, Boon J, Casadio F, Vermeulen M (2015b) Synergistic use of Py-THM-GCMS, DTMS, and ESI-MS for the characterization of the organic fraction of modern enamel paints. Herit Sci 3:30. https://doi.org/10.1186/s40494-015-0058-x

Lazzari M, Chiantore O (1999) Drying and oxidative degradation of linseed oil. Polym Degrad Stab 65:303–313. https://doi.org/10.1016/S0141-3910(99)00020-8

Mattiello JJ (1943) Protective and Decorative Coatings, vol. 3, ed. J. Mattiello. New York: John Wiley & Sons

McMillan G, Casadio F, Fiedler I, Sorano-Stedman V (2013) An investigation into Kandinsky's use of Ripolin in his paintings after 1930. J Am Inst Conserv 52:258–277. https://doi.org/10.1179/1945233013Y.0000000010

Meilunas RJ, Bentsen JG, Steinberg A (1990) Analysis of aged paint binders by FTIR spectroscopy. Stud Conserv 35:33–51. https://doi.org/10.1179/sic.1990.35.1.33

Muir K, Gautier G, Casadio F, Vila A (2011) Interdisciplinary investigation of early house paints: Picasso, Picabia and their "Ripolin" paintings. In: ICOM-CC preprints of the 16th triennial conference, Lisbon

Muir K, Langley A, Bezur A, Casadio F, Delaney J, Gautier G (2013) Scientifically investigating Picasso's suspected use of Ripolin house paints in Still Life, 1922 and The Red Armchair, 1931. J Am Inst Conserv 52:156–172. https://doi.org/10.1179/1945233013Y.0000000012

Osmond G (2014) Zinc white and the influence of paint composition for stability in oil based media. In: Van den Berg KJ, Burnstock A, de Tagle A, de Keijzer M, Heydenreich G, Krueger J, Learner T (eds) Issues in contemporary oil paints. Springer, Cham, pp 263–281

Osmond G (2019) Zinc soaps: an overview of zinc oxide reactivity and consequences of soap formation in oil-based paintings. In: Casadio F, Keune K, Noble P, Van Loon A, Hendriks E, Centeno S, Osmond G (eds) Metal soaps in art: conservation and research. Springer, Cham, pp 25–43

Osmond G, Boon J, Puskar L, Drennan J (2012) Metal stearate distributions in modern artists' oil paints: surface and cross-sectional investigation of reference paint films using conventional and synchrotron infrared microspectroscopy. Appl Spectrosc 66:1136–1144

Otero V, Sanches D, Montagner C, Vilarigues M, Carlyle L, Lopes JA, Melo MJ (2014) Characterisation of metal carboxylates by Raman and infrared spectroscopy in works of art. J Raman Spectrosc 45:1197–1206

Poli T, Piccirillo A, Zoccali A, Conti C, Nervo M, Chiantore O (2014) The role of zinc white pigment on the degradation of shellac resin in artworks. Polym Degrad Stab 102:138–144. https://doi.org/10.1016/j.polymdegradstab.2014.01.026

Poli T, Piccirillo A, Nervo M, Chiantore O (2019) Aging of natural resins in presence of pigments: metal soaps and oxalates formation. In: Casadio F, Keune K, Noble P, Van Loon A, Hendriks E, Centeno S, Osmond G (eds) Metal soaps in art: conservation and research. Springer, Cham, pp 143–152

Raven LE, Bisschoff M, Leeuwestein M, Geldof M, Hermans J, Stols-Witlox M, Keune K (2019) Delamination due to zinc soap formation in an oil painting by Piet Mondrian (1872–1944). In: Casadio F, Keune K, Noble P, Van Loon A, Hendriks E, Centeno S, Osmond G (eds) Metal soaps in art: conservation and research. Springer, Cham, pp 345–357

Robinet L, Corbeil M-C (2003) The characterization of metal soaps. Stud Conserv 48:23–40. https://doi.org/10.2307/1506821

Rogala D, Lake S, Maines C, Mecklenburg M (2010) Condition problems related to zinc oxide underlayers: examination of selected abstract expressionist paintings from the Collection of the Hirshhorn Museum and Sculpture Garden, Smithsonian institution. J Am Inst Conserv 49:96–113

Uebele C (1913) Paint making and color grinding. The Painters Magazine, New York. https://archive.org/details/cu31924003612268

van der Weerd J, Geldof M, Struik van der Loeff L, Heeren RMA, Boon JJ (2003) Zinc soap aggregate formation in Falling Leaves (les Alyscamps) by Vincent van Gogh. Z Für Kunsttechnol Konserv ZKK 17:407–416

van der Weerd J, Van Loon A, Boon JJ (2005) FTIR studies of the effects of pigments on the aging of oil. Stud Conserv 50:3–22. https://doi.org/10.2307/25487713

Part II
Innovative Approaches to the Characterization of Metal Soaps and Oxalates

Chapter 10
Tracking Metal Oxalates and Carboxylates on Painting Surfaces by Non-invasive Reflection Mid-FTIR Spectroscopy

Francesca Rosi, Laura Cartechini, Letizia Monico, Francesca Gabrieli, Manuela Vagnini, David Buti, Brenda Doherty, Chiara Anselmi, Brunetto Giovanni Brunetti, and Costanza Miliani

Abstract This chapter discusses the potential of reflection mid-FTIR spectroscopy for in situ identification and localization of metal oxalates and carboxylates in paintings. The infrared reflection profiles of the most common metal oxalates and carboxylates are discussed through the direct comparison with reference powders and model samples demonstrating the possibility of distinguishing the nature of the metal counterion. The whole spectral data here discussed represent a unique collection of reference infrared reflection spectra available to conservation scientists for monitoring occurrence and distribution of the most common alteration compounds affecting both ancient and modern paintings. An overview of case studies representing the most relevant situations where metal oxalates and carboxylates have been found in important artworks is presented. This broad vision of the phenomenon, based on the experience matured within the MOLAB transnational access and the ability to probe several points on a painted surface non-invasively, can significantly contribute to progress in the understanding of occurrence, evolution, and origin of such alteration compounds.

F. Rosi (✉) · L. Cartechini · L. Monico · F. Gabrieli · B. Doherty · C. Anselmi · B. G. Brunetti C. Miliani
Istituto CNR di Scienze e Tecnologie Molecolari (CNR-ISTM), Perugia, Italy

Centro di Eccellenza SMAArt, Università degli Studi di Perugia, Perugia, Italy
e-mail: francesca.rosi@cnr.it

M. Vagnini
Associazione Laboratorio di Diagnostica per i Beni Culturali, Spoleto, Italy

D. Buti
CATS-SMK (Centre for Art Technological Studies and Conservation, Statens Museum for Kunst), Copenhagen, Denmark

F. Casadio et al. (eds.), *Metal Soaps in Art*, Cultural Heritage Science,
https://doi.org/10.1007/978-3-319-90617-1_10

10.1 Introduction

Metal oxalates and carboxylates are widespread alteration products affecting both ancient and modern pictorial art as documented by the vast literature dealing with their identification on artworks through micro-destructive methods (such as micro-FTIR, micro-Raman, and ToF-SIMS imaging) (Spring et al. 2008; Nevin et al. 2008; Richardin et al. 2011; Sutherland et al. 2005) and references therein. Nevertheless, there is interest in broadening the knowledge about their occurrence in artworks subjected to different conservation conditions and/or different restoration processes, as well as in detailing their distribution within a specific painting in relationship with the chemical composition of different paints. To this aim, an analytical method satisfying the requirements of non-invasiveness, portability, specificity, and sensitivity to the identification of metal carboxylates and oxalates is needed. Among the currently available non-invasive and portable techniques (Brunetti et al. 2016), reflection mid-FTIR spectroscopy is a highly appreciated analytical tool for the identification of painting materials including not only pigments (Miliani et al. 2012; Buti et al. 2013; Legrand et al. 2014) and binding media (Legrand et al. 2014) but also surface deposits, cleaning residues (Ricci et al. 2006; Melchiorre Di Crescenzo et al. 2013), and alteration products (Monico et al. 2013; Rosi et al. 2016b). The technique provides a fingerprint of the investigated material demonstrating high specificity in the identification of the painting components. The main drawback is linked to the difficulty encountered in spectral interpretation. Reflection FTIR spectra, being originated from surface and volume component of the reflected light, result in complex profiles exhibiting derivative, inverted, and broadened bands with respect to conventional transmission FTIR spectra (Miliani et al. 2012; Buti et al. 2013; Rosi et al. 2016a, b; Ricci et al. 2006; Melchiorre Di Crescenzo et al. 2013; Monico et al. 2013). However, it has been already demonstrated that an extensive investigation of dedicated model paint samples strongly supports and improves the interpretation of reflection spectra recorded on real artworks (Miliani et al. 2012; Buti et al. 2013; Legrand et al. 2014; Ricci et al. 2006; Monico et al. 2013; Rosi et al. 2010a).

This contribution presents the systematic investigation by reflection mid-FTIR spectroscopy for the non-invasive identification and localization of metal oxalates and carboxylates in paintings. The information acquired by the study of reference materials and paint models is here first discussed and then exploited for interpreting reflection mid-FTIR spectra recorded on a wide number of ancient and modern paintings investigated by the mobile laboratory MOLAB within the access activity supported by European (Eu-Artech, CHARISMA, IPERIONCH.eu) and national (IPERIONCH.it) projects. In light of such long experience, a selection of representative case studies is presented emphasizing different situations where oxalates and carboxylates are present.

10.2 Experimental Materials and Methods

10.2.1 Reflection Mid-FTIR Spectroscopy

FTIR spectra were recorded by means of the spectrometer Alpha (Bruker Optics) which is equipped with a reflection module with optical layout of 22°/22°; spectra were acquired from an area of 6 mm in diameter in the range 7000–400 cm^{-1} with 180 scans. The same system is also equipped with an ATR module composed of a diamond crystal which has been used for analyzing the model paints. Part of the in situ FTIR reflection spectra were recorded by the portable spectrophotometer Jasco VIR 9500 which is equipped with chalcogenide glass fiber-optic probe; spectra were acquired from an area of 4 mm in diameter in the range 7000–900 cm^{-1}. With both systems, a spectral resolution of 4 cm^{-1} was used, and spectra are presented as pseudo-absorbance A' = log(1/R) where R = reflectance.

Reflection spectra have not been processed by Kubelka-Munk (KM) or Kramers-Kronig (KK) corrections. In fact, since the two contributions to reflected light (volume and surface reflection) may vary with both the optical properties and the morphological characteristics of the surface, any type of spectral correction is generally not advisable (Griffiths and De Haseth 2007). In paragraph 10.3.1, KK corrections is specifically tested for spectra acquired from oxalate reference compounds analyzed as pressed pellets.

10.2.2 Reference Materials

The following commercial powders of metal oxalates are reported as reference materials: $CaC_2O_4{\cdot}H_2O$ (Acros Organics), CuC_2O_4 (Alfa Aesar), $ZnC_2O_4{\cdot}2H_2O$ (Sigma-Aldrich), and $CaC_2O_4{\cdot}2H_2O$ (synthesized as reported elsewhere (Monico et al. 2013)).

The following mock-ups were prepared and used for the identification of metal carboxylates: lead white [hydrocerussite, $2PbCO_3{\cdot}Pb(OH)_2$] and linseed oil (4:1 by weight, spread on polycarbonate), malachite [$Cu_2CO_3(OH)_2$] and linseed oil (1:1 by weight, spread on glass), zinc white [ZnO] and linseed oil (1:1 by weight, spread on canvas), and ZnO, aluminum mono-stearate [$AlSt(OH)_2$], and oil (1:0.05:0.95, spread on canvas). Touch-dry mock-ups were artificially aged at 40 °C and 95–98% relative humidity (conditions which accelerate the hydrolysis of the lipidic binding media) and monitored by reflection FTIR at sequential steps till the development of stable spectral features ascribable to the formation of metal carboxylates. Since the reactivity of the selected pigments is different, different times of aging were required for the formation of metal carboxylates in each mock-up.

10.2.3 Artworks

The selection of easel and canvas paintings investigated by the MOLAB and cited in the present paper has been made as follows:

Four panel paintings: *Portrait of a Man* (also known as *Trivulzio's portrait*) (1476, Civic Museum of Ancient Art, Torino, IT) and *The Crucifixion* (1475, Royal Museum of fine Arts, Antwerp, BE) by Antonello da Messina and *Christ with Singing and Music-making Angels* (1485, Royal Museum of Fine Arts, Antwerp, BE) and *The Last Judgment* triptych (1465–1471, National Museum, Gdansk, PL) by Hans Memling

Ten canvas paintings: *San Pietro Celestino* (unknown attribution, 18th C., Palazzetto dei Nobili, L'Aquila, IT); *The Crucifixion of Christ* (1855–1893, Ethiopian painting, British Museum, London, UK); *Chemistry* (1909–1916, University Aula, Oslo, NO) and *Puberty* (1894, Munch Museum, Oslo, NO) by Edvard Munch; *Flora* (1925, Palazzo Penna, Perugia, IT) by Gerardo Dottori; *Victory Boogie-Woogie* (1942–1944, Gemeentemuseum, The Hague, NL) by Piet Mondrian; *Satyr, Faun and Centaur with Trident* and the *Still Life with Three Fish, Moray Eel and Lime on White Ground* (1946, Picasso Museum, Antibes, FR) by Pablo Picasso; *Alchemy* (1947, Peggy Guggenheim Collection, Venezia, IT) by Jackson Pollock; and *Bianco 1952* (Albizzini Collection, Città di Castello, IT) by Alberto Burri

10.3 Results

10.3.1 Identification and Localization of Metal Oxalates

Infrared spectra of (hydrate) metal oxalates are characterized by the following vibrational modes: (i) the CO antisymmetric and symmetric stretching mode [ν_{as}(CO) and ν_s(CO)] of the oxalate anion in the wavenumber range 1700–1600 cm^{-1} and 1400–1200 cm^{-1}, respectively (for hydrated compounds, the HOH bending mode also contributes in the ν_{as}(CO) region); (ii) the bending mode δ(OCO) in the range 830–770 cm^{-1}, and (iii) the OH stretching mode(s) [ν(OH) 4000–3000 cm^{-1}] and H_2O libration (770–598 cm^{-1}) (Monico et al. 2013 and references therein).

Although variably distorted by reflection effects, all these features are consistently recognizable in spectra acquired in reflection mode allowing for the distinction of different metal oxalates (Monico et al. 2013), as shown in Fig. 10.1 for Ca-, Cu-, and Zn-oxalate. Metal oxalate layers often show infrared reflection profiles characterized by derivative-like and/or inverted bands, since their morphology (smooth surface) as well as their optical properties (high absorption coefficients) is such to favor the surface component of the reflected light. Reflection profiles entirely due to the surface reflection can be converted to absorption-like spectra through the KK correction (Griffiths and De Haseth 2007). However, the KK algorithm does

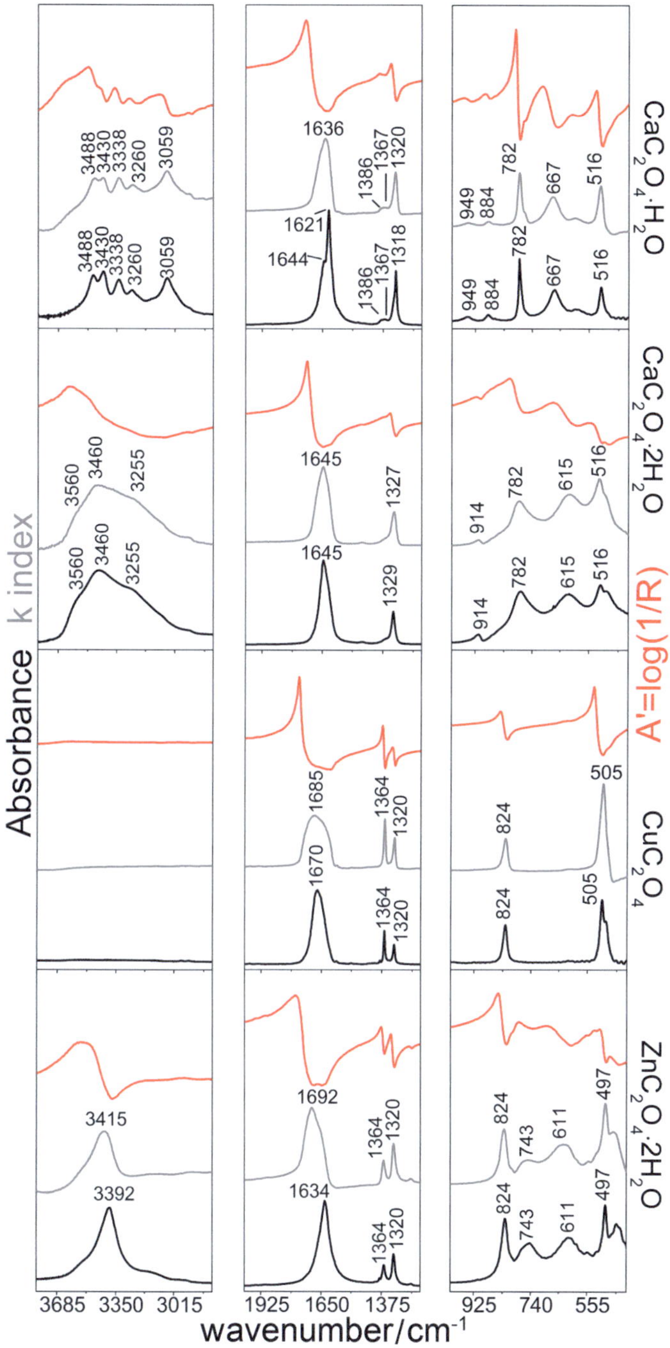

Fig. 10.1 Reflection mode FTIR spectra (red lines) of a selection of metal oxalates acquired from standards (prepared as pressed powder), corresponding *k* index resulting from the application of the KK correction (grey lines) and transmission profiles (black lines). (Modified from Monico et al. 2013)

not accurately correct all modes in spectra as has already been demonstrated in a previous paper (Monico et al. 2013).

Generally, the band shape and the wavenumber position of the ν_s (CO) and δ(OCO) (and water-OH modes for hydrate forms) enable metal oxalate speciation directly from the uncorrected reflection spectra. Furthermore, in the case of a dominant contribution of surface reflection, the application of KK provides for the distinction between different hydration phases (e.g., monohydrate and dihydrate calcium oxalate) (Monico et al. 2013). Conversely, the highly distorted ν_{as} (CO) is poorly corrected by the KK transformation and in any case is rarely considered for diagnostics of metal oxalate, also because it is in a spectral range where generally intense bands of both organic and inorganic components of artworks' materials are present.

From the experience matured in the last 15-year activity of MOLAB, the occurrence of metal oxalates on the surface of polychrome objects conserved indoor as well as outdoor is rather common, with calcium oxalate the one most frequently found. Two different situations have been observed: (i) metal oxalates distributed as uniform layer irrespective of the chemical composition of the paints underneath or (ii) metal oxalates with a peculiar zoned distribution typically related to specific pigments or paint formulations.

One common case encountered in ancient/historical paintings examined in the past 10 years is characterized by the presence of a uniform layer of calcium oxalate covered by one or more layers of restoration varnishes. As representative examples of this situation, Fig. 10.2 reports the oxalate distribution in two paintings by Antonello da Messina: *Trivulzio's portrait* (Fig. 10.2a) and *The Crucifixion* (Fig. 10.2b).

In the painting *Trivulzio's portrait* (analyzed after a conservation treatment aimed at the removal of an oxidized terpenic varnish), mid-FTIR revealed the presence of Ca-oxalate for all the areas measured (Fig. 10.2a, c). The oxalate patina was spread out all over the surface as an opaque and grey layer altering the painting aesthetic (Bellucci et al. 2010). It's worth reporting that, due to the hampering effect of the superficial terpenic varnish, no metal oxalates (or very weak signals) could be detected before the cleaning (Fig. 10.2c).

Similarly, in the painting *The Crucifixion*, which was analyzed after the removal of an external layer of natural varnish, non-invasive mid-FTIR spectroscopy indicated the widespread diffusion of Ca-oxalates (white asterisks in Fig. 10.2b) with the exception of few areas with overpaint and losses.

In the spectra acquired both from *The Crucifixion* (Fig. 10.2c C-01) and the *Trivulzio's portrait* (TR-cleaned), the shape of the ν_s(CO) mode reveals a mixed contribution of the volume and surface reflection resulting in a highly distorted band for which a proper KK correction was not possible. Thus, a differentiation between the monohydrate and the dihydrate forms of Ca-oxalate was not achievable in these two cases.

Another interesting example is the painting *Christ with Singing and Music-making Angels* (1487–1490, by Hans Memling) investigated by MOLAB in 2001 (Van der Snickt et al. 2011), for which non-invasive mid-FTIR spectroscopy

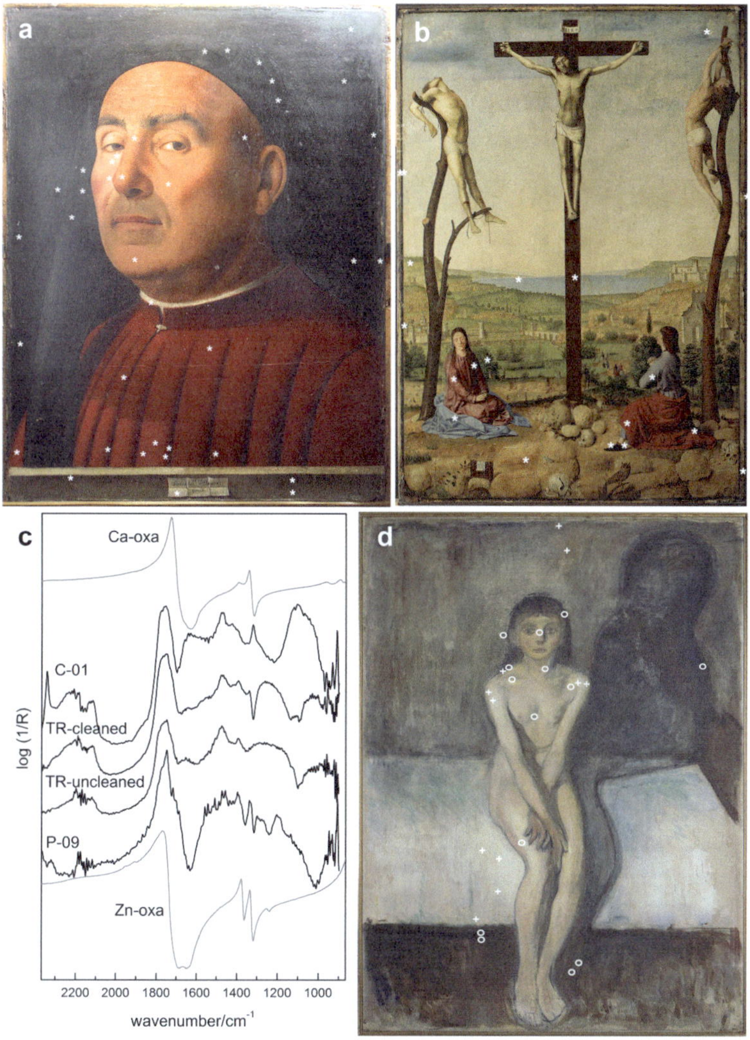

Fig. 10.2 Three case studies showing the presence of metal oxalate patinas: two ancient paintings (**a**) *Trivulzio's portrait* (1476) (Bellucci et al. 2010) and (**b**) *The Crucifixion* (1475) by Antonello da Messina and one modern painting (**d**) *Puberty* (1894) by Edvard Munch (Rosi et al. 2015). (**c**) Reflection mode FTIR spectra (black lines) compared with reference standards of zinc and monohydrated calcium oxalates (grey lines): spectra TR-uncleaned and TR-cleaned recorded on painting (**a**) and spectrum C-01 recorded on painting (**b**) showing example measurements for the identification of hydrated calcium oxalate on the points marked with asterisks in (**a**) and (**b**). Spectrum P-09 is representative of the measurements on painting (**d**) where zinc oxalate has been specifically identified on the measurements points marked with white plusses. On painting (**d**) white circles indicate areas where the acquired spectra show the presence of small amount of oxalates without providing for metal speciation

performed on uncleaned and cleaned areas revealed that a widespread Ca-oxalate patina is formed between two layers of varnish. A detailed investigation of the complex stratigraphy of the Memling's painting is discussed elsewhere in this volume (Klassen et al. 2019). Notably, the study of a micro-sample by ATR micro-FTIR spectroscopy taken from a green area revealed that, below these two varnish layers within which the oxalate patina is sandwiched, there are green paint layers (possibly made of verdigris) characterized by the presence of Cu-oxalates and Cu-carboxylates. These alteration compounds are localized on green areas and were not revealed by non-invasive in situ measurements due to the presence of various varnish layers.

The observation of metal oxalate patinas is not restricted to ancient and historic artworks, but sometimes modern paintings are also characterized by an extensive distribution of metal oxalates. As an example, the oxalate distribution non-invasively detailed in the painting *Puberty* (1894) by Edvard Munch (Rosi et al. 2015) is reported in Fig. 10.2d. Mid-FTIR spectra revealed the presence of oxalates in almost all the measured points (about 40 in total) with the exception of restoration retouches. In many areas, the signals of oxalates were rather weak, and the nature of the metal could not be speculated (white circles in Fig. 10.2d). In other points (denoted by white plus signs in Fig. 10.2d), the typical spectral profile of Zn-oxalate was observed (see Fig. 10.2c where one representative spectrum recorded from *Puberty* – P-09 – is compared with the reference spectrum of zinc oxalate). In *Puberty*, the presence of Zn has been related to the use of ZnO as a restoration material employed in the glue-paste lining treatment applied in 1951 (Rosi et al. 2015). Similarly, in another painting by Munch, *Chemistry (1914–1916)*, for which a similar Zn-based conservation treatment has been documented, Zn-oxalates along with zinc carboxylates have been non-invasively revealed (results not shown; for details see (Monico et al. 2013; Frøysaker 2015; Frøysaker et al. 2015)). The addition of zinc-based compounds (i.e., ZnO) in conservation treatments by some Norwegian restorers has been a common practice for more than 50 years until the end of the 1970s (Frøysaker 2015). In light of these unexpected recent findings concerns have been raised about conservation and restoration of the paintings subjected to this type of previous treatment.

Metal oxalate formation may also occur following different patterns. In what follows, other case studies are presented where the occurrence of metal oxalates is restricted to localized portions of the paintings, generally in correspondence of specific pigments or paint compositions. In the painting *The Last Judgment* (1466–1471) by Hans Memling, non-invasive mid-FTIR revealed that Cu-oxalates were present only in copper-rich green areas, while they were absent in the blue areas containing azurite (also a copper-containing pigment) (Szmelter et al. 2014). This finding is in agreement with literature data showing the higher reactivity with oxalic acid of copper green pigments (verdigris first and then malachite) with respect to azurite (Zoppi et al. 2010).

Observing one panel of Pablo Picasso's triptych *Satyr, Faun and Centaur with Trident* (1946) (Fig. 10.3a) under UV (Fig. 10.3b) (Casadio et al. 2013) shows that two different white pigments (of which only one is luminescent) are used

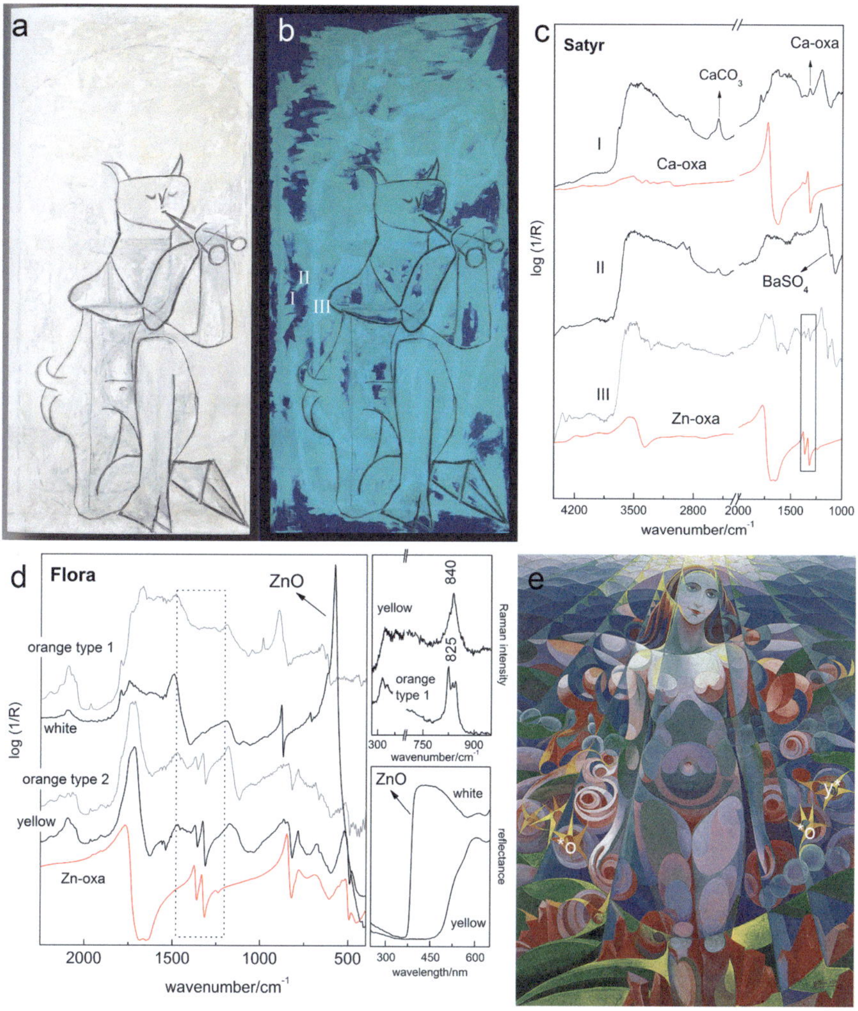

Fig. 10.3 Two case studies showing the presence of metal oxalates in localized areas: (**a**) *Satyr Faun and Centaur with Trident* (1946) by Pablo Picasso along with (**b**) the corresponding UV-induced fluorescence image (Casadio et al. 2013) and (**e**) *Flora* (1925) by Gerardo Dottori (Rosi et al. 2014a). (**c**) Reflection mode FTIR spectra recorded on three different white paints in (**a**) and shown in the fluorescence image (**b**) determined the following composition: (I) ($CaCO_3$ of the exposed white preparation of the fibrocement panel), (II) (lithopone and ZnO), and (III) (ZnO with minor amount of lithopone and strong signals of a lipidic binder). Spectral data show the presence of monohydrated calcium oxalate and zinc oxalate localized in paints I and III, respectively; (**d**) left, reflection mode FTIR spectra recorded on different orange, yellow, and white paints of (**e**) showing Zn-oxalates only in the paints containing the pigments chrome yellow and cadmium zinc sulfide; (**d**) top right, Raman spectra of the same yellow and orange paints where chrome yellow and chrome orange are detected; and (**d**) bottom right, UV-Vis reflection spectra of the same white and yellow paints showing the presence of ZnO only in the white. Reflection mode FTIR spectra of monohydrated calcium oxalate and zinc oxalate references are also shown for comparison (red lines)

for the background, while a third one (with a strong bluish emission) is used to hide a figure which remains underneath (*pentimento*). Non-invasive mid-FTIR spectroscopy in combination with UV-Vis spectroscopy revealed the composition of the three different types of white paints (Casadio et al. 2013). Specifically, (i) the dark blue non fluorescent background (I in Fig. 10.3b) is made of $CaCO_3$ and corresponds to the exposed white preparation of the fibrocement panel; (ii) the white background with the greenish fluorescence (II in Fig. 10.3b) is mainly composed of lithopone white and ZnO in a paint with a relatively low content of binder; and (iii) the paint employed for covering the previous composition (III in Fig. 10.3b) is still made of ZnO, but with minor amount of lithopone, and shows stronger signals attributed to the lipidic binder (as revealed by the corresponding IR bands in the 4500–4100 cm^{-1} range (Rosi et al. 2016a)). Notably, these three white paints manifested different propensities with respect to oxalate formation (Fig. 10.3c): the calcium carbonate background contains some calcium oxalates, and the lithopone-based background has no oxalates which, instead, have markedly formed as zinc oxalates in correspondence of the ZnO-based paint used by Picasso to cover up the *pentimento*.

Flora (1925) a painting executed by the Italian Futurist Gerardo Dottori (Fig. 10.3e) with a rather complex palette shows a similar situation (Rosi et al. 2014a). Very strong FTIR signals of zinc oxalates were detected on the yellows and some of the oranges (Fig. 10.3d left); in most of the other colors, bands ascribable to the presence of this class of compounds were too weak to enable the metal identification; in ZnO-based paints, no Zn-oxalates were present (Fig. 10.3d, the "white" spectrum). Both orange and yellow paints were complex mixtures of four pigments as identified by Raman and UV-Vis reflection spectroscopies (Fig. 10.3d, top and bottom right). Yellow paints resulted mainly composed of chrome yellow ($PbCr_{1-x}S_xO_4$, (Monico et al. 2014)) and a cadmium zinc sulfide pigment ($Cd_{1-x}Zn_xS$). The orange paints contained chrome orange and mercury sulfide and, in some cases, also the two cadmium- and chromium-based yellow pigments.

Excluding ZnO as the source of metal cations promoting the formation of Zn-oxalates (since no evidence of ZnO in the yellow/orange paints has been detected Fig. 10.3d bottom right), the cadmium-based yellow pigment ($Cd_{1-x}Zn_xS$ solid solution) has been identified as a possible source of zinc ions. According to several sources (Bayard and Fiedler 1986; Eastaugh et al. 2004) metal oxalates were added as reagent or additive in the early production of yellow cadmium sulfide pigments. Thus it is reasonable to expect that these residues might have reacted with the Zn-containing reagent used for the production of the pigment leading to zinc oxalates.

Investigation of several paintings demonstrated that the occurrence and formation of metal oxalates follow at least two different patterns: the formation of extensive and homogenous oxalate patinas and localized oxalate alterations. In the first case, the type of oxalate (i.e., the metal cation) is not related to the paint composition itself, but rather it derives from some additional materials that extend to the whole surface of the painting. Hence the formation of such patinas can be ascribed to the application of coating materials (either as original and/or from past restoration interventions) or to deposition from the atmosphere. The second type of occurrence reveals different processes contributing to oxalate formation, which are mainly driven by the specific/intrinsic reactivity of the paint constituting materials possibly exposed to specific environmental conditions.

10.3.2 Identification and Localization of Metal Carboxylates

Polymerized oil binding medium mixed with pigment has the tendency to form an ionomeric network where carboxylic acid functionalities of the polymer are linked to metallic cations (typically divalent lead, zinc, and copper) (Hermans et al. 2019). With aging the presence of these metal ions in the polymerized medium facilitates the progressive triglyceride ester hydrolysis liberating free fatty acids and leading to the formation of metal soaps (Hermans et al. 2019). From literature, microinvasive studies of soap protrusions in oil paintings mostly found the occurrence of metal carboxylate from stearic and palmitic acids (van der Weerd et al. 2005; Salvadó et al. 2009; Cotte et al. 2007; Keune and Boon 2004) which may dramatically affect the paint stability. Within this framework, the use of non-invasive techniques can be helpful to correctly evaluate the entity of the degradation phenomenon, permitting a broad overview of the whole painting surface to be complemented with microanalyses, which have to be ethically limited to few areas of the painting examined.

A systematic investigation of oil paint models artificially aged in order to promote the carboxylate formation was conducted by reflection mid-FTIR spectroscopy. The study of aged model paints (here the pigments lead white, malachite, and zinc oxide are presented) better accounts for the chemical composition of degraded paints after reaction of the pigments with the oil binder to form different metal carboxylates including also those which are not represented by the reference materials usually considered (Mazzeo et al. 2008; Otero et al. 2014). A noticeable example is given by the still debated and unresolved Zn-carboxylate form frequently observed in ZnO-oil paints. In fact, previous FTIR studies on oil paint samples have shown that there are two different types of metal carboxylate species (Hermans et al. 2015; Osmond et al. 2012; Clementi et al. 2013). One is characterized by a sharp carboxylate band matching the reference spectra of Zn palmitate and/or stearate (whose spectra are very similar); the other one shows a broader and up-shifted band with respect to the spectral profiles of the carboxylates expected from the reaction of Zn with the fatty acids resulting from oil hydrolysis (Hermans et al. 2015).

Metal carboxylates can be effectively identified by mid-FTIR spectroscopy by exploiting the relative position of the antisymmetric (ν_{as}) and symmetric (ν_s) stretching bands of the COO^- moiety (Otero et al. 2014; Robinet and Corbeil 2003) in the range 1800–1300 cm^{-1}. These bands are in fact sensitive to the nature of the metal and (to a lesser extent) of the organic acid chain (Otero et al. 2014). Although Raman spectroscopy demonstrated to be more appropriate for identification of the organic acid chain (Otero et al. 2014), the CH rocking modes (at about 700 cm^{-1}) and the CH twisting and bending modes (1300–1100 cm^{-1}) of the hydrocarbon

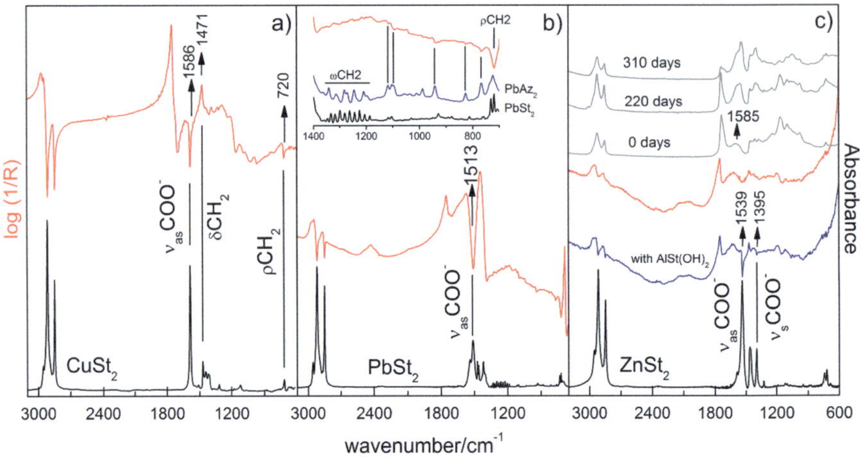

Fig. 10.4 Reflection mode FTIR spectra (red lines) of artificially aged oil model paints prepared with (**a**) malachite (aging time 270 days), (**b**) lead white (aging time 140 days), and (**c**) zinc white with (blue line) and without (red line) aluminum mono-stearate (aging time 30 days). For comparison transmission mode FTIR spectra of the respective metal stearate reference (black lines) are reported. In the inset of (**b**), the transmittance FTIR spectrum of lead azelate (blue line) is also shown. In (**c**) ATR-FTIR spectra of the ZnO paint sample aged at different times (grey lines) are also shown for comparison

chain (Otero et al. 2014; Catalano et al. 2015; Mesubi 1983) can be used for differentiating the fatty acids involved in soap formation.

The capability of FTIR to identify metal carboxylates on samples is retained in reflection mode, although the generally intense COO^- stretching bands tend to be distorted appearing as derivative or inverted peaks. Figure 10.4a–c report the reflection mode spectra (red lines) collected on artificially aged oil model paints prepared with malachite, lead white, and zinc white in comparison with the transmission mode spectra of the respective metal stearate (MSt_2) standards (black lines).

In the case of the malachite-oil model (Fig. 10.4a), it is possible to observe an inverted band at 1586 cm^{-1} matching the ν_{as} of COO^- common to Cu- palmitate, stearate, azelate, and oleate (Otero et al. 2014). The combined presence of the bands at 1471 cm^{-1} (δ CH_2) and at 720 cm^{-1} (ρ CH_2) enables the identification of Cu-stearate and/or palmitate in the aged malachite-oil model (Otero et al. 2014). This example clearly shows that band position in reflection is generally very close to that of the corresponding transmission mode spectrum, but the shape (inverted, absorption- and/or derivative-like) may significantly vary, due to surface optical properties. In fact, the Cu soap formed on the aged malachite-oil model sample shows the $\nu_{as}(COO^-)$ as a sharp inverted band and the CH_2 bending as a sharp absorption-like band.

In the case of the aged lead white oil model (Fig. 10.4b), the newly formed carboxylate shows a relatively broad inverted peak centered at 1513 cm^{-1} assigned to the ν_{as} of COO^- and common to Pb palmitate, stearate, and azelate (Otero et al. 2014). Unfortunately, the strong absorption band of the ν_{as} of CO_3^{-} of the pigment (Brunetti et al. 2016) hinders the ν_s of COO^- and the CH_2 bending modes, which usually enable the identification of the organic acid chain. On the other hand, the presence of a band at 767 cm^{-1} (Otero et al. 2014; Catalano et al. 2015) as well as of additional bands ascribable to the aliphatic chain suggests the preferential formation of lead azelate (inset of Fig. 10.4b).

The zinc white oil paint (Fig. 10.4c), as anticipated, shows a different behavior. In order to prove the possibility to non-invasively identify the two types of carboxylates above discussed, two ZnO-oil paint models were prepared with and without aluminum mono-stearate. This latter is a common jellifying agent added to paint formulations, which has already been demonstrated to strongly promote the formation of Zn soaps with sharp FTIR bands (Osmond et al. 2012; Osmond 2014; Gabrieli et al. 2016). The corresponding spectrum acquired in reflection mode (Fig. 10.4c, blue line) shows two sharp inverted bands at 1539 and 1395 cm^{-1} matching the position of the ν_{as} and ν_s of COO^- in Zn-stearate and/or palmitate as indicated by the comparison with the transmission mode spectrum recorded on $ZnSt_2$ (Fig. 10.4c, black line).

When investigating the ZnO-oil model without $AlSt(OH)_2$ (Fig. 10.4c, red line), no sharp FTIR Zn-carboxylate signals are observed but rather a broad band (also visible in the $AlSt(OH)_2$ containing model) falling in the range 1700–1400 cm^{-1} whose minimum is again matching the ν_{as} of COO^- in Zn- stearate. Additional ATR-FTIR measurements on paint samples artificially aged at different times (Fig. 10.4c, grey lines) confirmed that no sharp Zn-stearate bands are revealed, while the broad band centered at 1585 cm^{-1} is already observed before the aging and persists for long aging time (at least till 310 days). The correct evaluation of this broad band in reflection mode is quite complicated appearing as a broad and mixed inverted/derivative-like peak which may motivate misleading deductions about possible carboxylate detection.

Despite the fact that the literature frequently reports on the identification of lead and copper carboxylates in micro-samples, (Richardin et al. 2011; van der Weerd et al. 2005; Prati et al. 2016), a rather limited number of case studies investigated by MOLAB evidenced their presence. This is probably due to the common practice of applying surface varnishes on historical paintings (that are most likely to contain these types of pigment) which both optically and spectrally hamper the visualization of the carboxylate signals.

In Fig. 10.5 two examples of paintings characterized by the presence of lead and copper carboxylates (stearate and/or palmitate) are reported. The first one is an eighteenth-century painting (*San Pietro Celestino*, unknown attribution, Fig. 10.5a) in which lead soaps have been found in correspondence of the paints composed of lead white. FTIR reflection spectroscopy (Fig. 10.5c, grey lines) enabled the identification of the Pb soaps (specifically as lead stearate) in spite of the presence of a restoration material characterized by long aliphatic chains (i.e., natural wax or

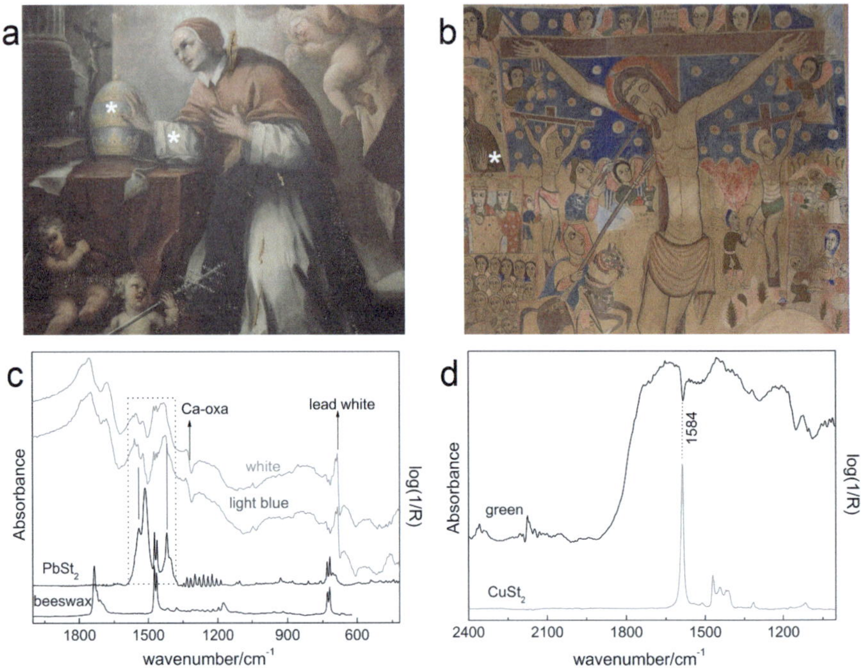

Fig. 10.5 Two case studies representative of ancient paintings showing the presence of metal carboxylates: (**a**) an eighteenth-century panel painting (*San Pietro Celestino*, unknown attribution) and (**b**) an Ethiopian painting on canvas (*The Crucifixion of Christ*, 1855–1893). (**c**) Reflection mode FTIR spectra (grey lines) of two paints of (**a**) containing lead white (marked with white asterisks) and showing the presence of lead soaps, identified as lead stearate and also Ca-oxalates. (**d**) Reflection mode mid-FTIR spectrum recorded on a preserved green area of (**b**) (white asterisk) showing a sharp band at 1584 cm^{-1} diagnostic for Cu soaps. The transmission mode FTIR spectra of Pb- and Cu-stearates and beeswax references are also shown for comparison

paraffin) with FTIR bands overlapping with the vibrational pattern of the aliphatic chain of carboxylates. Notably also Ca-oxalates have been detected.

The second case study is an Ethiopian painting on canvas, *The Crucifixion of Christ* (1855–1893, Fig. 10.5b), in which the original green pigment degraded to a brownish color. Non-invasive reflection mid-FTIR spectra recorded on preserved green areas revealed the presence of Cu-carboxylates evidencing a sharp band at 1584 cm^{-1} (Fig. 10.5d) (Buti et al. 2013; Cutts et al. 2010).

The metal carboxylates most frequently encountered by non-invasive mid-FTIR spectroscopy are those of zinc in modern paintings, probably because of the higher occurrence of unvarnished paintings and the frequent use of zinc white in modern artworks. In this regard, from the MOLAB activity, we may put in evidence additional interesting experiences inherent to carboxylate formation in Zn-based paints for further considerations.

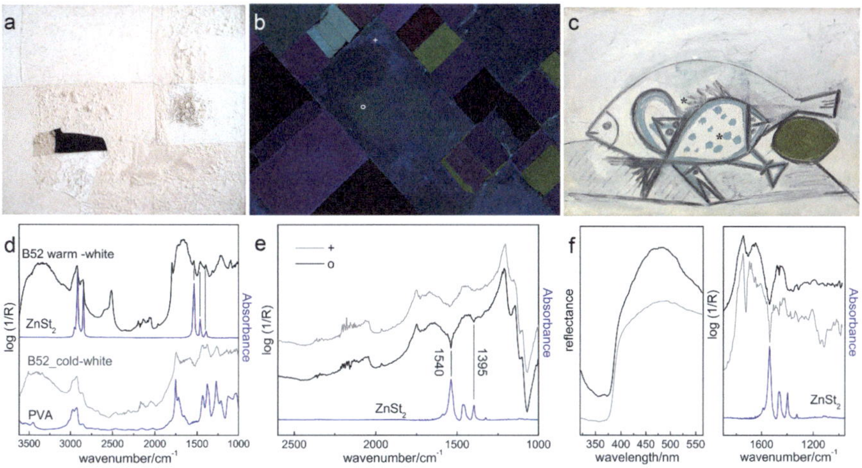

Fig. 10.6 Three case studies from modern art which show different behavior of some white paints with respect to metal soaps' formation: (**a**) *Bianco 1952* (1952) by Alberto Burri (Rosi et al. 2010b), (**b**) detail of UV-induced fluorescence image of *Victory Boogie-Woogie* (1944) by Piet Mondrian (Miliani et al. 2008), and (**c**) *Still Life with Three Fish, Moray Eel and Lime on White Ground* (1946) by Pablo Picasso (Casadio et al. 2013). (**d**) Reflection mode FTIR spectra collected on two different white paints of (a), one is cold white and the other warm white, both containing ZnO but with different binders: a vinyl-based resin and oil, respectively. (**e**) Reflection mode FTIR spectra of two white paints of (**b**) showing different emission properties at the UV light (white cross and circle in (**b**)). Only the white paint showing the green fluorescence of ZnO in (**b**) was found to contain Zn soaps (black line), while the other white paint (grey line) made of TiO_2 and lithopone has no carboxylates. (**f**) Left side: UV-Vis reflection spectra of two white paints of (**c**) (black asterisks) both showing the presence of ZnO; (**f**) right side, corresponding FTIR reflection mode spectra (black and grey lines) of which only one (collected on the smaller fish) exhibits the presence of Zn soaps. The transmission mode FTIR spectra of Zn-stearate and PVA references (blue lines) are also shown for comparison

The painting *Bianco 1952* (1952) by Alberto Burri (Fig. 10.6a) has been the subject of extensive investigation (Rosi et al. 2010b, 2014b).

The monochromaticity of the painting is modulated by the presence of two different tonalities of white: one brilliant bright and cold and the other rather opaque and warm. In both white hues, the main white pigment is zinc white (ZnO) along with additional minor compounds such as calcite ($CaCO_3$), gypsum ($CaSO_4{\cdot}2H_2O$), and barite ($BaSO_4$). The artist used a vinylic binder (PVA) for the cold white areas and a lipidic one for the warm white. Accordingly, zinc white gave evident zinc soap formations in the warm white (Fig. 10.6d, black line compared with the transmission mode spectrum of $ZnSt_2$, reference material) and not in the cold white (Fig. 10.6d, grey line compared with a reference spectrum of PVA acquired in reflection mode). It is still unsure if the difference in tonality of the white areas was the artist's deliberate choice, achieved by different combinations of binding media, or rather the unexpected result of the progressive degradation of the oil paint.

A detail in UV light of *Victory Boogie-Woogie* (1944) by Piet Mondrian (Fig. 10.6b) shows heterogeneous emission properties due to the contributions of different white pigments: ZnO, TiO_2, and lithopone (Miliani et al. 2008). ZnO resulted locally applied only in the green-yellow fluorescing areas (Miliani et al. 2008) where mid-FTIR spectroscopy revealed the presence of zinc soaps (Fig. 10.6e, black line). No soaps were, instead, observed in the other whites composed of TiO_2 and lithopone (Fig. 10.6e, grey line).

The painting *Still Life with Three Fish, Moray Eel and Lime on White Ground* (1946, Fig. 10.6c) by Pablo Picasso demonstrates that metal soap formation is strongly dependent not only on the nature of the pigment but also on the paint tube formulation – in terms of pigments' mixture, extenders, fillers, and additives. Here ZnO was identified as principal white pigment in all the white areas by UV-Vis reflection spectroscopy (Fig. 10.6f left) thanks to the typical sigmoidal shape of this semiconductor pigment positioned at about 380 nm. Notably, the mid-FTIR spectra recorded in the same two points as the UV-Vis revealed the presence of Zn-stearates only for one of the two measured white areas (Fig. 10.6f right, grey line), while for the other one, the broad band of Zn carboxylates (Fig. 10.6f right, black line) was, instead, observed. The localized occurrence of Zn soaps in only some of the white paints, all of which composed of ZnO, can be ascribed to different formulations of the zinc white paint tubes used by Picasso when painting the *Still Life with Three Fish, Moray Eel and Lime on White Ground*. In fact, according to the mid-FTIR study on model paints, the presence of paint additives, as the case of Al stearate, strongly affects the stability of ZnO-oil paints toward soap formation (Osmond 2014; Gabrieli et al. 2016). Moreover, the possibility that zinc stearate was included as an additive itself in the specific paint formulation can also be considered.

Recent analysis of Jackson Pollock's *Alchemy* (1947) (Rosi et al. 2016b; Gabrieli et al. 2016) with mid-FTIR reflection spectroscopy indicated the presence of zinc soaps (stearate and/or palmitate) as well as Zn-oxalates in correspondence with the thick white (made of ZnO and TiO_2), phthalocyanine blue, and cobalt phosphate violet paints (Fig. 10.7a–c).

Interestingly, investigating the drips of thick white paint running from the front to the edge of the painting (Fig. 10.7d) by FTIR analysis, it was revealed that zinc soaps are mainly localized at the painting front surface observing strong and sharp inverted band at 1539 cm^{-1} (Fig. 10.7e). No zinc carboxylates are observed at the edge of the painting (Fig. 10.7e) which is protected by a frame. No preferential distribution of Zn-oxalates was observed between the exposed and unexposed white paints, which show no evident variation in intensity for the FTIR bands characteristic of Zn-oxalates (Fig. 10.7e).

In this last example, the paint composition is not the discriminating factor for localized soap formation/aggregation. Most probably in the case of *Alchemy*, the preferential localization of zinc soaps at the surface of exposed drips was promoted by different concomitant factors: the high relative humidity provided by the environment (since 1948 the painting has been conserved at Palazzo Venier in Venice) and a discontinuous preparation layer of the canvas (very poor or totally

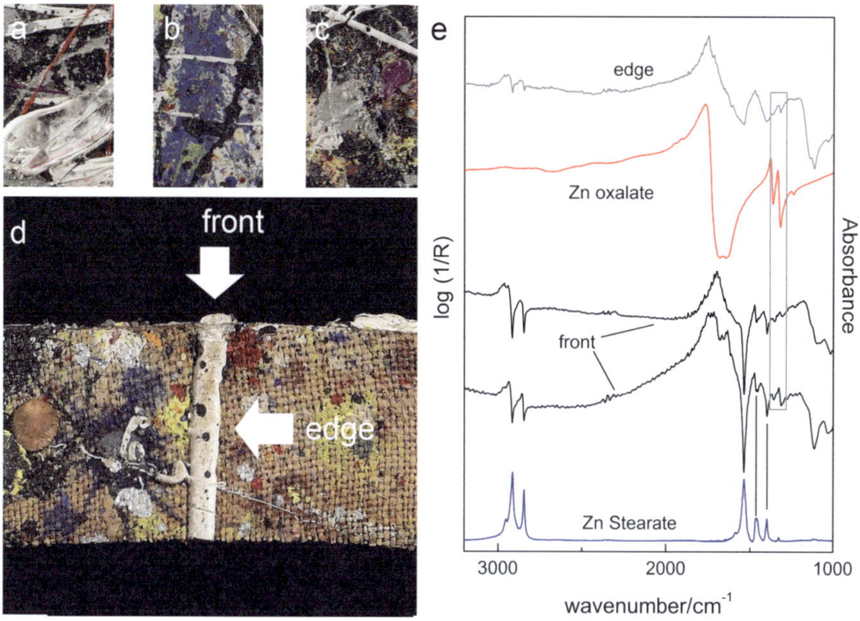

Fig. 10.7 (**a–c**) Details of some white, blue, and violet paints of *Alchemy* (1947) by Jackson Pollock that have all been found to contain zinc soaps (Rosi et al. 2016b). (**d**) A detail of the painting's edge where a thick drip of white paint containing ZnO runs laterally from the front. (**e**) Reflection mode FTIR spectra collected on the white drip on the front (black line) and the edge (grey line) compared with a Zn-oxalate reference (red line). The transmission mode FTIR spectra of Zn-stearate (blue line) is also shown for comparison

absent at the border) which may have locally affected the natural exchange of the paint with the environment and the diffusion/aggregation of zinc soaps.

10.4 Conclusion

The systematic investigation of paint model samples, complemented by a wide number of in situ studies of historical paintings summarized here, clearly demonstrates that reflection mid-FTIR spectroscopy is a powerful tool to carry out the characterization of metal oxalates and carboxylates in a non-invasive way. In fact, despite the limits in spectral interpretation of spectra, reflection mid-FTIR spectroscopy demonstrated to be able to provide sensitive and specific recognition of these two classes of degradation compounds typically affecting painting materials. The study of paint models allowed spectral features to be distinguished, useful for identification of the metal ion involved in the formation of oxalates (also distinguishing hydrated forms) and carboxylates. Furthermore, for this latter class of compounds, the technique is sensitive to the type of fatty acid which has reacted.

The experience gained in more than 10 years of MOLAB activity across Europe allowed us to collect a significant number of case studies to draw some general considerations about the form and the extent by which oxalates and carboxylates manifest themselves on painting surfaces.

Oxalates may be present as extensive patina often at the interface between the paint surface and the varnish or between different layers of varnish, suggesting a degradation phenomenon generalized to the entire surface, irrespective of the paint composition, and ascribable to different accidental or natural causes, as discussed in the chapter.

In some other specific cases, both oxalates and carboxylates form with preferential localizations as a result of specific paint formulations and/or environmental conditions which promote paint reactivity and thus local decay.

It follows that the non-invasive evaluation of the chemical nature and the distribution of metal oxalates and carboxylate is of crucial importance for monitoring the state of conservation of an artwork and addressing proper conservation procedures.

Acknowledgments This work was carried out through the support by the Laboratorio di Diagnostica di Spoleto and by the European Commission, through the project IIPERIONCH, H2020-INFRAIA-2014-2015 (Grant No. 654028).

References

Bayard MA, Fiedler I (1986) Cadmium yellow, orange and red. In: Feller RL (ed) Artist's pigments, A handbook of their history and characteristics, vol 1. Cambridge University Press, Cambridge, pp 65–108

Bellucci R, Bonanni P, Brunetti GB (2010) Il restauro del Ritratto Trivulzio di Antonello da Messina. In: OPD Restauro, vol 22. Centro Di Della Edifimi SRL, pp 15–54

Brunetti B, Miliani C, Rosi F, Doherty B, Monico L, Romani A, Sgamellotti A (2016) Non-invasive investigations of paintings by portable instrumentation: the MOLAB experience. Top Curr Chem (Z) 374:10–35

Buti D, Rosi F, Brunetti BG, Miliani C (2013) In-situ identification of copper-based green pigments on paintings and manuscripts by reflection FTIR. Anal Bioanal Chem 405: 2699–2711

Casadio F, Miliani C, Rosi F, Romani A, Anselmi C, Brunetti B, Sgamellotti A, Andral JL, Gautier G (2013) Scientific investigation of an important corpus of Picasso paintings in Antibes: new insights into technique, condition, and chronological sequence. J Am Inst Conserv 52:184–204

Catalano J, Murphy A, Yao Y, Yap GPA, Zumbulyadis N, Centeno SA, Dybowski C (2015) Coordination geometry of lead carboxylates-spectroscopic and crystallographic evidence. Dalton Trans 44:2340–2347

Clementi C, Rosi F, Romani A, Miliani C (2013) Photoluminescence properties of zinc oxide in paints: a study of the effect of self-absorption and passivation. Appl Spectrosc 66:1233–1241

Cotte M, Checroun E, Susini J, Walter P (2007) Micro-analytical study of interactions between oil and lead compounds in paintings. Appl Phys A Mater Sci Process 89:841–848

Cutts H, Harrison L, Higgitt C, Cruickshank P (2010) The image revealed: study and conservation of mid-nineteenth-century Ethiopian church painting. Brit Mus Tech Res Bull 4:1–17

Eastaugh N, Walsh V, Chaplin T, Siddall R (2004) Pigment compendium: a dictionary of historical pigments. Elsevier Butterworth-Heinemann, Oxford

Frøysaker T (2015) Unintended contamination? A selection of Munch's paintings with non-original zinc white. In: Frøysaker T, Streeton N, Kutzke H, Hanssen-Bauer F, Topalova-Casadiego B (eds) Public paintings by Edward Much and his contemporaries. Change and conservation challenges. Archetype Pubblication, London, pp 132–139

Frøysaker T, Miliani C, Grøntoft T, Kleiva I (2015) Monitoring surface blackening and zinc reaction products on prepared samples located adjacent to Munch's the source in the aula at the University of Oslo. In: Frøysaker T, Streeton N, Kutzke H, Hanssen-Bauer F, Topalova-Casadiego B (eds) Public paintings by Edward Much and his contemporaries. Change and conservation challenges. Archetype Pubblication, London, pp 126–131

Gabrieli F, Rosi F, Vichi A, Cartechini L, Pensabene L, Kazarian SG, Miliani C (2016) Revealing the nature and distribution of metal carboxylates in Jackson Pollock's Alchemy (1947) by micro attenuated total reflection FTIR spectroscopic imaging. Anal Chem. 89(2):1283–1289. https://doi.org/10.1021/acs.analchem.6b04065

Griffiths P, De Haseth JA (2007) Fourier transform infrared spectrometry, 2nd edn. Wiley, New York

Hermans JJ, Keune K, Van Loon A, Iedema PD (2015) An infrared spectroscopic study of the nature of zinc carboxylates in oil paintings. J Anal At Spectrom 30:1600–1608

Hermans JJ, Keune K, Van Loon A, Iedema PD (2019) Toward a complete molecular model for the formation of metal soaps in oil paints. In: Casadio F, Keune K, Noble P, Van Loon A, Hendriks E, Centeno S, Osmond G (eds) Metal soaps in art: conservation and research. Springer, Cham, pp 47–65

Keune K, Boon JJ (2004) Imaging secondary ion mass spectrometry of a paint cross section taken from an early Netherlandish painting by Rogier van der Weyden. Anal Chem 76: 1374–1385

Klaassen L, Van der Snickt G, Legrand S, Higgitt C, Spring M, Vanmeert F, Rosi F, Brunetti BG, Postec M, Janssens K (2019) Characterization and removal of a disfiguring oxalate crust on a large altarpiece by Hans Memling. In: Casadio F, Keune K, Noble P, Van Loon A, Hendriks E, Centeno S, Osmond G (eds) Metal soaps in art: conservation and research. Springer, Cham, pp 265–282

Legrand S, Alfeld M, Vanmeert F, De Nolf W, Janssens K (2014) Macroscopic reflection Fourier transformed mid-infrared (MA-rFTIR) scanning, a new technique for in situ imaging of painted cultural artefacts. Analyst 139:2489–2498

Mazzeo R, Prati S, Quaranta M, Joseph E, Kendix E, Galeotti M (2008) Attenuated total reflection micro FTIR characterisation of pigment–binder interaction in reconstructed paint films. Anal Bioanal Chem 392:65–76

Melchiorre Di Crescenzo M, Zendri E, Rosi F, Miliani C (2013) A preliminary FTIR-based exploration of the surfactant phase separation process in contemporary mural paintings. e-Prev Sci 10:10–18

Mesubi MA (1983) An infrared study of zinc, cadmium, and lead salts of some fatty acids. J Mol Struct 81:61–71

Miliani C, Kahrim K, Brunetti BG, Sgamellotti A, Aldrovandi A, van Bommel M, van den Berg KJ, Janssen H (2008) MOLAB, a mobile facility suitable for non-invasive in-situ investigations of early and contemporary paintings: the case-study of Victory Boogie Woogie (1942-1944) by Piet Mondrian, Preprints of the ICOM-CC 15th triennial meeting, Delhi, 15:857–864

Miliani C, Rosi F, Daveri A, Brunetti BG (2012) Reflection infrared spectroscopy for the non-invasive in situ study of artists' pigments. Appl Phys A Mater Sci Process 106:295–307

Monico L, Rosi F, Miliani C, Daver A, Brunetti B (2013) Non-invasive identification of metal-oxalate complexes on polychrome artwork surfaces by reflection mid-infrared spectroscopy. Spectrochim Acta A 116:270–280

Monico L, Janssens K, Hendriks E, Brunetti BG, Miliani C (2014) Raman study of different crystalline forms of $PbCrO_4$and $PbCr_{1-x}S_xO_4$ solid solutions for the non-invasive identification of chrome yellows in paintings: a focus on works by Vincent van Gogh. J Raman Spectrosc 45:1034–1045

Nevin A, Melia JL, Osticioli I, Gautier G, Colombini MP (2008) The identification of copper oxalates in a 16th century Cypriot exterior wall painting using micro FTIR, micro Raman spectroscopy and gas chromatography-mass spectrometry. J Cult Herit 9(2): 154–161

Osmond G (2014) Zinc white and the influence of paint composition for stability in oil based media. In: Van den Berg KJ, Burnstock A, de Tagle A, de Keijzer M, Heydenreich G, Krueger J, Learner T (eds) Issues in contemporary oil paints. Springer, Cham, pp 263–281

Osmond G, Boon JJ, Puskar L, Drennan J (2012) Metal stearate distributions in modern artists' oil paints: surface and cross-sectional investigation of reference paint films using conventional and synchrotron infrared microspectroscopy. Appl Spectrosc 66:1136–1144

Otero V, Sanches D, Montagner C, Vilarigues M, Carlyle L, Lopes JA, Melo MJ (2014) Characterisation of metal carboxylates by Raman and infrared spectroscopy in works of art. J Raman Spectrosc 45:1197–1206

Prati S, Bonacini I, Sciutto G, Genty-Vincent A, Cotte M, Eveno M, Menu M, Mazzeo R (2016) ATR-FTIR microscopy in mapping mode for the study of verdigris and its secondary products. Appl Phys A Mater Sci Process 122:10. https://doi.org/10.1007/s00339-015-9519-z

Ricci C, Miliani C, Brunetti BG, Sgamellotti A (2006) Non-invasive identification of surface materials on marble artifacts with fiber optic mid-FTIR reflectance spectroscopy. Talanta 69:1221–1226

Richardin P, Mazel V, Walter P, Leprevote O, Brunelle A (2011) Identification of different copper green pigments in renaissance paintings by cluster-TOF-SIMS imaging analysis. J Am Soc Mass Spectrom 22:1729–1736

Robinet L, Corbeil MC (2003) The characterization of metal soaps. Stud Conserv 48:23–40

Rosi F, Daveri A, Doherty B, Nazzareni S, Brunetti B, Sgamellotti A, Miliani C (2010a) On the use of overtone and combination bands for the analysis of the $CaSO_4$–H_2O system by mid-infrared reflection spectroscopy. Appl Spectrosc 64:956–963

Rosi F, Miliani C, Clementi C, Kahrim K, Presciutti F, Vagnini M, Daveri A, Cartechini L, Brunetti BG, Sgamellotti A (2010b) An integrated spectroscopic approach for the non-invasive study of modern art materials and techniques. Appl Phys A Mater Sci Process 100:613–624

Rosi F, Patti M, Cartechini L et al (2014a) Designs and colors. Gerardo Dottori through non invasive investigations. In: Duranti M, Baffoni A, Duranti F (eds) Gerardo Dottori the futurist view. Estorick Collection of Modern Italian Art, London, pp 33–45

Rosi F, De Cesare G, Iazurlo P, Daveri A, Vagnini M, Valentini F, Basile G (2014b) The Burri project: researches for technique and conservation. In: Sgamellotti A, Brunetti GB, Miliani C (eds) Science and art the painted surface Royal Society in chemistry. RSC, Cambridge, pp 499–521

Rosi F, Cartechini L, Romani A et al (2015) Non-invasive investigations of the materials and painting technique of Puberty (1894), anxiety (1894) and Vapire (1895) by Edward Munch. In: Frøysaker T, Streeton N, Kutzke H, Hanssen-Bauer F, Topalova-Casadiego B (eds) Public paintings by Edward Much and his contemporaries. Change and conservation challenges. Archetype Pubblication, London, pp 84–91

Rosi F, Daveri A, Moretti P, Brunetti B, Milliani C (2016a) Interpretation of mid and near-infrared reflection properties of synthetic polymer paints for the non-invasive assessment of binding media in twentieth-century pictorial artworks. Microchem J 124:898–908

Rosi F, Grazia C, Fontana R, Gabrieli F, Pensabene Buemi L, Pampaloni E, Romani A, Stringari C, Miliani C (2016b) Disclosing Jackson Pollock's palette in Alchemy (1947) by non-invasive spectroscopies. Herit Sci 4:18–31

Salvadó N, Butí S, Nicholson J et al (2009) Identification of reaction compounds in micrometric layers from gothic paintings using combined SR-XRD and SR-FTIR. Talanta 79:419–428

Spring M, Ricci C, Peggie D, Kazarian S (2008) ATR-FTIR imaging for the analysis of organic materials in paint cross sections: case studies on paint samples from the National Gallery, London. Anal Bioanal Chem 392:37–45

Sutherland K, Price B, Passeri I, Tucker M (2005) A study of the materials of Pontormo's "portrait of Alessandro de' Medici" Materials Research Society Symposium Proceedings 852, #OO2.2.1

Szmelter I, Cartechini L, Romani A, Pezzati L (2014) Multi-criterial studies of the masterpiece 'The Last Judgement' attributed to Hans Memling, at the National Museum of Gdańsk (2010–2013). In: Sgamellotti A, Brunetti GB, Miliani C (eds) Science and art the painted surface. RSC, Cambridge, pp 230–251

Van der Snickt G, Miliani C, Janssens K, Brunetti B, Romani A, Rosi F, Walter P, Castaing J, De Nolf W, Klaassen L, Labarque I, Wittermann R (2011) Material analyses of 'Christ with singing and music-making angels', a late 15th-C panel painting attributed to Hans Memling and assistants: part I. Non-invasive in situ investigations. J Anal At Spectrom 26:2216–2229

van der Weerd J, Van Loon A, Boon JJ (2005) FT IR studies of the effect of pigments on the aging of oil. Stud Conserv 50:3–22

Zoppi A, Lofrumento C, Mendes NFC, Castellucci EM (2010) Metal oxalates in paints: a Raman investigation on the relative reactivities of different pigments to oxalic acid solutions. Anal Bioanal Chem 397:841–849

Chapter 11
Identification and Distribution of Metal Soaps and Oxalates in Oil and Tempera Paint Layers in Fifteenth-Century Altarpieces Using Synchrotron Radiation Techniques

Nati Salvadó, Salvador Butí, Trinitat Pradell, Victòria Beltran, Gianfelice Cinque, and Jordi Juanhuix

Abstract The formation and distribution of metal soaps produced as a result of the reactivity and aging of the materials in a fifteenth-century egg tempera and oil paintings on wood are presented. The painting technique involves the application of several paint layers over a ground using, sometimes in the same paint layer sequence, drying oil and egg yolk binders. We show, with a selection of examples, how the use of thin sections and a combination of various micro-sensitive analytical techniques is adequate to obtain the high-quality data necessary for the unambiguous identification of metal soaps and metal oxalates as well as their distribution in the paint layers. The techniques include micro infrared spectroscopy (μSR-FTIR) and micro X-ray diffraction (μSR-XRD) with synchrotron radiation, optical microscopy (OM), and scanning electron microscopy with energy dispersive X-ray spectroscopy (SEM-EDS). The data obtained sheds light about the underlying reaction and aging mechanisms happening in each paint layer and among them. This helps to define the state of conservation of the artworks.

N. Salvadó (✉) · S. Butí · V. Beltran
Departamento d'Enginyeria Química, EPSEVG, and Center for Research in Nano-Engineering, Universitat Politècnica de Catalunya, Barcelona, Spain
e-mail: nativitat.salvado@upc.edu

T. Pradell
Center for Research in Nano-Engineering and Departament de Física, Universitat Politècnica de Catalunya, Barcelona, Spain

G. Cinque
Diamond Light Source, Oxfordshire, UK

J. Juanhuix
CELLS-ALBA Synchrotron, Cerdanyola del Vallès, Barcelona, Spain

F. Casadio et al. (eds.), *Metal Soaps in Art*, Cultural Heritage Science,
https://doi.org/10.1007/978-3-319-90617-1_11

Keywords μSR-FTIR · μSR-XRD · Egg tempera · Oil · 15th century · Metal soap · Metal oxalate

11.1 Introduction

The simultaneous presence of free fatty acids (a carboxylic acid with a long aliphatic chain) in the binding media and of metal ions (lead, calcium, aluminum, or copper, among others) in the pigments and ground layers results in the formation of long-chain metal carboxylates, known as metal soaps. Most paintings show a presence of metal soaps to some degree. In the last few years, the identification of metal soaps, that is, the determination of their molecular composition and crystal structures (as well as physical and chemical properties), has increased significantly (Catalano et al. 2014, 2015; Corbeil and Robinet 2002; Cotte et al. 2016; Helwig et al. 2014; Hermans et al. 2014, 2015, 2016; Martínez-Casado et al. 2014; Robinet and Corbeil 2003). This increased interest is principally due to conservation issues as these substances may produce efflorescence, spots, and protrusions and may cause the paint layers to separate or occasionally detachment, in particular for zinc and, in a lesser extent, lead soaps (Cotte et al. 2007, 2016; Higgitt et al. 2003; Keune and Boon 2007; Keune et al. 2011, 2016; Osmond et al. 2012; Sawicka et al. 2014; Salvadó et al. 2009). Nevertheless, in most of the cases, the main consequence of these soaps is a modification in color and texture and reduced opacity giving the artwork an aged aspect. Sometimes the presence of oxalates of the same metals has also been determined (Salvadó et al. 2005; Sotiropoulou et al. 2016).

One of the most important European collections of medieval paintings on wood (panels, altarpieces, wood carvings, and sculptures) in quantity, technical quality, and stylistic interest can be found in Catalonia. Among them, we focus on important fifteenth-century artworks painted with different materials and techniques (egg or glue tempera and oil) showing a variable state of conservation. Over the centuries, the reactivity between the original organic and inorganic materials in the paint with the atmosphere (weathering) and the natural aging of the compounds have affected the physical properties of the artwork (optical and mechanical properties).

Soaps are found both in egg tempera (Meilunas et al. 1990; Salvadó et al. 2005) and oil paintings (Higgitt et al. 2003; Keune and Boon 2004; Meilunas et al. 1990; Salvadó et al. 2005). This is due to the capability of both binders to produce free fatty acids. The most commonly used drying oils since medieval times are linseed and walnut oils. Linseed oil contains mainly triglycerides made up of about 70% of two polyunsaturated fatty acids, linoleic C18:2 and linolenic C18:3 acids, as well as ~20% monounsaturated oleic C18:1 acid. They also contain about 10% of saturated fatty acids, palmitic C16:0 and stearic C18:0. For these oils drying and aging involve the polymerization by uptake of oxygen of the unsaturated fatty acids, as well as the hydrolysis of the original ester bonds of the triglycerides. Saturated palmitic and stearic fatty acids are not involved in the cross-linking process, but in the triglyceride hydrolysis process. As a result, free fatty acids are produced (Bayrak et al. 2010; Erhardt et al. 2005; Mills and White 1994; Popa et al. 2012; de Viguerie et al. 2016).

Egg yolk initially dries up by evaporation of water. Dried egg yolk contains about 60% lipids and 33% proteins, each contributing to different drying and aging processes in the drying oil. The lipids, triglycerides, and phospholipids contain unsaturated acids (~40% oleic C18:1, ~17% polyunsaturated, mainly linoleic C18:2) and 33% saturated acids, mainly palmitic C16:0 (~28%) and stearic C18:0 (~5%). The lipids may provide some flexibility to the egg yolk paint. The aging of the lipids in the egg yolk is similar to those of drying oils, although there is less polymerization because there are fewer polyunsaturated fatty acids. Therefore, similar free fatty acids to those produced in drying oil are expected to occur in egg yolk. There is a higher saturated fatty acids proportion in egg yolk (Newman 1998; USDA National Nutrient Database for Standard References Legacy Release 2018).

This study focuses on the formation of metal soaps from free fatty acids present in drying oil or from egg yolk. This includes egg tempera and oil painting techniques commonly used during the Gothic period. The painting technique involves the application of several paint layers over a ground (gypsum/animal glue) using, sometimes in the same paint layer sequence, drying oil and egg yolk binders. Consequently, not only do the presence and nature of the soaps and other compounds in the paint layers need to be determined but also their distribution. The detection and characterization of metal soaps among the other compounds present in paint layers require a combination of various micro-sensitive analytical techniques. One essential requirement to obtain precise information on the composition and distribution of these reaction products across the painted submillimetric layers is the use of a very small beam (a few μm), attainable using synchrotron light. Good discrimination capability is the key to the correct interpretation of the data. In particular, μSR-FTIR together with μSR-XRD constitutes an excellent instrumental ensemble for the analysis of metal soaps (Salvadó et al. 2009). Their nondestructive nature is also a consideration for the selection of these techniques. Moreover, different sample preparation strategies are also required. In particular, the combination of thin-sections of the samples with synchrotron-based techniques has proven to be one of the most useful associations for paintings analysis (Beltran et al. 2015; Pouyet et al. 2015). The thin-sections are essential when working in transmission geometry, which provides maximum resolution and data quality. We show, with a selection of examples, how the use of thin-sections, prepared using a methodology developed by our group (Beltran et al. 2015) that is optimal for transmission μSR-FTIR and μSR-XRD analysis, is adequate to obtain high-quality data.

11.2 Methodology

IR analysis provides molecular specificity and structural information for the different compounds present in the paint layers. A high spectral quality (signal/noise ratio) at the highest spatial resolution (diffraction limited) can be achieved thanks to the high brilliance and collimation of synchrotron light. Synchrotron-based FTIR is nowadays as stable as globar. In addition, it allows one to work in confocal mode conveniently between 5 and 15 μm, i.e., accessing the true micron scale which is important in our studies to reveal both the chemical species and spatial distribution, with no compromise on the low wavenumber spectral range (an effect typical when closing slits with benchtop). Thus, μSR-FTIR is particularly useful due to its capability of mapping on a micrometric scale. In order to differentiate characteristic metal soap bands found between 1200 and 1800 cm^{-1}, $\nu_{as}(COO^-)$, $\delta(CH_2)$, and $\nu_s(COO^-)$, from the other compounds present in paints, we need maximum discrimination capability. Moreover, the same metals are also known to produce oxalates with strong absorption bands related to the carbonyl stretching $\nu(C{=}O)$ in the same region.

The setup used included a Hyperion 3000 IR microscope, a broadband high-sensitivity MCT detector, and a 36x condenser coupled to the Vertex 80v interferometer available at the end station of beamline MIRIAM B22 at Diamond Light Source, UK (Cinque et al. 2011). The spectra were obtained in transmission mode using a small beam spot of 15 × 15/12 × 12 square microns, 4 cm^{-1} resolution, co-adding 256 scans at scanner velocity 80 kHz (35 s), in the 4000 to 650 cm^{-1} wavenumber range. Molecular composition IR maps were obtained by scanning the sample via a micrometric resolution motorized X–Y stage.

Two strategies of sample preparation were used for μSR-FTIR analyses. A selected sample fragment was squeezed between two diamonds in a microcompression cell to obtain samples of adequate thickness for analysis. The various materials present in the sample were spread out, which allowed the different substances present to be singled out and, at the same time, maintained the layered structure of the paint. The second method consisted in obtaining a thin-section of the samples (Beltran et al. 2015). The samples were embedded in epoxy resin and subsequently microtomed. To prevent the penetration of the epoxy resin, a thin gold protection layer was applied before embedding the sample. The embedded sample had a typical size of a few hundred micrometers, and the thin section obtained of about 2 μm thickness was pressed between two KBr pellets in order to assure the flatness of the preparation.

μSR-XRD is used to resolve complex mixtures of crystalline compounds in micrometric layered samples where many substances are present in very low amounts, which is typical in paintings. μSR-XRD data were obtained at beamline XALOC at ALBA Synchrotron, Cerdanyola del Vallès (Barcelona) (Juanhuix et al. 2014). For the XRD measurements, 20 μm thick cross sections are microtomed from the same embedded sample used for the infrared measurements (Beltran et al. 2015). The thin sections were scanned vertically at steps of 3 μm through a 50 × 6 μm^2 (width × height FWHM) focused beam of 12.661 keV energy. X-ray diffraction patterns were collected at typical acquisition times varying between 0.2 and 1 s using a virtually noise free Pilatus 6 M (Dectris) detector with a large active area (424 × 435 square millimeters, 6 Mpixels) protected by a small beam stop. The powder diffraction patterns of the layers were obtained by radial integration of the images obtained by this two-dimensional detector. This setup provides a high sensitivity to poor diffracting compounds present in very low amounts and a good low angular limit/large d-spacing (~70 Å) for the metal soaps present. In particular, the diffraction peaks corresponding to the metal soaps are broad and can be clearly identified and separated from the other compounds due to the large size of those molecules (low diffraction angles).

11.3 Results

We present a selection of representative data obtained from the analysis of more than 200 samples taken from 25 fifteenth-century Catalan Gothic altarpieces. Similar to other paintings in Europe from the same period, various lead-based pigments were used: lead white $PbCO_3/2PbCO_3{\cdot}Pb(OH)_2$, minium "orange" Pb_3O_4, and lead-tin yellow (type I Pb_2SnO_4 or type II $Pb(Si,Sn)O_3$). The lead soaps were found in the white paints containing lead white and egg yolk. They are particularly difficult to discriminate by microscope in lead white paints as they appear uniformly distributed. Conversely, in lead-tin yellow paints, lead soap aggregates are found. For instance, the yellow paints from an egg tempera panel by Jaume Huguet, one of the most important fifteenth-century Catalan painters (Fig. 11.1), were made of mixtures of lead-tin yellow type I, lead white, and minium in various

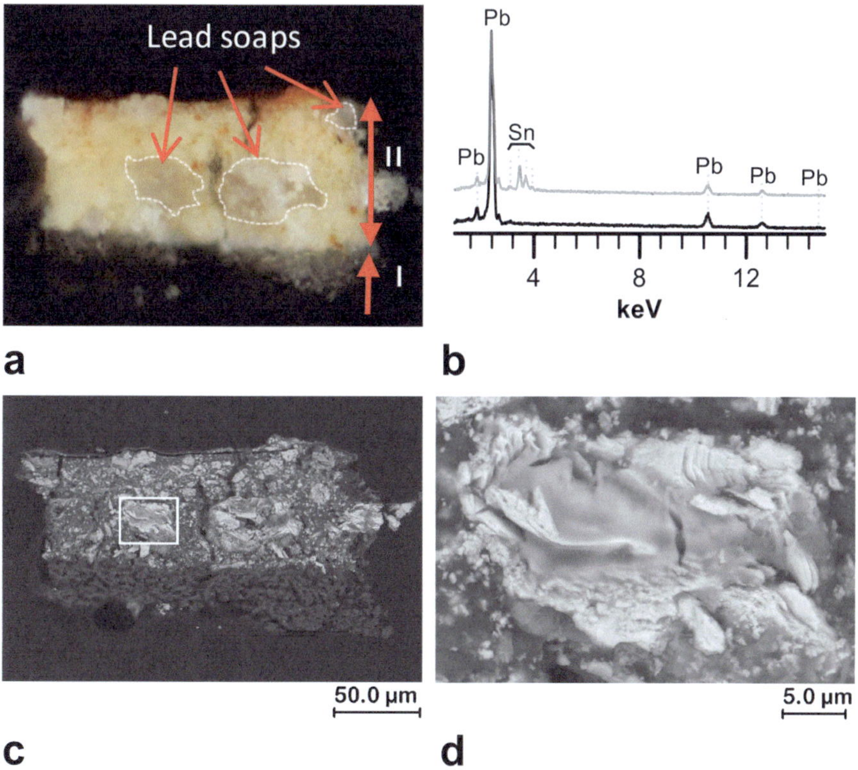

Fig. 11.1 Cross section of a yellow paint from the fifteenth-century egg tempera altarpiece "Sant Abdó i Sant Senen" by Jaume Huguet, St. Pere de Terrassa Church (Catalonia). (**a**) OM image: I, ground layer (gypsum and animal glue); II, yellow paint layer (lead-tin yellow Pb_2SnO_4, minium and egg yolk). (**b**) SEM-EDS spectra corresponding to the lead soap aggregate (black) and paint layer (gray). (**c**) Backscattered electron image. (**d**) Detail of backscattered electron image of a lead soap aggregate

proportions to obtain different yellow shades. The OM image (Fig. 11.1a) and the corresponding backscattered electron (BSE) image (Fig. 11.1b, c) from a cross section of the yellow paint show the presence of an aggregate containing only lead inside (determined by scanning electron microscopy with energy dispersive X-ray spectroscopy SEM-EDS). It is remarkable that from the data we have obtained so far and despite the presence of tin in the paint, tin soaps are not formed either in egg yolk or in drying oil.

The μSR-FTIR spectra from the yellow paint (Fig. 11.2a) also show the bands characteristic of lead soaps, the asymmetric COO^- stretching bands at 1540 cm^{-1} and 1515–1517 cm^{-1} and the symmetric COO^- stretching band at 1417 cm^{-1} (Fig. 11.2b, dark and light blue, top spectra). Lead soaps and protein bands overlap and are often not well discriminated (Fig. 11.2b, dark and light blue, top spectra). Due to the contribution of the asymmetric band at 1542 cm^{-1} from the δ(N-H) and ν(C-N) amide II of the protein from the egg yolk, the lead soap band at 1540 cm^{-1} appears more intense than those of reference lead soaps. This protein asymmetric broadband also contributes to a shift of the band maximum from 1512 cm^{-1} in reference lead soaps up to 1517 cm^{-1} in the egg yolk white paint (Fig. 11.2b, gray, middle spectrum). A series of weak CH_2 progression bands with wavenumbers between those of lead palmitate and lead stearate (Fig. 11.2c) are also seen. The wavenumbers of those bands agree with the data obtained from naturally aged lead white mixed with egg yolk (1348, 1334, 1316, 1296, 1274, 1253, 1232, 1210, 1188 cm^{-1}). The position of the CH_2 progression bands shows the predominance of palmitate according to the high palmitate/stearate ratio present in egg yolk. Nevertheless, the data suggest that mixtures with different proportions of palmitate and stearate, $Pb(C16)_{2x}(C18)_{2-2x}$, are produced. In complete agreement with this result, μSR-XRD data also shows that lead soaps with intermediate d-spacings between those corresponding to lead stearate and palmitate are formed.

Apart from lead-tin yellow, minium, and cassiterite (SnO_2), which are related to the synthesis of the yellow pigment, the μSR-XRD measurements of the yellow paint show the presence of lead soaps. The low diffraction angle region of a cross section sample is shown in Fig. 11.2d, e. The reflections corresponding to lead soaps show d-spacings that lie between those of lead stearate and lead palmitate, which is in good agreement with the IR data. This is also observed in other samples where the binding medium is drying oil (Salvadó et al. 2009) and found in the literature for other cations (changes in CH_2 progression bands in mixtures of binary zinc palmitate-stearate salts in oil) (Helwig et al. 2014; Hermans et al. 2015). Since the paint layers are heterogeneous, it is of great importance to measure more than one series (both cross section lines and maps) to obtain data which is as representative as possible of the sample. In Fig. 11.2d, e one representative cross section line is shown.

It is interesting to note that, in good agreement with other authors, among the samples analyzed lead soap aggregates seem to appear in relative higher amounts in lead-tin yellow or red lead paints than in white lead paints (Boon et al. 2005; Higgitt et al. 2003). Also, when litharge or massicot (lead (II) oxide) is present, lead soap formation is favored with respect to lead white (Erhardt et al. 2005; Van Loon et al. 2012).

Figure 11.3 shows another series of measurements obtained from a 20 μm thin-section of a yellow paint of the same painting with a similar composition. In this case, the presence of soaps of a different nature is suspected from the change observed in the d-spacings across the layers (Fig. 11.3d). In particular, the d-spacings at the ground/paint interface are smaller than those inside the paint. This suggests the presence of a different type of soap. Although it is difficult to determine, it is possible that the calcium from the ground intervenes in the formation of soaps at the interface (Salvadó et al. 2005, 2011). In this case, in addition to the formation of intermediate palmitate/stearate soaps, the presence of different cations further complicates the interpretation of the data.

The case of "The appearance of the Virgin to St. Francis at Porziuncola" altarpiece displayed at the Museu Nacional d'Art de Catalunya is particularly interesting because of the use of a mixed egg tempera and oil technique. The oil paint layer is applied over an egg tempera chromatic preparation. μSR-FTIR data taken from a white oil paint (layer II in Fig. 11.4a, b) applied over an ochre egg tempera preparation (layer III in Fig. 11.4a, b) show the presence of lead soaps in both layers. But, μSR-XRD data obtained from the same sample show the presence of crystalline lead soaps only in the chromatic preparation layer (Fig. 11.4a–d). Moreover, from the data we have obtained so far, which includes egg tempera, oil, and mixed techniques, we have observed that lead soaps tend to be more crystalline when the binder is egg yolk (we obtained similar results with the laboratory replica samples, naturally aged for 8–10 years) (Salvadó et al. 2005). This may be understood considering that the egg tempera technique uses an aqueous carrier and the water evaporates very slowly. The remaining water is expected to promote the complexation and nucleation of lead soaps (Pereira et al. 2012, 2014). In addition, porosity developed when water is evaporated (Mills and White 1994) can help the nucleation and growth of crystalline soaps. From our laboratory assays, the formation of crystalline lead soaps from lead white and egg yolk at room temperature is clearly detectable in less than 6 months. Nevertheless, the proportion of saturated fatty acids is larger (about two times) in egg yolk than in drying oil. It is also interesting to note that at the interface between layers where more nucleation sites are expected to happen, the lead soaps also appear more crystalline (Fig. 11.4d). Once again, this shows the complexity and difficulties found in the interpretation of historical data.

A sample extracted from the wing of an angel from the same altarpiece shows a series of different color painting layers. Over the gypsum ground layer, a very thin lead white mixed with egg yolk layer is followed by yellow, green, and red layers, all mixed with oil. The optical and backscattered electron images show the growth of lead soaps (only lead is detected in the EDS elemental analysis) originally formed in the yellow layer (containing lead-tin yellow type I and drying oil) appearing as a protrusion bursting between the green (copper acetate/basic copper acetate based pigment) layer and the red (minium pigment) layer (Fig. 11.4e–g). The formation of the protrusion produces the opening of small cracks on the surface. In the yellow layer, where lead soaps are formed, the infrared bands at 1630sh, 1587, 1311, 1290 cm^{-1} and 784, 770 cm^{-1} (Mancilla et al. 2009; Salvadó et al. 2009) characteristic

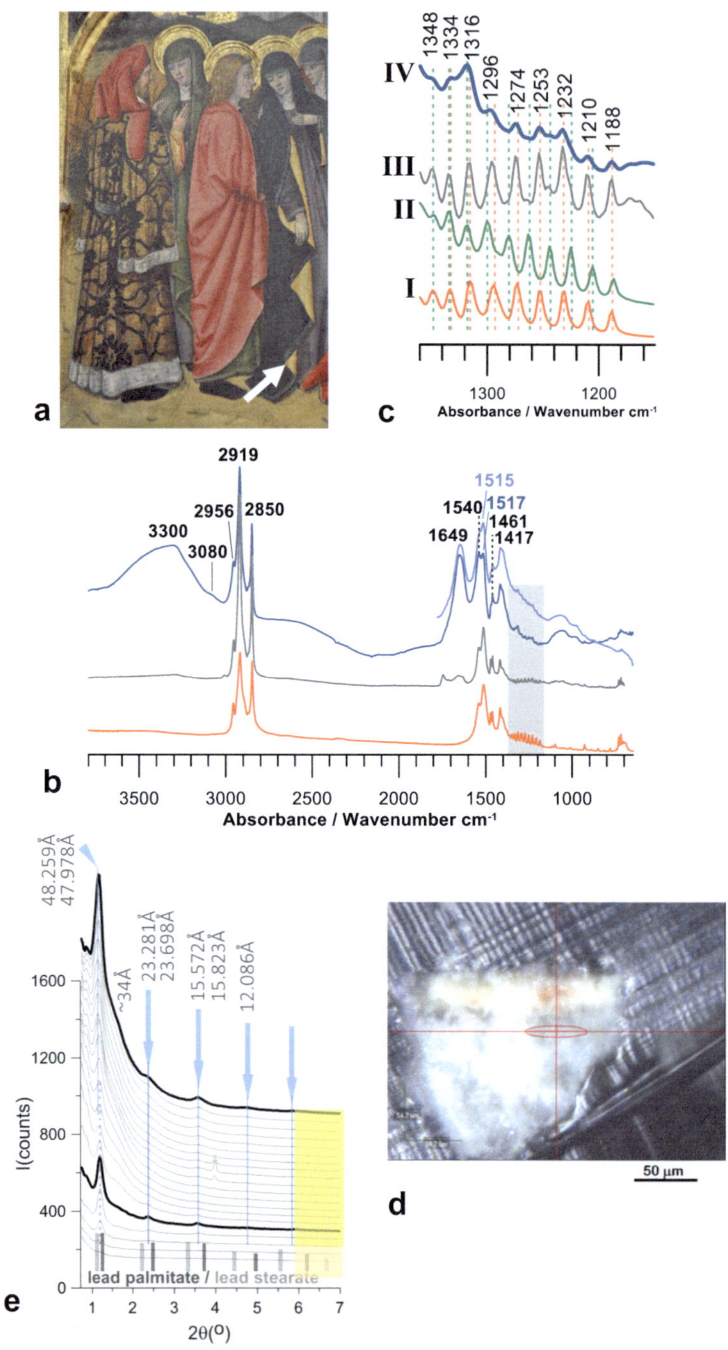

Fig. 11.2 (continued)

of lead oxalates are also determined. So far, lead oxalates have only been found in paint layers containing lead-tin yellow mixed with drying oil, never on the surface patinas (Monico et al. 2013).

Jaume Huguet is the painter of "El Conestable" altarpiece. This artwork is remarkable because it was a royal commission, has a high pictorial quality, and is still found in its original place, the Santa Àgata Chapel in Barcelona (this is highly uncommon). One of the conservation problems of the altarpiece is the loss of the dark red colored paint. Comparing several altarpieces by the same artist, we can see that this is generally happening when the dark red paint layer is thick (about 75–100 μm thick). The sequence of layers in the painting includes the ground layer (gypsum and animal glue), a chromatic preparation layer with lead white, cinnabar, and egg yolk, and a dark red, upper paint layer, containing a red lake pigment and egg white. The different degrees of contraction during the drying process of egg white compared to egg yolk are responsible for the lack of cohesion among the paint layers and the presence of micro-cracks (Bucklow 2012). Figure 11.5a–c shows the optical and backscattered electron images.

Figure 11.5d, e shows a series of 14 diffraction patterns from the cross section of the dark red paint and Fig. 11.5f a magnification of the low diffraction angle region. Again, the d-spacings of the lead soaps can be found in between those of the lead stearate and palmitate. Consequently, lead soaps are formed in the chromatic preparation layers, and it is possible that they grow in the micro-cracks and accelerate the detachment of the most superficial red layer. Figure 11.5c shows a backscattered electron image of the process.

Calcium oxalates (weddellite) are also found in the preparation layer made of gypsum and animal glue. In this case, the calcium oxalates are associated with the presence of organic binding media and are most probably formed due to the degradation of the organic matter. The formation of calcium oxalates has been

Fig. 11.2 (continued) (**a**) Detail of the fifteenth-century egg tempera altarpiece "Sant Abdó i Sant Senen" by Jaume Huguet, St. Pere de Terrassa Church (Catalonia); the sampling point is indicated with an arrow. (Photo Carles Aymerich. © CRBMC. Centre de Restauració de Bens Mobles de Catalunya). (**b**) μSR-FTIR spectra (transmission mode, diamond cell, spot size 12 × 12 μm) of lead palmitate (orange, bottom spectrum), lead soaps formed in a mixture of lead carbonate and egg yolk naturally aged for 8 years (gray, middle spectrum), and yellow layer from the paint sample containing lead-tin yellow Pb_2SnO_4, minium Pb_3O_4, and egg yolk binder (dark and light blue, top spectra). (**c**) Magnification of the 1170–1360 cm^{-1} region where the CH_2 bands related to the soaps are seen. The spectra of lead stearate and lead palmitate are essentially identical with the exception of this range: I, lead palmitate; II, lead stearate; III, lead soaps formed in a mixture of lead carbonate and egg yolk naturally aged for 8 years; IV, yellow layer from the paint sample. (**d**) OM image of a 20 μm thin cross section of the yellow sample from which μSR-XRD measurements were taken. The red ellipse corresponds to the beam footprint (50 × 6 μm) at the initial position. Measurements were taken along the vertical red line at steps of 3 μm covering 60 μm. (**e**) μSR-XRD patterns taken from the thin cross section shown in image (**d**). Those corresponding to the paint and the ground layers are marked in yellow and in orange, respectively. Some representative patterns of both layers are drawn with a thicker line. For some peaks two d-spacing values are indicated, which are the range of values obtained

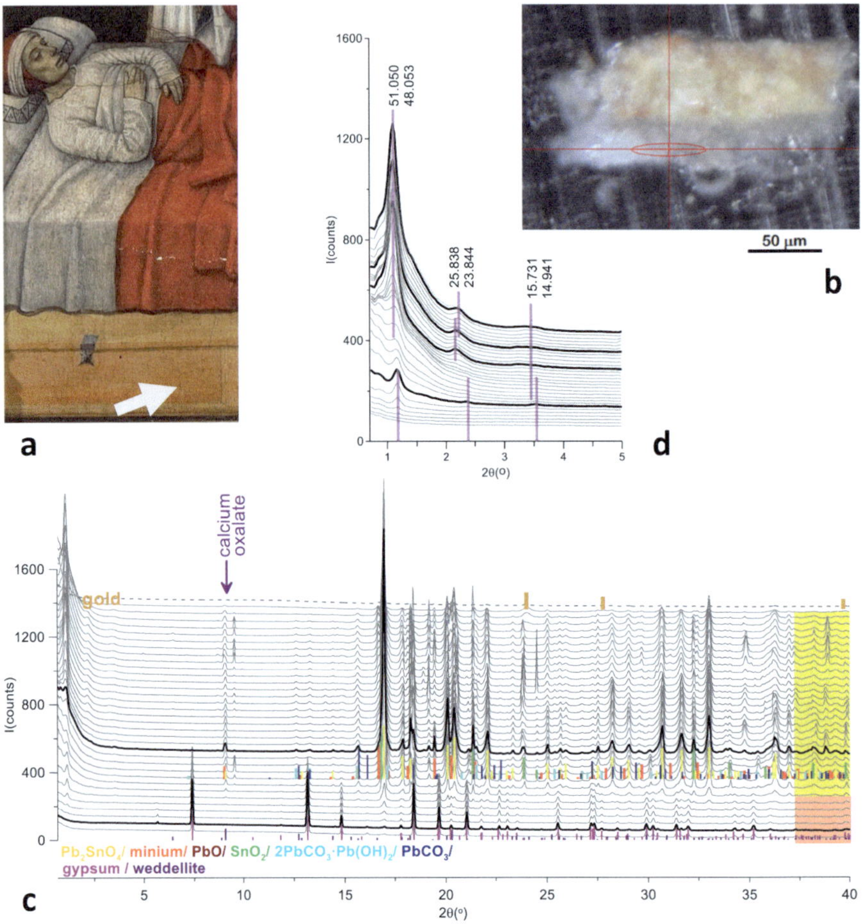

Fig. 11.3 (**a**) Detail of the fifteenth-century egg tempera altarpiece "Sant Abdó i Sant Senen" by Jaume Huguet, St. Pere de Terrassa Church (Catalonia); sampling point is indicated with an arrow. (Photo Carles Aymerich. © CRBMC. Centre de Restauració de Bens Mobles de Catalunya). (**b**) OM image of a 20 μm thin cross section of the yellow sample. The red ellipse corresponds to the beam footprint (50 × 6 μm) at the initial position. (**c**) μSR-XRD patterns series from the thin cross section of the paint layer. Measurements were taken along the vertical red line (in image **b**) at steps of 3 μm covering 100 μm. Those corresponding to the paint and the ground layers are marked in yellow and in orange, respectively. Some representative patterns of both layers are drawn with a thicker line. (**d**) Magnification of the μSR-XRD patterns at the low diffraction angle region

observed in all the ground layers analyzed containing gypsum mixed with animal glue (Salvadó et al. 2009, 2011).

The green paint layer applied over a gold foil from an altarpiece by Jaume Huguet (Fig. 11.6) is formed by a complex mixture of drying oil, egg yolk, a copper-based pigment, and tin-lead yellow type I pigment. The absorption band at 1585 cm^{-1}

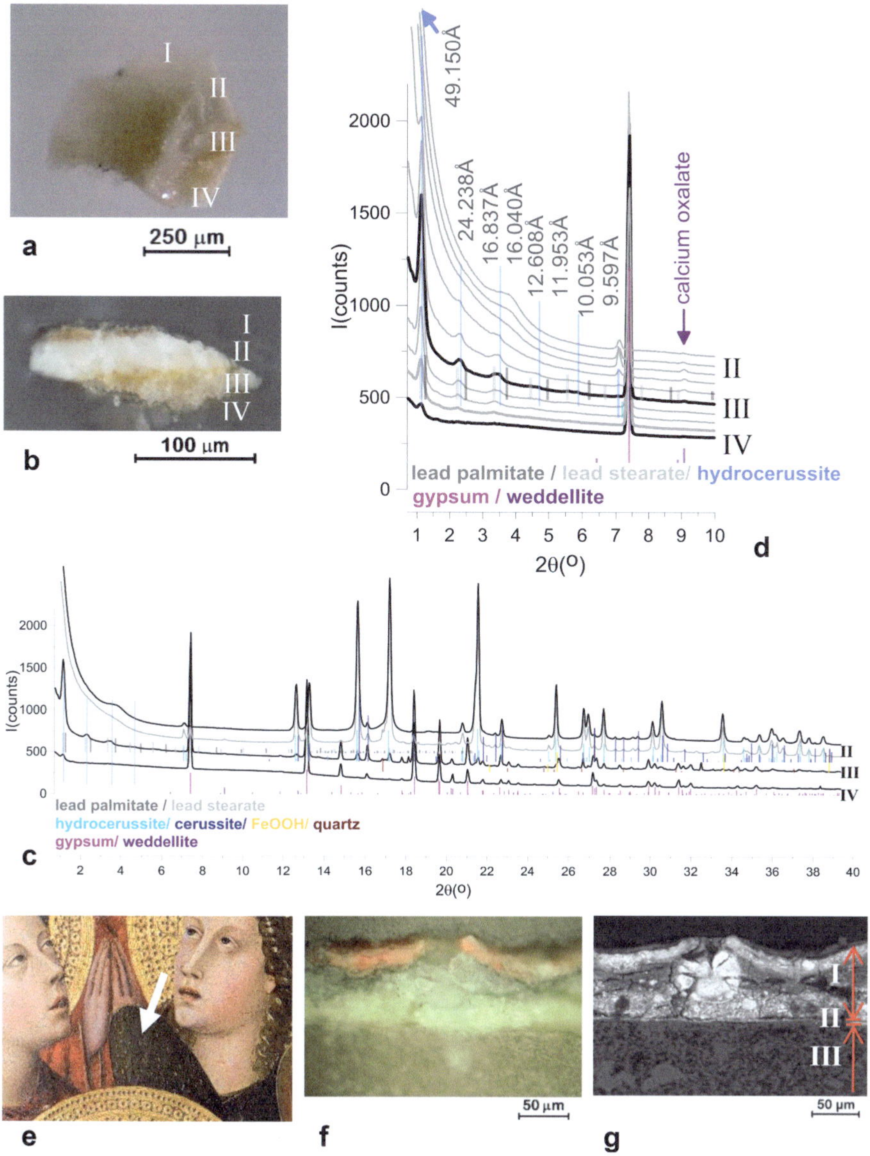

Fig. 11.4 (**a**) White sample from the fifteenth-century altarpiece "The appearance of the Virgin to St. Francis at Porziuncola", Museu Nacional d'Art de Catalunya MNAC (Barcelona). (**b**) Cross section of a white paint sample: I, superficial layer/varnish; II, lead white with oil paint layer; III, egg yolk tempera ochre preparation layer; IV, animal glue and gypsum ground layer. (**c**) µSR-XRD patterns series from the thin cross section. Some representative patterns of layers are drawn with a thicker line. (**d**) Magnification of the µSR-XRD patterns at the low diffraction angle region. Measurements were taken at steps of 3 µm covering 70 µm (ten patterns are shown). For some peaks two d-spacing values are indicated, which are the range of values obtained. (**e**) Detail of the altarpiece; the sampling point is indicated with an arrow. (Photo Calveras. © MNAC. Museu Nacional d'Art de Catalunya). (**f**) OM image of cross section of a lead soap protrusion. (**g**) Backscattered electron image of the same protrusion shown in (**f**): I, oil paint layers, red, green, and yellow; II, egg yolk tempera chromatic preparation layer; III, gypsum and animal glue ground layer

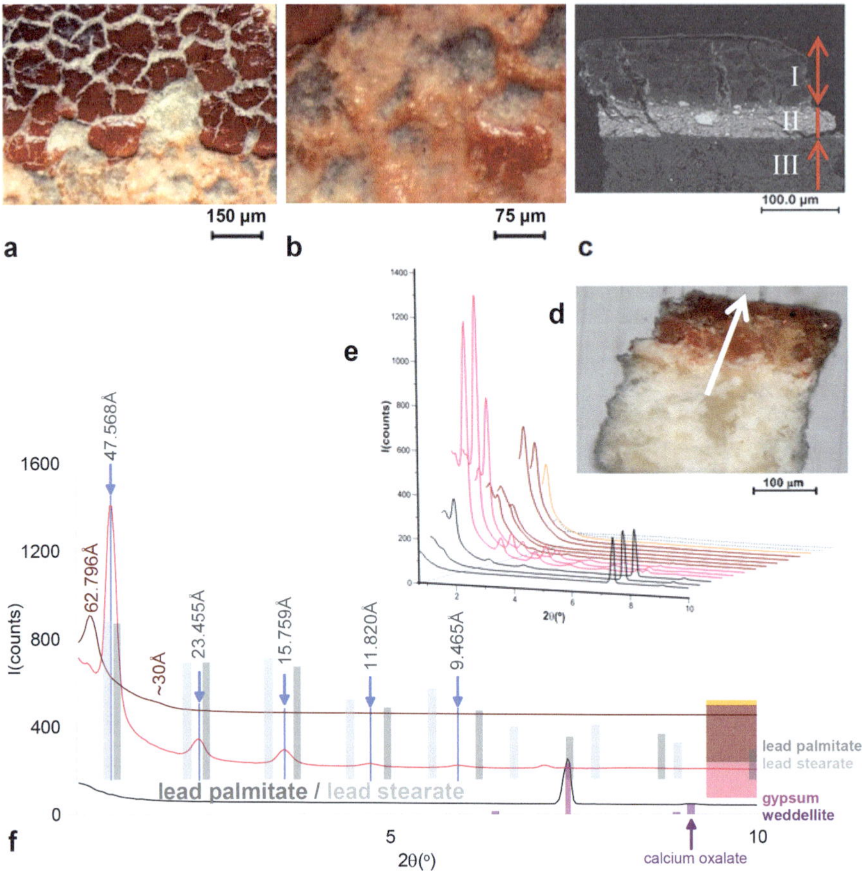

Fig. 11.5 (**a–b**) Surface details of dark red paint from fifteenth-century egg tempera altarpiece "El Conestable" by Jaume Huguet, Capella de Sant Àgata (Barcelona), showing the cracked and damaged red layer. (**c**) Backscattered electron image: I, dark red layers (red lake pigment and egg white); II, chromatic preparation layer (lead white, cinnabar and egg yolk); III, ground layer (gypsum and animal glue). The lead soap growth toward the surface through the cracks is observed. (**d**) OM image of the thin cross section of sample showing the line where the μSR-XRD patterns were obtained from. (**e**) Low diffraction angle μSR-XRD patterns corresponding to the thin cross section. (**f**) μSR-XRD patterns from the dark red layer (top), chromatic preparation layer (medium), and ground layer (bottom). Those corresponding to the chromatic preparation layer, dark red layer, and outmost surface gold protection layer are marked in pink, red, and yellow, respectively

corresponds to characteristic asymmetric stretching COO^- group of a copper soap (Fig. 11.6b) (Richardin et al. 2011; Salvadó et al. 2005, 2009, 2013).

Copper soaps are not determined by μ-XRD, maybe because they are not crystalline. We use X-rays with energy above that of the absorption edge of copper which is not the most adequate, and thus the analysis should be repeated with X-

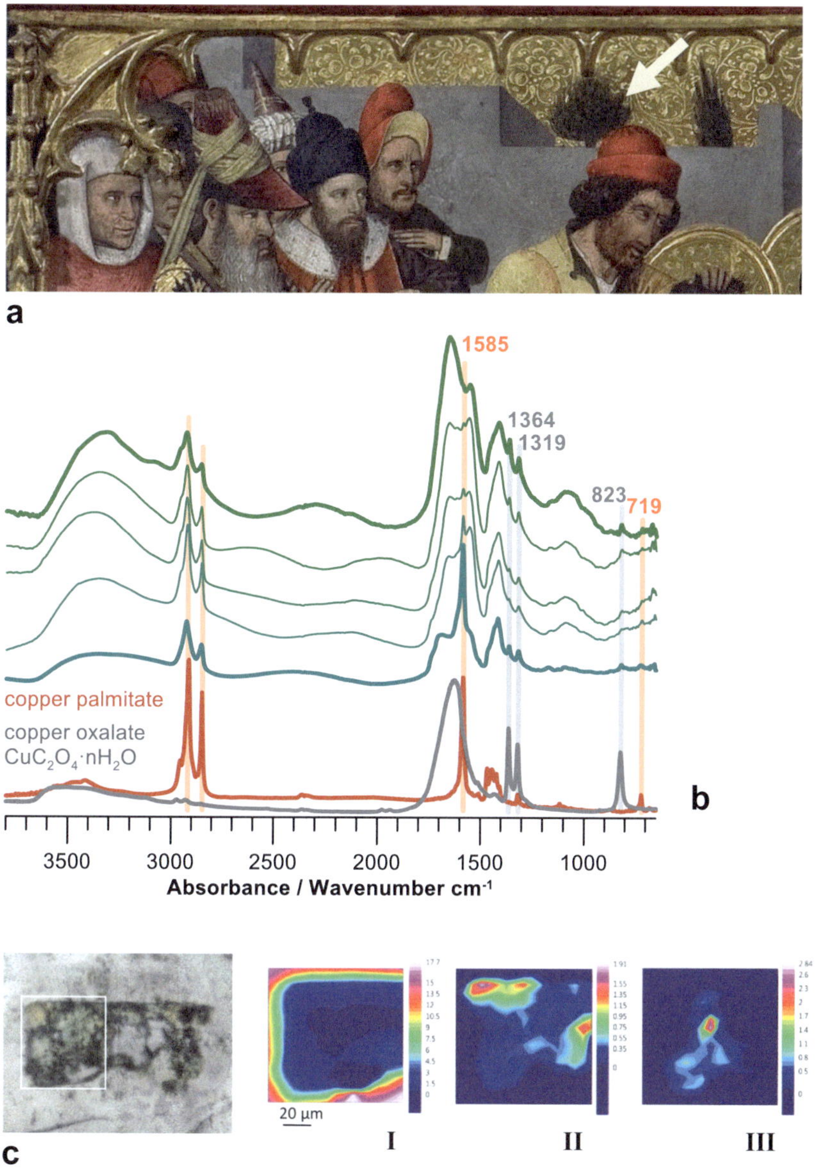

Fig. 11.6 (**a**) Detail of fifteenth-century egg tempera altarpiece "Sant Abdó i Sant Senen" by Jaume Huguet, St. Pere de Terrassa Church (Catalonia). (Photo Carles Aymerich. © CRBMC. Centre de Restauració de Béns Mobles de Catalunya). The sampling point is indicated with an arrow. A sample containing a drying oil, egg yolk, copper, and lead-tin yellow pigments was taken from the green paint applied over a gold foil. (**b**) μSR-FTIR spectra from the green layer compared to the reference spectra copper palmitate and copper oxalate. (**c**) SR-infrared microspectroscopic chemical maps obtained in transmission mode from a thin cross section (<2 μm) of the green sample. Spot size, 12 × 12 μm. Chemical maps of the band at I, 1247 cm^{-1} corresponding to the embedding epoxy resin; II, 1319 cm^{-1} corresponding to the copper oxalates; and II, 1585 cm^{-1} map corresponding to the copper soaps

rays of lower energy to verify this. In addition, the IR absorption bands present at 1364, 1319, and 823 cm^{-1} are related to copper oxalates (Salvadó et al. 2009, 2013). Copper oxalate and soap distribution maps were obtained by μFTIR from a thin section (<2 μm thick) showing areas with a high concentration of oxalates and low concentration of soap and vice versa. As metal oxalates are substances expected from the degradation of organic matter, it is not wholly unexpected to find the same metal forming oxalates and soaps.

11.4 Conclusions

We have studied the formation and distribution of metal soaps produced as a result of the drying, reactivity, and aging of the materials in fifteenth-century egg tempera and drying oil paintings. The formation of lead, calcium, and copper soaps is demonstrated. Lead soaps are always formed in relative higher amounts in lead-tin yellow or red lead paints than in white lead paints; and tin soaps are not found. Lead soaps appear to be mixtures with different proportions of palmitate and stearate and tend to form aggregates. Lead soaps appear more crystalline when formed in egg tempera paints, which has been explained because the technique uses an aqueous carrier which is expected to promote the complexation and nucleation of lead soaps; the porosity developed by water evaporation can help the nucleation and growth of crystalline soaps. Generally speaking, the presence of porosity, cracks, or interfaces favors the growth of lead soap crystals. Finally, lead, calcium, and copper oxalates are also found.

Acknowledgments The project received financial support from MINECO (Spain), Grant MAT2013-41127 R, and Generalitat de Catalunya, Grant 2014SGR-581. We acknowledge Diamond Light Source for time on beamline MIRIAM B22 under Proposal (SM10400) and the collaboration of Diamond staff. The μSR-XRD experiments were performed at BL13 XALOC beamline (Grant No. 2014071016) at ALBA Synchrotron with the collaboration of ALBA staff. Part of this work was carried out with collaboration between the Universitat Politècnica de Catalunya (UPC) and the "Museu Nacional d'Art de Catalunya" (MNAC) and the "Centre de Restauració de Béns Mobles de Catalunya" (CRBMC).

References

Bayrak A, Kiralam M, Ipek A et al (2010) Fatty acid compositions of linseed (Linum Usitatissimum L.) genotypes of different origin cultivated in Turkey. Biotechnol Biotechnol Equip 24(2):1836–1842

Beltran V, Salvadó N, Butí S et al (2015) Optimal sample preparation for the analysis of micrometric heterogeneous samples. Anal Chem 87:6500–6504

Boon JJ, Gore E, Keune K et al (2005) Image analytical studies of lead soap aggregates and their relationship to lead and tin in 15th C lead tin yellow paints from the Sherborne triptych. In: Picollo M (ed) Sixth infrared and Raman users group, Florence, Mar–Apr 2004, Il Prato, pp 66–74

Bucklow S (2012) The classification of craquelure patterns. In: Hill Stoner J, Rushfield R (eds) Conservation of easel paintings, Series in conservation and museology. Routledge, London/New York, pp 285–290

Catalano J, Murphy A, Yao Y et al (2014) ^{207}Pb and ^{119}Sn solid-state NMR and relativistic density functional theory studies of the historic pigment lead-tin yellow type I and its reactivity in oil paintings. J Phys Chem A 118:7952–7958

Catalano J, Murphy A, Yao Y et al (2015) Coordination geometry of lead carboxylates-spectroscopic and crystallographic evidence. Dalton Trans 44:2340–2347

Cinque G, Frogley M, Wehbe K (2011) Multimode InfraRed imaging and microspectroscopy (MIRIAM) beamline at diamond. Synchrotron Radiat News 24(5):24–33

Corbeil MC, Robinet L (2002) X-ray powder diffraction data for selected metal soaps. Powder Diffract 17(1):52–60

Cotte M, Checroun E, Susini J et al (2007) Micro-analytical study of interactions between oil and lead compounds in paintings. Appl Phys A Mater Sci Process 89:841–848

Cotte M, Checroun E, Nolf WD et al (2016) Lead soaps in paintings: friends or foes? Stud Conserv 62(1):2–23

de Viguerie L, Payard PA, Portero E (2016) The drying of linseed oil investigated by Fourier transform infrared spectroscopy: historical recipes and influence of lead compounds. Prog Org Coat 93:46–60

Erhardt D, Tumosa CS, Mecklenburg MF (2005) Long term chemical and physical processes in oil paint films. Stud Conserv 50(2):43–150

Helwig K, Poulin J, Corbeil MC et al (2014) Conservation issues in several twentieth-century Canadian oil paintings: the role of zinc carboxylate reaction products. In: Van den Berg KJ, Burnstock A, de Tagle A, de Keijzer M, Heydenreich G, Krueger J, Learner T (eds) Issues in contemporary oil paints. Springer, Cham, pp 167–184

Hermans JJ, Keune K, Van Loon A et al (2014) The molecular structure of three types of long-chain zinc (II) alkanoates for the study of oil paint degradation. Polyhedron 81:335–340

Hermans JJ, Keune K, Van Loon A et al (2015) An infrared spectroscopic study of the nature of zinc carboxylates in oil paintings. J Anal At Spectrom 30:1600–1608

Hermans JJ, Keune K, Van Loon A et al (2016) The crystallization of metal soaps and fatty acids in oil paint model systems. Phys Chem Chem Phys 18:10896–10905

Higgitt C, Spring M, Saunders D (2003) Pigment-medium interactions in oil paint films containing red lead or lead-tin yellow. Natl Gallery Tech Bull 24:75–93

Juanhuix J, Gil-Ortiz F, Cuni G et al (2014) Developments in optics and performance at BL13-XALOC, the macromolecular crystallography beamline at the ALBA synchrotron. J Synchrotron Radiat 21:679–689

Keune K, Boon JJ (2004) Imaging secondary ion mass spectrometry of a paint cross section taken from an early Netherlandish painting by Rogier van der Weyden. Anal Chem 76: 1374–1385

Keune K, Boon JJ (2007) Analytical imaging studies of cross-sections of paintings affected by lead soap aggregate formation. Stud Conserv 52(3):161–176

Keune K, Loon A, Boon JJ (2011) SEM backscattered-Electron images of paint cross sections as information source for the presence of the lead white pigment and lead-related degradation and migration phenomena in oil paintings. Microsc Microanal 17(5):696–701

Keune K, Kramer RP, Huijbregts Z et al (2016) Pigment degradation in oil paint induced by indoor climate: comparison of visual and computational backscattered Electron images. Microsc Microanal 22(2):448–457

Mancilla N, D'Antonio MC, González-Baró AC et al (2009) Vibrational spectra of lead(II) oxalate. J Raman Spectrosc 40:2050–2052

Martínez-Casado FJ, Ramos-Riesco M, Rodríguez-Cheda JA et al (2014) Short lead (II) soaps: from weakly fluorescent crystals to strongly phosphorescent and structurally varied vitreous phases. A thermal, structural and spectroscopic study. J Mater Chem C 2:9489–9496

Meilunas RJ, Bentsen JG, Steinberg A (1990) Analysis of aged paint binders by FTIR spectroscopy. Stud Conserv 35(1):33–51

Mills JS, White R (1994) The organic chemistry of museum objects, 2nd edn. Butterworth-Heinemann, Oxford
Monico L, Rosi F, Miliani C (2013) Non-invasive identification of metal-oxalate complexes on polychrome artwork surfaces by reflection mid-infrared spectroscopy. Spectroc Acta Pt A-Molec Biomolec Spectr 116:270–280
Newman R (1998) Tempera and other non-drying-oil media. In: Dorge V, Carey Howlett F (eds) Painted wood: history and conservation. The Getty Conservation Institute, Los Angeles, pp 33–63
Osmond G, Boon JJ, Puskard L (2012) Metal stearate distributions in modern artists' oil paints: surface and cross-sectional investigation of reference paint films using conventional and synchrotron infrared microspectroscopy. Appl Spectrosc 66:1136–1144
Pereira RFP, Valente AJM, Fernandes M et al (2012) What drives the precipitation of long-chain calcium carboxylates (soaps) in aqueous solution? Phys Chem Chem Phys 14(20):7517–7527
Pereira RFP, Valente AJM, Burrows HD (2014) The interaction of long chain sodium carboxylates and sodium dodecylsulfate with lead(II) ions in aqueous solutions. J Colloid Interface Sci 414:62–72
Popa VM, Gruia A, Raba D, Moldovan C et al (2012) Fatty acids composition and oil characteristics of linseed (Linum Usitatissimum L.) from Romania. J Agroaliment Process Technol 18(2):136–140
Pouyet E, Fayard B, Salomé M et al (2015) Thin-sections of painting fragments: opportunities for combined synchrotron-based micro-spectroscopic techniques. Herit Sci 3:3. https://doi.org/10.1186/s40494-014-0030-1
Richardin P, Mazel V, Walter P et al (2011) Identification of different copper green pigments in renaissance paintings by cluster-TOF-SIMS imaging analysis. J Am Soc Mass Spectrom 22:1729–1173
Robinet L, Corbeil MC (2003) The characterization of metal soaps. Stud Conserv 48:23–40
Salvadó N, Butí S, Tobin MJ et al (2005) Advantages of the use of SR-FT-IR Microspectroscopy: applications to cultural heritage. Anal Chem 77:3444–3451
Salvadó N, Butí S, Nicholson J et al (2009) Identification of reaction compounds in micrometric layers from gothic paintings using combined SR-XRD and SR-FTIR. Talanta 79:419–428
Salvadó N, Butí S, Labrator A et al (2011) SR-XRD and SR-FTIR study of the alteration of silver foils in medieval paintings. Anal Bioanal Chem 399:3041–3052
Salvadó N, Butí S, Cotte M et al (2013) Shades of green in 15th century paintings: combined microanalysis of the materials using synchrotron radiation XRD, FTIR and XRF. Appl Phys A-Mater 111:47–57
Sawicka A, Burnstock A, Izzo FC et al (2014) An investigation into the viability of removal of lead soap efflorescence from contemporary oil paintings. In: Van den Berg KJ, Burnstock A, de Tagle A, de Keijzer M, Heydenreich G, Krueger J, Learner T (eds) Issues in contemporary oil paints. Springer, Cham, pp 311–332
Sotiropoulou S, Papliaka ZE, Vaccari L (2016) Micro FTIR imaging for the investigation of deteriorated organic binders in wall painting stratigraphies of different techniques and periods. Microchem J 124:559–567
USDA National Nutrient Database for Standard Reference, Release 1 (April, 2018) Full Report (All Nutrients): 01137, Egg, yolk, dried
Van Loon A, Noble P, Burnstock A (2012) Ageing and deterioration of traditional oil and tempera paints. In: Hill Stoner J, Rushfield R (eds) Conservation of easel paintings, series in conservation and museology. Routledge, London/New York, p 229

Chapter 12
Photoluminescence Micro-imaging Sheds New Light on the Development of Metal Soaps in Oil Paintings

Mathieu Thoury, Annelies Van Loon, Katrien Keune, Joen J. Hermans, Matthieu Réfrégiers, and Barbara H. Berrie

Abstract This paper describes a photoluminescence imaging approach coupled to a synchrotron source (SR-PL imaging) that provides a new, powerful way to probe the processes involved in metal soap formation in paint films, processes that play an important role in the deterioration of paint films in traditional and modern oil paintings.

The technique couples multispectral PL mapping with sub-micrometer spatial resolution with the ability to tune the excitation radiation from UV-C to the visible range. False-color RGB images of the distribution of PL from three historical paint cross sections (two with lead soaps and one with zinc soaps) using a range of

M. Thoury (✉)
IPANEMA, CNRS, ministère de la Culture et de la Communication, Université de Versailles Saint-Quentin-en-Yvelines, Muséum National d'Histoire Naturelle, USR 3461, Université Paris-Saclay, Gif-sur-Yvette, France
e-mail: mathieu.thoury@synchrotron-soleil.fr

A. Van Loon
Conservation Department, Rijksmuseum, Amsterdam, The Netherlands

K. Keune
Conservation Department, Rijksmuseum, Van't Hoff Institute for Molecular Sciences, University of Amsterdam, Amsterdam, The Netherlands

J. J. Hermans
Van't Hoff Institute for Molecular Sciences, University of Amsterdam, Amsterdam, The Netherlands

M. Réfrégiers
Synchrotron SOLEIL, Gif-sur-Yvette, France

B. H. Berrie
National Gallery of Art, Washington, DC, USA

F. Casadio et al. (eds.), *Metal Soaps in Art*, Cultural Heritage Science,
https://doi.org/10.1007/978-3-319-90617-1_12

excitation and emission wavelengths were obtained. The resulting maps of PL from lead soaps within the cross sections provide new insight about their formation, which can be visualized occurring around lead white particles or flocs of particles as an initial step. Inhomogeneity within an individual lead soap aggregate is revealed and it can be seen that the aggregated forms have complex structures. A single aggregate in one sample was found to comprise five phases.

High spatial resolution images of the distribution of zinc soaps throughout a ZnO-containing paint film show that the periphery of small fissures containing soaps show luminescence of zinc oxide particles and no sign of a gradient in PL into the paint film, and thus allow inferring that Zn soap formation is not initiated at cracks, but rather soaps accumulate in them. Haloes of high luminescence around particles of emerald green implicate it in the development of Zn soaps.

These results show the potential of SR-PL imaging to provide improved characterization of Pb and Zn soap structures, leading to a better understanding of the kinetics of their formation and the development of macroscopic aggregates.

Keywords Lead soaps · Zinc soaps · Oil paintings · Degradation · Lead white · Zinc white · Photoluminescence · Synchrotron · Microscopy

12.1 Introduction

Many steps in the development of soaps in paintings remain incompletely understood, including the initiation factors, the rate of formation of soaps, and alteration and aggregation of the initial products. These are being studied using spectroscopic and chromatographic methods, but photoluminescence (PL) has not been used to a large extent, despite the observation that under UV-A irradiation aggregated lead and zinc soaps in paint films exhibit heterogeneous luminescence (Higgitt et al. 2003; Noble et al. 2002).

With the aim of performing an in-depth study of the luminescence properties of metal soaps and to establish the value of PL for investigating soap formation in oil paint films, at even very early stages, a novel synchrotron luminescence deep-UV imaging approach has been optimized and used to study the phenomenon. Developments in synchrotron deep-UV photoluminescence (SR-DUV-PL), micro-imaging implemented at the DISCO beam-line of synchrotron SOLEIL in collaboration with IPANEMA, have allowed the collection of unprecedented, detailed PL data on lead and zinc soaps at the submicron scale. The results rely on several factors in the set-up. Multi-spectral detection, optimized to collect emission from 300 to 1000 nm, in combination with a fully tunable deep-UV excitation wavelength allows unique spatial and spectral contrast between phases aggregated within metal carboxylate species that form as alteration products in oil paints. The full-field detection of luminescence makes it possible to obtain images of the distribution of PL in defined spectral ranges from large areas, even hundreds of square micrometers, with a spatial resolution of several hundred nanometers (Thoury et al. 2011). The high spatial dynamics attainable offers a new method for visualization and speciation within diverse structures and morphologies of metal soap alteration products found in typical cross-section samples, providing a better understanding of their evolution over time, from the earliest stages up to formation of large aggregates of soaps that are apparent on the surfaces of paintings.

To illustrate the potential of the PL full-field micro-imaging set-up for studying soap formation, this paper presents examples of its application to two lead soap-containing and one zinc soap-containing paint samples that present different manifestations of and stages in the development of metal soaps in oil paint films. A cross-section sample taken from a large painting on canvas by Jacob Jordaens, *Revolt of the Batavian against Roman Rule* (1661–66, Royal Palace, Amsterdam), shows an early stage of lead soap formation, at which the lead white particles are in the process of converting into soaps, while a paint sample taken from a seventeenth-century ceiling painting depicting *Justitia and Mars in a Warm Embrace* (1680, Room of Trustees, Burgerweeshuis, Zierikzee) illustrates a mature stage with aggregation of lead soaps within the paint film. The sample taken from a painting by Vincent van Gogh, *Roses* (1890, National Gallery of Art, Washington), reveals the formation of zinc soaps throughout the paint layer. The interpretation of the PL data is supported by other analytical techniques, including scanning electron microscopy backscatter electron imaging (SEM-BSE) and attenuated total reflection Fourier-transform infrared (ATR-FTIR) micro-imaging.

12.2 Experimental

12.2.1 Samples and Sample Preparation

The samples were prepared at the different institutions following the in-house protocols. The sample of the Jordaens painting, sample N103 CS9, was embedded in a polyester resin (Polypol PS230, Poly-Service Amsterdam). The sample of the historic ceiling painting from the Burgerweeshuis in Zierikzee, sample ZZ7210_x12, was embedded in a methacrylate mounting resin (Technovit 2000 LC, Heraeus Kulzer GmbH, Germany). The sample from van Gogh's *Roses* was embedded in Bio-Plastic™ resin (Ward's Science). The samples were wet-ground on a polishing wheel to expose the complete paint layer build-up, with the assistance of a sample holder. In the final steps, Micro-Mesh® sheets up to grit 12,000 (Micro-Surface Finishing Products Inc., Wilton, Iowa, USA) were used for dry polishing (Van Loon et al. 2005). Each sample is described more fully in the Results section.

12.2.2 UV/VIS Spectral-imaging Setup at Disco

SR-DUV-PL experiments were performed using the full-field micro-imaging TELEMOS set-up of the DISCO beamline (Bertrand et al. 2013; Giuliani et al. 2009). The monochromatic synchrotron beam is coupled to a modified Zeiss Axio Observer

Z1 microscope (Carl Zeiss, Germany) equipped with a 100×/NA 1.25 immersion objective and a Princeton PIXIS 1024B/BUV EM-CCD (1024 × 1024 pixels, 13 μm pixel size), in front of which a filter wheel is positioned. This configuration allowed collection of images of the luminescence from the samples with spatial resolution to 150 nm. For examination of the PL from the Jordaens and the *Roses* samples, excitation wavelengths were set at 280 and 390 nm. For the Zierikzee sample, several excitation wavelengths were used: 240, 260, 280, 340, 360 and 380 nm. All emission images were acquired in nine spectral bands between 327 and 870 nm using high transmittance band-pass interference filters and with two dichroic mirrors with 300 and 400 nm cut-off wavelength, respectively. Visualization of the distribution of emission was improved by assigning images of interest to red (R), green (G) and blue (B) channels. The intensities of each channel of the full-field RGB image were either stretched between the 2nd and 98th percentile or linearly stretched using the ENVI 5.0 software (EXELIS) to facilitate visual comparison and discrimination of the distribution of the materials having different emission spectra.

12.2.3 Light Microscopy, Scanning Electron Microscopy and ATR-FTIR Imaging

Light microscopy of the Jordaens sample was performed on a Zeiss Axioplan 2 microscope at the Rijksmuseum in Amsterdam. Luminescence images were obtained with a LED 365 nm light source and a filter cube composed of a 365 nm excitation filter (EX G 365), a beamsplitter at 395 nm (BS FT 395) and an emission long-pass filter at 420 nm (EM LP 420). The Zierikzee sample was examined using a Leica DMRX microscope at the FOM Institute AMOLF in Amsterdam. Luminescence images were obtained using an Osram HBO 50 high-pressure mercury lamp and Leica filter cube D, which has an excitation pass filter at 355–425 nm, with a dichroic cutoff filter at 455 nm and a long-pass suppression filter at 470 nm. The van Gogh sample was examined using a Leica DMRX microscope at the National Gallery of Art, Washington. In this case, three filter cubes were used: filter cube D has excitation band-pass filter 355–425 nm with dichroic cut-off filter at 455 nm and long-pass suppression filter 470 nm. Filter cube I3 has excitation band-pass filter 450–490 nm, dichroic filter at 510 nm and a long-pass suppression filter at 515 nm. Filter cube M2 has excitation at band-pass 546/14 dichroic filter at 580 nm and long-pass suppression filter at 590 nm.

SEM backscattered-electron (BSE) images of the Jordaens and Zierikzee samples were obtained on, respectively, an Evo 500 high vacuum electron microscope (Zeiss, Oxford) and on a Verios high-vacuum electron microscope (FEI, Eindhoven, Netherlands) at the FOM Institute AMOLF, Amsterdam.

The ATR-FTIR measurements of the Zierikzee sample were undertaken at the IPANEMA laboratory, Gif-sur-Yvette, France, on a Bruker HYPERION 3000 FTIR microscope using a focal plane array (FPA) detector and an ATR germanium hemisphere crystal.

12.3 Results and Discussion

12.3.1 Lead Soaps

12.3.1.1 Historical Sample (N103 CS9), Jacob Jordaens, *Revolt of the Batavian Against Roman Rule*, 1661–66, Royal Palace Amsterdam

One of the large-scale paintings in the gallery arches of the Citizen's Hall of the Royal Palace Amsterdam depicts *The Revolt of the Batavians against Roman rule* (Fig. 12.1a). The painting has recently undergone conservation and restoration treatment (2005–09). It was glue lined in the eighteenth century, and wax-resin lined in the 1960s. It demonstrates a dramatic darkening, which may have been caused by the combination of aging and past treatments (Froment and Van Eikema Hommes 2011). Extensive lead soap formation in the ground is thought to play a role, and impregnation of the paint with wax and resin from the lining adhesive likely also contributes to color change. The sample studied here was taken from a greenwash to the left side of the moon. The painting was prepared with a double ground. The first beige ground consists of mainly chalk with some (saponified) lead white, a little iron earth pigment with a few umber particles, while the second darker brown ground is composed of lead white, chalk, earth pigment and a relatively higher proportion of umber. On top of the ground layers there is a thin dark gray paint layer with carbon-based black and a little lead white pigment. The second ground is rich in small and large particles and agglomerates of lead white. The sample studied here illustrates a relatively high degree of saponification of the lead white pigment particles in the ground. The opaque lead white particles have converted into lead soaps that have a refractive index similar to the oil binder, causing increased translucency of the paint (Keune et al. 2011; Noble et al. 2008).

The appearance of the second ground layer is striking when observed using UV-A microscopy, which shows highly luminescent particles, as well as particles with a luminescent halo around them (Fig. 12.1c). The SEM backscattered electron image reveals differences in morphology of the lead white pigment particles, which point to lead soap formation. This is supported by the micro-ATR-FTIR imaging data (not shown here). Synchrotron-based photoluminescence (SR-PL) micro-imaging was employed to obtain more detailed information and to investigate the inhomogeneity of the luminescence behavior within the partially saponified lead white paint film.

Figure 12.1 shows the bright-field image, the UV-A-induced luminescence, the SEM backscatter electron image, and SR-PL images obtained upon excitation at 280 nm at two different emission wavelength ranges, 499–529 nm and 641–708 nm. In the SR-PL images collected using $\lambda_{exc} = 280$ nm, contrast is strongly enhanced compared to visualization using UV-A illumination which allows distinguishing the gray surface paint layer (layer II in Fig. 12.1b) and the second ground (layer I in Fig. 12.1b) from each other. It is observed that in the 499–529 nm emission image (Fig. 12.1e) the paint matrix appears much darker (shows less luminescence) in comparison to the 641–708 nm emission image (Fig. 12.1f). Various lead white

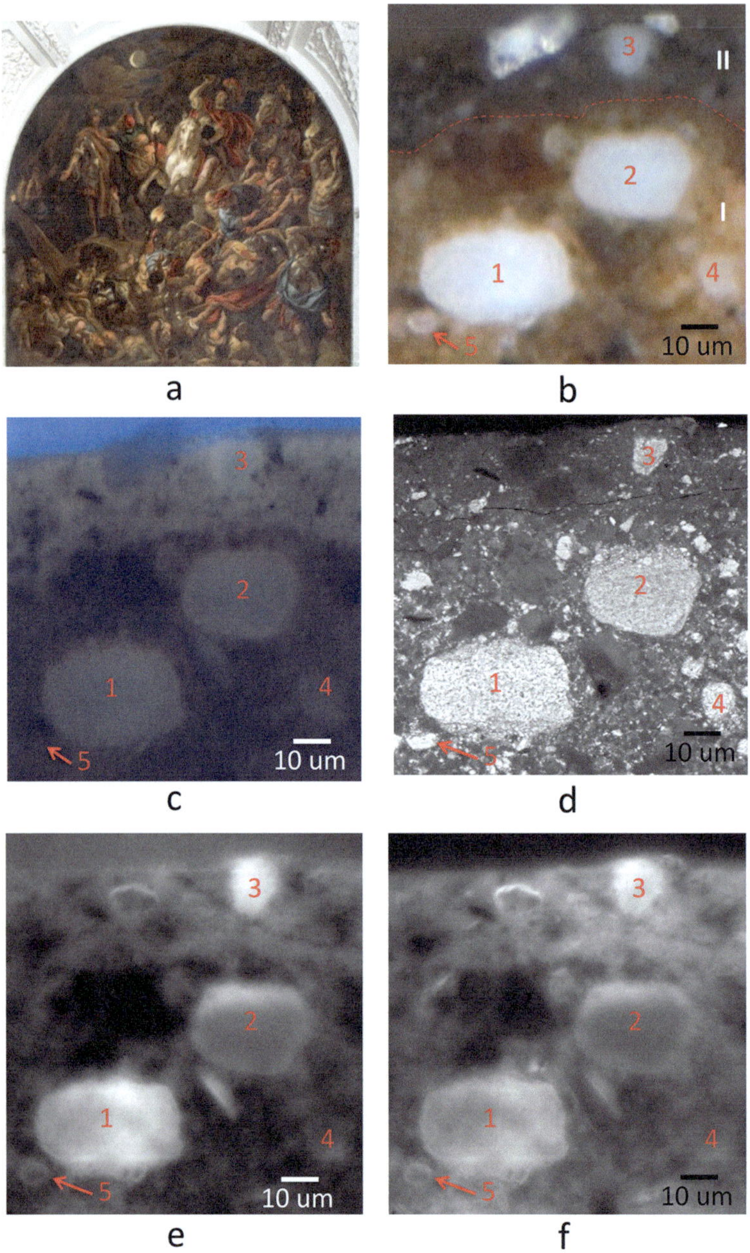

Fig. 12.1 (**a**) *Revolt of the Batavians against Roman rule*, Jacob Jordeans, 1661–66, Royal Palace, Amsterdam, the Netherlands, oil on canvas; (**b**) bright-field image of a sample cross section taken from a green-wash to the left side of the moon showing the gray surface paint layer (II) and the second ground (I); (**c**) UV-A induced microscopic image acquired using a fluorescence microscopy filter cube (filter block ex. 365 nm; em. > 365 nm); (**d**) SEM backscattered-electron image; (**e**) photoluminescence image collected at λ excitation = 280 nm and λ emission = 499–529 nm. Image histogram stretch, 2–98 percentile on the whole image; (**f**) photoluminescence image collected at λ excitation = 280 nm and λ emission = 641–708 nm; image histogram stretch, 2–98 percentile on the whole image

pigment particles/agglomerates (numbered 1–5 in the bright-field image, Fig. 12.1b) show differences in luminescence. Particles 1 and 3 are highly luminescent, while particles 2, 4 and 5 are hardly visible in the SR-PL images. Interestingly, the rim of these particles, however, does exhibit some luminescence. For example, the upper rim of particle 2 luminesces, while the rest of the particle appears dark. Differences in luminescence are visible within the large lead white agglomerate, particle 1. Both the upper and lower rims of this agglomerate exhibit some fluorescence. It is possible to relate the differences in luminescence to the condition/state of the lead white pigment. The SEM backscattered electron image (Fig. 12.1d) shows that the agglomerate of lead white particles (particle 1) has disrupted edges at the upper and lower rims, while the rest of the particle is solid and has clearly defined edges. The same is the case for the upper rim of lead white particle 2, which also looks degraded. Unaltered lead white does not show any detectable luminescence in this emission range (De la Rie 1982). Therefore it can be inferred that the luminescence in the lead white paint is associated with lead soap formation and is a sign of chemical change (MacDonald et al. 2016). The high spatial resolution that was achieved using the SR-photoluminescence imaging allowed visualization of the degradation within one lead white agglomerate. It is known from the literature and from our own experience that pure soaps of long-chain saturated fatty acids do not luminesce on excitation in the deep UV, but the photoluminescence behavior of metal soaps in oil paints may be caused by the presence of disordered states, as suggested by recent studies (Hermans et al. 2015; Martinez-Casado et al. 2014).

12.3.1.2 Historical Sample (ZZ7210_x12), Seventeenth-Century Painted Ceiling in the Room of Trustees, Burgerweeshuis, Zierikzee

In 2014, the seventeenth-century ceiling paintings of allegorical representations in the Room of Trustees, Burgerweeshuis in Zierikzee, underwent an intensive conservation treatment (Fig. 12.2a). The paintings are mounted directly on the wooden planks of the ceiling. In the eighteenth century, they were coated with a thick varnish and had been partly reworked while preserving the seventeenth-century picture beneath. As part of the recent treatment, the later layers of varnish and overpaint were removed revealing the original paint surface that dates to 1680. During the treatment, an earlier red-orange-colored decorative scheme was discovered below the 1680 picture. Various manifestations of lead soap-related degradation phenomena were observed in the multiple paint layers. The focus of this discussion is on the large lead soap aggregates that have formed in the red-orange paint, and how SR-PL was used to visualize multiple phases inside these soap aggregates.

The paint sample discussed here, ZZ7210_x12, was taken from the red sky of the allegorical representation of the reconciliation of Justice (Justitia) and War (Mars), on the west side ceiling located near the fireplace (Fig. 12.2a, right painting). The sample contains the complete layer structure before treatment,

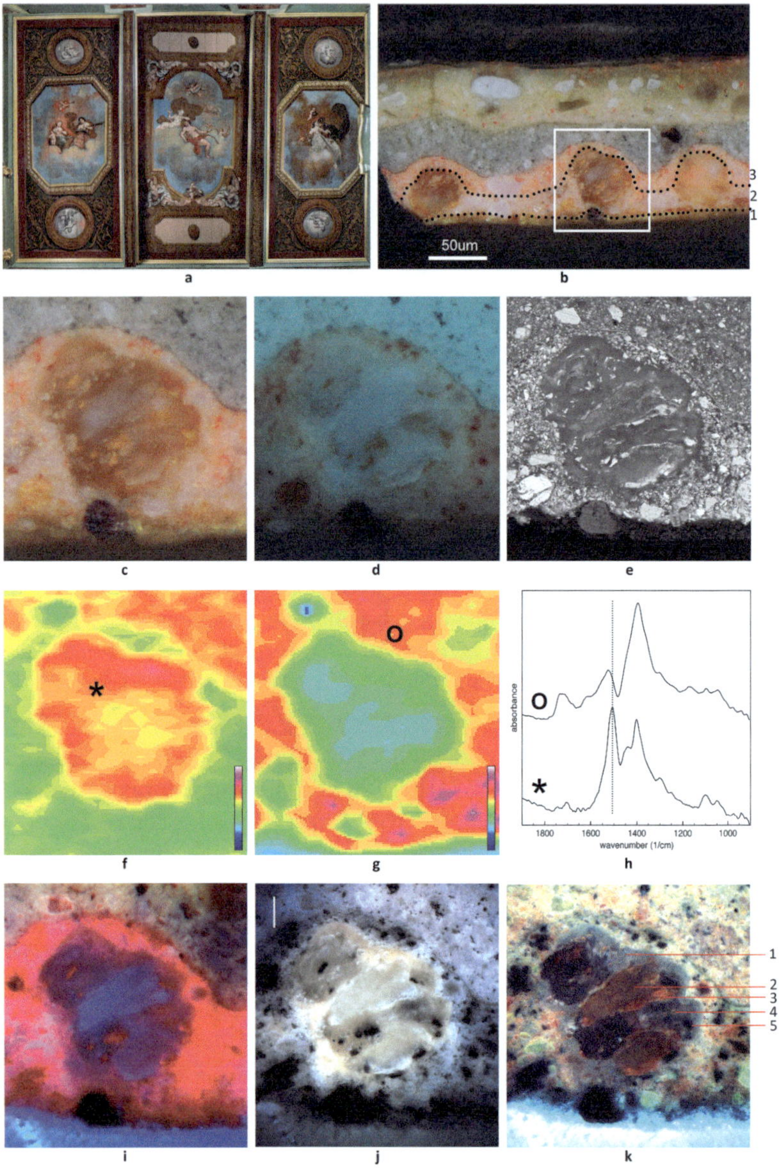

Fig. 12.2 (**a**) Seventeenth-century ceiling painting with allegorical representations, 1680, Room of Trustees, Burgerweeshuis, Zierikzee (Photo: Wim Ruigoord); (**b**) dark-field image of a sample cross section taken from the red sky of *Justitia and Mars in a Warm Embrace*. Photographed at 200×; (**c**) detail of the lead soap aggregate in dark field, corresponding to the white square in (**b**); (**d**) UV-A-induced image of the aggregate using fluorescence microscopy filter cube D (λ excitation = 355–425 nm; λ emission >470 nm); (**e**) SEM backscattered-electron image; (**f**) FTIR image of the *c.*1510 cm^{-1} band in false color (color ranging from blue-low intensity- to purple- high intensity; see inserted color bar) showing the distribution of lead soaps; (**g**) FTIR image of the *c.*1400 cm^{-1} band in false color showing the distribution of lead carbonates;

including the red-orange layers from the earlier decoration, the paint layers from the 1680 allegory, and the eighteenth-century additions (Fig. 12.2b). The red-orange decorative scheme consists of a thin, yellowish ground (layer 1) that contains chalk and earth pigments, followed by a red underlayer (layer 2) with lead white and red lead, and a red surface layer (layer 3) that contains lead white and vermilion. We conclude that the latter was indeed the final layer of a first decorative scheme, because it is covered with a thin surface haze (max. 1 μm thick). This superficial layer has a translucent brown appearance in visible light. EDX reveals it is rich in lead and sulfur. Thin surface layers such as this typically form as a result of interaction between the paint surface and atmospheric compounds (Van Loon 2008; see also the contribution in this Volume by Van Loon et al. on the Pellegrini paintings). Therefore, the red top layer (layer 3) must have been exposed to air for some time before the 1680 painting was commenced.

The lead soap aggregates originate in the red lead-containing layer (layer 2). They have a diameter up to 50 μm and have pushed up the red paint layer that lies on top (layer 3). Both light microscopy and SEM backscattered electron imaging reveal heterogeneity within the lead soap aggregates (Fig. 12.2c–e). Translucent regions visible in the dark-field image (Fig. 12.2c) correspond to the lower scattering (darker) regions in the BSE image (Fig. 12.2e), indicative of a higher organic (low Z) content. Less translucent, whitish regions show a slightly higher scattering (gray) in the BSE image, which points to an increase in lead content, and these regions may contain many dispersed nano-sized particles. In addition, the aggregate contains orange particles of red lead, as well as streaks of opaque whitish particles surrounding the semi-translucent whitish regions. The latter are lead-rich, high scattering (light gray or white) in the BSE. They are very likely re-mineralization products. The ATR-FTIR maps indicate the presence of lead carboxylates (map at *c.*1510 cm^{-1}) inside the aggregate, with a small contribution of carbonates (map at *c.*1400 cm^{-1}), and carbonates more abundantly present in the paint matrix surrounding the aggregate, indicative of unaltered lead white (Fig. 12.2f–g). The maps show some heterogeneity in the distribution of the carboxylate and carbonate bands inside the aggregate. In the FTIR spectrum of the surrounding lead white-containing paint, the position of the asymmetric stretch vibration of the lead carboxylate band ν_{as}(COO-) is shifted from *c.*1510 cm^{-1} to a slightly higher wavenumber (Fig. 12.2h, upper trace). This shift may be caused by lead ions bound

Fig. 12.2 (continued) (**h**) FTIR spectra of 1900–900 cm^{-1} region showing a shift in the position of the lead carboxylate band at *c.*1510 cm^{-1}. Lower trace: spectrum from aggregate. Upper trace: spectrum from surrounding paint matrix; (**i**) false-color photoluminescence emission image; λ excitation = 280 nm; λ emission = 698–766 nm (red), 452–486 nm (green), 370–410 nm (blue). Image histogram stretch, 2–98 percentile on the whole image; (**j**) false-color photoluminescence excitation image acquired using excitation in the UVA; λ emission = 452–486 nm; λ excitation = 380 nm (red), 360 nm (green), 340 nm (blue). Image histogram stretch, 2–98 percentile on the whole image; (**k**) false-color photoluminescence excitation image using excitation in the UVC; λ emission = 452–486 nm; λ excitation = 280 nm (red), 260 nm (green), 240 nm (blue); image histogram stretch, 2–98 percentile on the whole image to carboxylate groups on the polymerized oil network as in an ionomer (Hermans et al. 2016). The image obtained under UV-A illumination shows luminescence associated with the lead soap aggregate, but it is general and difficult to correlate to the different regions/phases within the aggregate (Fig. 12.2d).

Using the TELEMOS set-up at the DISCO beamline, RGB composite images of the photoluminescence at different emission ranges were obtained. Figure 12.2i shows 698–766 (red channel), 452–486 (green channel) and 370–410 nm (blue channel) for $\lambda_{exc} = 280$ nm. The false color of the aggregate observed in the image varies from red to blue-purple and shows the inhomogeneity of the emission throughout the aggregate. The red lead particles in the aggregate and the surrounding paint layer stand out in the emission image at 698–766 nm (red channel in RGB composite), as well as the vermilion particles present in the red top layer. The purple regions correlate to the translucent regions in the visible light image. The violet regions, which have a high contribution from emission in the 370–410 nm range (blue channel), can be correlated with the more opaque, whitish regions in the visible light image.

Interestingly, visualization of the extent of the inhomogeneity within the aggregate can be improved by using different excitation wavelengths while collecting images of emission over the same ranges. The RGB image collected at 452–486 nm using three different excitation wavelengths in the UV-A $\lambda_{exc} = 380$ (red channel), 360 (green channel) and 340 nm (blue channel) reveals that the aggregate comprises five phases (Fig. 12.2j). The regions at the left and top of the aggregate that appear yellowish could not be distinguished as a separate phase using light microscopy or BSE imaging. The two light gray masses (phase 2) at the center and bottom of the aggregate correspond to the semi-translucent, whitish regions observed in the light microscopy image. Their bright white luminescent rims (phase 3) have excitation-emission behavior similar to the lead-rich particles/streaks in the BSE image. Dark gray-brown regions (phase 4) are visualized in the PL image. These correspond to transparent regions in the light microscopy image. They contain particles of red lead that do not emit and appear black (phase 5).

Comparison of emission images obtained at 452–486 nm upon UV-C excitation $\lambda_{exc} = 280$ (red channel), 260 (green channel) and 240 nm (blue channel) with the UV-A images shows that the metal soap phases have less intense luminescence than on irradiation in the UV-A and, furthermore, the false-color images show more differentiation among phases in the aggregate (Fig. 12.2k). Using UV-C excitation, the core of the aggregate is surrounded by a gray rim (phase 1), which is low scattering (dark) in the BSE image. Brown-reddish masses (phase 2) are noticeable that correspond with the masses that appear light gray under the UV-A excitation image. They show orange-red rims (phase 3), indicating that the luminescence is highest at 280 nm excitation (red channel). Interestingly, the luminescence of the lead white particles in the paint outside the aggregate is the highest at 260 nm excitation (green channel). It can be hypothesized that the re-mineralized products that occur at the orange-red rims (region 3) may be associated with a different lead carbonate phase (possibly neutral lead carbonate) than lead white pigment (mostly basic lead carbonate). Some scattered reddish particles are also visible in the gray

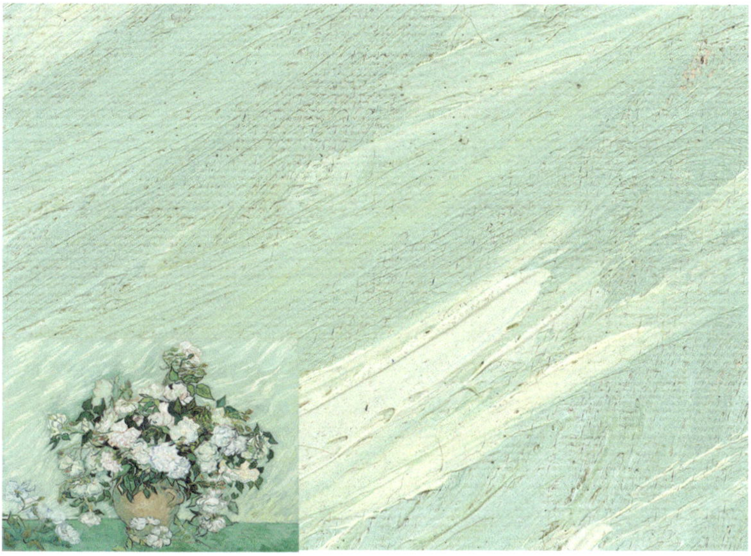

Fig. 12.3 *Roses*, Vincent van Gogh, 1890, National Gallery of Art, Washington. Detail from the background with overall image inset. A high resolution image is available at www.nga.gov

rim, as well as in the dark gray/black masses (phase 4). The red lead particles (phase 5) stand out more in the UV-A excitation image (black particles).

It can be concluded that the luminescence images obtained using different excitation energies have proved that there are a number of lead soap phases inside the aggregate and allow their distribution within the mass to be mapped. Work is under way to understand the origin of the variation in luminescence behavior that could lead to the identification of these phases.

12.3.2 Zinc Soaps

12.3.2.1 Historical Sample, Vincent van Gogh, *Roses*, 1890, Oil on Canvas, National Gallery of Art, Washington

Many paintings have condition problems associated with the use of zinc white paint. Writing to Vincent van Gogh (1853–1890), Paul Gauguin said, "The grape harvests are totally covered in scales as a result of the white which has separated" and he offered a remedy, which he suggested van Gogh might need for his own work (van Gogh Letters 1889). Indeed, the paint films in van Gogh's paintings are sometimes characterized by a chalky, fissured appearance (Osmond et al. 2005). In van Gogh's *Roses* (1890), the paint film in the mint green background is permeated by a network of small fissures that are aligned with the fibers of the canvas (Fig. 12.3). The vertical fissures seem to be more pronounced than the horizontal ones.

A sample from this painting was obtained to examine zinc soap formation using SR-PL imaging. The artist often used a high proportion of zinc white in mixtures, as is the case here. The green paint in the background of *Roses* is made from a mixture of zinc white with emerald green ($Cu(C_2H_3O_2)_2.3Cu(AsO_2)_2$) and a small proportion of a carbon-based black. Only the paint layer is present in the sample. The image obtained in bright field with crossed polars is shown in Fig. 12.4a. The emission of the sample on irradiation using UV-A/VIS excitation on the microscope is shown in Fig. 12.4b. Zinc oxide is a class II–VI semiconductor having a band gap of 2.9–3.45 eV (380–427 nm) and a strong luminescence centered at 384 nm, though particles with crystal defects or impurities have emission at longer wavelengths (Bertrand et al. 2013). In this paint sample, ZnO particles can be seen as small bright dots dispersed throughout the sample due to their strong emission in the visible range. Larger spherical particles of emerald green are dim, and the bulk of the paint, including the fissures, has a generally distributed green or blue-green glow that visually appears relatively homogeneous. Since ZnO absorbs only at wavelengths shorter than ca. 380 nm, the visible-induced emission, shown in Fig. 12.4c, allows observation of luminescence signals produced by species other than ZnO, such as the binding medium. Under these conditions, a diffuse emission from the paint film including in the cracks and fissures is observable; however it is not easy to distinguish by eye between emission from unaltered paint film and zinc soaps.

UV-C excitation and false-color multispectral imaging of the resultant emission clearly differentiate among and offer visualization of the phases that are present, shown by the results illustrated in Fig. 12.4d–f. The images in Fig. 12.4d–f were generated using $\lambda_{exc} = 280$ nm. In (d) emission from particles of ZnO dominate the image since the bad gap emission of ZnO is assigned to the red channel, but areas within the paint film have strong emission bands at 499–520 nm (blue channel) and at 641–708 nm (green channel). Additionally, it can be observed that the fissures are filled with a complex that emits in the UV.

In the zoomed detail of the false-color SR-PL image depicted in (e) the blue channel is used to map emission at 327–353 nm. The phase responsible for this emission completely fills the fissures and cracks in the paint film. It is apparent that this phase and/or another is present in "hot spots" that are magenta/purple in the image evident, for example, at the center of the field of view in the false-color image (e). A similar emission profile is present around individual particles of emerald green suggesting a chemical similarity between the hot spots and the periphery of the particles of emerald green. The highest zoom of the false-color SR-PL image (f) shows that the luminescence of zinc oxide particles (appearing red) seem to be unaltered. Therefore, zinc soap formation is not ongoing at these boundaries. These results suggest that soap formation does not begin at cracks and fissures, but once formed, soaps aggregate within fissures and likely cause their expansion.

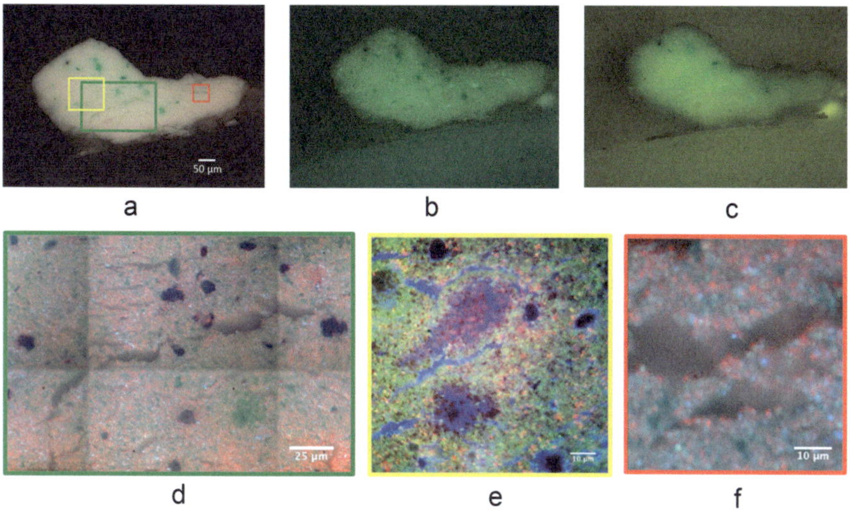

Fig. 12.4 Images of a sample cross section taken from the background of *Roses*, by Vincent van Gogh. (**a**) Bright-field (crossed polars) image showing that the paint is made from a mixture of zinc white, emerald green, and a carbon black; (**b**) the sample viewed using fluorescence microscopy filter cubes D (excitation band-pass 355–425 nm, dichroic filter 455 nm, and long-pass filter 470 nm); (**c**) fluorescence microscopy filter I3 (excitation band-pass filter 450–490 nm, dichroic filter 510 nm, and long-pass filter 515 nm); (**d**) detail corresponding to the green rectangle in **a** at 100×. λ excitation = 280 nm. λ emission of false-color image channels = 370–410 nm (red); G: 641–708 nm (green); B: 499–529 nm (blue). Image histogram stretch: square root on the whole image; (**e**) zoom of false-color PL image corresponding to the yellow rectangle in (**a**) at 100×. λ excitation = 280 nm. λ emission of false-color image channels = 452–486 nm (red); 370–410 nm (green); 327–353 nm (blue). Image histogram stretch: equalization on the zoomed area; (**f**) zoom of the false-color PL image corresponding the red rectangle in (a) at 100×. λ excitation = 280 nm. λ emission of false-color image channels = 370–410 nm (red); 698–766 nm (green); 641–708 nm (blue)

12.4 Conclusions

Synchrotron deep-UV photoluminescence micro-imaging has provided unprecedented information to help analyze the chemical complexity of metal soap formation and their aggregation. Taking advantage of the high spatial resolution of synchrotron PL imaging, we characterized early stages of reaction in a paint film occurring at the edges of individual lead white particles as well as at the interface of flocs of particles with the binder. The emission spectra of phases formed during the initial processes and later the development of aggregates of soaps have been shown to be different. The results indicate that there are many chemical phases present during the process of soap aggregate formation, and also within the final large lead soap protrusions. Here a larger number of phases than expected from traditional fluorescence microscopy or SEM-EDX was discovered. This finding was possible because the technique provides emission contrast at the sub-micron scale, allowing individual particles to be observed. The information obtained is distinct from and complementary to speciation using other techniques such as FTIR and SEM and elemental analyses.

Chemical alteration can be identified and located at an early stage of development and relatively low degree of degradation. In the case of zinc soap formation, it was observed that areas of alteration develop within the paint film and, in the case studied here, occurs around particles of emerald green. The alteration products do not appear to be developing at fissures, and soap formation is not active at their edges or at micro-cracks; however, soaps appear to aggregate within these cracks and this may be the cause of their expansion.

Ongoing work is aiming to use synchrotron PL to study the phenomena associated with soap formation in a much larger corpus of micro-samples to obtain information on the number of chemical species involved in the various processes of soap formation and robust chemical identification using PL characteristics and other techniques.

Acknowledgments This work is part of the PAinT project, supported by the Science4Arts program of the Dutch Organization for Scientific Research (NWO). We wish to thank Emilie Froment (University of Amsterdam) for providing the paint sample of Jacob Jordaens' *Revolt of the Batavian against Roman Rule*, Royal Palace, Amsterdam. We are also indebted to Roos Keppler and Annefloor Schlotter (independent conservators of historic interiors, the Netherlands) for providing the paint sample of the seventeenth-century painted ceiling in the Room of Trustees, Burgerweeshuis, Zierikzee.

References

Bertrand L, Réfrégiers M, Berrie B, Échard J-P, Thoury M (2013) A multiscalar photoluminescence approach to discriminate among semiconducting historical zinc white pigments. Analyst 138:4463–4469. https://doi.org/10.1039/c3an36874b

De la Rie RE (1982) Fluorescence of paint and varnish layers (part II). Stud Conserv 27(2):65–69

Froment E, Van Eikema Hommes MH (2011) The darkness of the nocturnal conspiracy of Claudius Civilis by Govert Flinck and Jürgen Ovens (1659–1662) in the Royal Palace Amsterdam. In: Bridgland J (ed) Preprints ICOM committee for conservation, 16th triennial meeting, Lisbon, 19–23 Sept

Giuliani A, Jamme F, Rouam V, Wien F, Giorgetta J-L, Lagarde B, Chubar O, Bac S, Yao I, Rey S, Herbeux C, Marlats JL, Zerbib D, Polack F, Réfrégiers M (2009) DISCO: a low-energy multipurpose beamline at synchrotron SOLEIL. J Synchrotron Radiat 16:835–841

Hermans JJ, Keune K, Van Loon A, Iedema PD (2015) An infrared spectroscopic study of the nature of zinc carboxylates in oil paintings. J Anal At Spectrom 30:1600–1608. https://doi.org/10.1039/c5ja00120j

Hermans JJ, Keune K, Van Loon A, Corkery RW, Iedema P (2016) Ionomer-like structure in mature oil paint binding media. RSC Adv 6:93363–93369. https://doi.org/10.1039/C6RA18267D

Higgitt C, Spring M, Saunders D (2003) Pigment-medium interactions in oil paint films containing red lead or lead-tin yellow. Natl Gallery Tech Bull 24:75–95

Keune K, Van Loon A, Boon J (2011) SEM backscattered-electron images of paint cross-sections as information source for the presence of the lead white pigment and lead-related degradation and migration phenomena in oil paintings. Microsc Microanal 17(5):696–701. https://doi.org/10.1017/S1431927610094444

MacDonald MG, Palmer MR, Suchomel MR, Berrie BH (2016) Reaction of Pb(II) and Zn(II) with ethyl linoleate to form structured hybrid inorganic–organic complexes: a model for degradation in historic paint films. ACS Omega 1(3):44–50. https://doi.org/10.1021/acsomega.6b00075

Martinez-Casado FJ et al (2014) Short lead(II) soaps: from weakly fluorescent crystals to strongly phosphorescent and structurally varied vitreous phases. A thermal, structural and spectroscopic study. J Mater Chem C 2:9489–9496. https://doi.org/10.1039/c4tc01645a

Noble P, Boon JJ, Wadum J (2002) Dissolution, aggregation and protrusion: lead soap formation in 17th century grounds and paint layers. In: Art matters, Netherlands technical studies in art, Waanders Publishers, Zwolle, pp 46–61

Noble P, Van Loon A, Boon JJ (2008) Selective darkening of ground layers associated with the wood grain in 17th-century panel paintings. In: Townsend J, Doherty T, Heydenreich G, Ridge J (eds) Postprints ICOM committee for conservation interim meeting, London, June 2007, pp 68–78

Osmond G, Keune K, Boon J (2005) A study of zinc soap aggregated in a late 19th century painting by R.G. Rivers at the Queensland art gallery. AIC CM Bull 29:37–46

Thoury M, Echard J-P, Réfrégiers M, Berrie B, Nevin A, Jamme F, Bertrand L (2011) Synchrotron UV-visible multispectral luminescence micro-imaging of historical samples. Anal Chem 83(5):1737–1745

Van Gogh Letters (734 Br. 1990: 739|CL: GAC 34 Date: Paris, between Tuesday, 8 and Wednesday, 16 January 1889). Les vendanges sont en totalité écaillées par suite du blanc qui s'est separé. http://vangoghletters.org/vg/letters/let734/letter.html

Van Loon A (2008) White hazes and surface crusts on dark oil paint films. In: Color changes and chemical reactivity in seventeenth-century oil paints. PhD dissertation, University of Amsterdam, Molart series (14), AMOLF, Amsterdam. Downloadable from http://www.amolf.nl/publications

Van Loon A, Keune K, Boon JJ (2005) Improving the surface quality of paint cross-sections for imaging analytical studies with specular reflection FTIR and static-SIMS. In: Proceedings of art'05 conference on non-destructive testing and microanalysis for the diagnostics and conservation of the cultural and environmental heritage, Lecce, Italy, May 15–19, 2005 (CD-ROM)

Chapter 13
Physicochemistry of Pure Lead(II) Soaps: Crystal Structures, Solid and Liquid Mesophases, and Glass Phases – Crystallographic, Calorimetric, and Pair Distribution Function Analysis

Francisco J. Martínez-Casado, José A. Rodríguez-Cheda, Miguel Ramos-Riesco, María Isabel Redondo-Yélamos, Fabio Cucinotta, and Alejandro Fernández-Martínez

Abstract A complete characterization of pure lead(II) soaps is crucial to their identification in the degradation of paintings. We present here our study on the physicochemical behavior and structure of the members of the lead(II) alkanoates series, from acetate to octadecanoate (stearate), in all the phases that they present: from the completely ordered phase (crystal) to the liquid phase. These soaps present two polymorphic structures in the crystal phase, now solved for all the compounds, one intermediate solid phase and a liquid crystal, prior to the isotropic liquid. In some members, different glass states (from different phases) are also found, with interesting photophysical properties. The compounds have been thoroughly analyzed as a function of temperature mainly by X-ray diffraction (powder and single crystal), differential scanning calorimetry and high-energy X-ray total scattering (for pair distribution function or PDF analysis), and other techniques like polarizing light microscopy, infrared and UV-Vis spectroscopy, and nuclear magnetic resonance. The use of new techniques that give information on the short-

---

F. J. Martínez-Casado (✉)
MAX IV Laboratory, Lund University, Lund, Sweden

Departamento Química-Física I. Facultad de Ciencias Químicas, Universidad Complutense de Madrid, Madrid, Spain
e-mail: francisco.martinez@maxiv.lu.se; fmarcas@quim.ucm.es

J. A. Rodríguez-Cheda · M. Ramos-Riesco · M. I. Redondo-Yélamos
Departamento Química-Física I. Facultad de Ciencias Químicas, Universidad Complutense de Madrid, Madrid, Spain

F. Cucinotta
School of Chemistry, Newcastle University, Newcastle Upon Tyne, UK

A. Fernández-Martínez
CNRS, ISTerre, Université Grenoble Alpes, Grenoble, France

F. Casadio et al. (eds.), *Metal Soaps in Art*, Cultural Heritage Science,
https://doi.org/10.1007/978-3-319-90617-1_13

range order (like PDF analysis or X-ray absorption) for the characterization of the disordered phases can help in the identification of lead(II) soaps and other compounds, in general, in the initial stages of formation prior to crystallization.

Keywords Lead soaps · Crystal structures · Calorimetry · Mesophases · Glass phases · Pair distribution function

13.1 Introduction

It was in 1997 when P. Noble and J. Wadum, conservators at the Mauritshuis in The Hague, when studying Rembrant van Rijn's *The Anatomy Lesson of Dr Nicolaes Tulp* (Keune 2005; Noble et al. 2000, 2003; van der Weerd 2002), found *protrusions* in the painting, later identified as lead(II) soaps. Lead soaps are anionic organic salts (or carboxylates) that are the cause of degradation of many famous oil paintings. Since then (Centeno and Mahon 2009; Hale et al. 2011; Higgitt et al. 2003; Keune 2005; Noble et al. 2003; Spring et al. 2008), not only lead soaps but others like zinc(II), copper(II), or potassium soaps have been found in many oil paintings. In fact, the presence of these soaps has been reported on the surfaces of metallic cultural artifacts (Mills and White 1999; Robinet and Corbeil 2003). As a consequence, the interest in these materials (which also behave as surfactants and mesogens such as thermo- and lyotropic liquid crystals) has increased dramatically in the area of conservation and restoration of cultural heritage.

The mechanism of metal soap formation has been proposed by Boon et al. (2007): metal soaps are formed when heavy metals in the pigments (e.g., *lead white* or *lead-tin yellow* pigments) react with the fatty acids that result from the hydrolysis of ester glycerides in the oil-binding medium (Boon et al. 2002; Higgitt et al. 2003; Noble et al. 2003; van der Weerd 2002). This mechanism of the ester hydrolysis consists of homogeneous acid-base catalysis due to acidic CO_2 in air moisture through the years that could penetrate into the painting. Soaps may also migrate to the surface of the painting, forming lead carbonates, hydroxychlorides, sulfates, etc., by reacting with carbon dioxide and other compounds in the environment (Boon et al. 2002; Higgitt et al. 2003; Keune 2005; van der Weerd 2002; Van Loon 2008).

The formation of metal soaps may affect a painting in different ways, such as forming aggregates that deform the paint surface or insoluble whitish surface hazes, making the paints drip or water sensitive, or even increasing the transparency of the paint (Boon et al. 2002; Centeno and Mahon 2009; Eibner 1909; Keune and Boon 2007; Noble et al. 2003, 2005; Van Loon 2008). In the latter case, for example, *undesired* transparency, which allows one to see the artist's preparatory drawing, alterations (*pentimenti*), and even seeing the support (wood, canvas, metal sheet, etc.) (Boon et al. 2002; Centeno and Mahon 2009; Eibner 1909; Van Loon 2008), has been attributed to lead soap formation, but without further explanation. On the other hand, physicochemical studies have shown that most of the pure lead alkanoates exhibit transparency in some phases at higher temperatures and some of them show these effects, even at room temperature due to the formation of glass states (Martínez-Casado et al. 2007a, 2012, 2014). Thus, the formation of these phases could explain the transparency in paintings.

Artworks can be considered "alive": with different components evolving and reacting over time and in different timescales. Therefore, the study of metal soaps in art requires a prior and thorough characterization of the pure compounds and all their phases, which is fundamental and necessary to understand their properties, characteristics, and behavior and to identify them in more complex systems, such as oil paintings. In the analysis of metal soaps, it is also important to group them in a series, so the particular behavior can be correlated and understood from the characteristics of the families (Cheda et al. 1999; Martínez-Casado et al. 2012; Ramos-Riesco et al. 2015), like in the case of similar phases and their properties.

Pure metal soaps, in general, and pure lead soaps, in particular, show several phases from the crystal to the liquid phase at increasing temperatures. For example, a thermotropic liquid crystal phase may exist under some conditions. Moreover, a lyotropic phase may be formed at lower temperatures, proving that the mixture of two or more components in complex systems, as in the case of paints (e.g., free acids from glyceride ester hydrolysis), can lead to the formation of different phases with diverse characteristics: opaque, translucent or transparent, fluid or solid, etc.

The unique nature of each of the phases makes the use of a wide variety of experimental techniques essential for proper characterization. Calorimetry, diffraction, spectroscopy, and microscopy are the most common ones and provide information about the thermal behavior, crystal structure, coordination, conformation of the alkyl chain, optical properties, etc. However, other techniques are necessary to describe the structure of disordered phases, like liquid crystal, liquid, and glass states. In this regard, pair distribution function (PDF) analysis and X-ray absorption (XAS) provide important information about the short-range order of those phases (bond distances and, in the case of XAS, coordination sphere as well), as will be shown in this chapter, with some PDF data. As in the case of the diffraction pattern for crystal structures, PDF analysis and XAS provide characteristic "fingerprints" that could be used in the later identification of metal soaps in paintings, even in the early stages of their degradation.

Soap formation chemistry is not yet fully understood, and it does not take place in the same way in all artworks containing the same potentially reactive materials. The activating factors and mechanisms are not fully known nor how to change or prevent them. There is a lot of work yet to be done, mainly in how phase structures and dynamic of the processes are involved.

13.2 Phases and Mesophases of Lead Soaps (and Their Glass States): From Crystal to Liquid

Lead soaps can be studied as a series of lead(II) alkanoates, as a function of the alkyl chain length of the alkanoate anion, e.g., from acetate to stearate, that is, from 2 to 18 C atoms in the alkyl chain. Lead soaps (**Pb(C*n*)$_2$**, where n is the total number of C atoms in the alkanoate anion) present different phases and mesophases

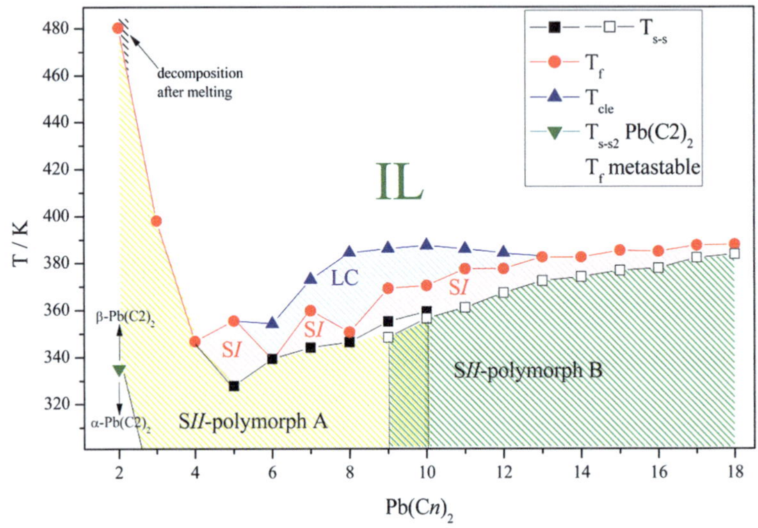

Fig. 13.1 Temperature vs. number of C atoms (n) diagram, showing the different phases and mesophases and the transition temperatures for the lead soap series (where, $\mathbf{T_{s\text{-}s}}$, solid *II*–solid *I*; $\mathbf{T_f}$, fusion, solid *I*–liquid crystal, in general; $\mathbf{T_{cle}}$, clearing or liquid crystal–liquid; $\mathbf{T_{s\text{-}s2}}$, solid α–solid β for $Pb(C2)_2$; $\mathbf{T_f}$**metastable**, metastable solid *I*–liquid or liquid crystal for $Pb(C4)_2$ or $Pb(C6)_2$, respectively)

(ones between crystalline solid and liquid), from low to high temperature: from total order to disorder. The sequential appearance of phases with temperature (*stepwise melting process*) (Martínez-Casado et al. 2012; Ramos-Riesco et al. 2015) is due to the fact that metal soaps contain two main components: one ionic (metal ions and carboxylate anions) and one organic (alkyl chain).

Depending on the specific member of the series, they may present different crystal phases or polymorphs (S*II*) and two mesophases, a solid *rotator* phase (S*I*) and a liquid crystal phase (LC), before they melt into the isotropic liquid phase (IL). The transition temperatures among the different phases are shown in Fig. 13.1.

The main characteristics of the four different phases and the glass states, summarized in Table 13.1, are detailed as follows:

Crystal Phase (Solid *II* or S*II*) This is the totally ordered phase at room and lower temperatures. The molecules have positional and orientational order, and the alkyl chains are in the *all-trans* conformation. Two polymorphic forms are found as a function of the alkyl chain length: polymorph **A** (from $\mathbf{Pb(C2)_2}$ to $\mathbf{Pb(C\mathit{10})_2}$) and polymorph **B** (from $\mathbf{Pb(C\mathit{9})_2}$ to $\mathbf{Pb(C\mathit{18})_2}$ or longer). $\mathbf{Pb(C\mathit{9})_2}$ and $\mathbf{Pb(C\mathit{10})_2}$ are the only members of the series showing both forms. Polymorph A shows the same molecular arrangement despite the fact that the members of the series belong to different crystal systems (monoclinic, $P2_1/m$, for $n \leq 4$, and triclinic, P-1, for $n \geq 5$) (Lacouture et al. 2001; Martínez-Casado et al. 2011, 2014, 2016, 2017). In the case of polymorph B, all members present a monoclinic unit cell

Table 13.1 Phases and mesophases in the lead soaps. Characteristics, orders, disorders, and types of glasses

Phase[a]			
Crystal (S*II*)	Rotator (S*I*)	Liquid crystal (LC)	Isotropic liquid (IL)
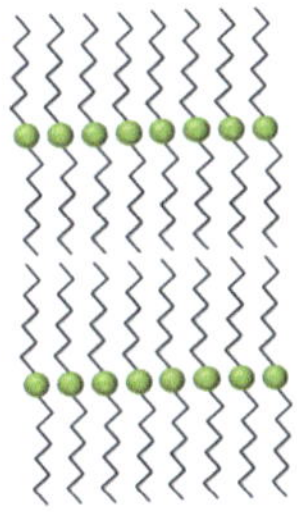	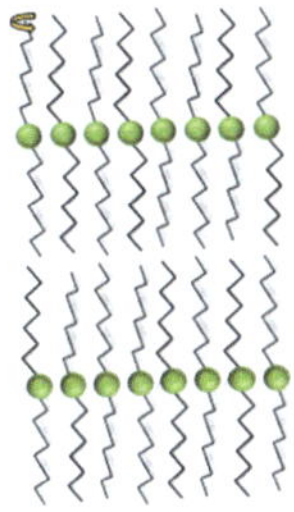	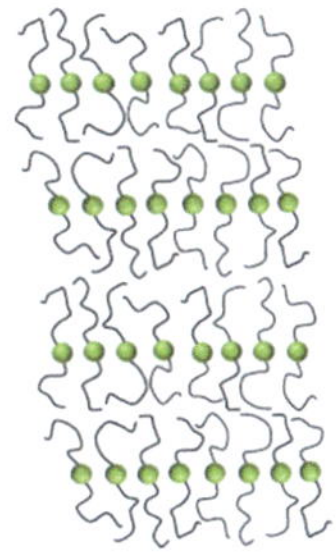	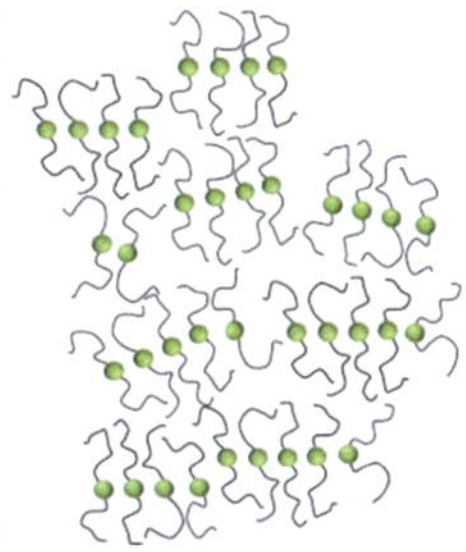
Characteristics of the (meso)phase			
Solid phase	Solid mesophase	Fluid mesophase	Fluid phase
Birefringent	Birefringent	Large domains ➔ birefringent	Small domains ➔ non-birefringent
Diffraction:	Diffraction:	Diffraction:	Diffraction:no
crystalline	crystalline	peaks (00 l)	peaks
Order			
All	Positional, orientational (directional), and conformational (*all-trans* alkyl chains)	Positional (domains) and orientational (direction vector perpendicular to the layers)	Short range
Disorder: dynamic (mesophase) or static (glass)			
None	Internal orientational (rotation of the alkyl chains)	Positional (fluid) and conformational (alkyl chains "melted")	All
Members of the $Pb(Cn)_2$ series presenting these phases			
All	Members with $n = 5$ and $n \geq 7$	Members with $6 \leq n \leq 12$	Members with $n \geq 3$; $Pb(C2)_2$ decomposes
Members of the $Pb(Cn)_2$ series presenting the corresponding glass state			
None	Rotator glass: G_I $Pb(C5)_2$, $Pb(C7)_2$	Liquid crystal glass: G_{LC} $Pb(C6)_2$	Regular glass (or amorphous): G_{IL} $Pb(C3)_2$, $Pb(C4)_2$, $Pb(C5)_2$

[a]In the figures, the carboxylates and lead atoms (ionic part) are represented with green balls, whereas the alkyl chains (lipidic part) with black lines

(monoclinic, *P2/c*) (Catalano et al. 2015; Martínez Casado et al. 2017). Particularly, $\mathbf{Pb(C2)_2}$ shows another polymorphic state (α), with an order-disorder transition to β, the high-temperature polymorph, which corresponds to the polymorphic form **A** (Martínez-Casado et al. 2016). All the structures of $\mathbf{Pb(C2)_2}$ were solved by single crystal and powder X-ray diffraction.

Both polymorphs, A and B, are two-dimensional coordination polymers (2D-CP), that is, Pb atoms and carboxylates grow in layers (2D), which are stacked and form plate-like crystals. For both as well, there are two different kinds of alkyl chains, both with *all-trans* conformation, and the planes formed by the C atoms in the hydrocarbon chains (*all-trans*) are arranged with a *herringbone* structure (*internal orientational* order) (Martínez-Casado et al. 2012). Although the coordination number of Pb is the same (7), the main difference between the two polymorphs lies in the different coordination of the Pb atoms: (a) *hemidirected* (with the O atoms directed throughout one hemisphere of the coordination sphere), and (b) *holodirected* (with the O atoms more regularly situated around the Pb atom) for polymorphs A and B, respectively. This different coordination was also observed by Catalano et al. using ^{207}Pb-NMR (Catalano et al. 2014, 2015), which can be used to distinguish between the two polymorphs.

The two polymorphic structures are shown in Fig. 13.2, showing the different coordination and arrangement of the molecules.

***Rotator* Phase (Solid *I* or S*I*)** The *rotator* phase is a solid mesophase with order in the position and orientation of the molecules (crystalline and birefringent) and with *all-trans* conformation of the alkyl chains (observed by infrared spectroscopy). However, ^{13}C-NMR spectroscopy shows a cooperative partial rotation (dynamic) of the alkyl chains (Martínez-Casado et al. 2007b, 2008), which causes the loss of the *internal orientational* order. This phase exists for members with $n = 5$ and $n \geq 7$, although it also appears as a metastable phase in the cases of $Pb(C\mathit{4})_2$ and $Pb(C\mathit{6})_2$ (Martínez-Casado et al. 2014).

PDF analysis in this phase indicates a shortening of the Pb-Pb distances with respect to both polymorphs of the S*II* phase, which explains that this one and the two remaining phases (liquid crystal and liquid) present interesting phosphorescence in comparison with the weak fluorescence of the crystal (Burrows 1988; Martínez-Casado et al. 2014).

Liquid Crystal (LC) The liquid crystal phase is a fluid mesophase with dynamic positional disorder (mobility) and orientational order which, in the case of the lead soaps, appears between $n = 6$ and 12. The molecules keep a layered structure in this phase as well, and the alkyl chains are melted (they present dynamic *trans-gauche* conformational disorder), but the average director vector is perpendicular to the layers, so this mesophase is *smectic-A*. In this sense, the LC phase is transparent to visible light but birefringent under polarized light, due to the orientation of molecules, organized in clusters or domains of a certain size.

The LC phase shows the same short-range arrangement of atoms as in the *rotator* phase, observed by PDF analysis.

Liquid Phase (Isotropic Liquid or IL) The liquid phase is characterized by the dynamic disorder of the molecules and does not show birefringence under polarized light. However, this phase presents a short-range order of the atoms (up to a distance of 10 Å) similar to the *rotator* and liquid crystal mesophases. This fact implies that

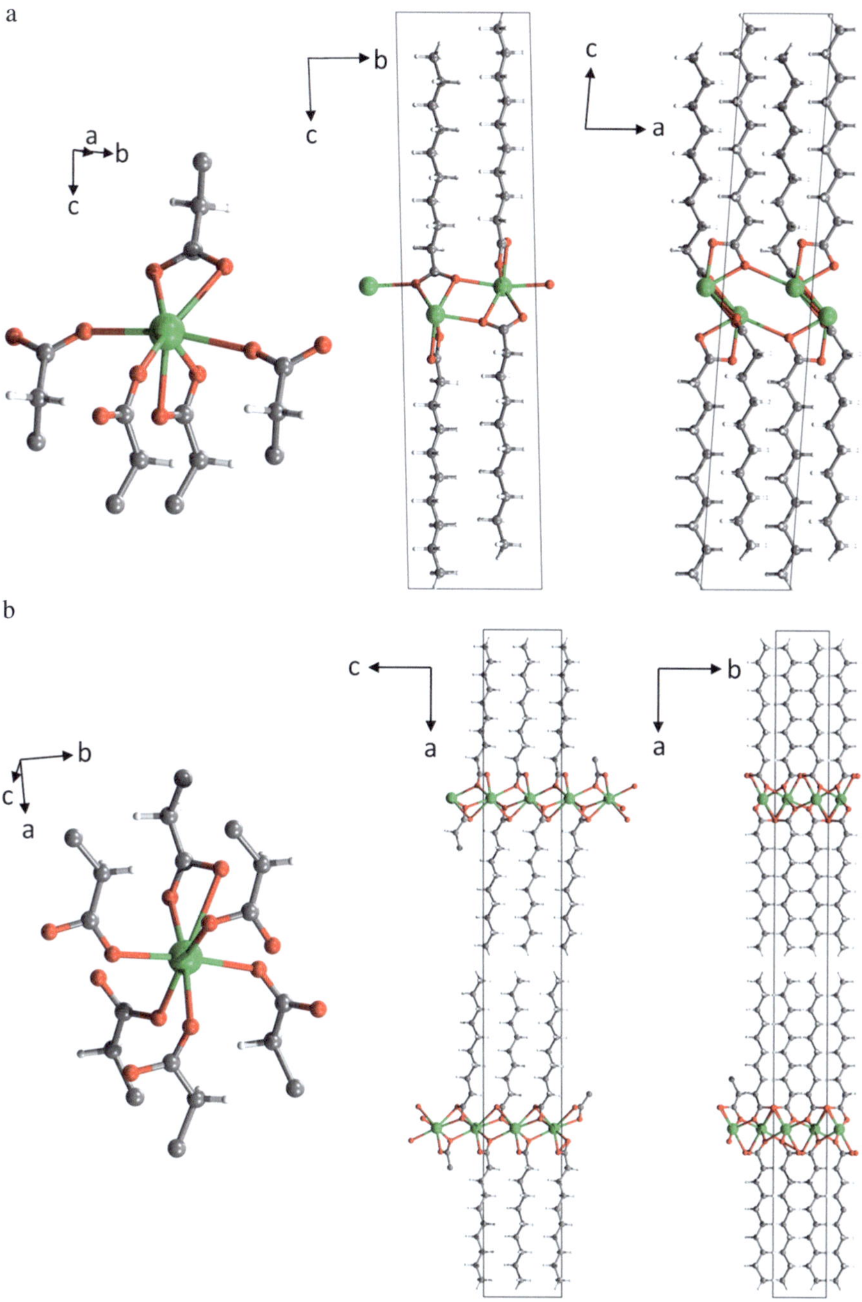

Fig. 13.2 Crystal structures of polymorphs A (a) and B (b) of Pb(C*10*)$_2$, showing the coordination of the Pb atoms (*hemidirected* and *holodirected*, respectively) and represented in the indicated directions in both cases. Pb atoms are represented in green, while O, C, and H are in red, grey, and white, respectively

the liquid is not completely disordered. Rather, in the liquid phase, the domain size is smaller than the one in the liquid crystal phase.

Three Different Glass States (G_{IL}, G_{LC}, and G_I) (Martínez-Casado et al. 2014) The last three phases (liquid, liquid crystal, and *rotator*) present different types of dynamic disorders that can be quenched, obtaining their corresponding glass states at low temperature, where the disorders are static. The main properties of these glasses remain similar to the properties of the phases from which they are obtained, like diffraction pattern and birefringence, since the molecular structure is maintained. These three types of glasses appear differently for some specific short members of the series. Thus, the regular glass (G_{IL}, from the liquid phase) is obtained for members with $n = 3–5$, liquid crystal glass (G_{LC}) only for $Pb(C6)_2$, and rotator glass (G_I) for $Pb(C5)_2$ and $Pb(C7)_2$.

As in their respective phases, these glasses show local structures with shorter Pb-Pb distances in comparison with both polymorphs of the S*II* phase, causing the strong phosphorescence at low and room temperate, in contrast of the weak fluorescent found in the S*II* phase.

It is worth noting that all these phases are referred to pure lead(II) alkanoates and appear at high temperatures or after thermal treatment. However, they may also appear in complex systems, like paintings, in their initial stages of formation (nucleation and growth) or by the effect of mixing with other components of those systems. Thus, lead soaps formed in a painting might not be necessarily in a crystalline state, so their presence in the form of a different phase or glass should be also sought, this last state being the possible cause of the formation of the lead soaps' transparencies found in some paintings (Boon et al. 2002; Centeno and Mahon 2009; Eibner 1909; Van Loon 2008).

13.3 New Tools for the Identification of Lead Soaps: PDF Analysis on Different Compounds and Their Phases

One of the main techniques used for the identification of metal soaps in paintings is Fourier transform infrared spectroscopy (FTIR), for its versatility and ease to adapt for in situ analysis, although other techniques, such as powder X-ray diffraction (P-XRD) or scanning electron microscopy (SEM), are employed as well. They are complementary but may have some limitations and fail to identify these compounds.

FTIR provides information about the vibration of chemical bonds. For example, C-O stretching modes are used to detect the presence of carboxylates and, therefore, of metal soaps. However, the bands depend on the environment of the bonds, and if the metal coordination changes, as it happens in a different phase, so do the vibration frequencies of the metal-ligand and carboxylate bonds. Then, those bands are not precise and reliable for the identification of carboxylate groups of a certain phase of a metal soap. On the other hand, P-XRD is used to analyze the diffraction pattern of a crystalline compound in a certain phase, which works like a *fingerprint* that can be used in the identification even for multicomponent mixtures. Nevertheless, the identification of metal soaps by P-XRD may not be possible for a different phase or for noncrystalline samples or phases. Finally, SEM requires a sample of the paint and studies its surface and identifies the chemical species, but not their coordination, and it does not give information about the core of the sample.

Other characterization techniques, like X-ray absorption (XAS) and pair distribution function (PDF) analysis, may complement the previous techniques, offering new possibilities in the field of metal soap identification in artworks. Both techniques use X-rays (PDF can also be obtained with neutrons), which are able to penetrate much deeper than low energy radiation and allow studying the inner part of the samples, not only the surface regions. Nowadays, large facilities for X-rays, like synchrotron sources, are optimal candidates to host these techniques due to the small size and intensity of the beam and the energy tunability.

XAS is used for determining the local geometric and/or electronic structure of matter. XAS data can be obtained tuning the photon energy to a range where core electrons of a certain element are excited. It gives information about the environment of that element: oxidation state, coordination number, and geometry of this coordination. The PDF describes the distribution of distances between pairs of atoms and can be obtained from high-energy X-ray (or neutron) total scattering, which includes Bragg scattering as well as diffuse scattering. Hence, PDF includes information about the periodicity or crystal structure (from the Bragg peaks) and about the short-range order (from the diffuse scattering). Both techniques are typically used for fluids, glasses, and disordered materials in general, but they can be used on crystal phases because they allow discrimination between short-range order (represented by finite nonrandom displacements from the ideal crystal structure) and random displacements of the atoms.

The PDF of a given compound in a certain phase shows the distance distribution unique for that phase. It can be used as a *fingerprint*, as in the case of the diffraction pattern, to discriminate between compounds or to find characteristic distances of one of them. In Fig. 13.3, the PDFs calculated for lead(II) stearate and *white lead* are shown.

Another important aspect of the PDFs consists of the source used to obtain it. The signal coming from the different atoms depends on the atomic scattering factor, which, in the case of X-rays, is a function of the atomic number (Z), so the signal of heavy atoms is stronger and this information dominates the PDF. On the other hand, the atomic scattering factor with neutrons does not vary in the systematic way as with photons, and the light atoms (H, C and O, for example) are more "visible" in the PDF.

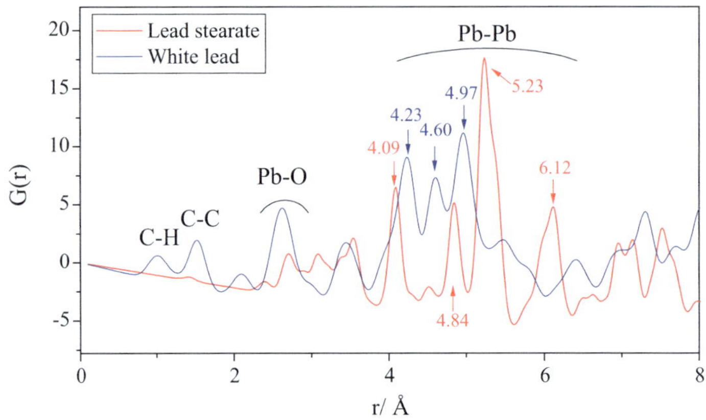

Fig. 13.3 Calculated PDF (for X-rays) of lead(II) stearate (or Pb(C*18*)$_2$, in blue) and *white lead* (basic lead carbonate, in red), showing some characteristic distances in the short range

13.3.1 P-XRD Versus PDF: Comparison of the Phases of Lead Soaps

The structural study of the crystalline phases of lead soaps has been carried out by single crystal or powder XRD, solving and refining the structures of the S*II* phase, as shown in the previous section. However, this has not been possible in the S*I* phase, and very little information is obtained in the more disordered phases, especially in the liquid crystal, liquid, or glass states. The diffraction patterns of the four phases of Pb(C*11*)$_2$ are given in Fig. 13.4a: the Bragg peaks are clear in the cases of the two solid phases, but only a few peaks (distance between layers) are observed for the liquid crystal, and no peaks at all in the liquid.

The information about the local short-range order is in the diffuse scattering, and it is accessible by PDF experiments. Thus, measurements of the same phases of Pb(C*11*)$_2$ were carried out to get their PDFs (Fig. 13.4b), where significant differences are detected from the S*II* phase and the remainder. In this sense, the first Pb-Pb distances are shorter in the mesophases than in the S*II* phase, explaining the enhanced photophysical properties in those cases (Martínez-Casado et al. 2014). But more importantly, the local structure (up to 5 Å) is similar for the S*I*, LC, and IL phases, although S*I* shows long-range order (due to its crystallinity). The patterns for the different phases, obtained from X-ray data, show mainly the contribution of the Pb-Pb distances (especially from 4 Å on), due to the high Z of Pb, as mentioned before.

On the other hand, experiments on other members of the series in all their phases show that the local structure is the same for the same type of phase (Martínez-Casado et al. 2017), so, for example, the S*I* phase presents the same PDF pattern independently of the alkyl chain length, and the same occurs for LC and IL. This means that any lead soap can also be identified in those phases using that

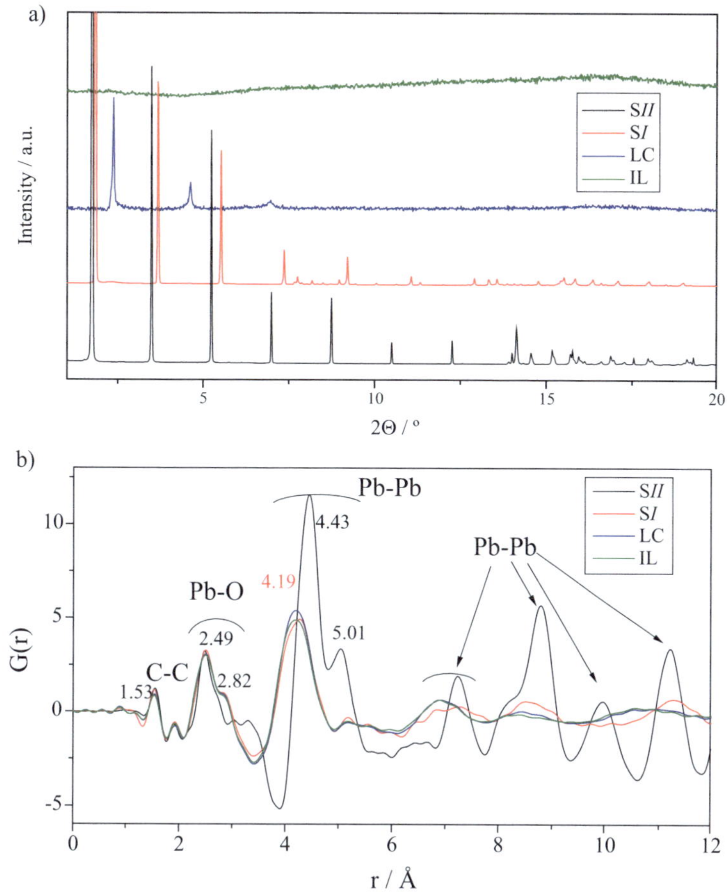

Fig. 13.4 P-XRD (**a**) and PDF, from X-ray, (**b**) data for the different phases of Pb(C*11*)$_2$. In the PDF analysis, some bond distances are highlighted and specified for a better comparison

PDF pattern as a *fingerprint*. In the particular case of *SII*, the short-range PDF is obviously different for polymorphs A and B, but again similar for the same polymorphic form in different members of the series. This observation shows that the use of a technique like PDF analysis helps significantly in the structural study of disordered, as well as ordered, phases, providing very useful information.

13.4 Conclusions

The analysis carried out on pure metal soaps and their phases shows the intricate behavior and helps understanding their properties. Four different phases and three glass states are found in this family of compounds. The role of temperature can be enhanced by mixture with other components, so the study of binary or more complex systems with metal soaps needs a previous knowledge of their pure phases. It is important to note that metal soaps may be present in paintings in different phases which would lead to different properties and, therefore, a correct diagnosis is necessary using complementary techniques. Thus, the use of FTIR or P-XRD should be complemented with others, such as PDF analysis or XAS, which allow detection of disordered phases of metal soaps. Moreover, PDF data of these pure phases can be used as *fingerprints* in the correct identification of metal soaps, allowing detection of them in the initial stages of formation and growth.

References

Boon JJ, van der Weerd J, Keune K, Noble P, Wadum J (2002) Mechanical and chemical changes in old master paintings: dissolution, metal soup formation and remineralization processes in lead pigmented ground/intermediate paint layers of 17th century paintings. In: Vontobel R (ed) ICOM committee for conservation, ICOM-CC: 13th triennial meeting, Rio de Janeiro, 22–27 Sept 2002. ICOM-CC, James and James 5, Rio de Jainero

Boon JJ, Hoogland F, Keune K (2007) Chemical processes in aged oil paints affecting metal soap migration and aggregation. In: Parkin HM (ed) AIC paintings specialty group postprints: papers presented at the 34th annual meeting of the AIC of historic & artistic works providence, Rhode Island, 16–19 June 2006, vol 19. AIC PSG Postprints, Washington, DC, pp 16–23

Burrows HD (1988) The luminescence of lead(II) decanoate. Mater Lett 6:191–193

Catalano J, Yao Y, Murphy A, Zumbulyadis N, Centeno SA, Dybowski C (2014) Nuclear magnetic resonance spectra and (207)Pb chemical-shift tensors of lead carboxylates relevant to soap formation in oil paintings. Appl Spectrosc 68:280–286

Catalano J, Murphy A, Yao Y, Yap GPA, Zumbulyadis N, Centeno SA, Dybowski C (2015) Coordination geometry of lead carboxylates-spectroscopic and crystallographic evidence. Dalton Trans 44:2340–2347

Centeno SA, Mahon D (2009) The chemistry of aging in oil paintings: metal soaps and visual changes. Metrop Mus Art Bull 67:12–19

Cheda JAR, Redondo MI, García MV, López-de-la-Fuente FL, Fernández-Martín F, Westrum EF Jr (1999) A thermophysical study of the melting process in alkyl chain metal n-alkanoates: the thallium (I) series. J Chem Phys 111(8):3590–3598

Eibner A (1909) Malmaterialienkunde als Grundlage der Maltechnik, vol 121. Springer, Berlin

Hale C, Arslanoglu J, Centeno SA (2011) Studying old master paintings. Technology and practice. The national gallery technical bulletin 30th anniversary conference postprints (ed Marika spring). Archetype Publications and the National Gallery, London

Higgitt C, Spring M, Saunders D (2003) Pigment-medium interactions in oil paint films containing red lead or lead tin yellow. Natl Gallery Tech Bull 75:75–91

Keune K (2005) Binding medium, pigments, and metal soaps characterised and localised in paint cross-sections. Ph.D. thesis. University of Amsterdam

Keune K, Boon JJ (2007) Analytical imaging studies of paint cross-sections illustrate the oil paint defect of lead soap aggregate formation. Stud Conserv 52:161–176

Lacouture F, Francois M, Didierjean C, Rivera JP, Rocca E, Steinmetza J (2001) Anhydrous lead(II) heptanoate. Acta Crystallogr C 57:530–532

Martínez-Casado FJ, Sánchez Arenas A, García Pérez MV, Redondo Yélamos MI, López de Andrés S, Cheda JAR (2007a) Short chain lead (II) alkanoates as ionic liquids and glass formers: a d.S.C., X-ray diffraction and FTIR spectroscopy study. J Chem Thermodyn 39: 455–461

Martínez-Casado FJ, García Pérez MV, Redondo Yélamos MI, Cheda JAR, Sánchez Arenas A, López de Andrés S, García-Barriocanal J, Rivera A, León C, Santamaría J (2007b) Intermediate rotator phase in lead(II) Alkanoates. J Phys Chem C 111:6826–6831

Martínez-Casado FJ, Ramos Riesco M, Sánchez Arenas A, García Pérez MV, Redondo MI, López-Andrés S, Garrido L, Cheda JAR (2008) A novel rotator glass in lead(II) Pentanoate: calorimetric and spectroscopic study. J Phys Chem B 112:16601–16609

Martínez-Casado FJ, Ramos Riesco M, da Silva I, Redondo MI, Labrador A, Cheda JAR (2011) Lithium and lead(II) Butyrates binary system. Pure compounds and an intermediate salt: from 2D to 3D coordination polymers. Cryst Growth Des 11:759–767

Martínez-Casado FJ, Ramos Riesco M, Redondo MI, Sánchez Arenas A, Cheda JAR (2012) The role of calorimetry in the structural study of mesophases and their glass states. J Therm Anal Calorim 108:399–413

Martínez-Casado FJ, Ramos Riesco M, Rodríguez Cheda JA, Cucinotta F, Fernández-Martínez A, Garrido L, Matesanz E, Marchese LJ (2014) Short lead(II) soaps: from weakly fluorescent crystals to strongly phosphorescent and structurally varied vitreous phases. A thermal, structural and spectroscopic study. J Mater Chem C 15(1):497–509

Martínez-Casado FJ, Ramos Riesco M, Rodríguez Cheda JA, Cucinotta F, Matesanz E, Miletto I, Gianotti E, Marchese L (2016) Unraveling the De-composition process of lead(II) acetate: anhydrous polymorphs, hydrates, and byproducts and room temperature phosphorescence. Inorg Chem 55(17):8576–8586

Martínez-Casado FJ, Ramos Riesco M, Rodríguez Cheda JA, Redondo MI, Garrido L, Fernández-Martínez A, García-Barriocanal J, da Silva I, Durán-Olivencia M, Poulain A (2017) PCCP 16(29):17009–17018

Mills JS, White R (eds) (1999) The organic chemistry of museum objects. Routledge, London

Noble P, Wadum J, Groen K, Heeren R, van den Beerg KJ (2000) Art et chimie, La couleur. CNRS Editions, Paris, pp 126–129

Noble P, Boon JJ, Wadum J (2003) Dissolution, aggregation, and protrusion. Lead soap formation in 17th-century grounds and paint layers. Art Matters 1:46–61

Noble P, Van Loon A, Boon JJ (2005) 14th triennial meeting of ICOM committee for con-servation. James and James, pp 496–503

Ramos Riesco M, Martínez-Casado FJ, Rodríguez Cheda JA, Redondo Yélamos MI, da Silva I, Plivelic TS, López de Andrés S, Ferloni P (2015) New advances in the one-dimensional coordination polymer copper(II) alkanoates series: monotropic polymorphism and mesomorphism. Cryst Growth Des 15:2005–2016

Robinet L, Corbeil MC (2003) The characterization of metal soaps. Stud Conserv 48:23–40

Spring M, Ricci C, Peggie D, Kazarian S (2008) ATR-FTIR imaging for the analysis of or-ganic materials in paint cross sections: case studies on paint samples from the national gallery, London. Anal Bioanal Chem 392:37–45

van der Weerd J (2002) Microspectroscopic analysis of traditional oil paint. Ph.D. thesis, University of Amsterdam

Van Loon A (2008) Color changes and chemical reactivity in seventeenth-century oil paintings. Ph.D. thesis, University of Amsterdam

Part III
Characterization and Treatment

Chapter 14
Taking Different Forms: Metal Soaps in Paintings, Diagnosis of Condition, and Issues for Treatment

Aviva Burnstock

Abstract This chapter introduces a selection of case studies of paintings that present a range of physical manifestations of metal soaps. The phenomena include surface spots, crusts, delamination, and textural features. The role of metal soaps in water-sensitive modern oil paints is introduced. Inferences based on a combination of technical and analytical evidence together with information about the history and context of the works are presented with hypotheses about the causes of metal soap formation or deterioration. Approaches to treatment of the works are introduced in a discussion of aesthetic, ethical, and practical options that address particular phenomena.

Keywords Metal soaps · Crusts · Spots · Delamination · Water sensitive oil paints · Treatment · SEM/EDX · FTIR · Chelating agents · EDTA

14.1 Introduction

The research summarized in this chapter aims to identify a broad range of phenomena related to metal soap degradation in paintings. It draws on evidence gathered from recent studies and inferences from a wide body of published research, comprising fundamental chemical investigations along with case studies that present similar or related metal soap phenomena. Recent advances in knowledge arising from molecular level and model studies of soap formation have informed an understanding of alterations observed in paintings. However, many of the key studies are published in scientific journals that are less accessible to conservators who encounter paintings with deterioration in the studio but may not recognize the range of phenomena. Furthermore, most conservators will not have the resources to undertake chemical analyses that provide specific characterization.

A. Burnstock (✉)
Department of Conservation and Technology, Courtauld Institute of Art, London, UK
e-mail: aviva.burnstock@courtauld.ac.uk

F. Casadio et al. (eds.), *Metal Soaps in Art*, Cultural Heritage Science,
https://doi.org/10.1007/978-3-319-90617-1_14

Selected examples are presented that exemplify different observed phenomena and highlight questions that might be usefully investigated in future studies. The cases discussed aim to provide a broad classification of manifestations of metal soaps in the paintings or to present instances where commonalities of cause point to the influence of environmental conditions, material composition, or conservation history. The role of metal soaps in water sensitivity of unvarnished oil paintings is also introduced.

The last section of this chapter considers the relevance of research into metal soaps for the treatment of paintings, discussing how an understanding of metal soap-related degradation is important in formulating appropriate treatment approaches. For instance, how such an understanding might impact upon the decision to remove coatings, surface crusts, or fatty acid efflorescence, the cleaning of water-sensitive paints, the consolidation of flaking paint, and the development of suitable preventive measures. Practical approaches to the treatment of deteriorated surfaces caused by metal soap formation are also discussed, alongside ethical considerations.

Different kinds of metal soaps have been identified in paintings and test samples using scanning electron microscopy with elemental analysis (SEM-EDX), Fourier transform infrared spectroscopy (FTIR), and other instrumental analytical methods. While a full review of published studies is beyond the scope of this chapter, key publications that have discussed lead and zinc soap aggregates, crusts, and hazes in paintings include Boon et al. (2002, 2007), Noble et al. (2002), Noble and Boon (2007), Keune and Boon (2007), Jones et al. (2007), Shimadzu et al. (2008), and Van Loon et al. (2012a, b). The five-year PAinT project (2012–2016)[1] sponsored by the Netherlands Organization for Scientific Research (NWO) has advanced our molecular understanding of metal soap formation in oil paint and paintings. Findings from this research project are included in other chapters of this volume.

14.2 Surface Phenomena

14.2.1 Efflorescence and Soap Crusts

The following case studies exemplify different visual manifestations of metal soaps that necessitated the development of specific approaches to treatment; the individual treatment challenges and decisions are detailed below in Sect. 14.5.

Portrait of Carolina Ewen, Circle of Thomas Hudson, c.1750 private collection, oil on canvas

Patches of insoluble white crystals were present on the paint surface of this mid-eighteenth-century British portrait.[2] A sample from the paint surface of an

[1] http://www.s4a-paint.uva.nl/research-topics/research.html

[2] Report CIA:2215 Department of Conservation & Technology, Courtauld Institute of Art, 2015.

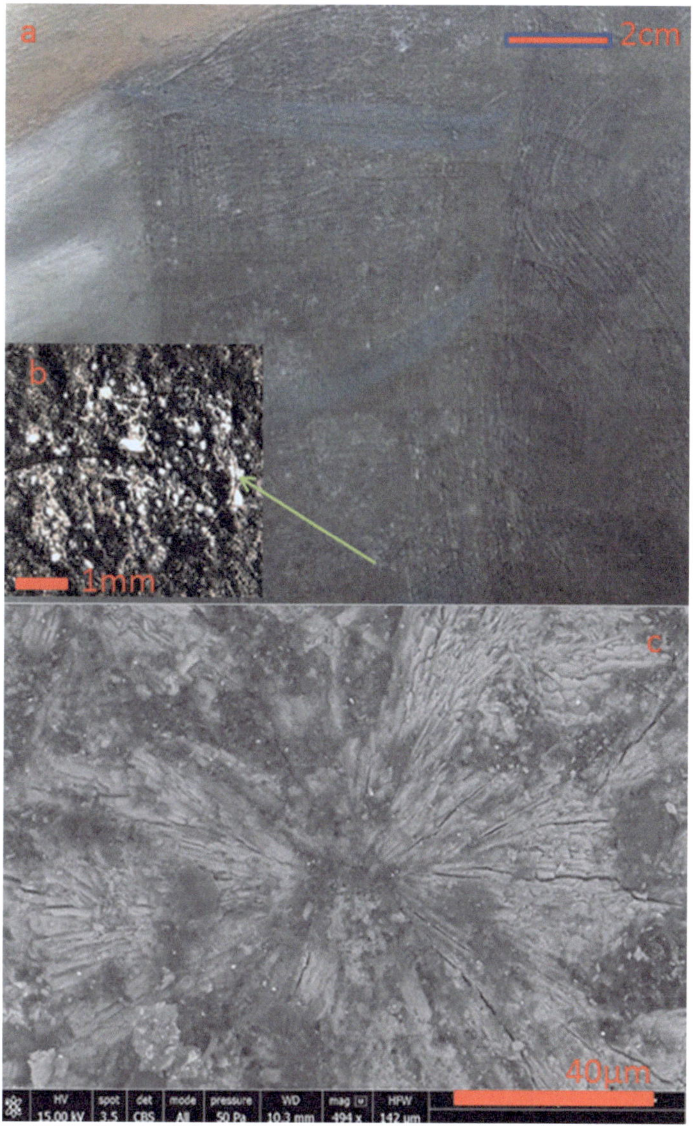

Fig. 14.1 (**a**) Detail of lead soap crystals on the surface of blue drapery in *Portrait of Carolina Ewen*. (**b**) Light microscope image of an aggregate of lead soaps. (**c**) SEM-BE image of the surface of a sample from the same area showing needle-shaped crystals of lead soaps

affected area viewed with SEM in a backscattered election image (BE) and analyzed using EDX showed inclusions of needle-shaped crystals containing elemental lead (Fig. 14.1) FTIR analysis of a surface scraping identified the crystals as lead stearate and re-mineralized lead soaps.

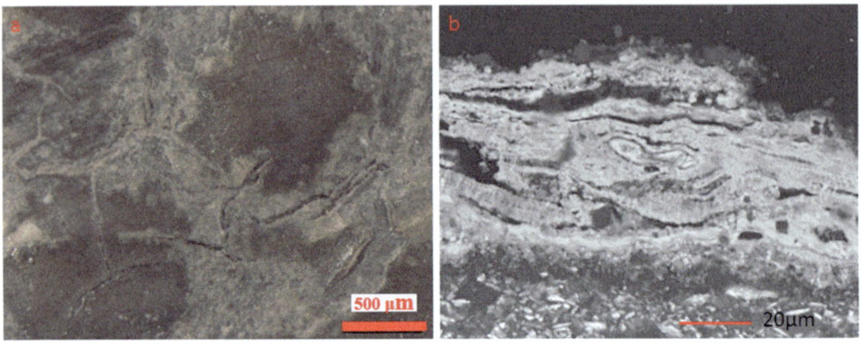

Fig. 14.2 (**a**) Detail from the foreground from *Return from the Front* showing hazy whitish soap crust. (**b**) A cross section from the same area the showing a laminated whitish crust of lead soaps

Return from the Front, Kate Olver, c.1915 private collection, oil on canvas

The foreground of this painting was covered by an uneven, laminated whitish crust of lead soaps with lead sulfate visible in a cross-sectional sample (Fig. 14.2) (Sawicka et al. 2014). The crust is particularly evident in the troughs of the paint following the canvas weave, although in other areas it completely covered the brown painted passages.

14.2.2 Paint-Coating Interactions

Two paintings examined present examples of soap formation associated with paint-coating interaction; the chemical relationship between paint and coating is further examined by Higgs in Chap. 7 in this volume. A third painting, *Female Nude* by Amedeo Modigliani, provided an example in which lead and zinc soaps had formed in the ground and paint in both varnished and unvarnished passages.

Wolmer Wood, Philip A. de László, d.1934 private collection, oil on wood panel

A technical examination of *Wolmer Wood* by Philip A. de László, a Hungarian artist working in Britain, identified the presence of insoluble zinc soaps at the surface of the painting[3] (Burnstock and Van den Berg 2014). The surface had developed a brown discoloration and the composition had become unreadable. Figure 14.3a shows a detail from the top left of the composition before treatment and 3b during removal of the coating. SEM-EDX examination of cross sections and FTIR analysis showed that zinc soaps had formed at the paint surface. These soaps appeared to be caused by the reaction between an oil-containing surface coating

[3]Report CIA:2093 Department of Conservation & Technology, Courtauld Institute of Art 2010.

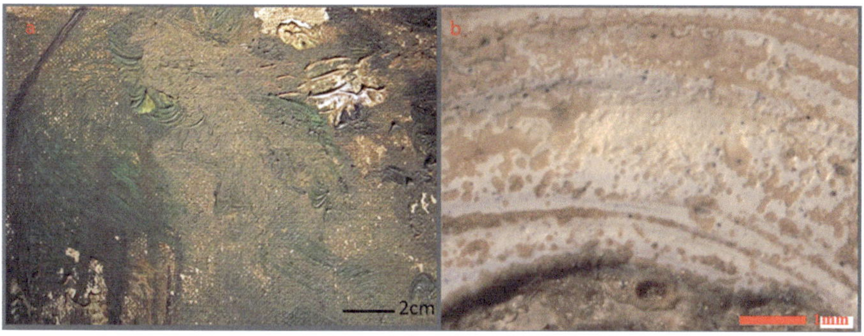

Fig. 14.3 (**a**) Detail of the surface of *Wolmer Wood* before treatment. (**b**) Detail of an area of zinc white-containing paint during removal of the brown layer with zinc soaps, showing pits in the paint surface where zinc soap aggregates have been removed

with zinc oxide-containing passages of paint. Notable in this case is the formation of zinc soap aggregates that are integral both to the insoluble coating and the paint surface. A brown surface residue containing zinc carboxylates was also noted in a painting by Shimadzu et al. (2008).

Cross sections from the painting show the artist's use of zinc white in a mixture with emerald green and cadmium yellow pigments applied over a ground that had been sealed with a protein-containing layer.[4]

Visible and SEM/BE images of a cross section from the zinc soap-containing paint and browned coating (Fig. 14.4a, b) clearly show crystals containing elemental Zn and O at the paint-coating interface (Fig. 14.4c).

In formulating hypotheses on the origin of the now brownish coating, it should be noted that application of an oil-containing varnish over oil paint, or simply applying oil, sometimes referred to as "oiling out," was recommended in artists' treatises in the nineteenth century and was common practice into the following century. If an oil-containing varnish of this sort was used to coat *Wolmer Wood* (whether or not by the artist), then a consequence would be the provision of free fatty acids or possibly acidic degradation products from resin if the applied coating was an oil-resin varnish. These free fatty or resin acids could then react with zinc ions to form soaps.

Portrait of Francisco de Saavedra, Francisco de Goya, 1798

Recent treatment of the full-length *Portrait of Francisco de Saavedra* by Goya (Courtauld Gallery, London) included removal of a browned, aged coating and revealed that the whole surface was covered with multiple shallow protrusions, each topped with a spot of dark brown material. While dense populations of protrusions were visible in brown painted areas, the visual effect of the brown spots was more pronounced in passages of light-colored paint. Figure 14.5a shows the low protrusions in an area of paint in the foreground, with one group shown at high

[4]Indicated in a cross section by positive Amido Black stain for protein.

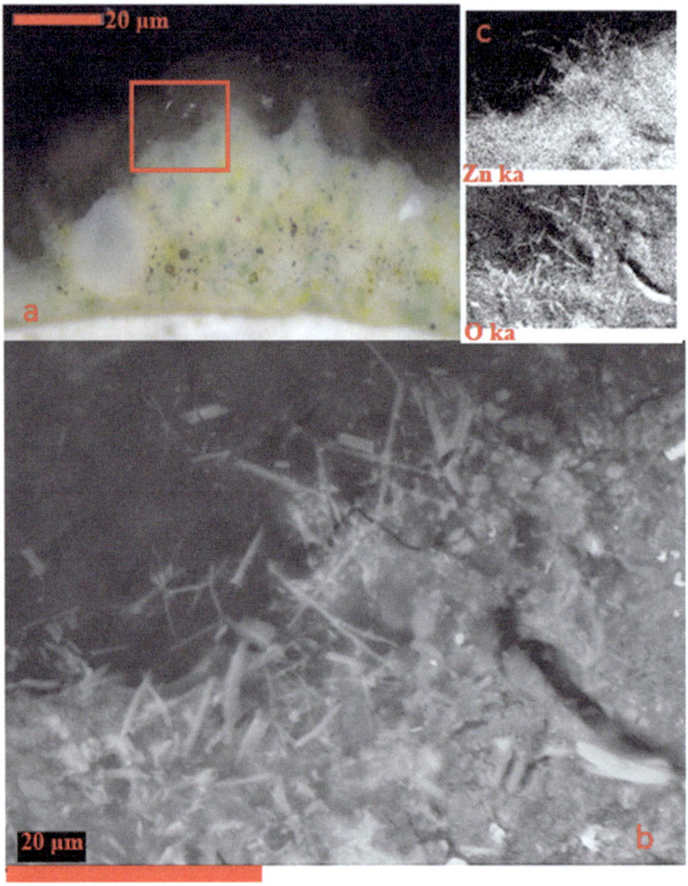

Fig. 14.4 (**a**) Cross section from zinc soap-containing paint and browned coating from *Wolmer Wood*. (**b**) SEM-BE image of the interface between the paint and the coating from the area of the cross section indicated in (**a**) by a red box. (**c**) Elemental mapping of the same area showing crystals containing elemental Zn and O at the paint-coating interface

magnification in 5b. A SEM secondary electron (SE) image of the surface of one protrusion is shown in 5c highlighting the globular shape of these protrusions, with convex or concave profiles. In cross section (Fig. 14.5d), the subsurface features of the protrusion become evident, with areas that are medium rich and devoid of pigment particles and areas of recrystallization of high atomic number species.

Analysis of paint cross sections from representative areas using light microscopy and SEM-EDX indicated iron oxide earth pigments, aluminosilicates, calcium sulfate, chalk, and lead carbonate in the ground. The cross sections included characteristic lead soap aggregates in various stages of recrystallization as reported in other studies of oil paintings. Similar protrusions with brown spotting have been observed on the surface of other portraits by Goya painted in the last decade of

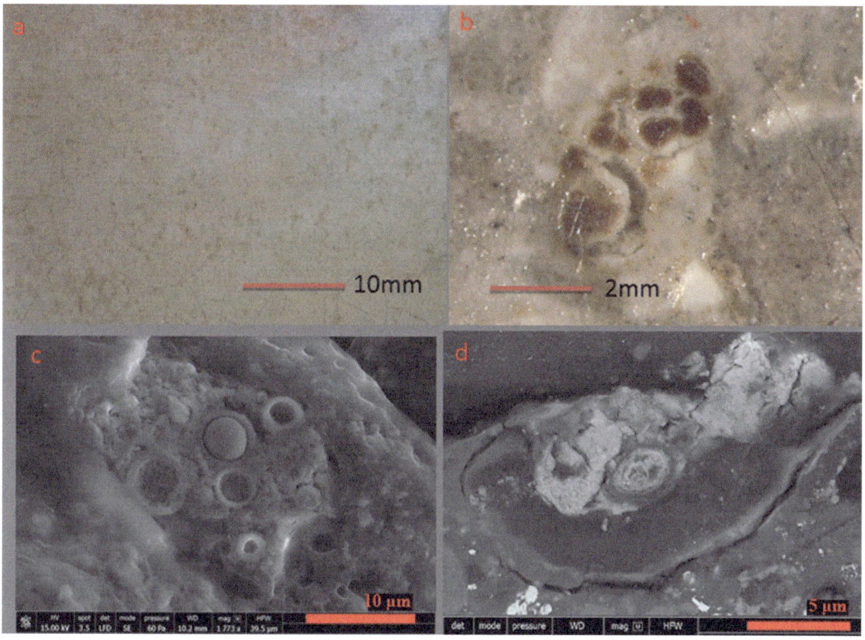

Fig. 14.5 (**a**) Detail of the foreground paint surface from *Portrait of Francisco Saavedra* showing spotting. (**b**) Detail from (**a**) showing a group of protrusions topped by brown spots. (**c**) SEM secondary electron image of the surface of a similar protrusion to that seen in. (**b**), (**d**) SEM-backscattered electron image of a cross section from a protrusion showing the subsurface features of the protrusion in (**c**)

the eighteenth and early in the nineteenth century[5] (Puig et al. 2016). The presence of identical surface degradation on works by Goya with different provenance and material histories points to factors related to the composition of the paint, ground or possibly the first coating applied after the work was completed.[6] There are records of the use of oil and resin varnishes in Spain from the early seventeenth century. In the case of Goya, apart from a comment he made in 1800 on the treatment of one of his paintings that allegedly included the use of egg white varnish (Veliz Bomford and Aterido 2016), there are no extant documents that refer to his vanishing practice.

When combined with empirical evidence from the conservation history of the works, condition of the surface and material analysis of both *Wolmer Wood* and the *Saavedra* portrait, the possibility arises that the application of an oil or oil-

[5]Enrique Quintana, Head of painting conservation, Museo Nacional del Prado, personal Communication 2015, and firsthand observation of contemporary portraits by Goya in the conservation studio of the Museo Nacional del Prado. Lead soap aggregates in paint samples from selected paintings by Goya in the Prado were confirmed by analysis carried out by Maria Dolores Gayo, head of the Museum's Scientific Department.

[6]Enrique Quintana, personal communication 2015.

resin varnish may contribute to the formation of metal soaps at the surface of the paintings. This hypothesis is plausible considering that oil-containing coatings – whenever they are applied in the lifetime of a painting – provide a source of free fatty acids that may react with metal ions at the paint surface. While the resin component of the coatings on these paintings may be removable using the range of solvents typically used for varnish removal, the oil component has the potential to form insoluble metal carboxylates. It is also plausible that acidic components in aged natural resin coatings may react with alkaline pigments in the underlying paint, or in a tinted varnish, with pigments that contain reactive ions to form soaps. This hypothesis requires further investigation.

Female Nude, Amedeo Modigliani, c.1917

Another physical manifestation of soaps on the surface of oil paintings is exemplified by *Female Nude*, by Modigliani (Courtauld Gallery, London) (Fig. 14.6). Under magnification the surface shows pinpoint losses exposing the white priming in dark-colored areas (Fig. 14.6b), while raking light observation highlights several scattered protrusions in the flesh tones (Fig. 14.6c). A cross section (6d) shows a layer of glue size, indicated by the arrow, followed by a layer of lead white and chalk (with blue pigments) superimposed by a blue-grey second ground layer containing barium sulfate, zinc oxide (or sulfide), and carbon black. The flesh paint contains lead white mixed with vermilion and shows a surface with aggregates of zinc soaps and lead soaps (typically 20–100 μm diameter) formed on the peaks of impasted

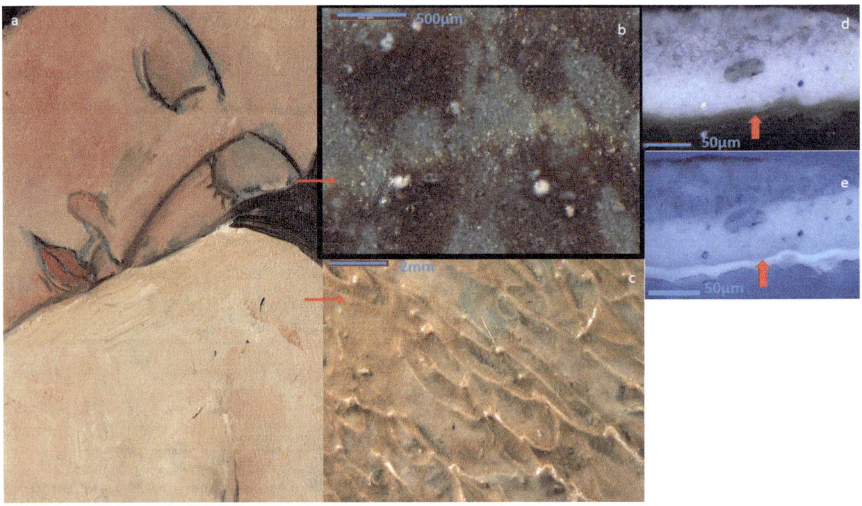

Fig. 14.6 (**a**) Detail of flesh paint from *Female Nude.* (**b**) Detail of the blue-grey ground with metal soap aggregates indicated by arrow. (**c**) Detail of flesh paint with lead soaps indicated by arrow. (**d**) A cross section from the double ground with a layer of glue size indicated by the arrow. (e) The same cross section in UV light

paint (indicated by a red arrow in 6c). Areas of exposed ground contained both lead and zinc soap aggregates.

The even distribution of zinc and lead soap aggregates in both varnished and unvarnished areas (the latter in a strip that had remained unvarnished and protected by the frame rebate) suggests that in this case the formation of lead and zinc soaps was not influenced by the surface coating.

14.3 Paint Delamination

Two works, painted on different supports, highlight conservation challenges posed by delamination of paint layers that can be caused by zinc soaps.

The Pilgrims Rest Hotel, Robert Brooks, d.1995, private collection, oil on cotton duck canvas

The painting primed with titanium white in oil has a history of blind cleavage and severe delamination between paint layers. It was first examined and treated at the Courtauld in 2009 and then treated again in 2013 and 2016.[7] The loss of cohesion between paint layers is caused by soaps formed from paint containing zinc oxide, confirmed by FTIR analysis. In the most pronounced areas of delamination, intralamellar fracture between layers containing zinc oxide was evident, rather than between the paint and the ground that did not contain zinc. Similar delamination phenomena in painted art have been observed by other authors including Van der Weerd et al. (2003); Rogala et al. (2010); Osmond G (2019); and Raven et al. (2018).

Man Clearing Snow Red Square, Ralph Lillford, 1986 private collection, acrylic and oil on hardboard

Similar delamination was observed in a painting by Ralph Lillford which was first examined and treated in 2009 and again in 2017. The composition was painted over another underlying scheme, on the smooth side of a hardboard support primed with titanium white, chalk, and kaolin. Like the painting by Brooks, delamination had occurred between and within paint layers containing zinc and/or calcium soaps (Fig. 14.7).

Cross sections indicated that the source of fatty acids for zinc soap aggregates visible in the upper paint layers could be a medium-rich underlayer of carbon black paint, or alternatively they may be formed from polar degradation products of hydrolysis of the oil medium at the paint surface. Metal soaps formed by reaction between zinc oxide and stearic acid, a chemical process that generates water, may also have favored ongoing soap formation in the paint film (Osmond 2014). The formation of a layer of zinc soaps that led to delamination, considered in the light of similar observations in published case studies, raises the question of what conditions

[7]Report CIA: 1998 2009, and notes on condition and re-treatment in 2009 and 2013, Department of Conservation & Technology, Courtauld Institute of Art.

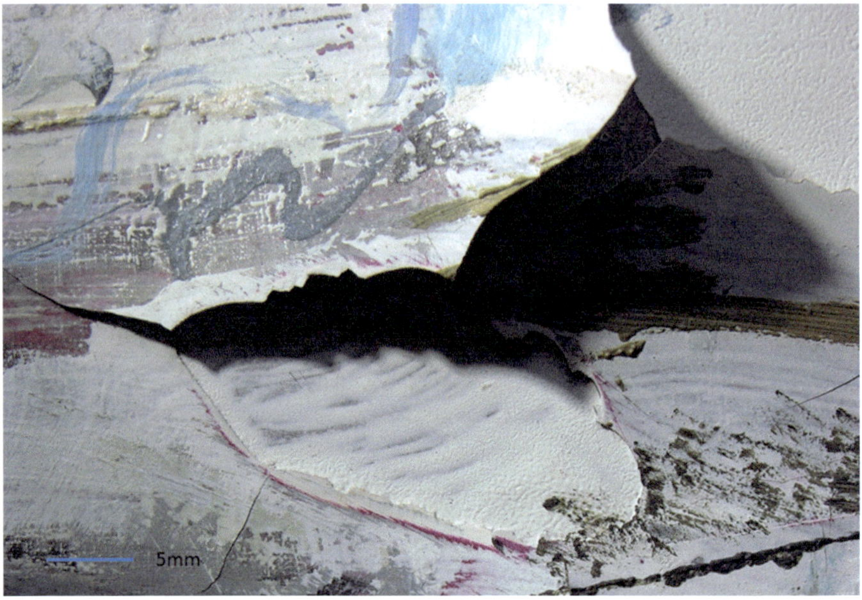

Fig. 14.7 Detail of flaking paint from *Man Clearing Snow Red Square*, showing delamination in areas of zinc oxide-containing paint

favor the formation of zinc aggregates or formations that result in delamination. The technical challenges for characterizing the coordination of fatty acids in zinc soaps within paint layers in works of art include identification of different kinds of oil media based on fatty acid profiles. This question is informed by investigation of model paint systems (Osmond 2014; Hermans et al. 2015).

14.4 Metal Soaps and Water Sensitivity in Unvarnished Modern Paints

Building site, Oxford Street, Frank Auerbach, 1959–1960, Tate Gallery, London [catalogue number T00418], oil on unprimed hardboard

Recent analytical studies of modern oil paintings that present water sensitivity on surface cleaning[8] raise questions about the role of metal soaps in relation to the response of paint to polar solvents. *Building site, Oxford Street* is thickly painted in oil on an unprimed hardboard support. The unvarnished surface includes passages of glossy and matte paint, reflecting different proportions of oil to pigment.

[8]Bay L (2015) Documenting water-sensitive oil paintings in the Tate Gallery. Final Year Project Department of Conservation & Technology, Courtauld Institute of Art, London.

Tests showed that pigment was removed from glossy paint after three to five rolls of a cotton swab with deionized water. Samples from the painting prepared as cross sections showed a skin of organic material at the surface together with fatty acid efflorescence. FTIR spectra from surface material exhibited split carbonyl absorptions at 1737 and 1713 cm^{-1} indicative of glycerides and free fatty acids in partially hydrolyzed oil and indicated the presence of polar azelaic acid.[9] Subsurface paint contained zinc stearate (indicated by absorption at 1539 cm^{-1}) and also iron oxide pigments extended with coarsely ground kaolin, gypsum, chalk, and barium sulfate. This combination of materials absorbs a significant quantity of oil medium and contains relatively few reactive metal ions, thus providing a source of fatty acids. However, the possibility that metal stearates including palmitates and free fatty acids were added to the paint by the manufacturer may also contribute to the availability of an excess of unreacted free fatty acids that could account for the observed surface phenomena (Burnstock and van den Berg 2014). The influence on water sensitivity of metal soaps added by the manufacturer and those formed in situ from inorganic pigments and fatty acids from the oil medium is the subject of current investigation.

14.5 Considerations for Treatment

The different forms of soaps and their relationships to the paint surface are an important consideration in formulating treatment approaches, examples of which are discussed below in relation to the case studies described in the preceding sections.

14.5.1 The Use of Aqueous or Hygroscopic Materials for Consolidation, Cleaning, and Lining

Where flaking or delamination is caused by zinc soaps, it is possible that the introduction of aqueous adhesives for consolidation may promote hydrolysis of oil media and further soap formation. This clearly conflicts with the objective to readhere delamination within a zinc soap-containing paint layer. A consideration of the treatment histories of the paintings by Brooks and Lillford (described in Sect. 14.3) supports this hypothesis.

The first treatment of the Brooks carried out in 2009 included consolidation of areas of flaking paint using a 20% solution of Aquazol 500 in deionized water applied with a small brush. The delaminating areas were gently massaged using a heated spatula at 45 °C and weighted overnight. The necessity for repeated treatment suggests that the paint was actively delaminating. Based on recent research that suggests that zinc oxide will readily form soaps in oil media and in an aqueous environment, it is possible that in this case the use of an aqueous adhesive or

[9] Analysis of the surface and bulk paints was carried out by PhD candidate Judith Lee using SEM-EDX and FTIR mapping and electrospray ionization mass spectrometry (ESI-MS).

hygroscopic resin for consolidation might have exacerbated the delamination by enhancing the formation of soaps (Arslanoglu 2004). In the case of the Lillford painting, areas of flaking retreated using BEVA 371 has proved effective.

Further investigation of the use of nonaqueous consolidants for this purpose may provide useful guidelines for conservation treatments. In practice, there may be a range of factors that influence the choice of adhesives for consolidation of flaking paint. They include the sensitivity of the paint to the carrier solvent, access to the cleavage that may require highly dilute solutions or pre-wetting with solvent, and risk of optical changes during removal of residual adhesive from unvarnished paint surfaces.

The influence of moisture on soap formation in paintings, including acute short-term contact with moisture as part of conservation treatment, has not been systematically investigated. In addition to consolidation using aqueous media, relevant procedures would include lining and the use of water-based cleaning methods. Also likely to be influential are aspects of artists' techniques and materials, such as canvas preparation (by artist or manufacturer) using a hygroscopic size layer, as was the case in Modigliani's *Female Nude* (Sect. 14.2.2). Other factors influencing soap formation include exposure to humid environmental conditions and absorption of moisture by the size layer or possibly the application of an aqueous protein interlayer (as was the case in *Wolmer Wood*, Sect. 14.2.2). All of these factors may provide a favorable environment for hydrolysis of the overlying oil paint (Osmond et al. 2014). *Female Nude*, painted in c1917, was lined (for the first time) with a wax-resin adhesive in the late 1950s; in the absence of experimental study, it is reasonable to speculate that this treatment has reduced the response of the size layer to increased humidity and thus reduced the risk of further hydrolysis of paint that may lead to soap formation. However, in this work the texture of the flesh paint and significant areas of exposed ground are pronounced due to the formation of soaps, whether intended or not, and this will be an important consideration in the plan for future treatment.

In cases where soap formation is the result of direct contact with moisture evidenced by water staining or its consequences (such as mold formation, in the case of the Ethel Walker described later in Sect. 14.5.3), it is clear that preventive conservation measures, such as dusting to remove fungal spores, the incorporation of backboards and glazing, or control of environmental moisture exposure by air-conditioning, reduces the risk of further deterioration.

14.5.2 Surface Cleaning of Unvarnished Water-Sensitive Paintings

Building site, Oxford Street (discussed in Sect. 14.4), in common with other twentieth-century unvarnished oil paintings, incorporates passages of paint that exhibit sensitivity to water. This precludes the use of water to remove dirt from their surfaces and limits conservators to dry cleaning. Alternative methods for cleaning using apolar solvents are ineffective in removing the polar fractions of soiling or dirt imbibed in the skin of medium at the paint surface.

Recent studies have examined the efficacy of rigid gels such as agarose for topical delivery of cleaning reagents. Such methods, together with the use of synthetic absorbent microfilament materials such as Evolon,[10] offer a method of cleaning with significantly reduced mechanical action compared with conventional cotton wool swabbing. Another approach is the use of microemulsions tailored for cleaning water-sensitive paints. Chung et al. (2017) examined a range of microemulsions for removal of artificial dirt from samples of moderately aged water-sensitive Winsor & Newton paints and found some formulations were moderately effective. However the clearance of non volatile components of these materials and the short or longer term effects of residues on the painting have not been systematically investigated.

New options for surface cleaning water-sensitive paints may be to introduce reactive ions to trigger the reaction between free fatty acids to form insoluble salts that might then be resistant to water swabbing. However, the ethics and practicalities of this approach considering the possibility of irreversible optical changes to the paint surface require careful consideration and further research.

14.5.3 Removal/Treatment of Metal Soap Efflorescence and Crusts

The removal or reduction of insoluble metal soap crusts can be achieved using aqueous solutions of chelating agents, formulated with specific concentration and pH, and applied using a range of methods. Sawicka[11] addressed the methods for removal of different forms of superficial white surface efflorescence in paintings using ethylenediaminetetraacetic acid (EDTA). She also formulated guidelines in diagram format for a sequence of treatment options that might be useful for conservators faced with treating paintings with metal soap hazes or crusts (Fig. 14.8). Likewise Van Loon et al. (2019) developed a procedure for determining the feasibility of removing these inorganic or semi-inorganic degradation layers. A first step would be SEM-EDX analysis of a cross section to determine the nature of the interface between the crust and the paint surface. The extent to which the coating or crust has become part of the paint surface will limit what can safely be removed. Another approach may be to try to saturate the surface using low molecular weight resins and/or to locally retouch the disfiguring areas of the painting. If the areas affected

---

[10] http://www.preservationequipment.com/Catalogue/Conservation-Materials/Materials-and-Fabrics/Evolon-Microfilament-Material

[11] SawickaA (2014) Clearing the "Haze" of Inorganic Efflorescence An investigation into the formation of lead soap efflorescence and the viability of its removal by means of the chelating agent Ethylenediaminetetraacetic Acid, Final year project, Department of Conservation &Technology, Courtauld Institute of Art, London.

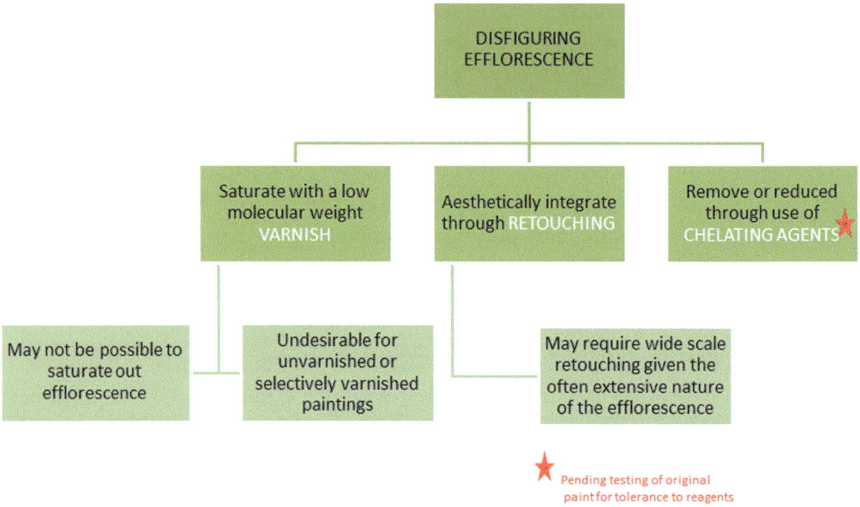

Fig. 14.8 Diagram of guidelines for a sequence of options for treating paintings with metal soap efflorescence and soap hazes (Adapted from Sawicka et al. 2014)

by efflorescence are too large, or too disturbing, the option to remove or reduce the crust can be achieved using an appropriate chelating agent. The choice of chelant depends on the preference for chelation of specific metal ions and lower chelation power for other metal ions present in the original paint. Their application should utilize solutions of pH range that optimize the chelation of specific metal ions in the soaps and present low risk of hydrolysis of the oil medium. Table 14.1 provides a list of stability constants which offers a guide for the choice of chelant, based on relative preferences for formation of stable metal-chelate complexes. Although the table presents a wider range of options, studies of the application of chelating agents for paintings have focused on the use of citrates for surface cleaning (Morrison et al. 2007) and EDTA for removal of soaps hazes (Sawicka et al. 2014). Both have shown that the pH, concentration of chelant, and method of application are critical in tailoring the reagent for purpose. Possible drawbacks include inefficient clearance of chelating agents, in particular where they are gelled (usually with a cellulosic thickener); the risk of affecting the equilibrium of the paint film by acute moisture exposure; and if the solution is not precisely tailored or applied, chelation of metal ions in the paint.

It was possible to reduce a brown layer containing zinc soaps from the surface of *Wolmer Wood* using EDTA solutions. Images of the surface taken during treatment illustrate the significant challenges involved in a kind of cleaning that necessitates the removal, or partial removal, of a coating that has become imbibed in the painting. There are pits left where soaps are removed, together with the residues of the coating that were not readily removable (Fig. 14.3b). EDTA solutions were also effective in removing lead soap efflorescence from the drapery of *Carolina Ewen* (Sect. 14.2.1).

Table 14.1 Stability/equilibrium constants for selected metal-chelant complexes (log K). The higher the log K value the more tightly the metal ion will be bound to the chelating agent, and the more likely that the complex will be formed (Martell and Smith 2003)

Metal ion	EDTA	GLDA	DTPA	HEDTA	STPP	Citric Acid	EDG
Al^{3+}	16.4	12.2	18.6	14.4		7.0	7.7
Ba^{2+}	7.9	3.5	8.7	6.2	3.0	2.8	3.4
Ca^{2+}	10.7	5.9	10.8	8.1	5.3	3.2	4.7
Cd^{2+}	16.5	9.1	19.0	13.7	6.5	4.2	7.4
Co^{2+}	16.5	10.0	18.8	14.5	6.9	4.4	8.0
Cu^{3+}	18.8	13.1	21.2	17.4	8.7	14.2	11.8
Fe^{2+}	14.3	8.7	16.2	12.2	2.5	3.0	6.8
Fe^{3+}	25.1	11.7	28.0	19.7		11.9	11.6
Hg^{2+}	21.5	14.3	26.4	20.1			5.5
Mg^{2+}	8.8	5.2	9.3	7.0	5.7	3.2	3.4
Mn^{2+}	13.9	7.6	15.2	11.1	7.2	3.5	5.5
Ni^{2+}	18.4	10.9	20.1	17.1	6.7	4.4	9.3
Pb^{2+}	18.0	10.5	18.8	15.6		6.5	9.4
Sr^{2+}	8.7	4.1	9.8	6.8	4.4	2.9	3.8
Zn^{2+}	16.5	10.0	18.2	14.6	7.6	4.5	8.4

As determined at an ionic strength of 0.1 M at 25C. *EDTA* ethylenediaminetetraacetic acid, *L-GLDA* glutamic acid N,N-diacetic acid, *DTPA* diethylentriaminetetraacetic acid, *HEDTA* hydroxyethylethylenediaminetriacetic acid, *STTP* sodium tripolyphosphate, *EDG* ethanoldiglycinic acid

However, in this case each area where an aggregate of lead salt crystals had been removed left a dark spot that required retouching. Using a combination of EDTA in cellulose-thickened gels, it was possible to reduce the lead soap crusts in the foreground of *Return from the Front* (Sect. 14.2.1) to some extent, but application of the reagent close to the paint surface proved too risky. In all these cases, as in all cleaning campaigns, the skill and judgment of the conservator carrying out the treatment resulted in an interpretation of the condition of the composition and texture of the surface with the ultimate aim to make the composition more legible. The success of treatments depended on the extent to which the coating or crust has become part of the painting surface, limiting what can safely be removed. When re-saturated with a new varnish, the composition of *Wolmer Wood* and the drapery of *Carolina Ewen* and the foreground of *Return from the Front* became more readable, although the compromises included exposing surface deterioration that necessitated aesthetic reintegration.

The influence of environmental moisture and the use of hygroscopic materials in paintings on the formation of lead soaps are highlighted in several of the case studies discussed in this chapter and more comprehensively in a recent review by Cotte et al. (2017). This link is important in relation to the approaches to both active and preventive conservation treatments, discussed below. Examples include soaps that formed in paintings exposed to humid environmental conditions, heat, and light (Keune et al. 2016). The moisture absorption properties of zinc oxide pigment may influence the tendency for soap formation in oil media by enhancing hydrolysis (Osmond 2012). This may have played a role in the formation of zinc and also lead soaps in *Female Nude*, which was painted on an open weave canvas prepared with a substantial layer of hygroscopic glue size. It is possible that this size layer may have provided a reservoir of moisture in damp conditions which, together with the hygroscopic character of zinc oxide, created the ideal conditions for formation of soaps. A similar stratigraphy has also been observed in a painting discussed by Osmond et al. (2014).

The Pilgrims Rest Hotel (Sect. 14.3) suffered cleavage between paint layers due to metal soap formation. The expansion and contraction of the canvas in response to changes in environmental moisture and to tapping out the stretcher keys had caused cracking in brittle zinc oxide-containing paints. The brittleness of zinc-containing paints poses a problem in particular where cupped and flaking paint is combined with shrinkage and deformation of the support.

Evidence for the effect of moisture on metal soap formation is exemplified by the Courtauld Gallery's *Draped Woman standing by a Mantlepiece* painted in 1938 by Ethel Walker (Burnstock et al. 2016).[12] The painting in oil on commercially primed canvas is unvarnished and displayed on the reverse of a stretcher with the unprimed side painted with another image facing outward. The painting has been stored and displayed with a backing board since its acquisition in 1973. That the painting had been exposed to moisture during the course of its history was evident when technical examination showed the presence of developed fungal hyphae. The identification of particularly concentrated areas of lead soaps in areas where fungal hyphae are present pointed to the favorable effect such conditions have on the development of both mold and lead soaps.

14.5.4 Treatment of Surface Aggregates

Numerous dark topped lead soap protrusions at the surface of the *Saavedra* portrait (Sect. 14.2.2) were particularly disturbing in the areas of the light-colored background paint, the flesh, and the blue and other pale-colored draperies. Tests showed that the brown spots were insoluble in the range of solvents used to remove varnish, but they could be reduced by repeated application of an oil-swelling solvent (such as n-methyl-2-pyrrolidone, NMP) in gelled form or removed together with some of the underlying lead soap mass using 2% EDTA in deionized water. The use of this solution applied in a gel was effective in removing the spot, by undercutting and removal of the surface of the underlying protrusion. The difficulties of limiting the application of the gel precisely to the spot and the clearance of the gel were challenging in this case. After consultation with conservators at the Prado, it was

[12] Report CIA:2050 Department of Conservation & Technology, Courtauld Institute of Art.

decided not to remove the material but to disguise the most disturbing dark spots with retouching.

The technical study of *Female Nude* was carried out in advance of future treatment. The characterization of soaps in the paint and ground and their significance in relation to the original paint texture, together with an understanding of the role of the varnish that is likely to have been applied at the same time as the wax-resin lining, will influence the approach to cleaning and interpretation of the work.

In all the cases discussed, removal or retention of metal soaps during cleaning required an interpretation of the paint surface that involved removal of surface material. Treatment decisions involve finding a balance between the risks to the paint film and the aesthetic gains. The interpretation of an image obscured by soap crusts or optically disturbing surface protrusions presents practical, aesthetic, and ethical challenges for conservators. Dark spots, similar to those found on the *Saavedra* portrait, are present on the surface of many paintings of different dates, including works in egg tempera and other media. These spots often remain on the paint as insoluble residues left during varnish removal using polar solvents and can sometimes be removed or reduced using an oil-swelling solvent, used together with a polar solvent or alkaline reagent in solution or gelled form. The question, however, of whether or not removal should be attempted is interesting to consider in the context of the acceptance, within contemporary conservation ethics, of other visual alterations such as fading of paint containing organic lakes, discoloration of smalt, and increased transparency of paint. More pertinent still is the decision generally taken not to remove areas of superficial degradation, such as the blackening of vermilion or browned copper-green paints, in order to reveal the better preserved material beneath.

14.6 Conclusion

The aim of this chapter has been to highlight various manifestations of metal soaps in paintings, add to the already significant body of published information and help to inform conservators' assessment of the condition of a work, and guide approaches to conservation. The strategies for treatment of the paintings introduced here present a range of options (and compromises) that require careful evaluation.

A combination of empirical and analytical evidence is presented that supports the notion that metal soaps form where fatty acids and metal ions are available; the idea that the introduction of new sources of fatty acids and/or metal ions may favor the mobility or reorganization of metal soaps within the painting structure; the suggestion that raised humidity and direct contact with water favor the formation of metal soaps; and the theory that the different physical manifestations of soaps are dependent upon chemical environment, the distribution of reactive species within the paint film, together with external environmental factors. Many of these ideas have been investigated in detail in previously published studies and in other contributions in this volume.

A number of questions arise from these studies that may influence approaches to treatment and would be informed by further research. The question of the time it takes for metal soaps to form in oil paint is pertinent to the question of whether or not artists anticipated changes in surface texture and in helping to date later interventions. Increased understanding of metal soap formation has a bearing therefore on interpreting artists' intentions and on the formulation of treatment rationales that consider the ongoing formation of soaps in paintings.

There may be some commonalities that can be identified regarding the predisposition of oil paints to specific kinds of soap formation, such as zinc soap aggregates, or lamellar formations that might lead to delamination or cleavage between paint layers. The role of metal soaps, formed in or added to modern oil paints, and their relationship to water sensitivity in unvarnished paintings is another area of current research.

Recognition of surface changes and their causal agents is essential for designing effective and appropriate conservation treatments. The removal of metal soap crusts and formations presents new ethical considerations for conservators because they constitute degradation products that have formed from original painting materials. However, there is currently limited research on effective methods for removal of residues of nonvolatile components of cleaning reagents including chelating agents, surfactants, and other materials applied in solution, poultice, or gel form. The long-term consequences of residual materials on works of art are a critical concern.

Good conservation practice requires chemical knowledge of the causes of deterioration, the design of an appropriate treatment, and the means to carry it out. The arsenal of methods and materials for cleaning paintings has been greatly enhanced in the last 30 years by practitioners such as Richard Wolbers (Angelova et al. 2017), and there is an awareness of the need to consider appropriate treatment for paintings with different physical histories and current contexts. Paramount is the judgment and skill of the conservator, who evaluates the evidence, takes the decisions, and undertakes the treatment of the work.

Acknowledgments Klaas Jan van den Berg (RCE/University of Amsterdam), Bronwyn Ormsby and Judith Lee (Tate, London), Jae Youn Chung (Hamilton Kerr Institute, Cambridge), Lucia Bay (Philadelphia Museum of Art), Alex Ball and Tomasz Goral (Natural History Museum London), William Luckhust (Kings College London).

References

Angelova L, Ormsby B, Townsend J, Wolbers R (2017) Gels in the conservation of art. Archetype Pubns, London

Arslanoglu J (2004) Aquazol as used in conservation practice WAAC Newsletter, Vol 26(No 1), pp 10–15

Boon JJ, Van der Weerd J, Keune K, Noble P (2002) Mechanical and chemical changes in old master paintings: dissolution, metal soap formation and remineralisation processes in lead pigmented ground/intermediate paint layers of 17th century paintings, In Vontobel R (ed) ICOM committee for conservation, 13th triennial meeting, Rio de Janeiro, 22–27 Sept 2002. James and James, London, pp 401–406

Boon JJ, Hoogland F, Keune K (2007) Chemical processes in aged oil paints affecting metal soap migration and aggregation. In: Post-prints for the AIC paintings speciality group 19. American Institute for Conservation of Historic and Artistic Works, Washington, DC, pp 16–23

Burnstock A, van den Berg KJ (2014) The interface between science and conservation and the challenges for modern oil paint research. In: Van den Berg KJ, Burnstock A, de Tagle A, de Keijzer M, Heydenreich G, Krueger J, Learner T (eds) Issues in contemporary oil paints. Springer, Cham, pp 1–20

Burnstock A, Lee J, van den Berg, KJ, Ormsby B (2016) Water sensitivity of modern paint films. In: Conference proceedings, international conference COLOUR AND CONSERVATION 2015, the association CESMAR7 and Il Prato publishing, pp 66–76

Chung JY, Ormsby, B, Lee J, Burnstock A, van den Berg, KJ (2017) An investigation of options for surface cleaning unvarnished water-sensitive oil paints based on recent developments for acrylic paints. In: Manuscript accepted for inclusion in the paintings group preprints for ICOM CC Copenhagen

Cotte M, Checroun E, De Nolf W, Taniguchi Y, De Viguerie L, Burghammer M, Walter P, Rivard C, Salomé M, Janssens K, Susini J (2017) Lead soaps in paintings, friends or foes? Stud Conserv 62:2–23

Hermans JJ, Keune K, Van Loon A, Iedema PD (2015) An infrared spectroscopic study of the nature of zinc carboxylates in oil paintings. J Anal Atomic Spectrom 30:1600–1608

Jones R, Townsend JH, Stonor K, Duff N (2007) Lead soap aggregates in sixteenth and seventeenth century British paintings. In: Postprints of the AIC paintings working group meeting 2006, pp 24–32

Keune K, Boon JJ (2007) Analytical imaging studies of cross-sections of paintings affected by lead soap aggregate formation. Stud Conserv 52:161–176

Keune K, Kramer RP, Huijbregts Z, Schellen HL, Stappers MH, van Eikema Hommes MH (2016) Pigment degradation in oil paint induced by indoor climate: comparison of visual and computational backscattered electron images. Microsc Anal 22(2):448–457

Martell AE, Smith RM (2003) NIST Critically selected stability constants of metal complexes (NIST standard reference database 46 Version 7.0)

Morrison R, Bagley-Young A, Burnstock A, van Keulen H, van den Berg KJ (2007) The effects of triammonium citrate and variable pH for surface cleaning unvarnished oil paintings. Stud Conserv 52(4):255–270

Noble P, Boon JJ (2007) Metal soap degradation of oil paintings: aggregates, increased transparency and efflorescence. In: Mar Parkin H (ed) AIC paintings specialty group, annual meeting in providence, vol 19, Rhode Island, 16–19 June 2006. Thompon-Shore, Dexter, pp 1–9. Postprints

Noble P, Boon JJ, Wadum J (2002) Dissolution, aggregation and protrusion: lead soap formation in 17th century grounds and paint layers. Art Matters 1:46–62

Osmond G (2012) Zinc white: a review of zinc oxide pigment properties and implications for stability in oil-based paintings. AICCM Bull 33:20–29 https://aiccm.org.au/sites/default/files/AICCMBulletin_33-Osmond.pdf

Osmond G (2014) Zinc white and the influence of paint composition for stability in oil based media. In: Van den Berg KJ, Burnstock A, de Tagle A, de Keijzer M, Heydenreich G, Krueger J, Learner T (eds) Issues in contemporary oil paints. Springer, Cham, pp 263–282

Osmond G, Ebert B, Drennan J (2014) Zinc oxide-centred deterioration in 20th century Vietnamese paintings by Nguyễn Trọng Kiệm (1933–1991). AICCM Bull 34:4–14

Osmond G (2019) Zinc soaps: an overview of zinc oxide reactivity and consequences of soap formation in oil-based paintings. In: Casadio F, Keune K, Noble P, Van Loon A, Hendriks E, Centeno S, Osmond G (eds) Metal soaps in art: conservation and research. Springer, Cham, pp 25–43

Puig I, Company X, Garrido C, Herreo MA (2016) Francisco de Goya Carlos IV Portrait of King Carlos IV, University of Leiden

Raven LE, Bisschoff M, Leeuwestein M, Geldof M, Hermans JJ, Stols-Witlox M, Keune K (2018) Delamination due to zinc soap formation in an oil painting by Piet Mondrian (1872–1944). Conservation issues and possible implications for treatment. In: Casadio F, Keune K, Noble P, Van Loon A, Hendriks E, Centeno S, Osmond G (eds) Metal soaps in art: conservation and research. Springer, Cham, pp 345–357
Rogala D, Lake S, Maines C, Mecklenburg M (2010) Condition problems related to zinc oxide underlayers: examination of selected abstract expressionist paintings from the collection of the Hirshhorn Museum and sculpture garden, Smithsonian Institution. J American Inst Conserv (JAIC) 49(2):96–113
Sawicka A, Burnstock A, van den Berg KJ, Izzo F, Keune K, Boon JJ, Kirsch K (2014) Metal soap efflorescence in contemporary oil painting. In: Van den Berg KJ, Burnstock A, de Tagle A, de Keijzer M, Heydenreich G, Krueger J, Learner T (eds) Issues in contemporary oil paints. Springer, Cham, pp 311–332
Shimadzu Y, Keune K, Boon JJ, Townsend JH, van den Berg KJ (2008) The effects of lead and zinc white saponification on surface appearance of paint. In: Bridgland J (ed) 15th triennial meeting of ICOM committee for conservation preprints, New Delhi, 22–26 Sept 2008. ICOM Committee for Conservation, Paris, pp 626–632
Van der Weerd J, Geldorf M, van der Loeff LS, Heeren RMA, Boon JJ (2003) Zinc soap aggregate formation in 'Falling Leaves (Les Alyscamps)' by Vincent van Gogh. ZKK Zeitschrift für Kunst Technologie und Konservierung 17(2):407–416
Van Loon A, Noble P, Boon JJ (2012a) The formation of complex crusts in oil paints containing lead white and smalt: dissolution, depletion, diffusion, deposition. In: Meeks N, Cartwright C, Meek A, Mongiatti A (eds) Historical technology, materials and conservation, SEM and microanalysis. The British Museum/Archetype, London, pp 207–209
Van Loon A, Noble P, Burnstock A (2012b) Ageing and deterioration of traditional oil and tempera paints. In: Hill Stoner J, Rushfield RA (eds) The conservation of easel paintings. Routledge, London, pp 214–241
Van Loon A, Hartman LE, van den Burg J, Haswell R, Pottasch C (2019) The development of an aqueous gel testing procedure for the removal of lead-rich salt crusts on the surface of paintings by Giovanni Antonio Pellegrini (1675–1741) in the "Golden Room" of the Mauritshuis. In: Casadio F, Keune K, Noble P, Van Loon A, Hendriks E, Centeno S, Osmond G (eds) Metal soaps in art: conservation and research. Springer, Cham, pp 285–297
Veliz Bomford Z, Aterido A (2016) The art of conservation V. Caring for the king's pictures: artists and restorers in the Spanish royal collection, 1576–1814. Burlingt Mag:447–459

Chapter 15
Characterization and Removal of a Disfiguring Oxalate Crust on a Large Altarpiece by Hans Memling

Lizet Klaassen, Geert van der Snickt, Stijn Legrand, Catherine Higgitt, Marika Spring, Frederik Vanmeert, Francesca Rosi, Brunetto Giovanni Brunetti, Marie Postec, and Koen Janssens

Abstract During the conservation treatment of Memling's *Christ with Singing and Music-making Angels*, three panel paintings that are among the most monumental works in early Netherlandish art, the conservators came across insoluble surface layers containing calcium oxalates. A very thin and irregular layer of this type, hardly visible to the naked eye, was spread across the surface of all three panels. A much thicker layer forming an opaque and highly disfiguring crust that obscured the composition (Figs. 15.1 and 15.7) was locally present on areas of dark copper-containing paint, where multiple layers of old discolored coatings and accretions remained in place before the most recent cleaning.

This article describes the application of a wide range of analytical techniques in order to fully understand the stratigraphy and composition of the crusts on the Memling paintings. FTIR spectroscopy in transmission and reflection mode, micro-ATR-FTIR imaging and macro-rFTIR scanning, SEM-EDX, mobile XRD,

L. Klaassen (✉)
Royal Museum of Fine Arts Antwerp, Antwerp, Belgium
e-mail: lizet.klaassen@kmska.be

G. van der Snickt
Department of Chemistry – AXES group, University of Antwerp, Antwerp, Belgium

Conservation Studies, University of Antwerp, Antwerp, Belgium

S. Legrand · F. Vanmeert · K. Janssens
Department of Chemistry – AXES group, University of Antwerp, Antwerp, Belgium

C. Higgitt · M. Spring
The National Gallery, London, UK

F. Rosi · B. G. Brunetti
Istituto CNR di Scienze e Tecnologie Molecolari (CNR-ISTM), Perugia, Italy

Centro di Eccellenza SMAArt, Università degli Studi di Perugia, Perugia, Italy

M. Postec
Independent Painting Conservator, Brussels, Belgium

F. Casadio et al. (eds.), *Metal Soaps in Art*, Cultural Heritage Science,
https://doi.org/10.1007/978-3-319-90617-1_15

and SR-μXRD showed that the crusts contained two related Ca-based oxalate salts, whewellite and weddellite, and were separated from the original paint surface by varnish, indicating that they did not originate from degradation of the original paint but from a combination of microbial action and a thick accumulation of dirt. Supported by the results from these different analytical techniques, which when used together proved to be very effective in providing complementary information that addressed this specific conservation problem, and aided by the presence of the intermediate varnish layer(s), the conservators were able to remove most of the crusts with spectacular results.

Keywords Memling · Christ with Singing and Music-making Angels · MA-rFTIR · ATR-FTIR · SR-μXRD · SEM-EDX · KMSKA · Oxalate · Whewellite · Weddellite

15.1 Introduction

One of the highlights in the collection of the Royal Museum of Fine Arts Antwerp is three large panel paintings entitled *Christ with Singing and Music-making Angels*, attributed to Hans Memling and assistants (Fig. 15.1). This paper deals with its

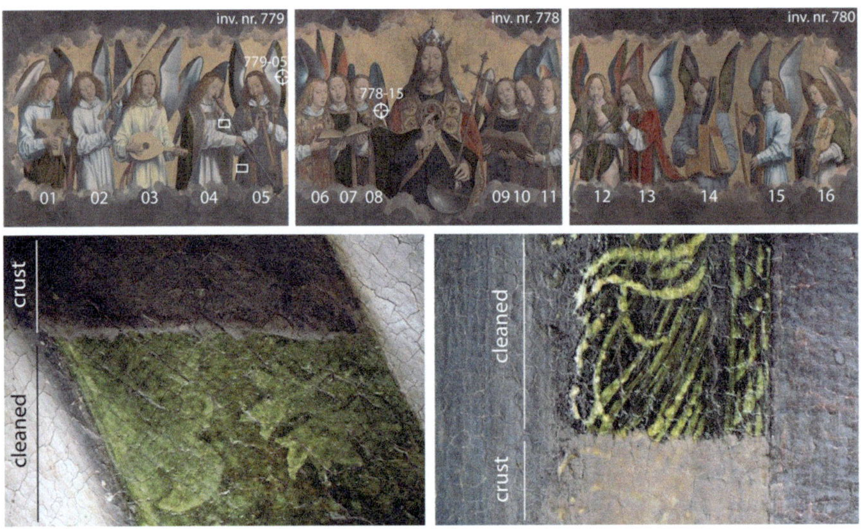

Fig. 15.1 Top row: *Christ with Singing and Music-making Angels* attributed to Hans Memling and assistants. Collection of the Royal Museum of Fine Arts Antwerp (KMSKA), with numbering of the angels and indication of sample locations (crosshair). The white rectangles indicate the location of the details shown in the bottom row. Bottom row: two macrophotographs showing two areas on panel 779 where oxalate crusts on green paint had been partly removed. Photos top row: KMSKA © Photo Rik Klein Gotink. Photos bottom row: KMSKA © Arcobaleno, Adri Verburg

extraordinary conservation treatment, which took from 2001 until 2017 and involved a team of conservators and scientists.

15.2 Memling's Retablo Major

Christ with Singing and Music-making Angels consists of three large panel paintings: the central panel measures 170 × 213 cm, and the two side panels are 170 × 231 cm each. The panels are attributed to Hans Memling and assistants and were made at the end of the fifteenth century, so late in Memling's career (Fransen 2018). Despite their sizeable dimensions, the panels are only a fragment of the original work of art. Originally, they formed the upper row of what must have been a very large polyptych, created for the high altar of the monastery of Santa Maria la Real in Nájera, Spain. The panels were described in 1795 by the Spanish writer Gaspard David Jovellanos who saw them, together with a large *Assumption of the Virgin*, two apostles, and the Saints Benedictus, Prudentius, Vitalis, and Agricola, on his visit to Nájera. According to Jovellanos, all of the artworks were by the same artist and belonged to the same altarpiece. Jovellanos saw the different parts in various places in the Monastery, the altarpiece having already been dismantled by 1795, but the *Assumption of the Virgin* was probably originally placed below Christ and flanked by the saints (Vandenbroeck 1985; Borchert 1993; De Vos 1994).

Although the Antwerp panels are undoubtedly among the most monumental works in early Netherlandish painting, not much is known about their history. As indicated above, the polyptych had already been dismantled by 1795, and when the panels were rediscovered around 1880 by Spanish art dealers, the central panel, *Christ with Singing Angels,* was placed above an altar, and the side panels, *Music-making Angels*, formed part of an organ decoration. After a stay of almost 400 years in Nájera, in 1886 the panels were sold to art dealers and finally bought by the Royal Museum of Fine Arts Antwerp in 1895.

The different retouching campaigns visible on the paint surface and the cradles on the back, probably attached in the 1950s or 1960s, indicated that there must have been several interventions in the past. However, there is very little documentation on the conservation history of the panels. Before they entered the art dealer circuit, they were brought to Madrid and "cleaned."[1] In the museum archives just a few, very short notes were found describing only minor interventions executed after their purchase.[2]

[1]"Llegó a Madrid, donde, tras su limpieza, se intentó que lo comprara la Reina Maria Christina," (Gómara 2013). Many thanks to Sara Mateu for providing the article.

[2]Flaking paint was noted in February and May 1901, in different areas. It was mentioned that the panels were covered by a layer of dust that adhered to the surface and diminished the brightness of the colors. Consolidation and surface cleaning with bread crumbs was proposed, but the charge made for the work was only for consolidation.
In 1952, 1977, and 1980, the panels occur in lists among other paintings that were treated, but

15.3 The Need for Intervention in 2001

The poor condition of the paintings was the impetus for the extensive conservation treatment that began in 2001. By that time there was severe flaking of the paint layers, the cradles were blocked, and the surface was extremely dirty. The consolidation of the paint layers and the unblocking of the cradles were the first steps of the treatment, and both could be executed without major problems. The cleaning, however, turned out to be problematic and took years to complete. Yellowed varnishes, discolored retouching, and layers of dirt subdued the bright colors and greatly reduced the legibility and the three-dimensionality of the composition. It was only through technical imaging, including infrared photography, infrared reflectography (IRR),[3] and X-ray radiography (XRR),[4] that the conservators became aware that underneath some of the plain, flat brown areas, decorative details, as well as modelling of the volume and folds in the draperies were present.

Cleaning tests showed that underneath the uppermost soluble varnish layers, a grey-brown opaque crust was present in the darkest, most obscured areas, in particular over the copper-containing blue and green paint and the grey clouds surrounding the figures, which also contain a copper-based pigment in the form of azurite.[5]

Unfortunately, this crust turned out to be highly insoluble in the extensive range of solvents and cleaning systems that were tested. It reacted to the cleaning tests by blanching but appeared impossible to dissolve. To gain more insight into its composition and with the hope of finding a safe method to remove it, scrapings of it were first analyzed by transmission FTIR microscopy and SEM-EDX. Further non-invasive investigations and detailed analysis of samples mounted as cross sections were also undertaken subsequently, in order to gain a more precise understanding of the chemical composition of the crusts, their location in the layer structure, and their distribution across the paint surface.

details were not specified. Although nothing is mentioned about the cradles, judging from their appearance, it is thought that they were attached in the 1950s or 1960s when the panels were already in the museum. In addition, a synthetic varnish (isobutyl acrylate) was found on the surface which must also have been applied in the museum.

[3]Infrared imaging was carried out twice: at the beginning of the project by Adri Verburg, using a phase I scanning back (4 × 5 inch) with spectral range 1000–1100 nm, and during the project by Claudia Daffara and Mattia Patti, using a multispectral scanner, spatial resolution 4 pt./mm (250 micron), InGaAs with spectral range 800–1700 nm, C.N.R.–I.N.O.A (Istituto Nazionale di Ottica Applicata).

[4]X-radiography was undertaken by Guido van de Voorde, Royal Institute for Cultural Heritage, Parc du Cinquantenaire 1, B-1000 Brussels.

[5]SEM-EDX detected copper in the blue and green paints, suggesting the use of azurite and a copper-containing green pigment (most probably originally verdigris). The grey paint layer depicting the clouds consists mainly of a mixture of lead white and a black pigment (probably a carbon-based black), mixed with a little azurite, identified by SEM-EDX and FTIR microscopy.

15.4 Experimental

Transmission Fourier transform infrared (FTIR) microscopy (mid-IR range) spectra were acquired using a Nicolet 710 Series FTIR spectrometer with a NicPlan infrared microscope, fitted with a mercury-cadmium-telluride (MCT) type A detector (cooled with liquid nitrogen). Small subsamples of the scrapings of the crusts were placed between the windows of a Spectra-Tech micro-compression diamond cell for analysis. Measurements were made in transmission mode over the range 4000–650 cm^{-1}, using a Spectra-Tech Reflachromat Cassegrain ×15 objective. One hundred twenty-eight or 256 scans were made at a resolution of 4 cm^{-1} with Happ-Genzel apodization. Both spectrometer and microscope were purged with water- and CO_2-free air.

Mobile reflection mid-infrared (reflection mid-FTIR) spectrometry point measurements were performed by means of a portable JASCO VIR 9500 spectrophotometer, made up of a Midac Illuminator IR radiation source, a Michelson interferometer, and a liquid nitrogen-cooled MCT detector. The probe was a 4 mm bifurcated cable containing 19 chalcogenide glass fibers allowing collection of spectra in the range 4000–900 cm^{-1} at a resolution of 4 cm^{-1}. The noncontact probe was kept perpendicular to the painting surface (0°/0° geometry) at a distance of about 3 mm. The total reflectance was collected over 400 scans using the spectrum from an aluminum mirror plate for background correction. The reflectance spectra (R) were converted to pseudo-absorbance (A' = log(1/R)).

The instrument used for the macroscopic FTIR scanning measurements in reflection mode (MA-rFTIR scanner) was a Bruker Alpha FTIR spectrometer, equipped with a frontal reflection module (20°/20° geometry) and a coaxial visual camera (Legrand et al. 2014). The spectrometer incorporates a globar IR source, Michelson interferometer, and a deuterated triglycine sulfate (DTGS) detector. Spectra acquisition and storage were carried out using the Bruker OPUS 6.0 software package. Before each scan a spectral background was recorded over 30 min by placing a gold-coated mirror in front of the reflection module. During scanning mode, spectra were recorded for 3 s on each point at a resolution of 4 cm^{-1} over the range 7500–375 cm^{-1}. For scanning purposes, the spectrometer was mounted on a motorized X (10 cm), Y (25 cm), and Z (10 cm) motor stage setup (Newport Corporation, Irvine, CA, USA). The X and Y stages are used during scanning, while the Z stage helps to bring the setup in focus (2 mm spot) prior to the scan. The processing of the hyperspectral datasets was performed by self-written routines using the data analysis software package IDL 8.3. For most chemical distributions and spectra, the recorded reflectance was converted to pseudo-absorbance. The resulting FTIR maps have a lateral resolution in the mm range.

For mobile X-ray diffraction (MXRD) measurements combined with XRF, a 30 W air-cooled iMOXS-MFR (IFG, Adlershof, Berlin, Germany) X-ray tube was used with a voltage of 40 kV and a current of 700 μA. The source is equipped with a polycapillary semi-lens providing a 4 mm (3.8–4.4 mm) diameter parallel X-ray beam (total exit divergence of 0.25° or 4.36 mrad). The copper anode provides, through a 0.1 mm beryllium window, polychromatic X-rays necessary for XRF measurements. The source is equipped with a 15 μm Ni filter to strongly attenuate the Cu K_β-line and avoid the presence of secondary diffraction peaks. XRD is therefore performed with the usual monochromatic radiation (Cu-K_α; E = 8.047 keV (= 0.154 nm)). Diffractograms are collected in reflection mode using an incident angle ω of 10° to the specimen surface allowing 2θ values larger than 10° to be reached. Fine-grain alumina samples were used for the calibration. The relatively prolonged acquisition times, amounting up to 30 min, limited the number of measurements. The FIT 2D software[6] was used to transform the two-dimensional diffraction images into 2θ spectra. Based on the latter and a database of X-ray powder diffraction patterns, the EVA[7] and XRDUA[8] software were used to determine the crystalline phases that were present. The technical features and analytical performance of this instrument are discussed elsewhere (Gianoncelli et al. 2008).

At the University of Antwerp, embedded paint cross sections were analyzed by means of a Jeol 6300 electron microprobe system equipped with an energy dispersive Si (Li) X-ray detector (SEM-EDX) (Princeton Gamma Tech). Elemental distribution maps were collected from the surface of paint cross sections at 20 kV, a magnification of 500×, a beam current of 1 nA, and a recording time of 3600 s.

The elemental analysis and mapping that was carried out at the National Gallery on paint cross sections and scrapings was done using a Carl Zeiss EVO® MA10 variable pressure scanning electron microscope (SEM) coupled to an Oxford Instrument X-Max 80 mm^2 energy dispersive X-ray spectrometer (EDX), using INCA 350 software. The operating parameters were 20 kV, 200 pA beam current, and 30 Pa chamber pressure (the minimum necessary to limit charging) with air as the chamber gas. Optical microscopy was carried out using a Leica DM4000 M microscope and a Zeiss AxioCam HRc camera for recording images. A filter system A (BP 340–380 nm, dichromatic mirror: 300 nm, suppression filter: LP 425 nm) was used for UV excitation.

Attenuated total reflection Fourier transform infrared (ATR-FTIR) microspectroscopic imaging spectra were acquired from embedded cross sections by using a Bruker Tensor 27 FTIR spectrometer connected to a Hyperion 3000 Series microscope, fitted with a 64 × 64 (4096 pixels) FPA detector (range = 4500–900 cm^{-1}), cooled with liquid nitrogen. The microscope was fitted with a CCD camera, X-Y stage (adjustment accuracy of 0.1 μm), and a dedicated ATR objective (20× magnification). The ATR had a germanium crystal with a tip size of 250 μm. Both spectrometer and microscope were purged with water- and CO_2-free air. One hundred twenty-eight scans were collected at a resolution of 4 cm^{-1}. The 64 × 64 pixel focal plane array collects image data from a 32 × 32 μm square. An effective (diffraction limited) lateral resolution of 3–5 μm is achieved.

[6]http://www.esrf.eu/computing/scientific/FIT2D/. Accessed February 2017.

[7]http://www.bruker-axs.de/eva.html. Accessed February 2017.

[8]http://xrdua.ua.ac.be/. Accessed February 2017.

Combined synchrotron radiation-based micro X-ray fluorescence and micro X-ray diffraction (SR μ-XRF/μ-XRD) imaging experiments were performed at the microprobe station of the P06 Hard X-ray Micro/Nano-Probe beamline (PETRA III, DESY, Germany). A photon energy of 21 keV was selected by means of a Si(111) double crystal monochromator. The beam was focused to 0.4 × 0.4 μm^2 (hor. × vert.) employing a Kirkpatrick-Baez mirror optic. Fluorescence radiation was recorded by a Vortex-EM silicon drift detector placed perpendicular to the incident X-ray beam. Simultaneously diffraction signals were recorded in transmission geometry using a Pilatus 300 K area detector at a distance of approximately 15 cm behind the sample. Initial calibration of the diffraction setup was performed using a LaB_6 reference sample. The software package PyMCA was used for spectral fitting of the fluorescence data (Solé et al. 2007), while whole pattern fitting of the diffraction data was performed using XRDUA (De Nolf and Janssens 2010). This software package provides several methods for obtaining crystalline-specific distributions from a large number of diffraction patterns typically obtained in μ-XRPD imaging experiments (De Nolf et al. 2014).

15.5 Analytical Results

15.5.1 Analysis of Scrapings of the Crust

Since rapid identification of the crusts was vital for the continuation of the conservation treatment, a number of samples in the form of scrapings were taken and analyzed after local removal of the upper varnish layers. These were collected from each panel, concentrating in particular on the crusts on the blue and green paint and on the grey clouds. The resulting powders were analyzed using FTIR microscopy in transmission mode and SEM-EDX at the National Gallery, London. All of the samples of the crusts had a similar composition, consisting essentially of calcium oxalate (bands at c.1650, 1327, and 783 cm^{-1}), with some calcium carbonate (c.1414, 875, and 712 cm^{-1}) and calcium sulfate (c.3401, 1120, and 671 cm^{-1}). In addition, when interpreted with the SEM-EDX results, the broad band at c.1090 (together with the doublet at c.800/780 cm^{-1}) indicates the presence of various silicates, and the band at c.1041 cm^{-1} shows the presence of calcium phosphate (and/or silicates such as phyllosilicates with bands at c.1040 and 915 cm^{-1}) (Farmer 1974). This composition suggests that the layer most probably resulted from an accumulation of dirt on the surface.[9]

[9]C. Higgitt and M. Spring, *Preliminary EDX and FTIR microscopic examination of samples*, National Galley, unpublished research report. For comparable spectra see the appendix in Higgitt and White (2005). For the chemical composition of dirt, see Van Grieken et al. (2000).

15.5.2 Non-invasive Analysis

In cooperation with the University of Antwerp and as part of a MOLAB campaign (Miliani et al. 2010),[10] the crusts were further investigated using mobile analytical equipment (Van der Snickt et al. 2011). The noncontact and non-invasive approach typical of MOLAB allowed a large number of measurements to be performed all over the vast paint surface of all three panels. In this way it was possible to assess the extent to which the local data obtained from a few scrapings was representative of the rest of the large surface.

Monico et al. demonstrated the usefulness and sensitivity of reflection mid-FTIR spectroscopy for the non-invasive identification of various oxalate salts on the surface of polychrome works of art (Monico et al. 2013). Interestingly, MOLAB reflectance FTIR point measurements, performed with the same instrument that was employed in that study, indicated the presence of oxalates across the entire paint surface. As exemplified by the spectrum in Fig. 15.2, characteristic calcium oxalate

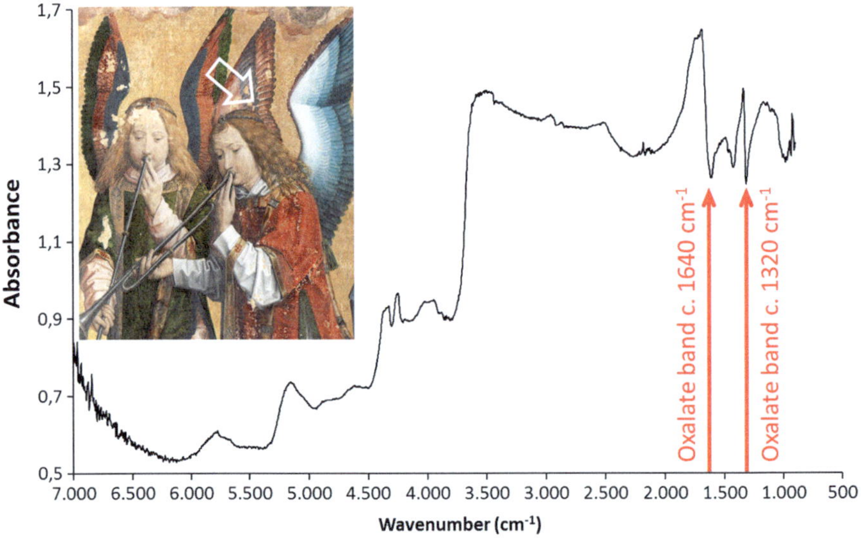

Fig. 15.2 Reflection mid-FTIR spectrum recorded on a reddish feather of angel 13 in an area where no crusts were observed. Insert: a white arrow indicates the exact spot where the spectrum was collected. Characteristic calcium oxalate bands around 1320 and 1640 cm^{-1} are indicated by red arrows

[10]MOLAB is a mobile laboratory composed of a unique collection of portable equipment which is available to cultural heritage researchers across Europe, including art historians, conservators, and conservation scientists. The MOLAB analyses presented here were funded under the EU CHARISMA project (FP7, grant no. 228330). MOLAB can currently be accessed through the Horizon 2020 IPERION project (www.iperionch.eu, Grant No. 654028).

bands at around 1320 and 1640 cm^{-1} were observed in all FTIR measurements, including those collected in areas where no crusts were visible to the naked eye. At the end of the campaign, it was evident that there was an oxalate-containing layer distributed across the whole paint surface of all three panels but so thin that it was hardly visible to the naked eye.

The numerous non-invasive reflectance FTIR measurements on the thicker crust present above areas of dark copper-containing paint also revealed exclusively calcium oxalates, with no evidence of the formation of the analogous compounds with copper, in agreement with the transmission FTIR analyses of the scrapings. Measurements performed on the crusts with the mobile X-ray diffraction (XRD) instrument from the C2RMF (Gianoncelli et al. 2008) confirmed the presence of two related Ca-based oxalate salts: whewellite and weddellite.[11]

15.5.3 Analysis of Paint Cross Sections

While the characterization of scrapings and non-invasive analysis of the paintings allowed identification of the surface materials, more detailed information about the layer structure of the paint and coatings was considered essential in order to assess the feasibility of removal of the crusts. The extensive non-invasive analysis guided the selection of sample locations and reduced the number of samples that were needed. The samples were taken both from areas where the thick, opaque crusts were evident and from areas where the oxalate-containing layer was hardly visible to the naked eye. The main objective was to compare these areas through the examination of paint cross sections and to gain insight into the location of the oxalate-containing layers within the overall layer structure.

Figure 15.3 shows the stratigraphy of the cross section prepared from a sample taken from the crust on top of the green feathers of the wing of angel 5 on panel 779. Figures 15.4 and 15.5 show the results of SEM-EDX analysis and micro-ATR-FTIR imaging of this sample. ATR-FTIR microspectroscopic imaging (Fig. 15.5) is a technique that was not available at the time of the first FTIR analyses on scrapings but since its introduction has proved very valuable for chemical imaging at high resolution (up to about 3 μm is achievable) directly on cross sections. It is especially suited to the characterization and location of oxalate-containing accretions.

At the bottom of the cross section are the ground layers on top of which are a thin brownish mordant and a layer of gold leaf (all layers labelled together as "1" in Figs. 15.3, 15.4, and 15.5). The ground in this sample consists of a first thin gypsum-rich layer followed by a thick chalk-rich layer. The gilded background

[11]In principle, whewellite and weddellite can also be distinguished on the basis of their FTIR spectra (although there is some disagreement in the literature about precise band positions), but in practice this is difficult, particularly when the salts are present in mixtures or in combinations with other compounds, see, for example, Garty et al. (2002), Conti et al. (2010), and Leroy (2016).

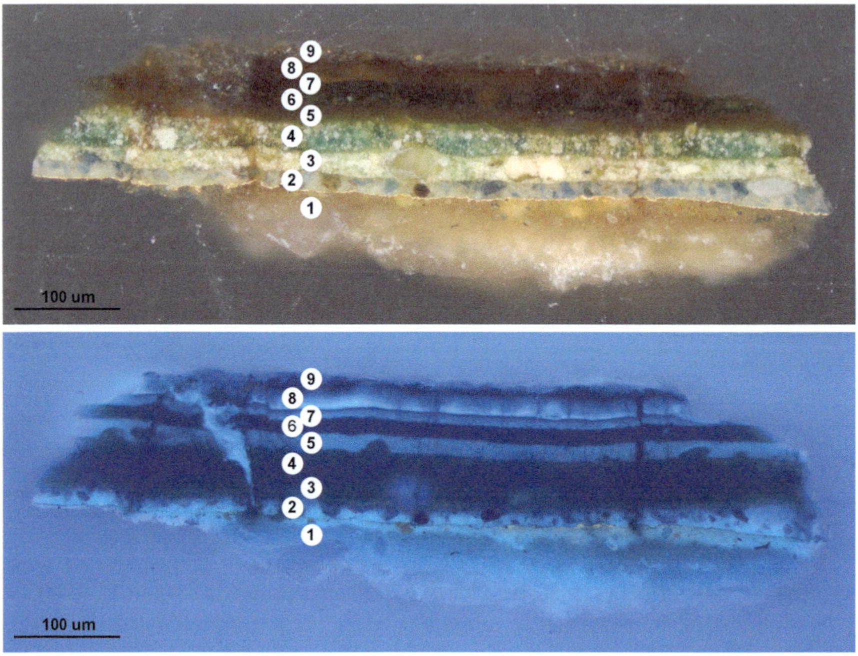

Fig. 15.3 Top: optical microscopy (OM) image of the cross section prepared from sample 779-05, removed from the green wing of angel 5, taken with visible illumination, and showing the sample stratigraphy. Bottom: photomicrograph of the same cross section taken under UV illumination

extends beneath the paint in this area. The next layer (labelled 2) contains a mixture of lead white and azurite and appears to correspond to the paint used to create the inner part of the wing. Over the blue paint is a light yellow-green paint (labelled 3) composed of a copper-containing pigment (probably verdigris or a related copper salt of an organic acid), lead white and lead-tin yellow (type I). This is covered with a more deeply colored green paint (labelled 4), containing a similar mixture to layer 3 but with rather less lead white and lead-tin yellow. Detailed examination of a series of FTIR spectra extracted from the ATR-FTIR imaging data confirmed that the copper-containing pigment in this layer is some form of (basic) copper acetate and revealed the presence of copper carboxylates (strong asymmetric carboxylate stretch at 1586 cm^{-1}), presumably from reaction of the pigment with the oil binder, and copper oxalates. Directly over the green paint is a very brown oil-resin varnish layer (labelled 5), which also contains some copper oxalate, suggesting migration of copper from the paint layer below. Over this varnish is a layer with a greenish color (labelled 6), which can be seen in the EDX maps to contain copper (both dispersed throughout the layer and in discrete copper-rich particles) and a little calcium. ATR-FTIR indicates that at least some of the copper in this layer is present in the form of copper carboxylates and oxalates. Since it lies over varnish, it is not completely

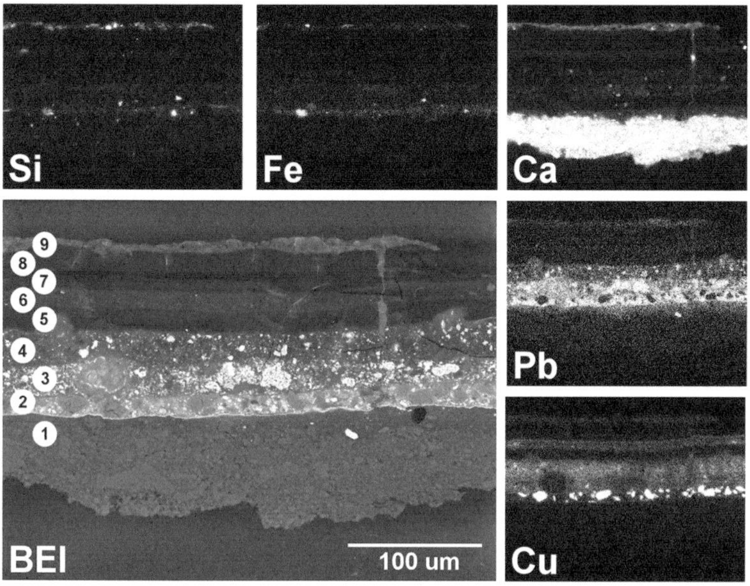

Fig. 15.4 SEM-EDX measurements on sample 779-05 taken from the green wing of angel 5. Bottom left: backscattered electron image (BEI) showing the sample stratigraphy. Top: elemental distribution maps of silicon (crust and mordant) and iron (crust and mordant). Right, top to bottom: elemental distribution maps of calcium (crust and ground, plus smaller quantities in the paint layers and the intermediate copper-containing layer), lead (paint layers), and copper (paint layers and intermediate layer between varnish layers)

clear whether this copper-containing layer is overpaint or an original glaze,[12] but it is covered by at least one, or possibly two, further oil-resin varnish layers (labelled 7 and 8). Copper oxalates were detected in the lower part of the varnish on top of the intermediate green layer (corresponding to layer 7) but not further up within the stratigraphy.[13] The calcium EDX map (Fig. 15.4) shows a thick calcium-rich layer (labelled 9) at the very top of the stratigraphy, corresponding to the calcium oxalate-containing crust, as can be seen in the ATR-FTIR imaging results (Fig. 15.5). The

[12]Although the conservators were of the opinion that the copper-containing paint on top of the varnish layer was probably overpaint, it was left on the surface during cleaning as no certain proof for this could be found in the cross sections or on the surface of the paintings. It was not present as a continuous layer but more as local spots.

[13]The location of the copper oxalates suggests that some degree of migration of copper ions into adjacent layers has occurred but not all the way up into the surface crust. These results clarify why only calcium oxalates were detected at the surface of the paintings in the scrapings and using the MOLAB non-invasive reflectance FTIR equipment. It should be noted that calcium oxalates may also be present in those layers containing copper oxalates, but this cannot be confirmed because of the overlap of the spectral bands and the 900 cm^{-1} cutoff for the micro-ATR-FTIR imaging equipment.

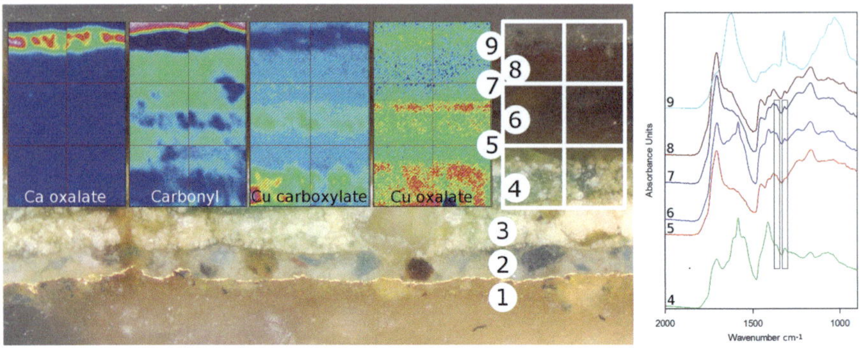

Fig. 15.5 Left: series of ATR-FTIR images produced by integration of IR spectral bands revealing the location of different materials within sample 779-05 taken from the green wing of angel 5. From left to right: the image produced by integration of the band between 1333 and 1302 cm^{-1} shows the distribution of calcium oxalates; between 1741 and 1674 cm^{-1} shows the distribution of carbonyl groups associated primarily with the oil-resin varnish layers, the paint, and the polyester mounting resin; between 1599 and 1564 cm^{-1} shows the distribution of copper carboxylates (and the green copper-containing pigment(s)); and between 1371 and 1345 cm^{-1} shows the distribution of copper oxalates. The scan window is 64 μm wide and 96 μm tall, i.e., 2 × 3 32 μm squares. Right: series of FTIR spectra from the different layers extracted from the ATR-FTIR imaging data. (9) Surface crust (calcium oxalate at c.1650 and 1321 cm^{-1}), (8) varnish layer directly below crust (oil-resin varnish at c.1733, 1711, 1435, 1414, 1318, 1254, and 1166 cm^{-1}), (7) varnish layer directly above copper-containing intermediate layer (possibly same layer as layer 8), (6) copper-containing intermediate layer (strong asymmetric carboxylate stretch at 1587 cm^{-1}), (5) varnish layer directly above paint (as above), and (4) upper green paint layer (copper-based pigment and copper carboxylates at c.1615, 1586, 1546, and 1414 cm^{-1}). Copper oxalates can be detected in layers 4–7 only (determined by the presence of the highlighted bands at c.1364 and 1319 cm^{-1})

elemental distribution maps reveal the heterogeneous nature of the crust which contains, in addition to calcium, distinct particles rich in silicon and iron.

Another sample was taken from the red lake-containing glaze in the feathers of the wing of angel 8 on panel 778. This was an area where the oxalate-containing surface layer was hardly visible to the naked eye. The sample was imaged using synchrotron radiation-based μ-XRF/μ-XRD mapping, and a number of the resulting chemical images are shown in Fig. 15.6. The XRD mapping shows a very thin layer (less than 1 μm) containing weddellite at the very top of the stratigraphy. Compared to the oxalate layers in the cross section that include the thick crust (see Figs. 15.3, 15.4, 15.5), this layer is much thinner, which explains why the oxalate deposits are hardly visible to the naked eye in this area. The oxalate layer might seem to be directly on top of the red lake-containing paint, but in fact – although this is not very clear in Fig. 15.6 – other chemical images of this cross section show that the weddellite-containing layer is separated from the original paint by an organic layer, presumably varnish (Janssens et al. 2016). At the very bottom of the cross section, the preparation layers are visible: the ground appears to consist of two layers, a layer rich in chalk applied over a layer rich in gypsum. The technique allowed the detected

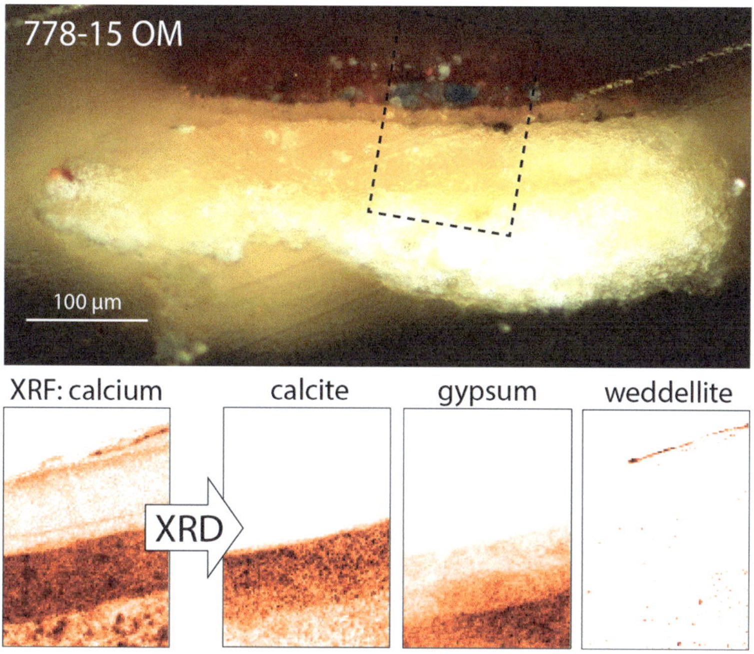

Fig. 15.6 Top: micrograph of sample 778–15 taken from the feathers of angel 8. The black rectangle indicates the area that was chemically imaged. Below: chemical images obtained by means of synchrotron radiation-based μ-XRF/μ-XRD. Bottom left: μ-XRF map of calcium collected in reflection mode. Bottom right: corresponding μ-XRD maps (transmission mode) of three calcium species: calcite, gypsum, and weddellite

calcium fluorescence signal to be resolved into four separate species: gypsum and calcite in the ground, hydroxylapatite on top of the ground (not shown), and a thin layer of weddellite at the top.

15.6 Discussion

The formation of metal oxalates, and particularly oxalate-rich surface layers, is commonly observed for a wide range of artworks including stone and bronze sculpture, wall paintings, and stained glass (Zoppi et al. 2010). Such oxalate-rich crusts, which are essentially inorganic in composition and highly insoluble, induce scatter and can be highly disfiguring. As oxalate and other salts form on surfaces, colored components of dirt (e.g., carbon-containing particles) can also become trapped within the crust, as observed on building stone (Potgieter-Vermaak et al. 2005). In recent literature, oxalate-containing layers on the surface of easel paintings are being reported with increasing frequency.[14] Although on easel paintings these layers are generally thin, their scattering effect gives an unsaturated appearance and can be optically very disturbing, especially in the dark areas. Nevertheless, during conservation treatments, these oxalate-containing layers are often left essentially undisturbed on the surface as their insolubility and strong adhesion to underlying layers make it impossible to remove them selectively.

The origin and formation of oxalate salts is complex and not fully understood. A number of different mechanisms may be involved depending on the exact materials and environment, but in general, oxalate salts form when oxalic acid reacts with metal ions. In paintings, the metal cations may originate from the paint layers (e.g., pigments or driers (Van der Snickt et al. 2012)), from the ground/substrate, or from particulate matter such as dirt deposited on the paint surface, but the source of the oxalic acid is still a subject of debate (Spring and Higgitt 2006). Microbiological activity and some form of extreme chemical deterioration of organic substances, either within the paint (e.g., the binder) or on the paint surface (e.g., protective or other coatings), have both been suggested in the literature (Mendes et al. 2008; Zoppi et al. 2010). It appears that either or both of these processes can be a source, but it is also clear that oxalate formation can occur even in the absence of an organic material (or where only small amounts are present).

In the case of Memling's *Christ with Singing and Music-making Angels,* a calcium oxalate-containing layer was found over the entire surface of the paintings, but in most areas, over light colors, this layer was so thin (less than 1 μm) that it was hardly visible to the naked eye. The oxalate-containing crusts on the darkest areas of the painting, however, were exceptionally thick, with an opaque grey-brown appearance, and highly disfiguring. Such crusts were present primarily over dark copper-containing paints, including the green garments painted with verdigris, and to a lesser extent over the azurite-rich garments and the grey clouds, which also contain a little azurite.

In the literature, copper-based pigments are noted to be one of the types of pigments with which oxalate formation seems to be particularly pronounced (Higgitt and White 2005; Spring and Higgitt 2006; Salvadó et al. 2013). Zoppi et al. reported that calcium- and copper-containing pigments show the highest reactivity toward oxalic acid, due to their high solubility in an acid environment (Zoppi et al. 2010). In this case, however, analysis showed that although some copper oxalates were detected lower in the stratigraphy in certain green passages, the surface crusts only contain calcium oxalates. Examination of cross sections and UV-induced visible fluorescence imaging on the painting during cleaning showed that the crust was not directly adjacent to the original paint surface but separated from it by one or more varnish layers, strongly suggesting that its formation is unrelated to the degradation of the underlying paint. The heterogeneous composition of the crust, which contains

[14]For examples of the detection of calcium oxalate in easel painting, see Matteini et al. (2002), Higgitt and White (2005), Noble and Van Loon (2007), Van Loon (2008), Kahrim et al. (2009), and Salvadó et al. (2009, 2013).

compounds that are typical of airborne particles, suggests instead that its formation is linked to dirt deposition and that the calcium is primarily of atmospheric origin. Supporting this suggestion is the fact that the panels spent a long period in the Spanish monastery in Nájera where – more than in a museum environment – they are likely to have accumulated high levels of calcium-containing dust and been exposed to microbial action (Kontozova-Deutsch et al. 2011). The thickest crust formation was found over the dark-colored copper-containing paint layers. Such areas are likely to have been left untouched during past selective cleaning when varnish(es) might only have been removed from the light colors to obtain a quick result, allowing dirt to build up most over these passages. These conditions would favor the formation of calcium oxalate salts which have effectively "cemented" the dirt to the surface of the painting, creating the thick, highly insoluble, and disfiguring surface crusts. The presence of multiple layers of very discolored and dirty oil-resin varnish and overpaint below the crust in the dark paint passages further obscured the composition. The dark passages may also have benefitted from local varnishing to (re)saturate the colors, with such additional organic layers perhaps further enhancing the oxalate formation, in a similar way as suggested by some authors for stone monuments (Lazzarini and Salvadori 1989; Rampazzi et al. 2004).

15.7 Removal

As the thick crusts concealed every hint of color and every pictorial detail over large areas, they were very disturbing, reducing the artistic quality and legibility of the paintings. To be able to fully appreciate again the original quality of the paintings, their removal was an essential part of the ongoing cleaning of the panels. The varnish layers in between the crust and the original paint indicated that the crusts were not part of the original paint stratigraphy. Together with the international advisory committee that oversaw the conservation process, it was therefore decided to undertake this time-consuming operation.

The presence of an intermediate varnish layer made it possible to separate the crusts from the original paint surfaces. In the green areas, where solvents might damage the vulnerable glazes, this could only be done mechanically, with a scalpel under the microscope. Working under magnification and progressing only millimeter by millimeter, the conservators were able to remove the disfiguring crust. Time-consuming as it was, the results were spectacular as can be seen in Figs. 15.1 and 15.7. Underneath the dull, plain brown surface appeared lush green draperies with three-dimensional folds, brocade patterns in different green hues, and decorative details such as fringes applied using lead-tin-yellow, as predicted by the imaging techniques. In the UV-induced visible fluorescence photograph (Fig. 15.7), it can be seen that the fluorescent varnish layer located between the crust and the original paint surface was revealed after removal of the crust, again confirming the observations made from paint samples.

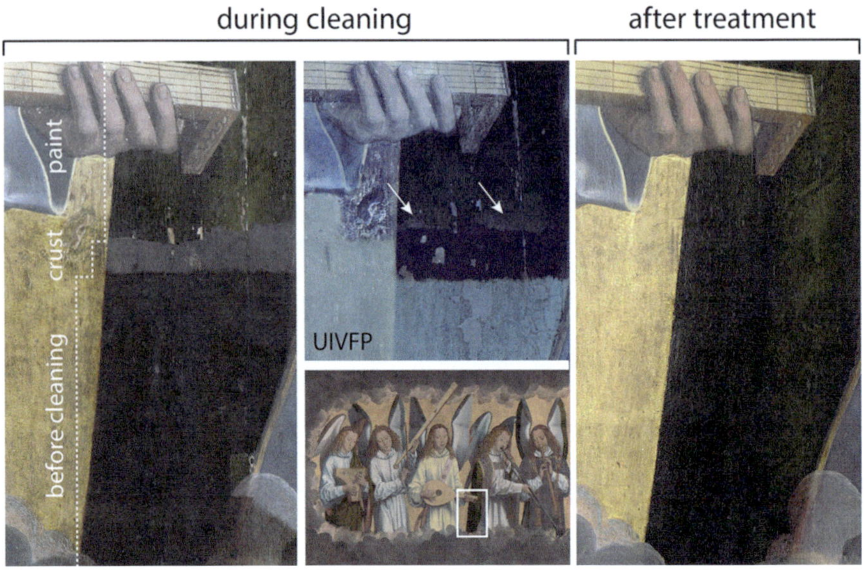

Fig. 15.7 Photo showing a detail of panel 779, as indicated by the white rectangle (middle, bottom). Left: cleaning staircase illustrating the gradual removal of the aged surface varnish(es) and the oxalate crust on the green paint. Middle, top: UV-induced visible fluorescence photograph (UIVFP) of the same area showing the typical fluorescence of the aged surface varnish(es) while the oxalate crusts appear dark in UV where exposed. Between the crust and the original paint surface, a fluorescent layer is visible (indicated by two white arrows), confirming the presence of an intermediate varnish layer. Right: the same area after treatment. Photo left: KMSKA © Arcobaleno, Adri verburg. Photo right: KMSKA © Rik Klein Gotink

Figure 15.8 illustrates how the gradual removal of the surface varnish layers and the underlying crust was documented in situ during the ongoing treatment by means of macroscopic FTIR scanning and spectroscopy in reflection mode.

The aim of these scans was to show the effectiveness of the (mechanical) cleaning procedure. As shown in the FTIR maps in Fig. 15.8, the distribution of the ν_S (CO) band of calcium oxalate clearly coincides with the crust, while the signal is absent where the green paint is revealed. The spectra in Fig. 15.8 were obtained in the latter cleaned area, proving what can be concluded from the oxalate distribution map: the intensity of the band at 1320 cm^{-1}, related to calcium oxalate, has dropped significantly after cleaning. Two paint losses appear as hot spots in the distribution maps for calcium carbonate, one of the materials employed in the preparation layers.

Although mechanical cleaning was possible on the smooth surface of the green colors, it was impossible on the more grainy blue colors and grey clouds. Removal of the crusts over the blue colors was possible to a certain degree by repeat applications of mixtures of dimethyl sulfoxide in ethyl acetate in different concentrations. On the grey clouds, a combination of solvent action and mechanical removal could be used. However, the vulnerability of the grey paint layer made it impossible to

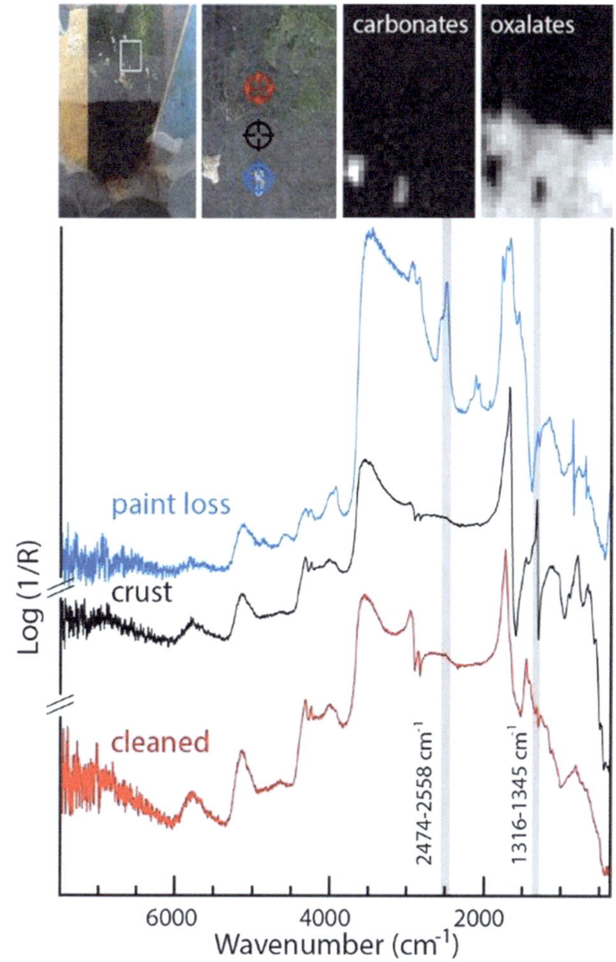

Fig. 15.8 Top, left: detail of the garment of angel 4 showing sequential cleaning tests. The white rectangle indicates an area where the crust was partially removed and imaged by means of macroscopic FTIR scanning in reflection mode. Top right: corresponding FTIR maps showing the distribution of carbonates in the paint losses and calcium oxalates in the crust. Bottom: FTIR spectra isolated from the hyperspectral data cube, with indication of the bands that were used to generate the chemical maps

remove the oxalate-containing crusts completely, and some material had to be left on the surface. Over the lighter colors, the oxalate-containing layers are very thin and hardly visible, for example, over the red feathers discussed above. In such areas, since the layers were not optically disturbing, they were left on the surface.

15.8 Conclusion

During the recent conservation of the three panels of Memling's *Christ with Singing and Music-making Angels,* an insoluble calcium oxalate-containing layer was found over the entire surface of the paintings. This crust was particularly thick, opaque, and disfiguring on the dark, copper-containing paint passages. The crust obscured the underlying composition but was extremely insoluble and posed a major conservation challenge.

A wide range of non-invasive and microinvasive analytical techniques were employed to determine the composition and distribution of the crusts and to establish the stratigraphy and nature of the underlying layers. Each contributed in a significant and complementary way to achieving as full an understanding as possible of the accretions on the paint surface, which was essential in informing the treatment of the paintings. The analytical results revealed that the crusts contained oxalate salts in the form of two related Ca-based oxalate salts, whewellite and weddellite, and a range of other components consistent with surface dirt deposition.

Oxalate salts have been found associated with a wide range of different artworks, and the mechanism of their formation is complex, possibly quite variable, and not fully understood. In the case of the Memling panels, the oxalates associated with the surface crusts did not appear to originate from degradation of the original paint but possibly from a combination of microbial action and a thick accumulation of dirt, particularly over the darker paint passages, where in addition a buildup of older varnishes had been left in place. The dirt deposits are likely to have accumulated during the period, while the panels were in the Nájera monastery.

The presence of intermediate varnish layer(s) between the crust and the original paint surface, as revealed by the analyses, indicated that the crusts were not part of the original paint stratigraphy and made it possible to remove most of them mechanically. The removal was a difficult and time-consuming process, and its progress was carefully monitored using macroscopic FTIR scanning and spectroscopy in the reflection mode. The treatment has, however, yielded spectacular results and has allowed the original colors and spatial depth of the composition to be recovered.

Acknowledgments A conservation project this large involves many people. Many thanks to Ineke Labarque, Régine Wittermann, Jean Albert Glatigny, Sara Matteu, Madeleine ter Kuile, Greta Toté, the advisory committee, and all colleagues from the conservation studio. We are indebted to the staff of the P06 beamline of the PETRA III facility and to DESY for providing the beam time for the synchrotron radiation-based measurements. KJ, GvdS, FV, and SL acknowledge financial support from the University of Antwerp Research Council via SOLARPAINT and other projects, the Fund Baillet Latour, and the Fund for Scientific Research (FWO), Brussels. CH and MS acknowledge the assistance of their colleague David Peggie. The conservation of the panels was supported by the Friends of the KMSKA and Interbuild NV, and the associated research was supported by the Baillet Latour fund.

References

Borchert T (1993) Memling's Antwerp "God the Father" with music-making angels. In: Verougstraete H, Van Schoute M (eds) Le dessin sous-jacent dans la peinture, Colloque X, September 1993, vol 1995. Le dessin sous-jacent dans le processus de creation, Louvain-la-Neuve, pp 153–168

Conti C, Brambilla L, Colombo C et al (2010) Stability and transformation mechanism of weddellite nanocrystals studied by X-ray diffraction and infrared spectroscopy. Phys Chem Chem Phys 12:14560–14566

De Nolf W, Janssens K (2010) Micro X-ray diffraction and fluorescence tomography for the study of multilayered automotive paints. Surf Interface Anal 42:411–418

De Nolf W, Vanmeert F, Janssens K (2014) XRDUA: crystalline phase distribution maps by two-dimensional scanning and tomographic (micro) X-ray power diffraction. J Appl Crystallogr 47:1107–1117

De Vos D (1994) Hans Memling. Het volledige oeuvre. Mercatorfonds, Antwerpen

Farmer V (1974) The layer silicates. In: Farmer V (ed) The infrared spectra of minerals. Mineralogical society monograph, vol 4. Mineralogical Society of Great Britain and Ireland, London, pp 331–363

Fransen B (2018) Hans Memling's Nájera altarpiece: new documentary evidence, The Burlington Magazine. Feb 2018. CLX 101–105

Garty J, Kunin P, Delarea J et al (2002) Calcium oxalate and sulphate-containing structures on the thallial surface of the lichen Ramalina lacera: response to polluted air and simulated acid rain. Plant Cell Environ 25:1591–1604

Gianoncelli A, Castaing J, Ortega L et al (2008) A portable instrument for *in situ* determination of the chemical and phase compositions of cultural heritage objects. X-Ray Spectrom 37:418–123

Gómara E (2013) Las tablas Najerinas del Muro de Amberes. Arte e Historia 21:16–19

Higgitt C, White R (2005) Analyses of paint media: new studies of Italian paintings of the fifteenth and sixteenth centuries. Natl Gallery Tech Bull 26:88–97

Janssens K, Legrand S, Van der Snickt G et al (2016) Virtual archaeology of altered paintings: multiscale chemical imaging tools. Elements 12:39–44

Kahrim K, Daveri A, Rocchi P et al (2009) The application of *in situ* mid FTIR fibre-optic reflectance spectroscopy and GC-MS analysis to monitor and evaluate painting cleaning. Spectrochim Acta A 74:1182–1188

Kontozova-Deutsch V, Cardell C, Urosevic M et al (2011) Characterization of indoor and outdoor atmospheric pollutants impacting architectural monuments: the case of San Jerónimo Monastery (Granada, Spain). Environ Earth Sci 63:1433–1445

Lazzarini L, Salvadori O (1989) A reassessement of the formation of the patina called 'scialbatura'. Stud Conserv 34:20–26

Legrand S, Alfeld M, Vanmeert F et al (2014) Macroscopic Fourier transform infrared scanning in reflection mode (MA-rFTIR), a new tool for chemical imaging of cultural heritage artefacts in the mid-infrared range. Analyst 139:2489–2498

Leroy C (2016) Oxalates de calcium et hydroxyapatite: des matériaux synthétiques et naturels étudiés par techniques RMN et DNP. PhD thesis in Chemistry, Université Pierre et Marie Curie, Paris. https://tel.archives-ouvertes.fr/tel-01443727. Accessed Feb 2017

Matteini M, Moles A, Lanterna G et al (2002) Characteristics of the materials and techniques. In: Ciatti M, Seidel M (eds) Giotto. The crucifix in Santa Maria Novella. Edifir, Florence, pp 387–403

Mendes N, Lofrumento C, Migliori A et al (2008) Micro-Raman and particle-induced X-ray emission spectroscopy for the study of pigments and degradation products present in 17th century coloured maps. J Raman Spectrosc 39:289–229

Miliani C, Rosi F, Sgamellotti A et al (2010) In situ noninvasive study of artworks: the MOLAB multitechnique approach. Acc Chem Res 43:728–738

Monico L, Rosi F, Miliani C et al (2013) Non-invasive identification of metal-oxalate complexes on polychrome artwork surfaces In reflection mid-infrared spectroscopy. Spectrochim Acta A 116:270–280

Noble P, Van Loon A (2007) Rembrandt's Simeon's song of praise, 1631: pictorial devices in the service of spatial illusion. Art Matters 4:19–35

Potgieter-Vermaak S, Godoia R, Van Grieken R et al (2005) Micro-structural characterization of black-crust and laser cleaning of building stones by micro-Raman and SEM techniques. Spectrochim Acta A 61:2460–2467

Rampazzi L, Andreotti A, Bonaduce I et al (2004) Analytical investigation of calcium oxalate films on marble monuments. Talanta 63:967–977

Salvadó N, Butí S, Nicholson J et al (2009) Identification of reaction compounds in micrometric layers from gothic paintings using combined SR-XRD and SR-FTIR. Talanta 79:419–428

Salvadó N, Butí S, Cotte M et al (2013) Shades of green in 15th century paintings: combined microanalysis of the materials using synchrotron radiation XRD, FTIR and XRF. Appl Phys A 111:47–57

Solé V, Papillon E, Cotte M et al (2007) A multiplatform code for the analysis of energy-dispersive X-ray fluorescence spectra. Spectrochim Acta B 62:63–68

Spring M, Higgitt C (2006) Analyses reconsidered: the importance of the pigment content of paint in the interpretation of the results of examination of binding media. In: Nadolny J (ed) Medieval paintings in northern Europe: techniques, analysis, art history. Studies in commemoration of the 70th birthday of Unn Plahter. Archetype, London, pp 223–229

Van der Snickt G, Miliani C, Janssens K et al (2011) Material analyses of 'Christ with singing and music-making Angels', a late 15th-C panel painting attributed to Hans Memling and assistants: Part I. non-invasive *in situ* investigations. J Anal At Spectrom 26:2216–2229

Van der Snickt G, Janssens K, Dik J et al (2012) Combined use of synchrotron radiation based micro-X-ray fluorescence, micro-X-ray diffraction, micro-X-ray absorption near-edge, and micro-Fourier transform infrared Spectroscopies for revealing an alternative degradation pathway of the pigment cadmium yellow in a painting by Van Gogh. Anal Chem 84:10221–10228

Van Grieken R, Gysels K, Hoornaert S et al (2000) Characterization of individual aerosol particles for atmospheric and cultural heritage studies. Water Air Soil Pollut 123:215–228

Van Loon A (2008) Color changes and chemical reactivity in seventeenth-century oil paintings. Dissertation, University of Amsterdam. MOLART Reports, vol. 14

Vandenbroeck P (1985) Catalogus schilderijen 14e en 15e eeuw. Ministerie van de Vlaamse Gemeenschap/Koninklijk Museum voor Schone Kunsten, Antwerpen, pp 138–143

Zoppi A, Lofrumento C, Mendes N et al (2010) Metal oxalates in paints: A Raman investigation on the relative reactivities of different pigments to oxalic acid solutions. Anal Bioanal Chem 397:841–849

Chapter 16
The Development of an Aqueous Gel Testing Procedure for the Removal of Lead-Rich Salt Crusts on the Surface of Paintings by Giovanni Antonio Pellegrini (1675–1741) in the "Golden Room" of the Mauritshuis

Annelies Van Loon, Laura Eva Hartman, Julia van den Burg, Ralph Haswell, and Carol Pottasch

Abstract During the treatment of the decorative ceiling and chimney paintings by Giovanni Antonio Pellegrini (1675–1741) located in the Golden Room of the Mauritshuis, a visually disturbing surface haze was encountered on the paintings following varnish removal. SEM-EDX, FTIR, and DTMS analysis demonstrated that this layer is primarily inorganic and complex in composition, with major elemental components including lead and sulfur followed by a lower proportion of potassium and calcium. The major elements are present in the form of sulfates of lead and lead-potassium, in addition to calcium oxalates. This salt-containing crust layer appears as a distinct layer located directly above the paint and does not seem to be intimately bound to the paint, but is composed of elements coming from the paint as well as environmental contaminants. As this layer significantly altered the artist's original intent, its removal was desirable and justifiable. Based on this analysis, an aqueous gel testing procedure was developed in order to determine a safe cleaning approach for the removal of the surface crust.

A. Van Loon (✉)
Conservation Department, Rijksmuseum, Amsterdam, The Netherlands
e-mail: a.vanloon@rijksmuseum.nl

L. E. Hartman · J. van den Burg · C. Pottasch (✉)
Royal Picture Gallery Mauritshuis, The Hague, The Netherlands
e-mail: c.pottasch@mauritshuis.nl

R. Haswell
Shell Global Solutions International, Amsterdam, The Netherlands

F. Casadio et al. (eds.), *Metal Soaps in Art*, Cultural Heritage Science,
https://doi.org/10.1007/978-3-319-90617-1_16

Keywords Efflorescence · Metal soaps · Giovanni Antonio Pellegrini · Aqueous cleaning gels · Chelating agents

16.1 Introduction

With the closing of the Royal Picture Gallery Mauritshuis for expansion and renovation (2012–2014), the opportunity was taken to study and treat the decorative painting ensemble, c. 1718, by Giovanni Antonio Pellegrini (1675–1741) located in the Golden Room of the Mauritshuis (Fig. 16.1). The ensemble consists of three large rococo ceiling paintings, *Apollo*, *Aurora*, and *The Dwindling Night*; two chiaroscuro chimney paintings, one recently identified as *Venus and Vulcan* and the other as a possible *Raison d'état* or *Republica*; four grisailles representing the four elements; and six flower *tondos*. The ceiling paintings underwent their last extensive treatment in situ in 1925; the chimney paintings, grisailles, and flower *tondos* in 1962–1965 under the guidance of restorer Luitsen Kuiper. For the 2012–2014 treatment, the paintings were taken out and transported to a temporary studio where the treatment was carried out. During examination of the paintings, a thin, but visually disturbing, grey surface haze was noticed beneath the varnish. This became more pronounced and disfiguring after the varnish was removed, appearing as if the paintings were covered with a grey veil. The grey surface haze was uniformly present on the painting ensemble, apart from the borders that had been protected by the frame and had retained their original color. The visual impact was strongest on the ceiling and chimney paintings (Fig. 16.1a, b).

This paper focuses on the identification and origin of the grey surface haze, as well as the consequences for the conservation treatment of the Pellegrini paintings. First, we needed to know what this grey layer was, before being able to make further decisions about whether or not to remove it. In order to determine its nature, samples were taken and examined with the light microscope. Scanning electron microscopy coupled with elemental analysis (SEM-EDX), Fourier transform infrared (FTIR) spectroscopy, and direct temperature-resolved mass spectrometry (DTMS) analyses were carried out in collaboration with the laboratories of Shell Global Solutions International. It could be established that the grey layer is a consequence of several degradation processes, of which metal soap formation is one. As the grey layer significantly altered the artist's original intent – Pellegrini is known for his use of bright pastel colors that were now completely disguised by the grey veil – its removal was desirable, in particularly on the three ceiling and two chimney paintings. The grey layer was completely insoluble in water or organic solvents. Therefore, an aqueous cleaning test protocol was developed, based on analysis results and review of the gel cleaning literature by Richard Wolbers and Chris Stavroudis (Wolbers and Stavroudis 2012). The aqueous cleaning gel was designed to precisely target the surface crust and influence its dissolution. This paper also addresses the methodology of testing, which involves the use of test panels as well as both empirical testing and chemical analysis to evaluate the effectiveness of the

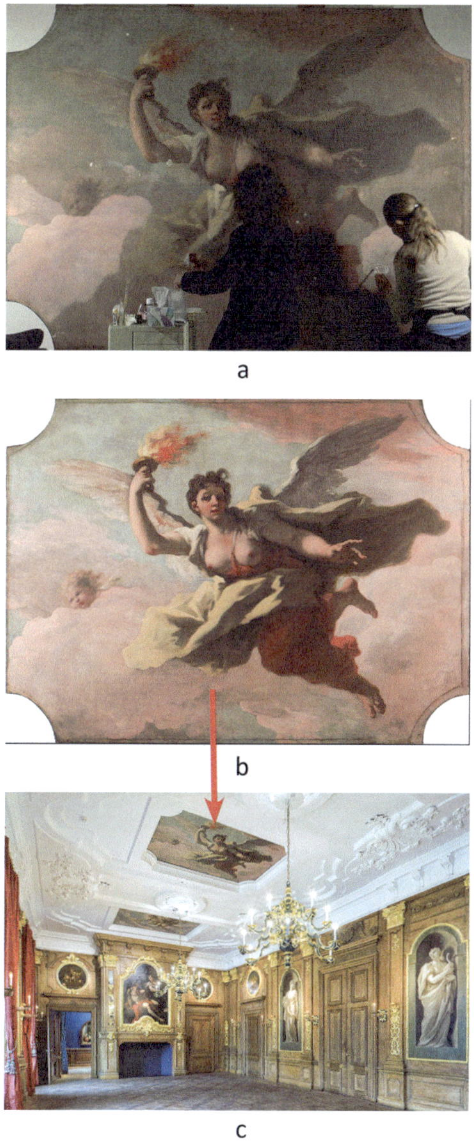

Fig. 16.1 Giovanni Antonio Pellegrini, *Aurora*, 1718, oil on canvas, 191.6 × 262 cm, ceiling painting inv. no. 1136, part of the decorative painting ensemble in the Golden Room, Mauritshuis, The Hague. (**a**) During treatment: after varnish removal and during removal of grey surface crust. The crust is still present on most of the surface, but the lower left side of the painting has been cleaned revealing the pink color of the clouds. (**b**) After treatment. The painting has regained its intended and spectacular bright pastel palette. (**c**) Current situation. The painting (see red arrow) has returned to the Golden Room after the recent renovation of the museum

cleaning gels and their effect on the original paint surface. It shows how the research enables us to come to a better understanding of the salt crust and the underlying original paint and ultimately make better-informed treatment decisions during the conservation treatment of the Pellegrini paintings.

16.2 Identification of the Grey Surface Layer

Paint cross sections reveal the presence of a thin, discrete grey layer (1–3 μm), containing tiny dispersed black particles, that has formed on the surface of the paints. Figure 16.2a shows a representative sample taken from the pink sky in *Aurora*, one of the ceiling paintings. It contains a single pink paint layer (layer 2) consisting of lead white, smalt (partly discolored), and vermilion, which is applied over a brown ground (layer 1) with remarkably coarse particles of quartz (SiO_2) and feldspar ($KAlSi_3O_8$), in addition to lead white, earth pigments, chalk, and a little charcoal black (Pottasch et al. 2015). The pink layer is covered with a thin grey layer. In the SEM backscattered electron image, this layer is noticeable as a thin, regular light grey (indicating a relatively high atomic mass) layer that closely follows the uneven surface texture of the pink paint below (Fig. 16.2b, c). It is located directly above the pink paint as a superficial layer, and not incorporated into the paint structure. Elemental analysis using EDX identifies lead and sulfur as the main elements present in the grey surface layer, with smaller, variable percentages of potassium and calcium (Fig. 16.2f).

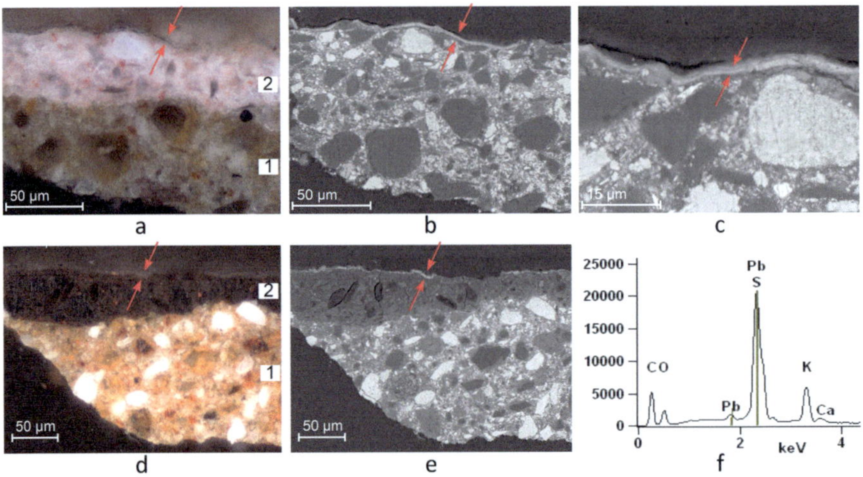

Fig. 16.2 Paint cross-sectional analyses. (**a**) Cross-sectional sample from the pink sky in *Aurora* (sample 1136 × 01), dark-field illumination: the red arrows indicate the thin grey layer that has formed on the surface of the pink top paint layer. (**b**) SEM backscattered electron image (SEM-BSE) of same cross section as (**a**), the red arrows indicate the grey layer that shows up as a highly scattering (light grey) layer in the BSE image due to the presence of lead, a high atomic weight element. (**c**) Detail of (**b**), recorded at 2500× magnification. (**d**) Cross-sectional sample from dark paint area in *Raison d'état* (sample 1143 × 08), close to the left edge, dark-field illumination: the red arrows indicate the thin crust that has formed on the surface of the dark paint layer. (**e**) SEM-BSE of (**d**). (**f**) EDX spectrum of grey surface layer (sample 1136 × 01) revealing the presence of lead, sulfur, potassium, and calcium

Isolated sample material of the grey surface layer was collected from several areas of each of the ceiling and chimney paintings for FTIR transmission spectroscopy and DTMS. FTIR spectra show overall similarities in absorbance peaks present, with slight variations in intensity, shape, and position between the different sampled areas. Figure 16.3a upper trace (i) represents a FTIR spectrum acquired from the grey layer from an area of pink sky (*Aurora*). Main peaks are to be observed at around 1051–1045 and at around 1392 wave numbers (cm^{-1}). The peaks at 1051–1045 cm^{-1} suggest the presence of sulfates (S-O stretch). In relation to elemental data acquired through SEM-EDX analysis of cross sections taken from a neighboring location and based on relative spectral matches from reference databases, sulfates of lead and lead-potassium are likely possibilities. Reference spectra of lead sulfate ($PbSO_4$, anglesite) and lead-potassium sulfate ($K_2Pb(SO_4)_2$, palmierite) (Fig. 16.3a lower trace (iv)) show intense absorption bands at around 1048 cm^{-1} (with minor features at 1160 and 968 cm^{-1}) and 1051 cm^{-1}, respectively. The observed broad band at around 1392 cm^{-1} suggests the presence of lead carbonate. This is confirmed by the presence of peaks at around 3550 (weak, sharp, OH stretch), 1050 (sharp, carbonate stretch), 837 (sharp, carbonate stretch), 776 (OH bend), and 681 (strong, sharp, carbonate stretch), which are associated with lead white, a basic lead carbonate. It is possible that the lead white comes from the pink paint as an impurity in the sample. The small band at around 1530 cm^{-1} may be attributed to lead carboxylate groups incorporated in the ionomeric structure of the oil paint medium (Van der Weerd et al. 2005; Hermans et al. 2016a). Minor peaks in the C-H stretching region at 2922 and 2851 cm^{-1} and in the carbonyl stretching region at 1740 (ester) and 1710 cm^{-1} (free carboxylic acid) are also present that could be indication of an organic material (likely coming from the oil paint medium).

The FTIR spectra acquired from the grey layer from an area of blue sky rich in smalt of the same painting (*Aurora*) show a broad peak located at around 1066–1064 cm^{-1} (Fig. 16.3a second trace (ii)). Although slightly shifted in position, this band most likely indicates sulfates of lead and lead-potassium. Silicates from the smalt from the paint beneath (present as an impurity in the sample) may also have contributed to this broad peak. They exhibit an intense absorption band in the same region as the sulfates. Similar absorption bands have been observed in samples of crust layers formed on smalt-rich paints (Van Loon 2008; Van Loon et al. 2011; Spring et al. 2005).

The dark paint areas of the chimney pieces exhibit a more brownish/blanched appearance. Paint cross sections reveal the presence of a thin, discrete lead-rich surface layer on top of the dark paint (layer 2) (Fig. 16.2d, e). The dark paints are found to be mixtures of earth pigments, carbon blacks with additions of vermilion, lead white, chalk, and smalt. The FTIR spectra collected from the surface layer from dark paint areas demonstrate the presence of bands at around 1637 cm^{-1} (broad, asymmetric carboxylate stretch), 1320 cm^{-1} (sharp, symmetric carboxylate stretch), and 780 cm^{-1} (OCO bend) that are characteristic of calcium oxalates (Cariata et al. 2000), in addition to sulfate bands in the 1100–1000 cm^{-1} region (Fig. 16.3a third trace (iii)).

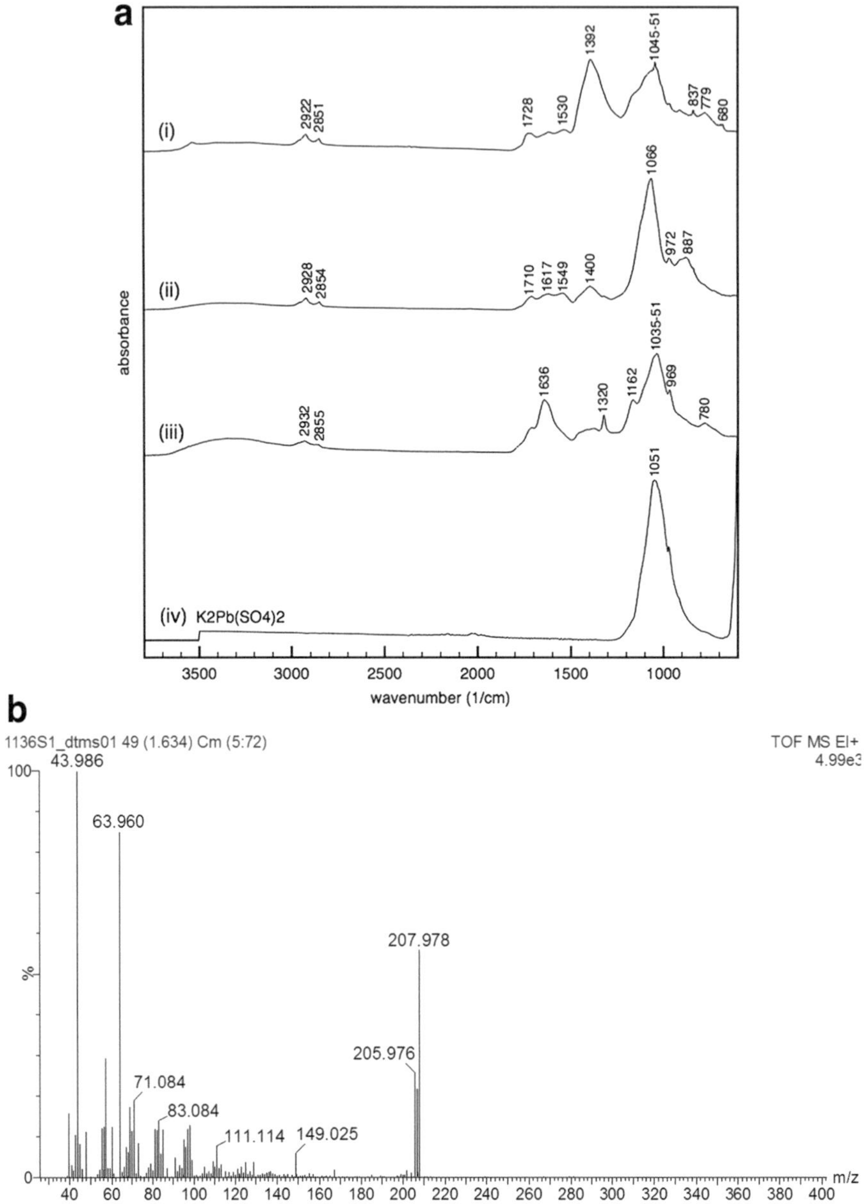

Fig. 16.3 FTIR and DTMS analyses. (**a**) ATR-FTIR spectra (3800–600 cm^{-1}) of (i) the thin surface crust from an area of pink sky in *Aurora* (sample 1136S05), (ii) the thin surface crust from an area of blue sky in *Aurora* (sample 1136S06), (iii) the thin surface crust from a dark paint area in *Raison d'état* (sample 1143S07), and (iv) reference spectrum of palmierite $K_2Pb(SO_4)_2$ (synthesized by Joen Hermans, University of Amsterdam, according to Tissot et al. 2001). (**b**) Summed DTMS mass spectra of isolated material of the thin surface crust from an area of pink sky in *Aurora* (sample 1136S01)

The DTMS spectra of the surface layer are dominated by mass peaks of lead isotopes at m/z 205.976–207.978, sulfur dioxide (SO_2) at m/z 63.960, and carbon dioxide (CO_2) at m/z 43.986 (Fig. 16.3b). The absence of organic components in the spectra indicates that the crust has a strongly inorganic nature.

From the above, we can conclude that the grey or brown layer that has formed on the surface of the paintings is a complex inorganic crust rich in lead sulfates (anglesite) and lead-potassium sulfates (palmierite) with minor contributions of calcium oxalates. It also contains tiny black particles of dirt or soot. The composition of the salt crusts was found to vary slightly, depending on the composition of the paint layers beneath the crust and possibly on the location of the painting in the room. Comparable lead-rich surface hazes and crusts have been identified in historical paintings, both easel painting and interior decoration (Noble and Van Loon 2007; Van Loon 2008; Van Loon et al. 2011).

16.3 Origin of the Grey Surface Layer

The complex composition of the surface crust suggests the involvement of various reaction processes in the development of the crust. Diffusion or migration of lead and potassium ions, originating from pigments in underlying paint layers, through the paint system appears to play an important role in its development. Their reaction with atmospheric compounds/pollutants, in particular sulfur dioxide (SO_2), then leads to the formation of a discrete salt crust on the paint surface. Comparable reaction processes and deterioration products may occur at the surface of historic glass, which has been more extensively researched than surface crusts on paintings (Woisetschläger et al. 2000; Spring et al. 2005). The primary source of potassium here is the smalt pigment, a blue cobalt-containing potash glass. Pellegrini used this pigment in large amounts in the ceiling and chimney paintings, in the blue and pink sky paints as well as in dark paint mixtures. Smalt is not a stable pigment in oil paint. Upon degradation, potassium is leached from the glass and reacts with fatty acids from the oil medium to form potassium soaps (Spring et al. 2005). Potassium soaps are water-soluble and diffuse easily through the paint film. Other sources of potassium in these paintings may be earth and lake pigments (Van Loon 2008). The percentage of potassium in the crust varied, depending on the paint composition. The source of lead is the lead white pigment used in the paint layers or the ground. Lead is solubilized from the pigment by interaction with the oil binder, either forming free lead soaps or binding to carboxylic acid groups of the oil network (Hermans et al. 2016a). Thus, the thin oil skin present on a lead white-containing paint will be enriched in lead that is directly exposed to atmospheric compounds. In the case of the dark paints of the chimney pieces, lead comes from the ground and has diffused into the upper paint layers and formed deposits at the surface (Fig. 16.2d, e). Recent crystallization studies have shown that lead soaps are insoluble in oil and easily crystallize (Hermans et al. 2016b). Therefore it is highly unlikely that lead diffuses in the form of lead soaps, as was initially thought, but rather through a mechanism of "ion hopping" via the carboxylic acid groups in the oil medium (Hermans et al. 2015).

The reaction into sulfates is likely the result of past environmental conditions when the paintings were still in an unvarnished state and when the galleries were heated by coal stoves. An old picture from the Mauritshuis archives shows that coal stoves heated the galleries of the Mauritshuis until the beginning of the twentieth century (Fig. 16.5). Sulfur dioxide (SO_2) is produced during the combustion of coals, which contain sulfur and combine with oxygen in the air. Sulfur dioxide quite easily reacts with further oxygen in the air to form sulfur trioxide gas (SO_3) (Thomson 1994). In the presence of water or moisture, it immediately forms then sulfuric acid (H_2SO_4), a strong acid that will react with the lead and potassium compounds into sulfates of lead and potassium. Potassium sulfates are still soluble in water, while the sulfates of lead and lead-potassium are highly insoluble in water or organic solvents. Lead seems to trap the potassium sulfates leading to the formation of insoluble crusts. The paintings were for the first time varnished, either in 1888 or 1925, as is known from the treatment history of the paintings. This means that the surface crust had developed over a period of more than 100 years. This corresponds with observations by Van Eikema Hommes and her team, who investigated the original and later color schemes of the frames and architecture in the Golden Room (Van Eikema Hommes et al. 2018). While a paint sample from *Aurora*, from an area covered by brass paint of the first frame from 1718, did not show a crust layer yet, in other samples from areas covered by paint from the 1810 and 1850 frames, a crust layer was indeed present (Van Eikema Hommes 2016 personal communication).

The surface crusts formed on the dark paints of the chimney pieces also contained relatively high amounts of calcium oxalates. These dark colors are richer in binding medium and therefore may trigger the formation of oxalates. Calcium oxalates can arise either from the environment, or from oxidative degradation of the oil binding medium, where oxalic acid is formed, and subsequent reaction of oxalic acid with calcium ions from calcium carbonate materials (Higgitt et al. 2005). Chalk is present as (minor) constituent in both the ground and dark paint.

16.4 Development of an Aqueous Gel Testing Procedure

16.4.1 An Aqueous Cleaning Gel

The salt crust layer was completely insoluble in water or organic solvents. This had been noted during varnish removal when the solvents did not affect the crust layer in any way. In order to remove the inorganic salt crust, a cleaning system was established, based on analysis results and review of the gel cleaning literature by Richard Wolbers and Chris Stavroudis (Wolbers and Stavroudis 2012). The desired gel would allow for control of pH, conductivity, and chelation to best target the salt crust while allowing adjustments to ensure safe parameters for the paint

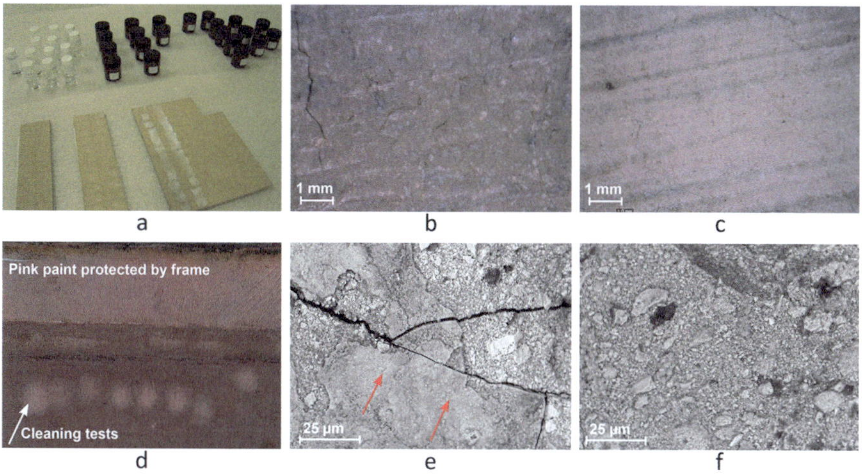

Fig. 16.4 Development of an aqueous gel testing procedure. (**a**) Complete set of aqueous cleaning gels used for testing, and the test panels prepared with the saturated solutions of the salts ($PbSO_4$, K_2SO_4, $CaC_2O_4 \cdot H_2O$) that were identified in the surface hazes. (**b**) Photomicrograph of an area of pink sky, in *Aurora*, before removal of the grey salt crust and (**c**) shown after cleaning. (**c**) After removal of the grey salt crust with a Pemulen gel at pH 7.5 containing 0.5% DTPA. The grey color visible in the thinner areas of paint is from the underlying grey/green ground showing through. (**d**) Detail image of *Aurora*, the upper right edge, showing the tests done with the Pemulen gels selected (see white arrow). Testing was done at the edge of the painting, adjacent to an area of paint which was protected by the frame and did not have this grey layer, for comparison. (**e**) SEM-BSE image of paint surface of an area of pink sky, before cleaning, recorded at 1000× magnification. The red arrows show the salt crust. In the adjacent area, the crust has chipped off the surface showing the underlying original paint, for comparison to (**f**). (**f**) SEM-BSE image of paint surface of an area of pink sky, after cleaning with Pemulen gel pH 7.5 containing 0.5% DTPA, for comparison to (**e**)

layers. It was decided to test three chelating agents that would target the main components in the salt crust and encourage their dissolution: triammonium citrate (TAC), ethylenediaminetetraacetic acid (EDTA), and diethylenetriaminepentaacetic acid (DTPA). EDTA and DTPA have high affinity for lead salts (Wolbers and Stavoudris 2012: 506).[1] Aqueous solutions of the chelating agents (0.004 M) were buffered at pH of 5.5, 7.0, 7.5, and 8.5 and gelled with two different gelling agents, Pemulen TR-2 and methyl cellulose (Fig. 16.4a).[2]

[1]The stability constant of TAC with Ca(II) is 4.68; with Pb(II) 6.50. The stability constant of EDTA with Ca(II) is 10.96; with Pb(II) 18.04. The stability constant of DTPA with Ca(II) is 10.9; with Pb(II) 18.9.

[2]Drops of concentrated acetic acid and 25% ammonia solution were used to obtain a buffered solution at pH 5.5; triethyl amine (TEA) and dilute HCl were used to obtain buffered solutions at pH 7.0, 7.5 and 8.5.

16.4.2 Test Panels

Test panels were prepared to gain information regarding the solubility of the salts (Fig. 16.4a). Three salts were chosen based on analysis and included calcium oxalate monohydrate (CaC_2O_4•H_2O), lead sulfate ($PbSO_4$), and potassium sulfate (K_2SO_4).[3] Saturated solutions of these salts were prepared and applied onto painted test panels (Fig. 16.4a). Since calcium oxalate monohydrate and lead sulfate have low solubility in water, the saturated solutions were heated to increase dissolvability and stirred slightly, so that the precipitate would disperse through the solution. These mixtures were applied to the boards with a pipette with approximately the same amount of solution applied to each space. The painted test panels are artist board panels with pre-primed canvas and an extra layer of flesh-colored alkyd paint. The board was divided into three sections for the three different salts. Each section was again divided into 24 spaces, for each different solution/gel that was to be tested. Based on solubility results (for a complete overview, see Van den Burg 2013), the most effective gels were then tested in discrete areas directly on the paintings (Fig. 16.4d).

16.4.3 Testing of the Effectiveness of the Cleaning Gels on the Paintings

Considering the composition of the salt crust layer varied slightly, different gels were tested depending on the area and painting in question. Test areas on the paintings were evaluated with the naked eye, with UV light, and with high magnification under the stereomicroscope and digital microscope (Dino-Lite) (Fig. 16.4b, c). In addition, samples were taken before and after testing and examined with light microscopy and SEM-EDX, in order to evaluate the effect of the gel cleaning on the painting at the micro level. Before embedding the samples, the surface of the paint samples was imaged and examined with high magnification and BSE imaging (Fig. 16.4e, f). The results of the testing indicated that the safest and most effective gel for removing the crust from the paintings was the Pemulen pH 7.5 with 0,004 M EDTA. By changing the chelating agent and pH, the effectiveness of the gels, depending on the major salt present, could be optimized. For areas where the crust was thicker, a Pemulen pH 7.5 0,004 M DTPA gel could be used locally. Importantly, the concentration, type of chelating agent, pH, and conductivity of the gels used were determined not only by the solubility of the salt but also the sensitivity of the paint beneath. Some of the brown and red areas, painted with earth pigments and vermilion, respectively, were particularly sensitive to the gel conditions.

[3]Solubility of CaC_2O_4 in water is 6.7 mg/l at 25 °C; $PbSO_4$ 4.25 mg/L at 25 °C; K_2SO_4 120 g/L at 20 °C.

16.5 Removal of the Grey Haze: Decision-Making Process

In the decision-making process whether or not to remove the grey haze from the ceiling and chimney paintings, both ethical and aesthetic aspects were considered. Previous research shows cases in which the crust was found to be ingrained and interlaced with the paint layer, preventing (investigation of) its removal (Van Loon 2008; Van Loon et al. 2011). On the Pellegrini paintings of the Golden Room, the crust was found to sit primarily on the surface of the paint (Fig. 16.2b, c, e). This meant that it was at least theoretically possible to remove it without damaging the paint layer below it. The testing procedure and the analysis of the samples taken from the tests conducted indicated that the method we had chosen to remove the crust was safe. Moreover, the paintings would be returned to a stable climatological environment (Fig. 16.1c). The other elements needed for the formation of the crusts, the sulfur dioxide, and other pollutants in the atmosphere had been significantly reduced if not eliminated with the removal of the coal fireplaces that had previously heated the rooms (Fig. 16.5). A new air filtration system was also installed in the Mauritshuis, meaning that atmospheric pollutants from outside the building would also be reduced to a minimum. In combination with the planned varnishing of all the paintings, which would act as a filter between the paint layer and the atmosphere, it was thought that re-formation of the crusts would be minimal, if possible at all. This was a large factor in the decision to remove the crusts.

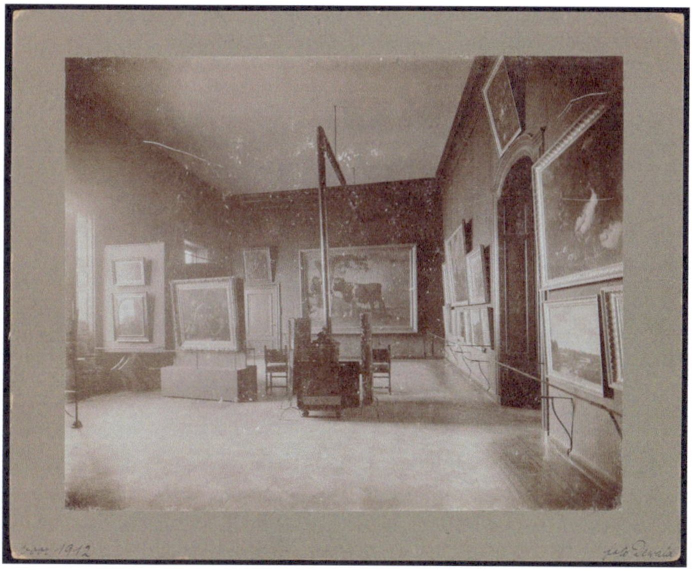

Fig. 16.5 The old museum climate situation: old photograph of the "Potter" Room showing the coal stove in the middle of the room (Mauritshuis archives, with thanks to Q. Buvelot)

Aesthetically, the removal of the crusts was considered desirable. Various tests, conducted along the edges and in various areas of the paintings, indicated that the crusts caused significant disfigurement to the images and decreased the visual pleasure when viewing the Pellegrini paintings. The rococo palette of the ceiling paintings was dulled, and the nuances on both the ceiling and chimney paintings were hidden. By removing the crust, the color palette, depth, and tonality would be partially restored and the paintings easier to read and returned closer to their intended appearance. These considerations – the safety of the method used to remove the crusts, the stable environment into which the paintings were returning preventing/minimizing the re-formation of the crusts, and the aesthetic benefits of removal – eventually weighed stronger than the risks in the decision to remove the crusts.

The treatment of the ceiling and chimney paintings, including removal of the surface crusts, took place over a period of 1 year. The work was finished in 2013, after which the paintings were returned to the Golden Room at the beginning of 2014.

16.6 Conclusions

This research enabled us to come to a better understanding of the nature of the salt crust and the underlying original paint and make scientifically supported treatment decisions. Through the removal of this grey salt crust layer, the paintings returned closer to their intended and spectacular bright pastel palettes. From the test results, we were able to tailor the gels to the needs of our paintings. A range of gels were prepared and modified during cleaning, by adjusting the pH and conductivity levels, type of chelating agent, and its concentration and application to attain the most efficient removal possible while reducing the risk of swelling of the paint surface.

16.7 Experimental Details

Microscopic paint samples were embedded in Technovit 2000 LC mounting resin (Heraeus Kulzer GmbH, Germany) and polished using a sample holder and Micro-Mesh sheets up to grade 12,000 (Micro-Surface Finishing Products Inc., Wilton, Iowa, USA) (Van Loon et al. 2005). A Leica DM2500 light microscope equipped with a Leica DFC490 digital camera was used to analyze the polished cross sections under visible light and by ultraviolet fluorescence (Leica filter block "A": BP 360/40 RKP 400-LP425).

Scanning electron microscopy coupled with energy dispersive spectroscopy (SEM-EDX) was performed on the prepared cross sections and unembedded samples using the JEOL 7000F field emission gun (FEG) high-pressure electron microscope, with a Thermo Scientific Silicon drift X-ray detector and Noran

System Seven (NSS) software. We used a 20 kV primary beam energy and a beam current between 1 and 5 nA. The samples were carbon coated to improve surface conductivity, prior to analysis. The spectrum fitting tool of the NSS software was used to confirm elements which have overlapping peaks, such as Pb and S.

Fourier transform infrared spectroscopy (FTIR) equipped with an attenuated total reflection (ATR) crystal (Ge) was performed using a VARIAN 3100 FTIR spectrometer coupled to an UMA600 microscope. Spectral resolution is 2 cm^{-1} and the number of scans is 128. Spectra have not been corrected for the varying path length inherent to ATR measurements.

Direct temperature-resolved mass spectrometry (DTMS) was undertaken on a Waters Micromass GCT Premier time-of-flight instrument with a mass resolution $M/\Delta M$ of 6000 at 200 Da (Van Loon et al. 2016). Samples were applied to the Pt/Ir filament of a direct insertion probe. Electron ionization was used for ion generation with a low electron energy (16 eV) to reduce ion fragmentation.

The salts used on the test panels, $PbSO_4$, K_2SO_4, and $CaC_2O_4 \cdot H_2O$, were purchased from Sigma-Aldrich. Palmierite $K_2Pb(SO_4)_2$ was synthesized by Joen Hermans, University of Amsterdam, according to Tissot et al. 2001, and confirmed with X-ray powder diffraction.

Acknowledgments FTIR and DTMS analyses were carried out by Fred Singelenberg and Wim Genuit of Shell Netherlands. We thank Joen Hermans (*PAinT* project) for his help with preparing the gels and salt solutions. We are grateful to Richard Wolbers for his advice regarding gel cleaning. We also thank colleagues of the Mauritshuis, especially Petria Noble for her critical comments on the poster, as presented at the Metal Soaps in Art Conference, Rijksmuseum Amsterdam, March 2016. This research took place as part of the "Partners in Science" collaboration with Shell Netherlands, which is chaired by Bob van Wingerden, and the *PAinT* Project supported by the *Science4Arts* program of the Netherlands Organization for Scientific Research (NWO).

References

Cariata F, Rampazzi L, Toniolo L, Pozzi A (2000) Calcium oxalate films on stone surfaces: experimental assessment of the chemical formation. Stud Conserv 45(3):180–188

Hermans JJ, Keune K, Van Loon A, Iedema PD (2015) An infrared spectroscopic study of the nature of zinc carboxylates in oil paintings. J Anal Atom Spectrom 30:1600–1608. https://doi.org/10.1039/c5ja00120j

Hermans JJ, Keune K, Van Loon A, Corkery RW, Iedema P (2016a) Ionomer-like structure in mature oil paint binding media. RSC Adv 6:93363–93369. https://doi.org/10.1039/c6ra18267d

Hermans JJ, Keune K, Van Loon A, Iedema PD (2016b) The crystallization of metal soaps and fatty acids in oil paint model systems. Phys Chem Chem Phys 18:10896–10905. https://doi.org/10.1039/c6cp00487c

Higgitt C, Spring M, Saunders D (2005) Analyses of paint media: new studies of Italian paintings of the fifteenth and sixteenth centuries. Natl Gallery Tech Bull 26:75–95

Noble P, Van Loon A (2007) Rembrandt's Simeon's Song of Praise, 1631: pictorial devices in the service of spatial illusion. In: Hermens E (ed) Art matters Netherlands technical studies in art, vol 4. Waanders, Zwolle, pp 21–36

Pottasch C, Smelt S, Haswell R (2015) Breaking new ground: investigating Pellegrini's use of ground in the golden room of the Mauritshuis. In: Evens H, Muir K (eds) Studying 18th-century paintings and works of art on paper, CATS proceedings II, CATS. Archetype, Copenhagen/London, pp 16–30

Spring M, Higgitt C, Saunders D (2005) Investigation of pigment-medium processes in oil paint containing degraded smalt. Natl Gallery Tech Bull 26:56–70

Thomson G (1994) The museum environment, second edition. Elsevier Butterworth-Heinemann, Oxford

Tissot RG, Rodriguez MA, Sipola DL, Voigt JA (2001) X-ray powder diffraction study of synthetic palmierite, $K_2Pb(SO_4)_2$. Powder Diffract 16(2):92–97

Van den Burg JM (2013) Treatment of the Raison d'Etat of the Golden Room ensemble, the Royal Picture Gallery Mauritshuis. Post-Graduate Thesis Conservation & Restoration Programme, University of Amsterdam (unpublished)

Van der Weerd J, Van Loon A, Boon JJ (2005) FTIR studies of the effect of pigments on the aging of the oil medium. Stud Conserv 50:1–20

Van Eikema Hommes M, Keune K, Jongsma R, Pottasch C (2018) Determining the early eighteenth-century colour scheme of the Golden Room in the Mauritshuis, The Hague: interpretation issues caused by changes to paint chemistry. In: Macro to micro: examining architectural finishes, postprints of the 6th architectural paint research conference, New York, Mar 2017. Archetype, London

Van Loon A (2008) White hazes and surface crusts on dark oil paint films. In: Color changes and chemical reactivity in seventeenth-century oil paints. PhD dissertation, University of Amsterdam, Molart Series (14), AMOLF, Amsterdam. Downloadable from http://www.amolf.nl/publications

Van Loon A, Keune K, Boon JJ (2005) Improving the surface quality of paint cross-sections for imaging analytical studies with specular reflection FTIR and static-SIMS. In: proceedings of Art'05 conference on non-destructive testing and microanalysis for the diagnostics and conservation of the cultural and environmental heritage, Lecce, 15–19 May 2005 (CD-ROM)

Van Loon A, Noble P, Boon JJ (2011) White hazes and surface crusts in Rembrandt's Homer and related paintings. In: Bridgland J (ed) Preprints ICOM committee for conservation 16th triennial meeting, Lisbon, 19–23 Sept 2011. Critério – Produção Gráfica Lda, Almada (CD-Rom)

Van Loon A, Genuit W, Pottasch C, Smelt S, Noble P (2016) Analysis of old master paintings by direct temperature-resolved time-of-flight mass spectrometry: some recent developments. Microchem J 26:406–414. https://doi.org/10.1016/j.microc.2015.12.025

Woisetschläger G, Dutz M, Paul S, Schreiner M (2000) Weathering phenomena on naturally weathered potash-lime-silica-glass with medieval composition studied by secondary electron microscopy and energy dispersive micro-analysis. Microchim Acta 135:121–130

Wolbers R, Stavroudis C (2012) Aqueous methods for the cleaning of paintings. In: Hill Stoner J, Rushfield R (eds) Conservation of easel paintings. Routledge, London, pp 500–523

Chapter 17
The Formation of Calcium Fatty Acid Salts in Oil Paint: Two Case Studies

Kate Helwig, Élisabeth Forest, Aimie Turcotte, Wendy Baker, Nancy E. Binnie, Elizabeth Moffatt, and Jennifer Poulin

Abstract Two case studies of calcium fatty acid salts in oil paint are presented. The first is a nineteenth-century oil on canvas painting, and the second is a decoratively painted ceiling. In both cases, the calcium soaps have formed in the lower layers of the paint systems and are associated with significant delamination and paint loss. The results of analysis of samples from the two works using FTIR, SEM-EDX, Raman, PLM, and Py-GC-MS are presented. The components of the paint and ground involved in the soap formation, the conditions that may have led to their formation, and the treatment strategies that were chosen are described.

Keywords Calcium soaps · Calcium fatty acid salts · Delamination · FTIR

17.1 Introduction

High concentrations of metal fatty acid salts (metal soaps) in oil paint layers can contribute to various types of deterioration, including efflorescence or surface crusts, protrusions, cracking, and delamination (Noble and Boon 2007). The vast majority of published studies on metal soap-related degradation involve either lead or zinc soaps. Only a few examples of other cations reacting with fatty acids in oil paint to produce soap degradation products have been documented.

Two works showing the presence of calcium soaps in oil paint are presented in this article: the first is a mid-nineteenth-century oil on canvas painting by Théophile Hamel entitled *Le Martyre de saint Pierre de Vérone*, and the second is a decoratively painted ceiling in the National Research Council (NRC) building in Ottawa. The formation of calcium fatty acid salts in paint is unusual; to our

K. Helwig (✉) · W. Baker · N. E. Binnie · E. Moffatt · J. Poulin
Canadian Conservation Institute, Ottawa, ON, Canada
e-mail: kate.helwig@canada.ca

É. Forest · A. Turcotte
Centre de conservation de Québec, Québec, QC, Canada

F. Casadio et al. (eds.), *Metal Soaps in Art*, Cultural Heritage Science,
https://doi.org/10.1007/978-3-319-90617-1_17

knowledge, there is only one previously published identification of calcium soaps in an oil painting (Ferreira et al. 2011). In the two case studies described here, the calcium soaps have formed in the lower layers of the paint systems and are associated with significant delamination and paint loss. The results of the analysis of the paint and ground in the two works are presented, and the components involved in the calcium soap formation are described. The conditions that may have led to their formation and the treatment strategies chosen are also summarized.

17.2 Methods of Analysis

A similar analysis protocol was utilized for both case studies. Cross sections were examined to determine the overall stratigraphy and to compare intact and delaminating areas. The cross sections were prepared using standard grinding and polishing techniques and examined by light and fluorescence microscopy as well as by scanning electron microscopy using a backscattered electron detector. Unmounted fragments of selected layers were analyzed in order to identify the pigments and binding medium. Pigments were identified using a combination of the following techniques: scanning electron microscopy-energy dispersive X-ray spectroscopy (SEM-EDX), Fourier transform infrared (FTIR) spectroscopy, and, in some cases, Raman spectroscopy or polarized light microscopy (PLM). The binding medium was identified using FTIR spectroscopy and in a few cases by pyrolysis-gas chromatography-mass spectrometry (Py-GC-MS).

For FTIR, samples were mounted on a low pressure diamond anvil microsample cell and analyzed using a Bruker Hyperion 2000 microscope interfaced to a Tensor 27 spectrometer. Spectra were collected in transmission mode from 4000 to 430 cm^{-1} using a wide band MCT detector. Py-GC-MS was performed after derivatization with tetramethylammonium hydroxide (TMAH, 2.5% in methanol) using an Agilent thermal separation probe assembly with an Agilent 7890 gas chromatograph interfaced to an Agilent 5975 mass spectrometer (for experimental details, see Moffatt et al. 2015). SEM-EDX was undertaken using a Hitachi S-3500 N scanning electron microscope integrated with a lithium-drifted silicon, light element X-ray detector, and an Oxford Inca X-ray microanalysis system. The SEM was operated at an accelerating voltage of 20 kV and a working distance of 15 mm in high vacuum mode using a backscattered electron detector. Prior to analysis, cross sections were carbon coated to ensure conductivity. Raman spectra were collected with a Bruker Senterra dispersive Raman microscope using a 785 nm wavelength laser. A 50× objective lens was used to produce an analysis area of approximately 2 μm in diameter. Selected samples were dispersed in Cargille Meltmount ($n = 1.66$) for examination by PLM using a Leica DMRX microscope.

17.3 Case Study: *Le Martyre de saint Pierre de Vérone*

17.3.1 Background

The mid-nineteenth-century oil painting on canvas by Québec artist Théophile Hamel (1817–1870), *Le Martyre de saint Pierre de Vérone*, is a copy of an original work by Titian, which was located in the Chapel of the Rosary at the Dominican Church SS. Giovanni e Paolo in Venice. During his 1-year stay in Venice, Hamel produced a small format copy of Titian's altarpiece, which, many years after his return to Québec in 1846, he used as a model for the large format painting examined here (Béland 2014). The painting, shown in Figs. 17.1 and 17.2, depicts a dramatic narrative of Peter of Verona's assassination.

Le Martyre de saint Pierre de Vérone remained in Hamel's collection until his death in 1870. In 1862, it was rescued from a fire in his workshop and exhibition room along with 21 other of his paintings, as well as some engravings and drawings in his personal collection (Vézina 1975). It is likely that smoke and heat from this fire have had an effect on the condition of the painting. Following his death, his estate put the painting into storage at the Grand Séminaire de Québec until 1919. The painting was then acquired by the Dominican Order in Québec City and was hung in the old chapel-monastery, which was converted into a convent in 1934. A fire destroyed the attic of this building in 1939, but there is no evidence that the painting was damaged. Around the mid-1940s, the painting was marouflaged onto an interior wall of the Chapter room.

In 2004, the Hamel painting was examined and subsequently demarouflaged by a private conservation firm (A. E. Henry, Ltd., Montréal); lifting paint and paint losses, principally in the upper half of the work, were already apparent at this time. Following the demarouflage, which included consolidation and facing with 5% w/v rabbit skin glue, the painting was stored flat, facing up, on a sheet of plywood for 10 years. The painting was donated to the Musée national des beaux-arts du Québec in 2013, and conservation treatment was initiated at the Centre de conservation du Québec (CCQ) in 2014.

17.3.2 Conservation Issues

When the painting arrived at the CCQ in 2014, its surface was obscured by the facing tissues and the verso was covered by the marouflage adhesive, as well as with plaster and cement in certain areas. Removal of the facing tissues and adhesive on the verso at the outset of the conservation treatment allowed the painting to be examined closely for the first time (Fig. 17.1).

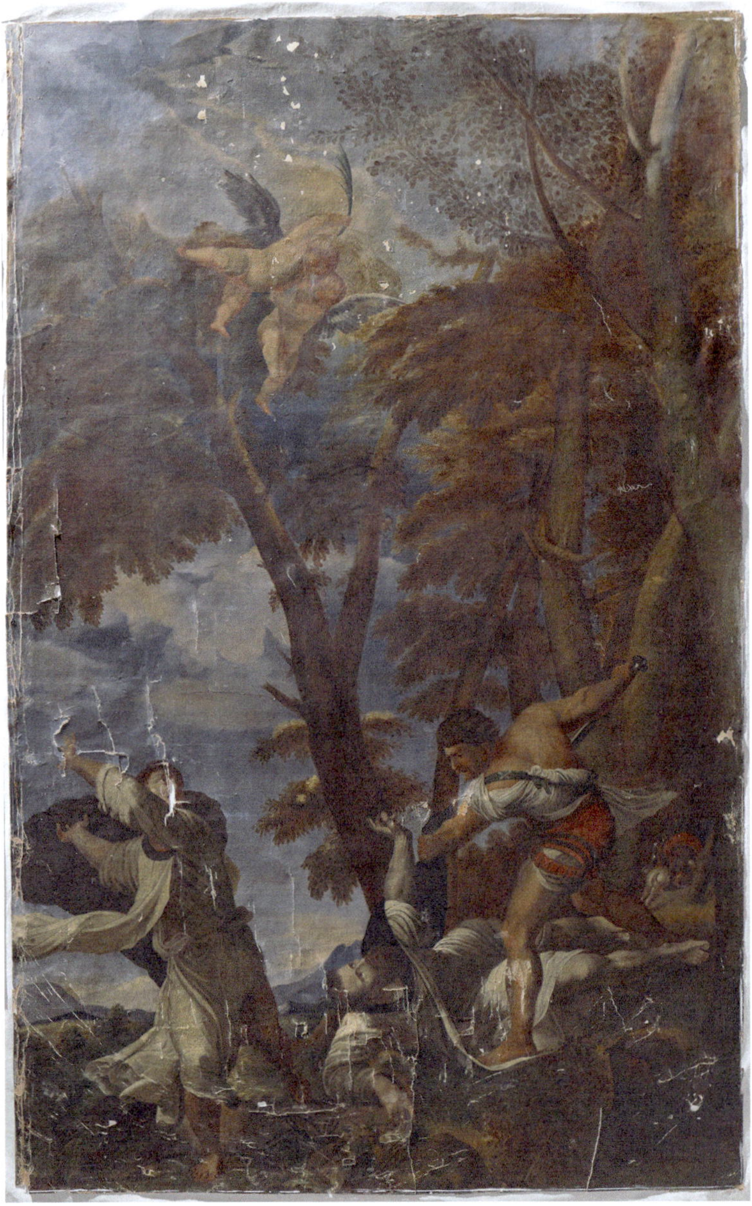

Fig. 17.1 *Le Martyre de saint Pierre de Vérone,* circa 1855 by Théophile Hamel, after a work by Titian; oil on canvas, 197 × 125 cm, collection of the Musée national des beaux-arts du Québec, gift of the Corporation des Dominicains de la cité de Québec (2013.21). Photographed during conservation treatment at the Centre de conservation du Québec, © M. Élie/CCQ (2014)

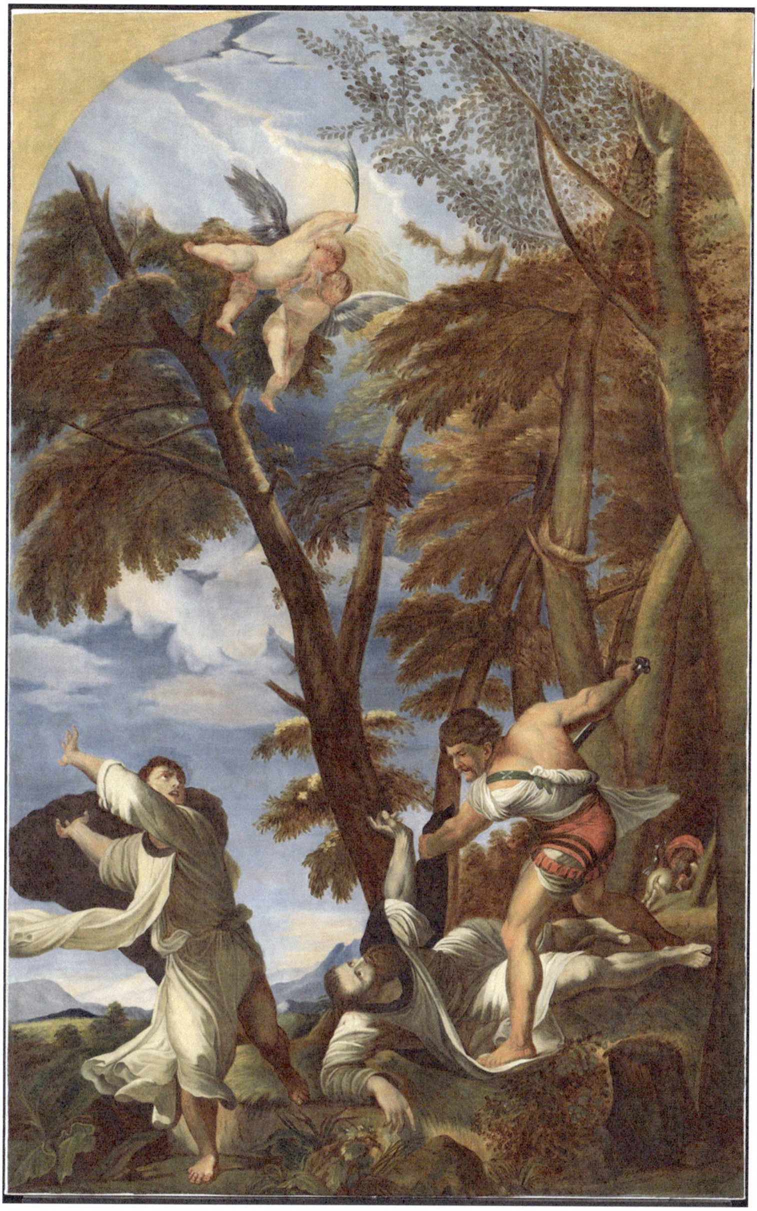

Fig. 17.2 *Le Martyre de saint Pierre de Vérone*, photographed after conservation treatment at the Centre de conservation du Québec, © M. Élie/CCQ (2016)

The paint surface was covered with many layers of grime and varnish. A thin black layer, located between two layers of varnish, had the appearance of soot and could have been deposited during the fire at Hamel's workshop in 1862. However, no blistering or burn marks were noted. The upper spandrels were covered with oil overpaint that continued the landscape over the original monochrome paint.

It was also noted that there was cleavage and delamination within the ground layers and significant paint losses, which did not relate to the demarouflage or to the surface tension created by the rabbit skin glue of the facings. In some paint losses, translucent brown material was visible within the thin layer of ground that remained on the linen canvas (Fig. 17.3a).

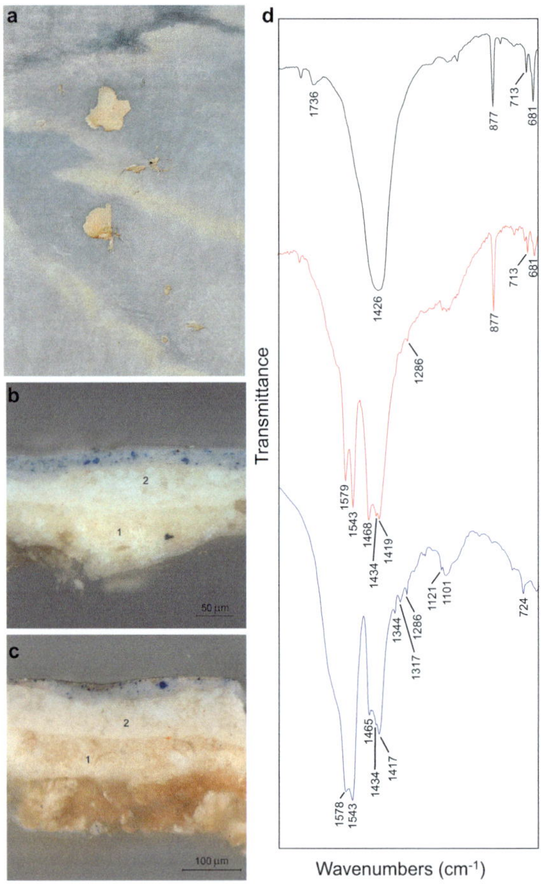

Fig. 17.3 (**a**) Detail of the sky during treatment showing losses and flaking paint, © A. Turcotte/CCQ (2015); (**b**) cross section from an intact area showing the two applications of white ground (labeled 1 and 2); (**c**) cross section next to a loss showing translucent, brown material intermixed with the first ground layer; (**d**) FTIR spectra of the first ground layer (upper, black spectrum); interface between the ground and transparent brown material (middle, red spectrum); and brown material (lower, blue spectrum). (Images (**b–d**) © CCI)

17.3.3 Analysis Results

Cross sections from several areas of the painting showed that it has two white ground layers. The first (lower) ground layer is a mixture of chalk, lead white, and a few scattered particles of barium sulfate in a drying oil medium. The second (upper) ground layer has a similar composition, but with a higher proportion of lead white. Analysis by Py-GC-MS showed that the oil in the ground layers is heat pre-polymerized linseed oil and contains a small amount of Pinaceae resin. The paint layers have a similar binding medium. Overall, the following pigments and associated constituents were identified in the paint: lead white, calcium carbonate, kaolin, quartz, ultramarine blue, Prussian blue, vermilion, iron oxides, bone black, carbon black, and a yellow lake.

Figure 17.3b is a cross section from an intact area of the sky showing the two applications of white ground followed by the ultramarine blue paint of the sky. Figure 17.3c is a cross section from an area next to a loss in the sky, where translucent, brown material was visible within the thin layer of ground remaining on the canvas. This cross section shows the same two white ground layers as seen in the other cross sections, but the lower ground is intermixed with a translucent brown substance that presumably corresponds to the translucent brown material observed in the adjacent paint loss.

Selected portions of an unmounted fragment of the sample illustrated in Fig. 17.3c were analyzed by FTIR spectroscopy. The upper spectrum (black) in Fig. 17.3d is from the first application of white ground and confirms the presence of calcium carbonate, lead white, and drying oil. No metal fatty acid salts were identified, and the weak ester band at 1736 cm^{-1} suggests a relatively low concentration of binding medium.

The middle spectrum (red) is from the interface between the first white ground and the translucent brown material. The spectrum indicates the presence of calcium monocarboxylic fatty acid salts, such as calcium palmitate or stearate (Bio-Rad Sadtler 2017; Ferreira et al. 2011), as well as calcium carbonate and lead white. The strong doublet at 1579 and 1543 cm^{-1} corresponds to the asymmetric carboxylate stretch of the calcium salts, while the doublet at 1434 and 1419 cm^{-1} (partially obscured by the broad carbonate band) corresponds to the symmetric carboxylate stretch. The band at 1468 cm^{-1} is a CH_2 deformation mode.

The lower spectrum (blue), from the translucent brown material, shows similar bands at 1578, 1543, 1465, 1434, and 1417 cm^{-1} from calcium fatty acid salts, although they are less well resolved than those in the middle spectrum. Some of the weak CH_2 progression bands, which are sensitive to the fatty acid chain length, are also present (1344, 1317, 1286 cm^{-1}). Sample size precluded the determination of the fatty acid profile in the soaps using Py-GC-MS. The SEM-EDX analysis showed the presence of some lead as well as calcium in the translucent brown material, suggesting that in addition to the calcium fatty acid salts identified by FTIR, a smaller amount of the corresponding lead salts may also be present.

17.3.4 Treatment

The thick, white adhesive layer on the back of the painting prevented flattening of the deformations and realigning the many tears inevitably caused by the demarouflage. Based on visual observation, solubility, and testing for lead with Plumbtesmo® Test Paper at the CCQ during the conservation treatment, the adhesive was determined to be lead white in a drying oil medium. After implementing a protocol for the safe treatment of objects containing lead, the material on the verso was removed with a small belt sander until the interface of the canvas and lead white adhesive layer was reached. The painting was then impregnated with BEVA 371® in Stoddard solvent (1:3 v/v), applied warm from the back. After complete evaporation of the solvent, the BEVA 371® was reactivated with heat to ensure complete adhesion between the paint and support. The facing tissues and rabbit skin glue were then removed using warm deionized water and sponges. The consolidation from the back stabilized much of the delaminating paint. However, after varnish, dirt and overpaint removal, remaining areas of lifting paint, located primarily in the upper part of the sky, also required local consolidation from the front with BEVA 371® in solution.

Following lining with BEVA film 371® on sailcloth (Dacron) fabric and mounting onto a stretcher, new areas of lifting paint were observed. These were most likely areas of blind cleavage caused by the presence of calcium soaps within the lower ground layer. Various adhesives were tested, including 5% w/v rabbit skin glue, a wax-resin mixture (beeswax, Cosmolloid 80H® microcrystalline wax and Ketone Resin N®), a Jade 403® emulsion, and 15% w/v Aquazol 200® in a 1:1 v/v mixture of ethanol and water. Aquazol 200® was found to give the best results and was used to consolidate these new areas of lifting paint.

17.3.5 Discussion

The results of the analysis indicate that the formation of calcium soaps, and possibly lead soaps, within the lower ground layer of the painting has caused a lack of cohesion, leading to delamination and paint loss. The fact that the soap formation occurred in the lower part of the ground, next to the canvas, suggests that the source of the free fatty acids for the soap formation may be the drying oil-based adhesive applied to the back of the canvas during the marouflage procedure.

No unusual components that would have promoted the formation of calcium soaps were identified in the ground and paint layers. Ground layers composed of a mixture of lead white and calcium carbonate in drying oil were very common in the nineteenth century, and no other instances of calcium soap formation in this type of ground have been published.

Although the composition of the ground is the same over the entire painting, the calcium soaps are confined primarily to the upper part of the work. This suggests that external factors that preferentially affected the top of the painting may be responsible for the conservation issues. Effects from the 1862 fire and the high humidity or water damage that may have occurred during the 1939 fire may have played a role. One hypothesis is that heat from the fire could have caused thermal conversion of some portion of the calcium carbonate to calcium oxide (Halikia et al. 2001), which could further react to form calcium hydroxide in the presence of moisture. Both calcium oxide and calcium hydroxide would be much more prone to soap formation than calcium carbonate.

While we can only speculate as to the specific conditions that led to the formation of the calcium soaps in the painting, it is clear that *Le Martyre de saint Pierre de Vérone* is an artwork that remains fragile. Periodic examination will be necessary to assess the stability of the ground and paint layers. The unknown element is whether the calcium soaps will continue to evolve and expand or will remain dimensionally stable in the painting's current environment.

17.4 Case Study: National Research Council Auditorium

17.4.1 Background

Designed by Sproatt & Rolph Architects and constructed between 1930 and 1932, the National Research Council (NRC) building in Ottawa, Canada was meant to be "a symbol of the Canadian government's recognition of the value of scientific research as a basis for economic planning in the 20th century" (FHBRO 2016). While the laboratory planning focused on "an attention to safety and utilitarianism," the overall building was inspired by the grand Beaux-arts style with rich interior finishes including a number of coffered and decoratively painted ceilings (FHBRO 2016). The auditorium was the largest in the city and was under constant use by different groups until the early 1970s (Mortimer 2002). The decoratively painted ceiling was laid out in a three-dimensional geometric pattern of coffered squares surrounded by hexagonal coffers. There is a combination of monochromatically painted and then stenciled features, fine hand-painted elements with or without the addition of stenciling, oil-rich glazing layers, and gilded and scumbled surfaces (Fig. 17.4a).

Specifications for plastering, preparation of the plaster, and details pertinent to the quality of the paint were provided by the architectural firm Sproatt & Rolph. These directives indicate that the final plaster application should be composed of lime putty mixed with plaster of Paris. They also specify that the plaster on the walls and ceilings should be covered with linen cloth prior to painting (Sproatt and Rolph 1929).

By 2013, the auditorium ceiling was in very poor condition. Canadian Conservation Institute (CCI) conservators were asked to undertake an examination to determine the stratigraphy and composition of the paint, to identify possible causes

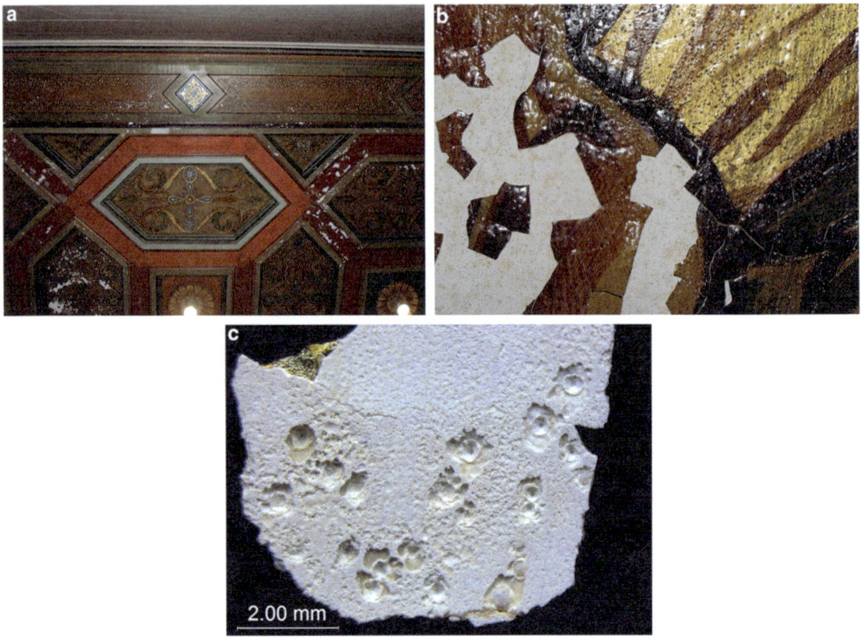

Fig. 17.4 National Research Council Auditorium, Ottawa, Canada; (**a**) detail of the decoratively painted coffered ceiling; (**b**) detail showing metal soap aggregates producing small lumps at the surface; (**c**) detail of the verso of a detached paint flake showing craters in the priming layer where soap aggregates have formed. (All images © CCI)

of damage to the paint layers, and to propose a viable treatment for stabilization and restoration.

17.4.2 Conservation Issues

The preliminary examination of the ceiling revealed a sporadic but massive failure of adhesion between the priming layer and the underlying plaster. Paint and priming layers had cracked and delaminated. Sizeable segments had subsequently peeled away from the plaster in large, brittle flakes, often several centimeters in width and length. It was apparent that the specification to apply linen to the plaster prior to painting had not been followed and that the priming and paint layers were applied directly to the plaster ceiling.

It was also noted that small lumps or inclusions were present over much of the painted surface, regardless of paint color or location (Fig. 17.4b). When paint flakes were viewed from the priming side, these lumps appeared to have absorbed surrounding material and expanded toward the top surface of the paint (Fig. 17.4c). The vast majority of the aggregates have not broken through the overlying paint, but rather have deformed the paint layers.

As there was no known history of water leaks, initially, the conservation issues were thought to be related to unstable humidity in the auditorium; data loggers were installed, and very wide fluctuations were observed between winter and summer months. It was not until 2015, when highly damaged areas were accessed via scaffolding by the CCI team, that signs of water damage were observed. It was suspected that this water had penetrated through roof leaks in the library room above, run down inside and on the surface of columns in this room, through the floor, and out and across the auditorium ceiling.

17.4.3 Analysis Results

In most areas of the ceiling, analysis showed that the plaster is composed of gypsum, calcium carbonate, calcium hydroxide, and magnesium hydroxide.[1] The presence of both gypsum and calcium carbonate is consistent with the specification that a mixture of lime putty and plaster of Paris be used for the final plastering. Cross sections confirmed the visual observation that the priming and paint layers were applied directly on the plaster. In all areas of the ceiling, there is a white to pale yellow priming layer over the plaster, followed by the decorative paint layers. The priming is composed of lead white, with smaller amounts of titanium white and barium sulfate, in a drying oil medium. Analysis by Py-GC-MS showed that the drying oil is heat pre-polymerized and contains a small amount of Pinaceae resin. The subsequent decorative, colored paint layers were all lead white-based with a drying oil medium. Pigments include red and yellow iron oxides, burnt umber, chrome yellow, Prussian blue, and emerald green. Bleached shellac varnish was applied in one to two layers followed by a thin layer of an oil-resin varnish (heated linseed oil, Pinaceae resin and copal).

Figure 17.5a is a cross section from an intact area of the ceiling, showing the plaster, oil-based priming and decorative paint layers. Figure 17.5b is a cross section showing one of the agglomerates that has protruded into the paint layers. Analysis of the material in the protrusion by FTIR revealed that it contains calcium fatty acid salts along with gypsum, calcium carbonate, and a small amount of drying oil (Fig. 17.5c). As in the previous case study, calcium monocarboxylic fatty acid salts were identified by the characteristic bands at 1578, 1541, 1470, 1434, and 1420 cm^{-1}. The same weak CH_2 progression bands at 1342, 1319, and 1286 cm^{-1} were also observed in the spectrum.

While the FTIR spectrum shows clear evidence of calcium fatty acid salts in the protrusion, the corresponding magnesium salts may also be present; some magnesium was identified in the protrusion along with calcium using SEM-EDX, and the strongest magnesium fatty acid salt bands (at 1576 and 1470 cm^{-1}) overlap with those of the calcium fatty acid salts (Bio-Rad Sadtler 2017). Further work could

[1] In one area, on a flat, vertical surface of a hexagonal coffer, the plaster was found to be composed of gypsum.

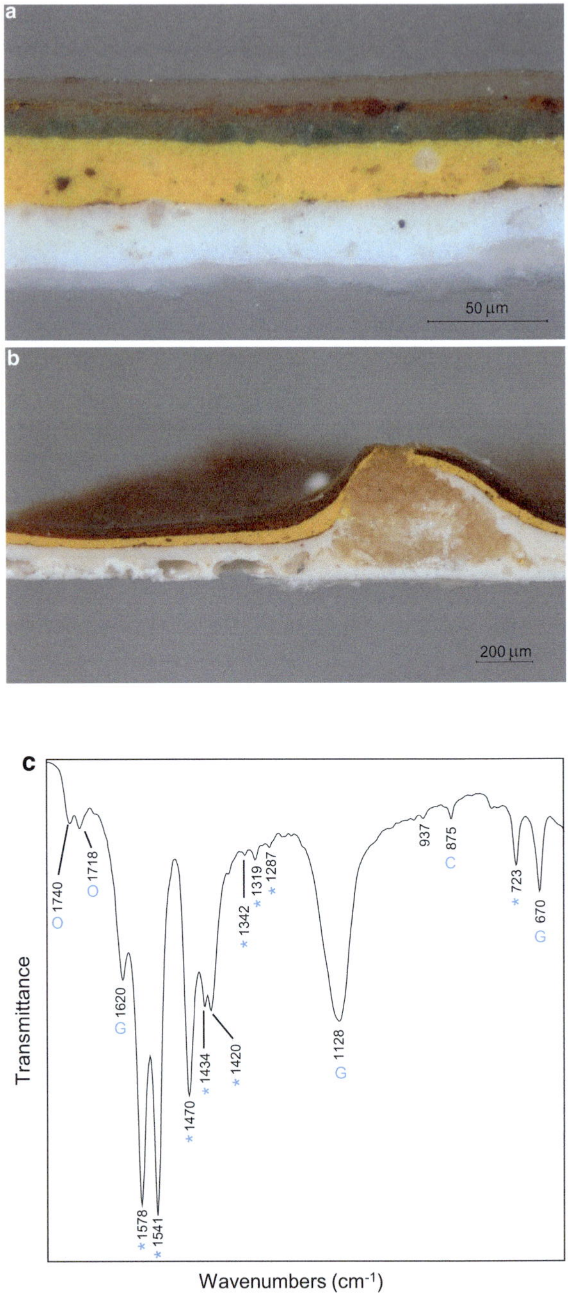

Fig. 17.5 (**a**) Cross section from an intact area of the ceiling showing the upper part of the plaster and the oil-based priming; (**b**) cross section from a problematic area showing a protrusion; (**c**) FTIR spectrum of material in a protrusion (* = calcium fatty acid salts; O = drying oil [ester at 1740 cm^{-1} and acid at 1718 cm^{-1}]; G = gypsum; C = calcium carbonate). (All images © CCI)

include Py-GC-MS to determine the fatty acid profile in the protrusions and FTIR mapping to study the spatial distribution of the soaps.

17.4.4 *Treatment*

In 2015, a prototype treatment was undertaken to determine the viability of an overall treatment of the flaking paint. One of the most severely damaged hexagonal panels and surrounding beams were chosen in order to establish methodologies and time estimates.

A number of thermoplastic adhesives diluted with water were tested for consolidation: diluted BEVA D8® dispersion (1:1 v/v), Lascaux consolidation medium (full strength), Plextol B® 500 (full strength and diluted 1:1 v/v), Aquazol 200® (5% and 10% concentrations w/v), and rabbit skin glue (5–7% w/v with several drops of ethanol). While synthetic adhesives worked reasonably well in holding the flakes in place, adhesive residues on the surface were difficult to remove. Rabbit skin glue appeared to hold the flakes with the smallest quantities of dispersed adhesive, the least number of repeat applications, and the lowest reactivation temperatures where heat was required. It was primarily for these reasons that rabbit skin glue was chosen as the consolidant.

The heavily discolored varnish layers were removed with solvents. In order to minimize risk to the unstable paint layers, exposure times were reduced by choosing solvents that were able to dissolve the varnish layers rapidly. Reconstruction of missing elements followed consolidation and varnish removal, bringing this inset panel and surrounding moldings back to a closer-to-original appearance. Details of the prototype treatment can be found in Binnie et al. (2017).

17.4.5 *Discussion*

The results of the analysis indicate that the protrusions originated in the plaster. They were formed by the reaction of alkali components in the plaster (calcium hydroxide and possibly magnesium hydroxide) with fatty acids from the oil paint above. Over time, the aggregates increased in size and penetrated into the overlying paint layers.

Although the exact mechanism of delamination is unclear, the presence of the calcium soap agglomerates has undermined the integrity of the bond between the plaster and the priming layer. Both uncontrolled environment, water condensation, and water infiltration may have played a role in the formation of the metal soaps. The alkali hydroxides in the plaster, compounds that are not commonly found in paint systems, would also be relatively susceptible to forming fatty acid salts.

The long-term effect of the conservation treatment on the stability of this compromised decorative surface remains uncertain. The introduction of water and alcohols, which could potentially exacerbate metal soap formation (Noble and Boon

2007), was balanced against the imminent and severe loss of decorative surfaces. The treated area will be monitored for at least a year before any further conservation treatment is considered. This will allow for verification of the efficacy of the prototype treatment as well as identification of potential problems associated with the chosen consolidation method. The NRC is aware of the need to maintain a stable environment in the auditorium in order to reduce further damage.

17.5 Conclusions

The oil on canvas painting, *Le Martyre de saint Pierre de Vérone*, contained calcium fatty acid salts in the translucent brown material intermixed with the ground layer. No unusual components that would have promoted the formation of these calcium salts were identified in the ground and paint. It was noted that the calcium soaps are confined primarily to the upper part of the work. External factors (exposure to fire and high humidity) may have led to their formation in this painting.

The decoratively painted NRC auditorium ceiling exhibits numerous calcium fatty acid salt protrusions that originated at the level of the plaster support. In this case, the alkali hydroxides in the plaster, compounds that are not commonly found in paint systems and are susceptible to forming soaps, have reacted with fatty acids from the oil paint above.

These two case studies illustrate that calcium fatty acid salts can form in oil paint systems in certain circumstances. In both cases, the calcium soap formation is associated with delamination and paint loss. Both environmental factors and the specific materials present in each of the works have likely played a role. More research and case studies are needed to clarify the factors leading to the formation of calcium soaps in oil paint, as compared to more commonly observed lead and zinc soaps.

References

Béland M (2014) D'après Le Titien. Au Musée national des beaux-arts du Québec Cap-aux-Diamants 116:50–51

Binnie N, Baker W, Moffatt E, Veal M-A, Helwig K, Poulin J (2017, in press) Investigation of paint instability and conservation requirements for decorative auditorium and library ceilings. In: Powers of ten – architectural paint research conference papers, postprints of the 6th international architectural paint research conference, New York, 15–17 Mar 2017. Archetype Publications, London

Bio-Rad Sadtler (2017) IR – Lubricant Additives 1. Accessed through Bio-Rad KnowItAll® Informatics System, vibrational spectroscopy edition. Bio-Rad Laboratories, Inc, Philadelphia. Spectra consulted: LA1X 80, 1088, 1819 and 1828

Ferreira ESB, Boon JJ, Stampanoni M, Marone F (2011) Study of the mechanism of formation of calcium soaps in an early 20th-century easel painting with correlative 2D and 3D microscopy. Preprints of 16th triennial ICOM-CC conference, Lisbon, pp 1604–1608

FHBRO Statement of Significance-100 Sussex Drive, NRC Laboratories (2016) http://www.historicplaces.ca/en/rep-reg/place-lieu.aspx?id=4030&pid=0. Accessed 16 Aug 2016
Halikia I, Soumpoulakis L, Christodoulou E, Prattis D (2001) Kinetic study of the thermal decomposition of calcium carbonate by isothermal methods of analysis. Eur J Mineral Process Environ Prot 1(2):89–102
Moffatt E, Salmon A, Poulin J, Fox A, Hay J (2015) Characterization of varnishes on nineteenth-century Canadian furniture. J Can Assoc Conserv 40:3–18
Mortimer DC (2002) 100 Sussex Drive, The Temple of Science. A concise story of the building which was the National Research Laboratories. National Research Council of Canada, Ottawa, pp 3–8
Noble P, Boon JJ (2007) Metal soap degradation of oil paintings: aggregates, increased transparency and efflorescence. AIC PSG Postprints 19:1–15
Sproatt H, Rolph E (1929) Sussex, original specifications, July 1929. Selection of pages from resource document provided by the National Research Council of Canada, Administrative Services and Property Management Branch, Ottawa, pp 86, 87, 126, 187
Vézina R (1975) Théophile Hamel, Peintre national (1817–1870), Tome I. Éditions Élysée, Ottawa, p 214

Part IV
Case Studies in Nineteenth and Twentieth Century Art: Artists and Paint Makers

Chapter 18
Everything Old Is New Again: Revisiting a Historical Symposium on Zinc Oxide Paint Films

Dawn V. Rogala

Abstract On June 6, 1949, members of the Victorian Branch of the Oil and Colour Chemists' Association gathered in Melbourne, Australia, to share their concerns regarding a sudden increase in problems associated with zinc oxide oil paints, problems which Association members posited were tied to industry-wide changes in processing methods for the zinc oxide pigment. The issues raised during this meeting inform and parallel current studies regarding metal soap behavior. This paper reviews the topics discussed at the 1949 symposium, distilling the historical research and its relationship to modern conservation and scientific inquiry. Topics include existing research literature, pigment processing methods and observed relationships between particle morphology and paint film behavior, and zinc oxide paint film failure patterns. Related metal soap research from the period is also discussed, including investigations of zinc oxide soap formation, the effect of environment on soap formation and film failure in zinc oxide oil paints, and the impact of zinc oxide-specific failure mechanisms on single- and composite-paint systems.

Keywords Zinc oxide · Zinc white · Oil paint · Zinc soap · Metal soaps · Paintings conservation

---

D. V. Rogala (✉)
Paintings Conservator, Museum Conservation Institute, Smithsonian Institution, Washington, DC, USA
e-mail: RogalaD@si.edu

F. Casadio et al. (eds.), *Metal Soaps in Art*, Cultural Heritage Science,
https://doi.org/10.1007/978-3-319-90617-1_18

18.1 Introduction

The Victorian Branch of the Oil and Colour Chemists' Association held a meeting in Melbourne, Australia, in the summer of 1949 to discuss a sudden and marked increase in problems associated with house paint formulations containing zinc oxide pigment and to investigate any possible relationship between this increased failure rate and recent changes in manufacturing methods for the pigment. Ten papers from the meeting were published shortly thereafter in a special issue of *Paint Notes: A Journal of Paint Technology* and are still a valuable resource for modern-day researchers investigating the aging behaviors of oil paint films made with zinc oxide and other metal soap-producing pigments.

The 1949 symposium on zinc oxide highlights how shifts in commercial production can impact the conservation study and treatment of cultural heritage objects. Modern art objects are often composed of materials produced by industries with manufacturing processes in a state of near-constant flux in response to internal and external market pressures. Production methods play an integral role in material behavior, yet this information is proprietary and therefore rarely accessible to researchers from outside the manufacturing industry. Successful production relies on an in-depth knowledge of material behavior; industrial research literature from this period details competing production methods, links processing techniques to specific material behaviors, and pinpoints shifts in manufacturing method that may help preservation professionals assess the behavior of artwork under their care.

What follows is a review of topics presented at the 1949 symposium. Papers presented at the meeting offer an "insider's" view of the historical research, provide updates on ongoing work, and outline trends in scientific inquiry. Topics include pigment processing methods and observed relationships between particle morphology and paint film behavior. Symposium papers are supplemented here with information from related period research, including examinations of zinc oxide soap formation, the effect of environment on soap formation and film failure patterns in zinc oxide paints, and the impact of zinc oxide-specific failure mechanisms on single- and composite-paint systems. This paper concludes with a summary of the historical zinc oxide research findings of interest to modern conservation and scientific inquiry.

18.2 Symposium Background and Opening Presentations

The reproduction of papers presented at meetings and symposia is common in research literature, and it is not unusual to see papers in late nineteenth- and early twentieth-century research journals cite one of the several regional association meetings at which the paper would be presented. Meetings typically addressed a range of topics and were attended by a mixed group of material producers, suppliers, and end users. The 1949 Melbourne meeting is unique in its focus. The

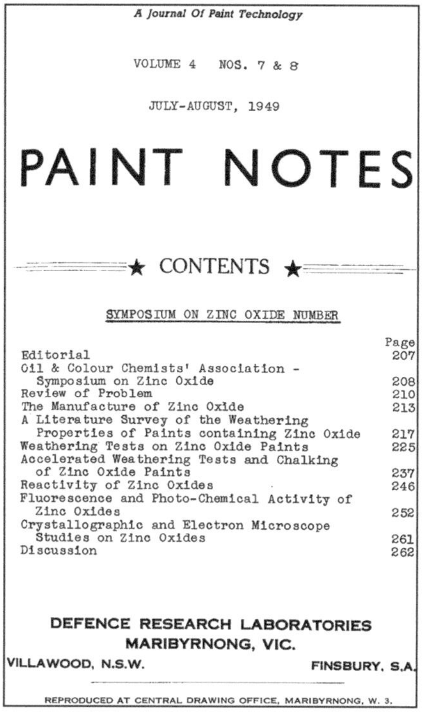

A Journal Of Paint Technology

VOLUME 4 NOS. 7 & 8

JULY-AUGUST, 1949

PAINT NOTES

★ CONTENTS ★

SYMPOSIUM ON ZINC OXIDE NUMBER

	Page
Editorial	207
Oil & Colour Chemists' Association - Symposium on Zinc Oxide	208
Review of Problem	210
The Manufacture of Zinc Oxide	213
A Literature Survey of the Weathering Properties of Paints containing Zinc Oxide	217
Weathering Tests on Zinc Oxide Paints	225
Accelerated Weathering Tests and Chalking of Zinc Oxide Paints	237
Reactivity of Zinc Oxides	246
Fluorescence and Photo-Chemical Activity of Zinc Oxides	252
Crystallographic and Electron Microscope Studies on Zinc Oxides	261
Discussion	262

DEFENCE RESEARCH LABORATORIES
MARIBYRNONG, VIC.
VILLAWOOD, N.S.W. FINSBURY, S.A.

REPRODUCED AT CENTRAL DRAWING OFFICE, MARIBYRNONG, W. 3.

Fig. 18.1 Contents page from the 1949 Zinc Oxide Symposium postprints, Victorian Branch (Australian Section) of the Oil and Colour Chemists Association. (© OCCA. Reprinted with permission)

introduction to the *Postprints from the Zinc Oxide Symposium of the Victorian Branch (Australian Section) of the Oil and Colour Chemists Association* states that while "no zinc oxide problem existed in Australia prior to the war," the introduction of new manufacturing methods designed to meet increased wartime demand for the pigment resulted in paint film problems "noticeable throughout the industry," with a resulting push for further investigation which prompted the meeting (Introduction 1949, 207).[1] Presentations at the symposium were delivered by representatives from paint makers and suppliers; the vice-chairman of the Victoria Branch "regretted that no paper was to be presented by any of the zinc manufacturers, although they had been invited to do so" (Geary 1949, 208) (Fig. 18.1).

[1]A relative definition of "no problem" should be applied to speakers' remarks throughout the symposium papers. The preferred material characteristics of house paint films differ from those of fine art materials. Regular repainting and short life expectancies allowed greater latitude in acceptable behavior for paint films, while some aging behaviors at odds with long-term preservation goals may not appear until after industrial paints reach the end of their anticipated commercial lifespan. Interpreting historical house paint literature from a preservation perspective has been covered previously by this author (Rogala 2011).

18.2.1 Manufacturing Methods

Coatings manufacturers opened the symposium program with a review of pigment production methods and existing scientific theory regarding the behavior of zinc oxide pigment in oil paint films. The work of pigment suppliers received even-handed treatment; the vagaries of production and market pressures were understood by paint manufacturers and researchers and accepted as a variable to be included in any study of paint film behavior. C.H.Z. Woinarski, then Senior Chemist at Hardie Trading Ltd., outlined the shift in paint production and performance observed during World War II, when new methods of pigment manufacture emerged – at newly built or converted facilities – to meet an increased demand for zinc oxide pigment (Woinarski 1949). Direct Process (also called American Process) methods of pigment production were replaced by Indirect Process (sometimes called French Process) methods, and this manufacturing shift was accompanied by a precipitous rise in failure rates of oil-based house paints containing zinc oxide pigment produced using the new method.[2] These disparate manufacturing processes were further delineated by D.G. Davidson, Works Manager at Goodlass Wall & Co., Pty. Ltd., who summarized the substantive differences as (1) production through simultaneous reduction and oxidation (Direct Process) versus oxidation alone (Indirect Process) and (2) particles whose size and shape are largely inconsistent (Direct Process) or consistent (Indirect Process).

18.2.2 Pigment Morphology

Particle shape and size were the focus of a literature review provided by K.R. Bussell, who noted a prevalence of marketing bias in research that "postulated theories explaining superior durability of certain types of zinc oxide with little or no experimental evidence to justify them," and were therefore of little value in accurately assessing the role of particle morphology in the observed increase in film failure of oil-based zinc oxide paint (Bussell 1949, 224). This literature does, however, offer useful descriptions of particle shapes and manufacturer expectations. At the time of the symposium, zinc oxide pigment was available in three shapes – amorphous, acicular, and irregular. Amorphous particles are defined in the period literature as crystalline, mostly spherical, and of consistent shape, while acicular particles are needle-shaped and less prone to clumping during mixing because of their uneven surfaces. Irregular particles (the term nodular is used by some authors) can be a mixture of sizes and shapes and include both amorphous and

[2]Indirect Process production was also an economical alternative due to higher pigment purity, lower lead content, lower water-soluble content, greater hiding power and tinting strength, and improved storage properties (Schmutz 1935; Morley-Smith 1950a).

acicular forms.[3] Both Bussell and Davidson reported that Direct Process zinc oxide particles tend to be larger and acicular, characteristics believed to be less reactive and cause fewer problems, while zinc oxide produced using the Indirect Process resulted in smaller, more reactive particles with fewer acicular components. Film-forming behaviors attributed to acicular pigments that would be undesirable in artwork include the development of "innumerable microscopic failures" as gaps between binder and large pigment particles created during drying of the paint film (Bussell 1949, 221). Bussell's review included work by authors whose names appear throughout the historical zinc oxide literature, including Kekwick and Pass (1938), Werthan (1941), and Calbeck et al. (1941).

A 1940 *Chemical Industries* article by H.A. Nelson (Fig. 18.2a) referenced by Bussell provides more information on the pigment structures resulting from various production methods. Zinc oxide is typically a crystalline, close-packed lattice structure of alternating zinc and oxygen atoms. Nelson confirms that Direct Process oxides contained various percentages of acicular variations on the crystalline form, some joined to form "twins" and "threelings" (referred to as "brush-heap" formations by Bussell), while Indirect Process zinc oxides (often marketed under the term "Seal" oxides) were typically irregularly shaped particles of uniform size distribution (Nelson 1940, 509). Nelson suggests that the acicular particles found in Direct Process zinc oxides delay film failure but are more likely to cause localized, longitudinal failures in oil-based house paints, while the film failures related to Indirect Process zinc oxides appear more quickly and are primarily useful, "self-cleansing" failures (surface-layer shedding/sloughing) of even distribution. (Nelson was at this time in charge of the paint research division at the New Jersey Zinc Company, part of the industry using the Indirect Process.)

18.3 Zinc Oxide Research Prior to the Symposium

The remaining presentations at the Zinc Oxide Symposium were devoted to early-stage and continuing research that examined the relationship between pigment characteristics and zinc oxide paint film behavior, with a particular focus on differences between oil-based films containing Direct Process and Indirect Process pigments. J.R. Rischbieth and W. Griffiths updated attendees on early results from weathering tests that supported observations of the superior performance of Direct Process over Indirect Process zinc oxide, the slight but not substantive improvement provided by the addition of boiled oils (Bussell), and the acceleration of film failure related to increased exposure to moisture (Woinarski). Just as industrial materials used by artists are formulated for behaviors that may conflict with preservation goals, the needs of the paint industry are complicated by their reliance on materials

[3]Present-day texts simplify zinc oxide particle shapes into two primary categories: nodular and acicular, with the former used in reference to characteristically fine, rounded particles.

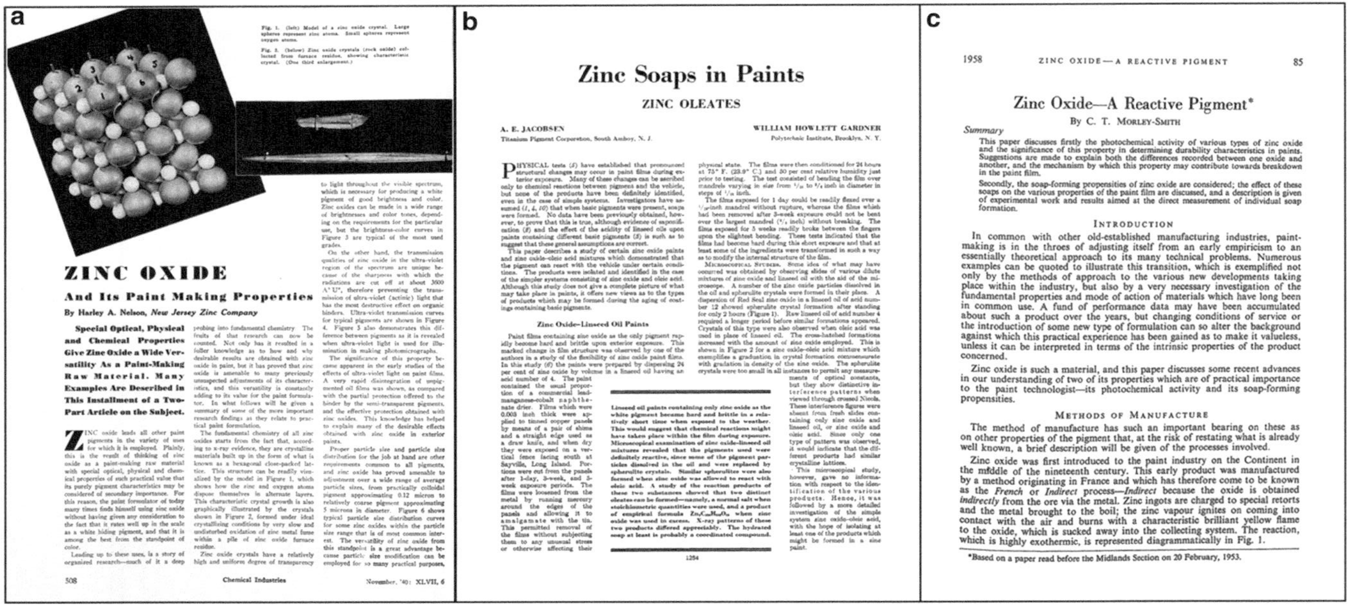

Fig. 18.2 (**a–c**) Examples of period research on zinc oxide pigment in oil paint vehicles (l-r): "Zinc oxide and its paint making properties" (Nelson 1940), "Zinc soaps in paints: Zinc oleates" (Jacobsen and Gardner 1941), and "Zinc oxide—a reactive pigment" (Morley-Smith 1958). (© NJ Zinc Co., ACS, OCCA. Reprinted with permission)

manufactured for other purposes. Griffiths points out that the wartime increase in zinc oxide production was directly related to an expanded need for rubber, with more than 50 percent of zinc oxide sales supplying the rubber industry (Griffiths 1949, 244). Acceptable behaviors for the newly produced pigment were therefore not the same as those for the zinc oxide primarily manufactured for use in paint films. Both Rischbieth and Griffiths noted the lack of research on the effects of new pigment production techniques on affiliated industries and called for further investigation into particle morphology and zinc oxide reactivity.

18.3.1 Metal Soaps

The first Zinc Oxide Symposium paper to address the role played by metal soaps in paint film failure was presented by H.A. Laurie and D.K. Box, who discussed surface area and particle shape in relation to the formation and control of zinc oxide soaps. Their review of existing research revisited problems associated with marketing bias but found useful information on the influence of surface area on metal soap formation, with the reduced surface of acicular particles deemed less reactive. The potential for continued formation of crystalline metal soaps throughout the lifetime of a paint film was noted, but the presenters downplayed the seriousness of this situation as the articles addressing this behavior were primarily written by industry competitors (lead pigment companies). Laurie and Box did alert attendees that their own examination of commercial zinc oxide determined that pigment labeled as Direct Process or Indirect Process often contained particle mixtures and impurities that could skew research results. They concluded that while experimental data was still limited, "the effect of zinc soaps on the properties of paint films... [is] of interest and point[s] to a further line of investigation" (Laurie and Box 1949, 251). Familiar authors mentioned by Laurie and Box included Robertson (1942), Dunn (1946), and Dunn and Baier (1948).

In related literature, an early paper by Nelson and G.W. Rundle suggested that paints be formulated with oil binders of low acid number to control soap production, along with limited use of hygroscopic materials such as glycerol to control the strongly moisture-absorbing characteristics of zinc oxide paint films, a behavior "not directly proportional to the relative humidity" and unique among other pigments in the length of time required to return to equilibrium (Nelson and Rundle 1923, 365–367). Work published in the early 1940s by D.G. Nicholson and his team from the University of Illinois found that continued soap production – beyond that necessary for film formation – had a negative impact on film strength and moisture sensitivity of the paint film. Acicular particles were again viewed favorably in terms of strength and durability. Nelson's claim that acicular particles show longitudinal failure patterns was supported by an observation that "acicular films tend to crack in lines which parallel the stroke used in their application" (Nicholson and T.W. Mastin 1942, 999). Other points of interest in this research include reports that (1) zinc oxide pigment can react with moisture prior to paint formulation and become more basic (a reference to Eide and Depew 1936), (2) zinc oxide pigment treated with carbon dioxide gas makes a more durable paint film, and (3) exposure to atmospheric contaminants can vary drying times and behaviors (from Nicholson 1941). Like other researchers, Nicholson and Mastin found incomplete or inaccurate labeling of commercial Direct and Indirect Process pigments.

18.3.2 Performance in a Layered System

The compatibility of zinc oxide oil paints and adjacent pigmented paint films arose as a topic of research interest prior to the symposium, with the emphasis placed on layering paints that exhibit well-matched material characteristics. Research from the US Department of Agriculture summarized the argument against zinc oxide oil paint in composite-paint systems by stating that the behavior of zinc oxide paint films differed so greatly from other paint films that it was not recommended as a priming layer for any paint system that did not also contain zinc oxide pigment (Browne 1936, 1941).[4] Similar research by industry competitors is mentioned at the 1949 meeting by Laurie and Box (a reference to Dunn), and more than a decade before the Zinc Oxide Symposium, a presentation to the Baltimore, Chicago, Northwestern, Philadelphia, Pittsburgh, and Western New York Paint and Varnish Clubs by Titanium Pigment Company representative D.W. Robertson noted that differences in moisture absorption between adjacent oil paint films could also lead to failure through the release of water-soluble compounds between paint layers, resulting in loss of adhesion in the interface between paint layers or between paint and support layers. Interface failure was noted by Robertson in paints that formed zinc oxide soaps, because "metal-oil compounds developed by the interaction of linseed oil and metallic salts have poor adhesive properties compared to the original oil film" (Robertson 1935, 229). A 1941 paper from Titanium Pigment Company representatives A.E. Jacobsen (who also published with Robertson) and W.H. Gardner (Fig. 18.2b) looked at zinc oleate soaps and reported finding (1) an increase in soap production in the presence of basic pigments and (2) differences in structure between salts formed under normal (stoichiometric) conditions and those produced under conditions with an excess of zinc oxide (sometimes forming complex salts of layered zinc and oleic acid).[5] Twenty-five percent zinc oxide is suggested in the pre-symposium literature (Keckwick and Pass 1938; Nelson 1940) without evidentiary support. Pigment volume concentrations for zinc oxide paint films became the subject of further research following the symposium.

[4] Attendees at a 1949 meeting of London Section of the Oil and Colour Chemists' Association also noted the chalking of zinc whites in simple zinc-and-varnish layering (Morley-Smith 1950b, 268).

[5] Industry research on the poor adhesion of stoichiometric salts in oleic acid appears as early as 1929, in a presentation at the Philadelphia Club by J.T. Baldwin which reported preliminary testing results suggesting that metal soaps appearing in zinc oxide paint remained separate from the body of the paint film under all testing conditions (Baldwin 1929, 866).

18.3.3 Particle Fluorescence

The 1949 symposium closed with two papers that reflected increasing interest in developing new analytical techniques for evaluating paint film characteristics and behavior. R.I. Garrod (1949) suggested a possible correlation between ultraviolet-induced visible fluorescence and pigment particle size, and G. Winter and R. N. Whittem (1949) posited a relationship between fluorescence and pigment reactivity. Winter and Whittem noted that research in this area was just beginning, and while fluorescence microscopy was a promising technique, results could be compromised by the impurities and mixed particle types found in commercial pigments, by altered pigment morphology (related to manufacturing), and by disrupted crystalline structures (e.g., due to zinc soap formation). Further study was recommended before the usefulness of fluorescence as an analytical technique could be properly assessed. Authors cited by Winter and Whittem include Kekwick (1938) and Robertson (1942).

18.4 Post-symposium Research

Research following the Symposium on Zinc Oxide built on promising hypotheses from the 1949 meeting. The fluorescence of zinc oxide was further explored as attempts were made by pigment manufacturers to modify Indirect Process pigments to better mimic the stability and aging characteristics observed in Direct Process zinc oxides. In an announcement made to the 1950 meeting of the Oil and Colour Chemists' Association (New South Wales Branch), a representative of the Zinc Pigment Development Association at Durham Chemicals acknowledged the roles played by moisture and particle size in zinc oxide film failure but hypothesized the existence of another, more controllable variable with influence over the performance of zinc oxide oil paints (C.T. Morley-Smith 1950a). Fluorescence microscopy was used to differentiate between zinc oxide particles exhibiting green fluorescence (said to be produced in an oxidizing atmosphere) and particles exhibiting purple-blue fluorescence (produced under reducing conditions). The "green" particle was reported to contain stoichiometric proportions of zinc and oxygen, while the "blue" particle showed excess zinc and interstitial zinc atoms capable of initiating photochemical reactions and accelerated film failure (Morley-Smith 1958, 89) (Fig. 18.2c). Despite the mixed reduction-oxidation conditions of Direct Process production, the Direct Process zinc oxide pigment showed more of the stable, less reactive "green" particles. (Indirect Process particles tended toward dull green-brown and purple-blue fluorescence.) Morley-Smith reported that efforts were underway to reduce the inclusion of ultrafine particles (more surface area) in zinc oxides and to modify Indirect Process particles to exhibit green fluorescence (and the attendant performance characteristics).

F.L. Browne revisited the high swelling rate of zinc oxide oil paints and confirmed responsiveness of the paint films to high humidity and applied water, with paints incorporating Indirect Process pigments showing greater sensitivity than those made with Direct Process pigments (Browne 1957). Browne's work supported earlier suggestions that stored pigment could become more basic through carbonation and that paint films containing zinc oxide retained some amount of swelling after drying (Browne 1955, 1957). Additional research from the US Department of Agriculture corroborated reports that even small amounts of zinc oxide pigment "exert a determining influence on [paint film] properties" (Eissler and Princen 1966, 19). Increased swelling was attributed to the basic nature of zinc oxide, as "the weakly acid character of linseed oil is insufficient to overcome excess alkalinity contributed by certain pigments to a film in the presence of water." Equivalent swelling was reported in zinc oxide oil paint films containing 1 percent and 30 percent volume concentration of the pigment (Eissler and Princen 1970, 155). Continued weathering research by Rischbieth showed that problems related to moisture often appeared months after exposure to water (1957). Research from the University of Stuttgart built on rubber industry literature citing a possible link between the oil and resin vehicle and swelling behavior in zinc oxide paint films (Funke 1967) and related tests by Morley-Smith on the influence of fatty acids in zinc oxide reactivity suggested that soap formation slowed with increasing acid chain length and noted that "a marked difference was apparent between the behavior of the saturated and unsaturated acids, unsaturation reducing the rate of soap formation appreciably" (1958, 94).

18.5 Summary of Themes and Relation to Contemporary Research

There has recently been renewed interest in the behavior of metal-based pigments in paint films, and historical literature can be a valuable resource. Papers from, and related to, the 1949 Symposium on Zinc Oxide present substantive research regarding pigment production and the behavior of oil paint films containing zinc oxide pigment. Table 18.1 outlines the comparative pigment characteristics and behaviors as presented in the period literature. Direct Process zinc oxide is reported to be produced through a combined reduction-oxidation process in stoichiometric conditions that results in larger, acicular particles with reduced surface area. Direct Process zinc oxide is said to be less reactive (on a relative scale) than other zinc oxides. Indirect Process zinc oxide, by comparison, is produced solely through oxidation, resulting in smaller particles with more reactive surface area and the potential for excess zinc and interstitial zinc atoms available for further reaction. Period research associates Direct Process zinc oxide with localized, longitudinal failures in oil paint films and with the potential for localized soap formation at groupings of pigment particles spaced unevenly throughout the paint film. Indirect Process zinc oxide is reported to show a tendency to form soaps composed of layered pigment and fatty acids, positioned throughout the paint film. The symposium postprints also link the shift

Table 18.1 Comparative zinc oxide pigment characteristics as reported c. 1949 in the paint research literature

Direct/American Process	Indirect/French Process
Established production in the US mid-nineteenth century; other markets (e.g. Australia) from early twentieth century	Industrially produced in France in the 1840s; introduced in other markets (e.g. Australia) during WWII
Manufactured from ores via simultaneous reduction and oxidation processes; heavy metal impurities	Manufactured from zinc metal via oxidation; high relative purity
Larger, variable particle size	Fine, relatively uniform particle size
Smaller reactive surface area	Larger reactive surface area
Predominantly acicular particle shape	Predominantly nodular particle shape
Stoichiometric crystal structure	Non-stoichiometric crystal; interstitial zinc atoms
Predominantly green UV fluorescence	Predominantly purple-blue UV fluorescence
Forms relatively durable films	Poor relative durability
Associated with longitudinal cracking in paint film	Associated with widespread/uniform chalking of paint film
Favors localized soap formation in paint film	Favors layered soap formation in paint film
Paint films experience high, prolonged swelling response to water; pigment is less sensitive to RH than Indirect Process	Paint films experience high, prolonged swelling response to water; pigment is sensitive to RH

by regional manufacturers from Direct to Indirect pigment processing methods with increased wartime demand for zinc oxide, which may be of interest to conservation and material science researchers investigating the appearance of lamellar zinc oxide soap behavior in mid-twentieth-century paintings.[6]

Zinc oxide oil paint films were reported to exhibit high swelling behavior. Even small amounts of zinc oxide in a paint film were said to induce swelling. Exposure to moisture was reported to enhance film failure, with some failures appearing long after the exposure period. The research suggested that moisture remained in the paint film and prolonged swelling, thereby extending moisture-related reactivity and alteration of the paint film's physical properties. Water-soluble compounds and metal soaps could also become trapped between paint layers, leading to adhesion loss. Zinc oxide pigment was reported to be sensitive to moisture while in its raw state, prior to paint formulation. Several papers from the 1949 meeting claimed that water exposure prior to mixing made the pigment more basic and that alkaline pigments could exhibit increased soap production, which in turn made the zinc oxide paint film more moisture sensitive. It is important to note that the historical research related to moisture sensitivity utilized levels of liquid and lengths of exposure

[6]Previous work on this topic by this author includes Rogala et al. (2010) and Maines et al. (2011).

unlikely to be replicated in standard conservation treatment; further research is needed to determine what impact moisture exposure (and duration of exposure) may have on the preservation of art materials containing zinc oxide pigment.[7]

A strong grounding in industrial research is central to the preservation of modern artworks. For modern-day researchers of aging behaviors and metal soap formation in zinc oxide oil paints, mining this literature is not as simple as dating a shift from Direct to Indirect pigment production or using fluorescence microscopy. Terminology can be confusing, or vary from author to author.[8] Product labeling can be inaccurate or incomplete. Advancing technology can also confuse physical characteristics, as in the attempts to modify Indirect Process pigments to mimic the behavioral and analytical markers of zinc oxide pigments made using Direct Process methods. The behavior of art materials such as zinc oxide pigment will be affected by the changing needs of other markets, from wartime increases in rubber production to the contemporary call for semiconductor coatings on electronics. The network of influences apparent in historical research relates to era-specific manufacturing trends just as modern-day market pressures direct contemporary material science research and innovation. With careful study, historical research can be a valuable resource in the preservation of cultural heritage objects.

References

Postprints from the Zinc Oxide Symposium of the Victorian Branch (Australian Section) of the Oil & Colour Chemists Association in Melbourne on June 6, 1949. In: Paint Notes: A Journal of Paint Technology 4(7–8), in order of publication. Author affiliations, when available, appear in brackets.

Unknown. Introduction, p 207

Geary RJ. Minutes, Oil & Colour Chemists' Association—Symposium on Zinc Oxide, pp 208–209

Woinarski CHZ. Review of problem, pp 210–213

Davidson DG. The manufacture of zinc oxide, pp 213–217. [Goodlass Wall & Co. Pty. Ltd.]

Bussell KR. A literature survey of the weathering properties of paints containing zinc oxide, pp 217–224

Rischbieth JR. Weathering tests on zinc oxide paints, pp 225–237

Griffiths W. Accelerated weathering tests and chalking of zinc oxide paints, pp 237–246. [Technical Director, Glazebrooks Paints Australia Ltd.]

[7]Weathering tests and lab experiments used to evaluate house paints far exceed the exposure conditions anticipated from a single conservation treatment, but the material response patterns revealed in these studies are relevant to cultural heritage preservation interests, e.g. the cumulative effects of conservation treatment over the lifetime of an artwork and the long-term exposure of art materials to unregulated environments.

[8]Examples of confusing language pairings include Direct/American, Indirect/French, and irregular/nodular. In one audience exchange at the 1949 symposium, "Mr. Sutton asked whether the various phenomena of erosion, cracking, chalking, etc. could not perhaps be merely manifestations of one and same thing" to which Mr. Rischbieth "admitted that there probably was… some connection between these phenomena" (Discussion 1949, 265).

Laurie HA, Box DK. Reactivity of zinc oxides, pp 246–251
Winter G, Whittem RN. Fluorescence and photochemical activity of zinc oxides, pp 252–261
Garrod RI. Crystallographic and electron microscope studies on zinc oxides, pp 261–262
Various. Discussion, pp 262–265

Other Papers Mentioned

Baldwin JT (1929) Philadelphia Club—preliminary report on the soap investigation. Am Paint Varnish Manuf Assoc Newsl 356:858–882
Browne FL (1936) Paints as protective coatings for wood. Ind Eng Chem 28(7):798–809. [Forest Products Laboratory, United States Department of Agriculture]
Browne FL (1941) Two-coat system of house painting. Ind Eng Chem 33(7):900–910. [Forest Products Laboratory, United States Department of Agriculture]
Browne FL (1955) Swelling of paint films in water, III: absorption and volumetric swelling of bound and free films from air of different relative humidities. For Prod J 5:92–96. [Forest Products Laboratory, United States Department of Agriculture]
Browne FL (1957) Swelling of paint films in water, XI: Mixed-pigment paints in linseed oil. For Prod J 7:248–252 [Forest Products Laboratory, United States Department of Agriculture]
Calbeck JH, Eide AC, Easley MK (1941) Acicular zinc oxide. Paint Ind Mag 56(9):300–313. [American Zinc Sales Company]
Dunn EJ Jr (1946) Chemical reaction in metal protective paints. Am Paint J 30(43):56–67
Dunn EJ Jr, Baier CH (1948) Effect of white pigments on physical properties of paint films. Am Paint J 32(52):42–46. 76–104
Eide AC, Depew HA (1936) Evaluation of zinc oxide for paint. Am Paint J 20(27):7–9. 20(28):51–56. [American Zinc Sales Company]
Eissler RL, Princen LH (1966) Effect of some pigments on tensile and swelling properties of free linseed oil films. Papers Presented – American Chemical Society, Division of Organic Coatings and Plastics Chemistry 26(1):16–23. [Northern Regional Research Laboratory, United States Department of Agriculture]
Eissler RL, Princen LH (1970) Swelling of linseed oil films in acid and alkaline environments. J Paint Technol 42(542):155–158
Funke W (1967) On the relation between the pigment-vehicle interaction and liquid water absorption of paint films. J Oil Colour Chemists' Assoc 50(10):942–975. [Research Institute for Pigments and Paints, University of Stuttgart
Jacobsen AE, Gardner WH (1941) Zinc soaps in paints: zinc oleates. Ind Eng Chem 33(10):1254–1256. [Titanium Pigment Corporation; Polytechnic Institute]
Kekwick LO, Pass A (1938) Acicular zinc oxide. J Oil Colour Chemists' Assoc 21(215):118–139
Maines C, Rogala D, Lake S, Mecklenburg M (2011) Deterioration in abstract expressionist paintings: analysis of zinc oxide paint layers in works from the collection of the Hirshhorn Museum and Sculpture Garden, Smithsonian Institution. In: Vandiver PB et al (eds) Materials research society symposium proceedings 1319, materials issues in art and archaeology, vol 9. MRS, Warrendale, pp 275–286
Morley-Smith CT (1950a) The development of anti-chalking French Process zinc oxides. J Oil Colour Chemists' Assoc 33(365):484–490. [Zinc Pigment Development Association; Durham Chemicals Ltd.]
Morley-Smith CT (1950b) The tint retention of coloured paints based on white pigments. J Oil Colour Chemists' Assoc 33(360):249–269. [Durham Chemicals Ltd.]
Morley-Smith CT (1958) Zinc oxide—a reactive pigment. J Oil Colour Chemists' Assoc 41:85–97. Durham Chemicals Ltd.
Nelson HA (1940) Zinc oxide and its paint making properties. Chem Ind 47(6):508–512. [New Jersey Zinc Company]

Nelson HA, Rundle GW (1923) Further studies of the physical properties of drying-oil, paint and varnish films. In: American society for testing materials proceedings. ASTM, Philadelphia, pp 356–368. [New Jersey Zinc Company]
Nicholson DG (1941) Drying of linseed oil paint: effect of atmospheric impurities on rate of oxygen absorption. Ind Eng Chem 33(9):1148–1153. [University of Illinois]
Nicholson DG, Mastin TW (1942) Durability of soap-treated zinc oxide paints. Ind Eng Chem 34(8):996–1002. [University of Illinois]
Rischbieth JR (1957) Weathering characteristics of zinc oxide. J Oil Colour Chemists' Assoc 40:212–220
Robertson A (1942) Zinc oxide as a paint pigment. J Oil Colour Chemists' Assoc 25:53–64
Robertson DW(1935) Exterior house paint pigment combinations in relation to durability and type of failure. Official Digest – Federation Paint Varnish Production Clubs 146:228–253. [Titanium Pigment Corporation]
Rogala D (2011) Industrial literature as a resource in modern materials conservation: zinc oxide house paint as a case study. In: Paintings specialty group postprints, American institute for conservation 39th annual meeting, Philadelphia. AIC, Washington, DC, pp 78–90
Rogala D, Lake S, Maines C, Mecklenburg M (2010) Condition problems related to zinc oxide underlayers: examination of selected abstract expressionist paintings from the collection of the Hirshhorn Museum and Sculpture Garden, Smithsonian Institution. J Am Inst Conserv 49(2):96–113
Schmutz FC (1935) Exterior house paints custom built with zinc pigments. Paint. Oil Chem Rev 97(8):26–28. [New Jersey Zinc Company]
Werthan S (1941) Zinc oxide—as you like it. Paint Ind Mag 56(6):198–204. [The New Jersey Zinc Company]

Chapter 19
Notes on Metal Soap Extenders in Modern Oil Paints: History, Use, Degradation, and Analysis

Klaas Jan van den Berg, Aviva Burnstock, and Michael Schilling

Abstract The long-term behavior of metal stearates used as additive in paints is not well known, and their detection is challenging. This paper presents some aspects of the production of technical metal stearates from stearin in twentieth-century oil paint and its use. Two fast direct analytical mass spectrometry (MS) techniques, evolved gas analysis MS and electrospray ionization MS, are introduced as useful tools for the detection of low (<2%) quantities based on the different palmitate/stearate (P/S) ratios of the stearates, oils, and other lipids in the paint. Early occurring fatty acid efflorescence on twentieth-century oil paints is caused primarily by metal stearates, as shown in one example from a modern painting by Mia Tarney. Metal stearates are not stable in the course of time; artificial aging of a set of paints shows the fast hydrolysis of aluminum stearates under elevated relative humidity.

Keywords Metal stearate additives · P/S ratio · Stearin · Hydrolysis · Evolved gas analysis · Electrospray ionization · Mass spectrometry

19.1 Introduction

Metal soaps have been an important ingredient of industrial and artists' oil paints throughout the twentieth century (Tumosa 2001) and are still commonly used. Metal soaps are used as additives that function to disperse particles in the medium and as stabilizers that prevent the paint components in the tube or tin from separating out.

K. J. van den Berg (✉)
Cultural Heritage Agency of the Netherlands, Amsterdam, The Netherlands

University of Amsterdam, Amsterdam, The Netherlands
e-mail: k.van.den.berg@cultureelerfgoed.nl

A. Burnstock
Department of Conservation and Technology, Courtauld Institute of Art, London, UK

M. Schilling
The Getty Conservation Institute, Los Angeles, CA, USA

F. Casadio et al. (eds.), *Metal Soaps in Art*, Cultural Heritage Science,
https://doi.org/10.1007/978-3-319-90617-1_19

Used in higher concentrations, the soaps also function as an extender in cheaper paints, as they create a gel with the oil component, reducing the proportion of pigment in the formulation (Tempest et al. 2013; Mayer 1970). The gel-like behavior of metal soaps influences the rheology of the paint making it more smooth and buttery.[1] This follows the general development in twentieth-century oil paints to shorter paints with higher medium content. The recommended proportion of metal stearate addition from many manufacturers is typically in the range of 1–5% of the volume of paint (Heaton 1947; Tumosa 2001).

The long-term behavior of metal stearates in paints is not well understood although several researchers have presented evidence that metal stearates may play a role in a range of physical manifestations of deterioration in artist's paints and paintings. Problems with paints formulated with high amounts of aluminum stearate are frequently mentioned in the literature; in particular, it was found to impede the drying time of the paint film (Heaton 1947; Gettens and Stout 1966). Furthermore, higher concentrations would cause the paints to become brittle and deteriorate (Heaton 1947).

More recently, aluminum stearate has been shown to create problems when used in zinc oxide white paints due to excessive formation of zinc soaps. These problems are related to zinc soap protrusions and the reduction to the binding capacity of the medium (Osmond 2014). Fatty acid efflorescence that is often found on twentieth-century oil painting may be caused by metal soap additives (Tempest et al. 2013). The role of metal stearates has been mentioned in relation to the cause of dripping paints (Boon and Hoogland 2014) and medium exudations (Bayliss et al. 2016) although the link between these physical manifestations and the presence of the stearates may be mostly indirect. It is therefore of interest to gain more information about the role of metal stearates in oil paints either added by the manufacturer or formed during curing and ageing of the paint.

The present study investigates the use of metal stearates in artists' paints by review of notes made in previous investigations by Tumosa (2001) and other hitherto unpublished research on methods for analysis of metal stearates. The long-term stability of particularly aluminum soaps in artists' paints is also addressed in the analytical study.

19.2 Manufacture

19.2.1 Historical Developments and Use

It is not known when aluminum stearate was first used as a component of paint. In his painting treatise of 1892 Vibert mentions "salts of aluminum" used in commercial paint for certain pigments (Carlyle 2001), but its specific purpose is not specified. Aluminum stearate added as a dispersing agent in commercial paints

[1] Saltmarsh, P., van den Berg K.J., Burnstock, A. 2008. "Making Paint Without a Recipe: Modern Paint Reconstructions," Poster presented at ATSR conference in Glasgow, June 2008.

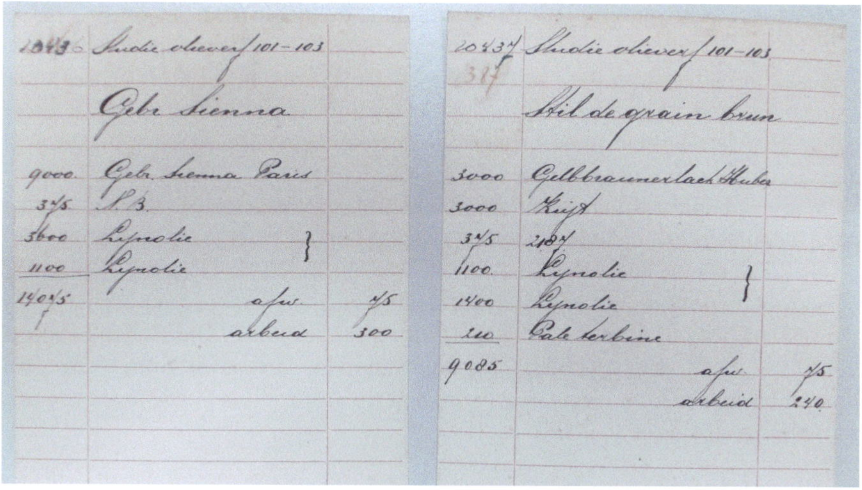

Fig. 19.1 Paint recipes from the Talens student oil colors range from the 1930s. Left: burnt sienna with sodium stearate (N.B.) and linseed oil; right: *Stil de grain brun* with a yellow/brown lake (*Gelbbrauner Lack*), chalk, zinc stearate (2187), and linseed oil (Van den Berg et al. 2016)

was first patented in 1920 in the United States (Tumosa 2001). The additive was subsequently widely incorporated into paints due to cheapness and light weight. It can be assumed that it was used in the manufacture of artists' as well as industrial oil paints from this time.

Written sources include discussion of the benefits of including aluminum stearates in paint making. It was found to help suspend the pigments in the medium, creating a more stable dispersion and improving the handling properties and body of the paint (Weiss 1957; Turner 1958). It also aids grinding by creating a more viscous mixture at lower pigment concentrations, thus reducing the amount of grinding time required (Tumosa 2001) which reduces the cost of manufacture.

To this date, manufacturers do not state the ingredients on paint tubes other than the pigments and binding medium. However, much information on historical manufacture has become available in the past few years from paint manufacturers' archives. Dialogue with manufacturers specializing in research and development combined with the results of analysis of tube paints has shown that aluminum, magnesium, and zinc stearates are still used on a large scale by most manufacturers of artists' paints.

Since the patenting of aluminum stearate, other metal soaps were probably also soon introduced. Talens, for example, used sodium and zinc stearates in the 1930s and possibly before (Fig. 19.1). Whereas aluminum and zinc stearates were obtained commercially, sodium stearates were produced in-house from sodium hydroxide and stearin (Van den Berg et al. 2016). In the late 1960s, aluminum stearates were used, replacing the sodium stearates and possibly the zinc stearates (Van den Berg et al. 2016; Mills et al. 2008).

Winsor & Newton has used magnesium and aluminum stearates in their paints throughout the second half of the twentieth century.[2] Weber used aluminum stearates in paints from the 1940s to the 1960s (Phenix et al. 2017). These three companies used stearates in all their paint quality ranges, although not in every color. Michael Harding uses magnesium stearate dispersion agent in his paints,[3] and Grumbacher has been reported to employ aluminum stearates (Osmond 2014).

Many examples of the use of aluminum and zinc stearates by these and other paint manufacturers, including Schmincke and Maimeri, exist in the literature (Mills et al. 2008; Boon and Hoogland 2014; Izzo et al. 2014; Osmond 2014). The conclusion must therefore be that metal stearate extenders are or have been used by the vast majority of oil paint producers, with few exceptions: Old Holland appears not to use metal soap extenders; this company adds fatty acids from hydrolyzed linseed oil as a wetting agent and hydrogenated castor oil as a stabilizer in some formulations (Mills et al. 2008).

19.2.2 Metal Stearates Are Produced from Stearin

The metal stearates added to paints are in general of a technical grade, applied widely in industry for their lubricating properties, separating properties, water repellence, gelling capacity, stabilizing effect, foam inhibition, etc. Technical-grade stearates are relatively inexpensive and are not pure.

Metal soaps are produced by reaction between basic metal oxides or hydroxides and lipidic fractions. Although there are exceptions (e.g., palmitates), as paint *additives*, most soaps used are made from stearin.

Stearin is a waxy material produced from different sources, one of the main being tallow, which is a cheap by-product from the processing of beef and other types of meat. Tallow contains a mixture of glycerides with a typical composition for, e.g., beef: saturated fatty acids, palmitic acid (C16:0), 24–33%; stearic acid (C18:0), 14–29%; myristic acid (C14:0), 2–8%; monounsaturated fatty acids, oleic acid, 39–50%; and polyunsaturated fatty acids, linoleic acid, 1–5% (Schumann and Siekmann 2000).

Stearin is separated from the low-temperature melting tallow which is high in oleic acid, through (1) "dry fractionation" by pressing the fatty mixture, leading to separation of the higher melting stearin-rich material from the liquid or (2) hydrogenation with nickel catalysts (US patent US4158666 A 1979). Stearin is a hard, waxy triglyceride containing c. 60% stearic acid and c. 35% palmitic acid.[4]

[2] Alun Foster and Ian Garrett, personal communications 2005–2015, reconfirmed by Ian Garrett March 2017.

[3] M. Harding, personal communication 2007.

[4] Van den Berg and Schilling (2008) unpublished results.

Technical-grade aluminum stearate consists mostly of distearates ($Al(OH)St_2$) in mixture with mono- and tristearates. Aluminum tristearate cannot or is very difficult to be produced pure (Alexander 1954). Technical-grade stearates contain up to 20% free fatty acids. They have a very low palmitate/stearate (P/S) ratio, which can be used to distinguish the stearates from other fatty components in the oil paints, using mass spectrometry (see below). The free fatty acids add acidity to the paint and thus enhance reactivity to certain alkaline pigments such as zinc white (Osmond 2014).

19.3 Experimental Setup

19.3.1 Experimental Paint Samples

A number of oil paint reconstructions were employed in this study. The composition of these paints has been described elsewhere (Mills et al. 2008; Tempest et al. 2013). At the time of analysis, paints were cured and aged for 3–5 years, respectively, under natural conditions, whereas some had undergone additional artificial light aging (Table 19.1).

The painting, *Rununculas* (2007), by the contemporary British artist Mia Tarney was made using Winsor & Newton artists' oil paint. Black paint was sampled and analyzed in 2011.

Winsor & Newton (W&N) paints were used from swatches of naturally aged artists' and students' quality oil paints now stored at Tate (Cooper et al. 2014).

Table 19.1 Experimental paints analyzed in this study

Paint	Date of paint-out	Stearates if present	Date paint analyzed	Reference
Ultramarine, linseed oil, and various proportions of zinc stearate (Fig. 19.2)	2008	Zn	2011	Tempest et al. (2013)
Talens Rembrandt Cadmium Yellow (Fig. 19.3)	2006 (light aged eq. 45 years)	Al	2011	Mills et al. (2008)
Michael Harding ultramarine (Fig. 19.3)	2008 (light aged eq. 45 years)	Mg[a]	2011	Tempest et al. (2013)
W&N ultramarine student quality (Fig. 19.3)	1964	–	2011	Izzo et al. (2014)
Talens Rembrandt Cadmium Yellow with added linseed oil (Fig. 19.5)	2006	Al	2016	Mills et al. (2008)

[a]Stearate found in this study – Mg – M. Harding, personal information

19.3.2 Analytical Techniques

19.3.2.1 EGA-MS

Evolved gas analysis mass spectrometry (EGA-MS) was carried out using a Frontier Laboratories Ltd. PY-2020D microfurnace pyrolyzer. The microfurnace was mounted onto the split inlet of an Agilent Technologies 5975C inert MSD/7890A gas chromatograph/mass spectrometer via a heated syringe needle interface. A deactivated transfer line (Frontier Ultra ALLOY DTM EGA tube, 2.5 M × 0.15 mm) connected the GC inlet to the MS detector via a Frontier Vent-Free adaptor. The injector was set to 320 °C. For analysis, a 50 μL stainless steel eco-cup loaded with c. 100 μg sample ground in the cup was placed into the cool upper zone of the microfurnace where it was purged with helium for three minutes prior to heating. Sample vapor was generated by ramping the temperature of the microfurnace from 100 to 700 °C at 20 °C /min, followed by a 5 min isothermal period. The mass spectrometer was scanned from 10–600 amu.

Data were recorded and analyzed using Excalibur software.

19.3.2.2 ESI-MS

Electrospray ionization mass spectrometry (ESI-MS) data were recorded on a Micromass QTOF2 quadrupole time-of-flight hybrid mass spectrometer (Manchester, UK), equipped with a nano probe and ESI source.

Paint samples were extracted with ethanol (EtOH) and made into a solution of 10 mM NH_4Ac in EtOH after homogenization and centrifuging. Samples were delivered to the MS system with a Micromass CapLC system, comprised of a ternary pump and autosampler. 1 μL sample was injected with a flow of 0.20 μL/min 10 mM NH_4Ac in EtOH. The effluent of the CapLC was delivered to the ion source with a picotip. Operating conditions: desolvation gas, nitrogen, 150 °C, 2 L/min; nebulizer gas, nitrogen, 1.5 L/min; cone gas, nitrogen, 2 L/min; and collision gas, argon. The source temperature was set at 80 °C, cone voltage 30 V, capillary voltage 3.0 kV, collision energy 10 V. The mass axis was calibrated using phosphoric acid.

Data were recorded and analyzed using Masslynx software.

19.4 Results and Discussion

19.4.1 Detection of Metal Stearates in Artists' Oil Paints

Characterization of metal soaps is straightforwardly done by Fourier transform infrared analysis (FTIR) through their absorptions around 1390–1610 cm^{-1}. However, low concentrations of stearates generally used in paints are often below the detection limit of c. 5%. Nevertheless, FTIR microscopes may be able to detect pockets of metal soaps in protrusions or undispersed metal stearates (Mills et al. 2008).

Both added stearates and soaps formed by reaction between metal ions from inorganic pigments and fatty acids from the medium and other sources are detected which may make it difficult to distinguish between the two types. This holds especially for zinc soaps, which are formed from fatty acids from, e.g., aluminum stearates (Osmond 2014) or from the hydrolyzed oil medium and zinc white pigment. MS techniques have been used to explore the difference in P/S ratio of added stearates and the lipidic binders.

Analysis of metal stearates from binding media of a variety of historical commercially produced oil paint samples was undertaken by Izzo et al. 2014 by extraction of samples of oil paints using a solvent mixture of chloroform/methanol (1/4 v/v), leaving the metal soaps in the residue. The presence of stearates in the formulation was indicated by a significantly lower P/S ratio in the residue than in the extract of the oil paint. With gas chromatography-mass spectrometry (GC-MS), technical stearates were shown to consist of c. 35% palmitic and 60% stearic acids, as well as traces of, e.g., oleic and myristic acids. The resulting P/S ratio of 0.6 is much lower than any drying oil which is, despite exotic rare exceptions, always higher than 1.0 (Mills and White 1994).[5] Using classical preparation methods for GC-MS, all the fatty acids that derive from oil binders and added stearates are analyzed simultaneously, which results in a ratio that includes stearates from both sources. Since oil binders contain low amounts of palmitic and stearic acids (in linseed oil this is less than 10%), metal stearates add high relative amounts of P and S to the paint. In a typical paint with c. 60% binding medium and 4% aluminum stearates, as much as 40% of the P/S ratio will be made up from the additive.

These are the reasons why assignments of binders from artists' oil paints from after 1920 based on data from general GC-MS procedures should be reviewed.

19.4.2 EGA-MS of Artists' Oil Paints

Evolved gas analysis mass spectrometry (EGA-MS) is a direct mass spectrometry technique which gives temperature-resolved information of evolving gases, in a manner similar to the more established direct temperature-resolved MS (DTMS) technique (Van den Berg 2002). With EGA-MS, samples are heated from room temperature to 700 °C. In Fig. 19.2a, EGA analyses of naturally and artificially aged paints (5 and 45 years under fluorescent light conditions, respectively) containing ultramarine, linseed oil, and varying amounts of zinc stearate are presented (Tempest et al. 2013). In the sample without added zinc stearate, only a broad peak in the high temperature (pyrolysis) region III (c. 400–500 °C) is visible, showing the

[5]See e.g. http://www.chempro.in/fattyacid.htm

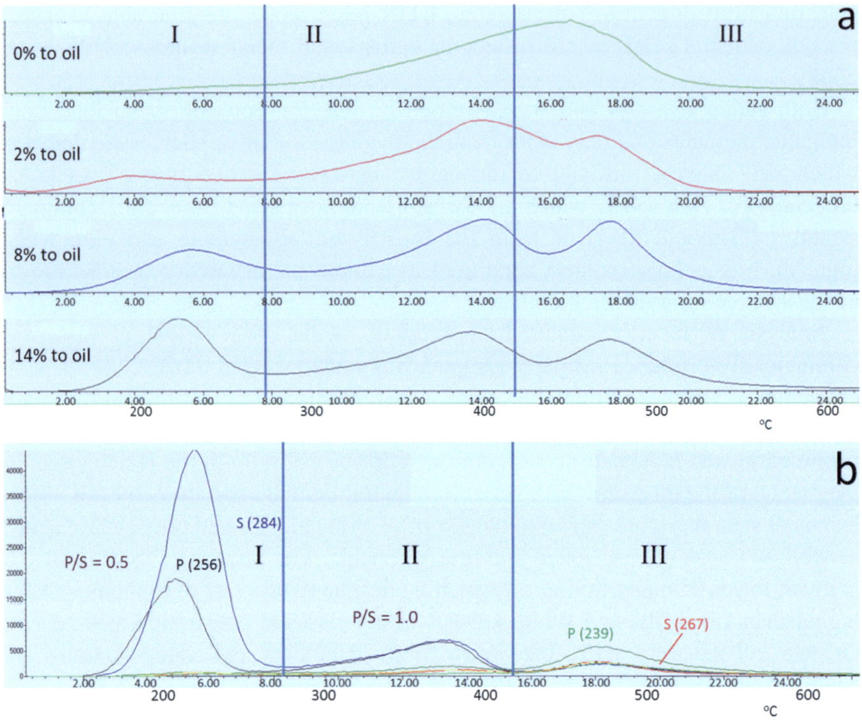

Fig. 19.2 Evolved gas analysis MS of test paints containing ultramarine, linseed oil, and various proportions of zinc stearate: (**a**) total ion currents, (**b**) single ion chromatograms of four ions related to palmitic acid (m/z 256 and 239), and stearic acid (m/z 284 and 267). Paints were analyzed after 5 years of aging and artificial light aging equivalent to 45 years under museum conditions. I: free fatty acids; II fatty acids formed from metal stearates, oligomers of oil; III fatty acid moieties from oil network

presence of negligible amounts of free fatty acids or oligomers, evolving at lower temperatures in regions I and II, respectively. This indicates that the hydrolysis of the oil binder has been negligible. Only when zinc stearate is present in the paint are free fatty acids and metal soaps detected in the temperature regions I (c. 150–270 °C) and II (c. 300–400 °C), respectively. The signal in the lower temperature region I is almost exclusively related to molecular ions of palmitic and stearic acids and their fragment ions. At pyrolysis temperatures, the spectrum is dominated by a broad range of pyrolysis products related to the oil network.[6]

In Fig. 19.2b, single ion chromatograms show molecular ions for palmitic and stearic acids at m/z 256 and 284, whereas at higher temperatures fragment ions dominate at m/z 239 and 267, respectively. These fragment ions are formed from

[6]Van den Berg and Schilling, manuscript in preparation.

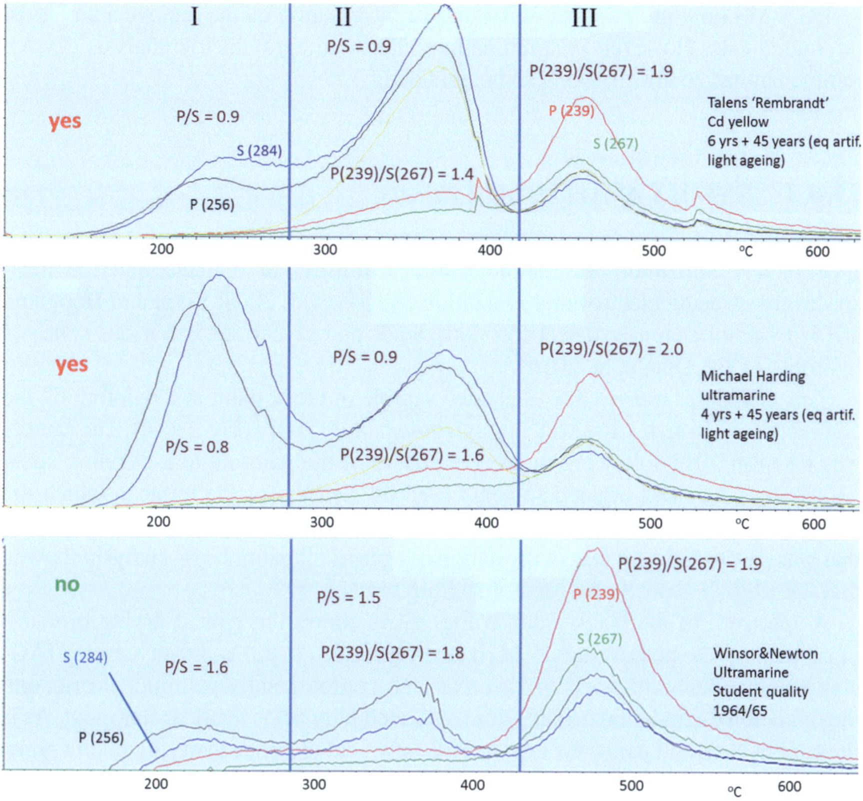

Fig. 19.3 EGA-MS of three naturally aged commercial paints, showing three temperature regions: I evaporation of low molecular weight fractions (e.g., free fatty acids), II low molecular weight molecules formed from metal soaps, and III products from pyrolysis. In the relatively young Talens and Michael Harding paints, indication of metal stearates is found; in Winsor & Newton paint none are detected

the palmitic and stearic acid containing moieties from soaps and the binder network. The ions show a variation in P/S ratios from 0.5 in region I to 1.0 in region II and 2.0 in region III. This variation is related to the free acids in the added metal stearates and the higher P/S ratio in the oil medium.

Analysis of cured commercial paints of different ages with EGA-MS (Fig. 19.3) shows the detection of low amounts of metal stearates in the Talens and Michael Harding paints; again, in the low temperature region, the P/S ratios are significantly lower than in the pyrolysis region, where palmitic and stearic acids are only formed from the oil binder.

The Winsor & Newton paint does not show any indication of the presence of metal stearates since in all temperature regions the P/S ratios are more or less similar.

EGA-MS only provides semiquantitative information on the relative amounts of the ingredients. However, in combination with thermogravimetric analysis (TGA), more quantitative information can be obtained.[7]

19.4.3 ESI-MS of Artists' Oil Paints

Electrospray ionization MS employs direct infusion of extracts into the mass spectrometer using electrospray ionization (Keune et al. 2008; Boon and Hoogland 2014) to simultaneously detect free fatty acids and glycerides in various states of hydrolysis (Van Dam et al. 2016).[8]

This technique was used to analyze a sample of black paint in a painting by the contemporary British artist Mia Tarney, *Rununculas* (2007) (Fig. 19.4a). The sample was taken in 2011 following indirect exposure of the painting to a cleaning agent containing water and organic solvents used by cleaners of the room in which the work was displayed. The acute exposure led to the development of efflorescence that was particularly evident in the dark passages of oil paint. FTIR analysis showed that the whitish deposit consisted of sodium soaps.[9]

A spectrum of the black paint in Fig. 19.4b shows the typical drying products of oil paint in the positive mode, such as diglyceride (DAG) and triglyceride (TAG) compounds which are esters of glycerol with predominantly palmitic, stearic, and the oxidation degradation product of unsaturated fatty acids in oil, azelaic acid. As is the case in cured oil paint, the unsaturated fatty acid moieties in drying oil are gone. The molecular (ammonium adduct) ions of the glycerides containing one saturated fatty acid are used to calculate the P/S ratio of the oil binder, e.g., from m/z 518/546 and m/z 688/716. Both show a P/S ratio of 1.5, which is consistent with that of linseed oil. Linseed oil is to date the most common drying oil employed for artists' oil paints other than in white, blue, and several light colors where less yellowing oils such as safflower or sunflower are preferred.

In the negative mode, a P/S ratio of 0.5 is detected for the free fatty acids ($M\text{-}H^-$ ions) which indicates that these free acids originate from a metal stearate source.

Since the glycerides only derive from the drying oil (and possible other lipid components), and the free fatty acids are mostly originating from stearates, the P/S ratio of both the negative and positive ion modes can thus distinguish oil binder from metal stearates in young paints.

[7] Van den Berg and Schilling, manuscript in preparation.

[8] K.J. van den Berg, A. van den Doel, J. Jansen (2016), An Investigation of twentieth-century oil paint chemistry aided by ASCA chemometric analysis. Unpublished manuscript.

[9] K.J. van den Berg and S. de Groot, Investigation of overnight efflorescence development on Mia Tarney's Rununculas (2007). Internal RCE report 2011–013, April 2011.

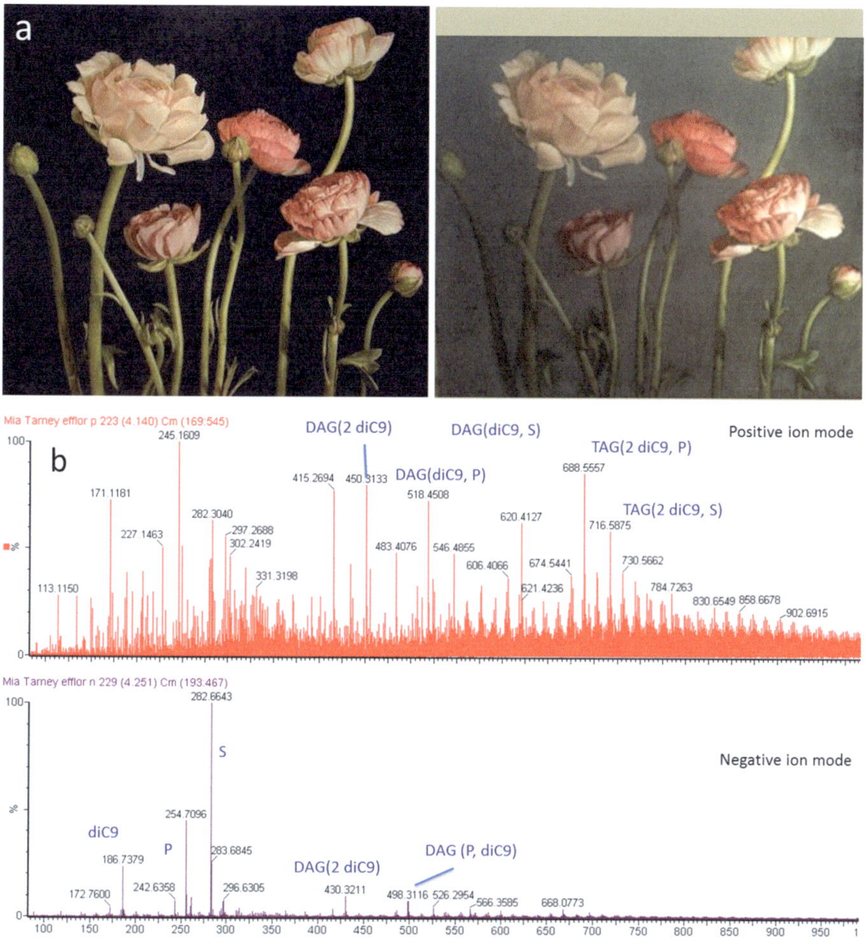

Fig. 19.4 (**a**) Mia Tarney, *Rununculas*, d.2007, before and after the event. (Photographs: Ms. Melanie Caldwell). (**b**) ESI-MS of black paint showing the presence of free fatty acids from technical metal stearates in a linseed oil matrix

19.4.4 Hydrolysis of Metal Stearates Due to Aging

In a series of EGA-MS analyses carried out at The Getty Conservation Institute and The Cultural Heritage Agency of the Netherlands on naturally aged Winsor & Newton paint swatches dating from the 1950s to the 1990s (Cooper et al. 2014), P/S ratios of free fatty acids and metal soaps of higher than 1.0 were consistently found, and differences with P/S ratios of the polymer network were small.[10] Since the use

[10]Van den Berg and Schilling (2010), unpublished results.

of aluminum and magnesium stearates was quite common in Winsor & Newton paints (vide supra), this indicates that extensive hydrolysis of the paints has taken place in the course of time. As a result, the reservoir of free fatty acids from added stearates has been supplemented with fatty acids released from the oil binder, giving a mixed P/S ratio.

To substantiate this, a series of semiquantitative experiments was carried out to study the development of the free acid content in 10-year-old naturally aged paints. Paints were exposed for up to 9 weeks to high (75%) or low (6%) relative humidity, at 60 °C. [11]

Figure 19.5 shows the results of ESI-MS analysis in the negative ion mode of a sample of Talens cadmium yellow paint, to which 2% linseed oil was added. The paint was analyzed previously and contains aluminum stearate as was detected with FTIR and SEM-EDX (Mills et al. 2008). The results of ESI-MS show the development in P/S ratios within a few weeks of artificial aging. The figure shows a steady decrease of the P/S ratio for both relative humidities. The decrease in itself can be a reflection of a relative depletion of palmitic acid under the elevated temperature due to its relative low vapor pressure (Schilling et al. 1999). However

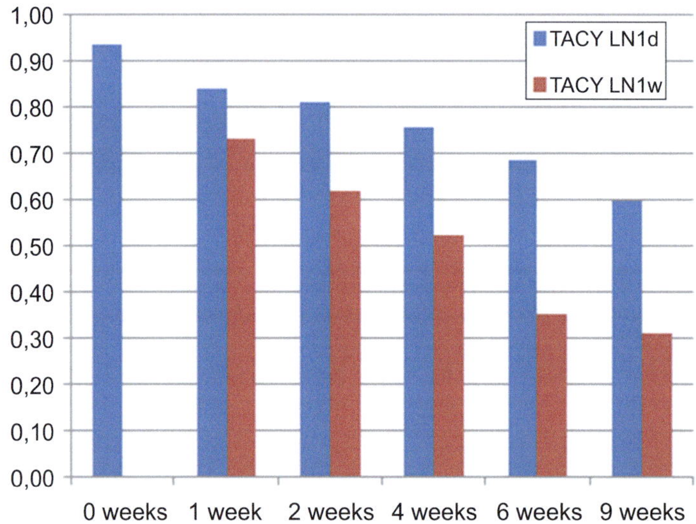

Fig. 19.5 ESI-MS analysis of a 10-year-old naturally aged Talens Rembrandt Cadmium yellow oil paint, with 2% added raw linseed oil. The paint is additionally aged up to 9 weeks in 60 °C under 6% (blue bars) and 75% (red bars) relative humidity. The graph shows P/S ratios taken in the negative ion mode from the palmitic and stearic acid ions at m/z 255 and 283, respectively. (Figure courtesy of Fabiana di Gianvincenzo)

[11]Results of an extensive analytical study on the ageing of a number of oils paints aged under different relative humidities are forthcoming (Di Gianvincenzo et al. manuscript in preparation).

the relative humidity is clearly responsible for the faster decrease of the P/S ratio, indicating hydrolysis of the aluminum stearate present in the paint, enriching the paint with free stearic acid.

19.5 Conclusion

This study describes aspects of the development and history of use of metal stearates in artists' oil paints in the twentieth century. The stearates are a waxy material manufactured from stearin, which has a composition that is enriched in stearates compared to the oil binders generally used in oil paint. This feature, and the resulting low P/S ratio, makes it possible to detect low quantities of stearates in oil paints.

EGA-MS and ESI-MS are presented for the detection of metal stearates in artists' oil paints. These techniques provide a fast and effective alternative to the current methods that necessitate time- consuming extraction and/or derivatization procedures employing GC-MS.

The study of black paint in a modern painting showed that technical-grade stearates used by manufacturers can cause efflorescence early in its lifetime. It was speculated that free fatty acids present in the stearates may be responsible, supporting a previous study (Tempest et al. 2013). However, the present study also supports evidence that aluminum stearates hydrolyze in the course of time and that this is promoted by atmospheric water in and exposure to high relative humidity. The question of whether other metal stearates show the same instability as aluminum stearate remains unclear and will be addressed in future studies. Indications are, however, that as relatively low quantities of metal stearates contribute to relatively high amounts of saturated fatty acids, much of the free fatty acids in oil paintings will derive from metal stearates, and therefore these additives are largely responsible for fatty acid efflorescence in twentieth-century oil paintings.

Acknowledgments Jaap Boon, Katrien Keune, Frank Hoogland; Thomas Learner, Alan Phenix; Polly Saltmarsh; Bronwyn Ormsby, Judith Lee; André van den Doel, Eliane van Dam; Suzan de Groot; Lindsey Doyle; Fabiana di Gianvincenzo, Federica Parlanti. Ian Garrett (formerly W&N), Bert Klein Ovink (formerly Talens), Wim v.d. Zwan (Old Holland). Gillian Osmond for editorial comments.

References

Alexander AE (1954) The chemical nature and physical properties of aluminium soaps. JOCCA 37(7):378–382

Bayliss S, van den Berg KJ, Burnstock A et al (2016) An investigation into the separation and migration of oil in paintings by Erik Oldenhof. Microchem J 124:974–982

Boon JJ, Hoogland FG (2014) Investigating fluidizing dripping pink commercial paint on van Hemert's seven-series works from 1990–1995. In: Van den Berg KJ et al (eds) Issues in contemporary oil paint. Springer, Switzerland, pp 227–246

Carlyle L (2001) The Artist's assistant. Archetype Publications Ltd, London

Cooper A, Burnstock A, van den Berg KJ et al (2014) Water sensitive oil paints in the twentieth century: a study of the distribution of water-soluble degradation products in modern oil paint films. In: Van den Berg KJ et al (eds) Issues in contemporary oil paint. Springer, Switzerland, pp 295–310

Gettens RJ, Stout GL (1966) Painting materials: a short Encyclopaedia. Dover Publications, Inc, New York

Heaton N (1947) Outlines of paint technology. Charles Griffin & Company Ltd, London

Izzo FC, van den Berg KJ, van Keulen H et al (2014) Modern oil paints – formulations, organic additives and degradation: some case studies. In: Van den Berg KJ et al (eds) Issues in contemporary oil paint. Springer, Switzerland, pp 75–104

Keune K, Hoogland F, Peggie D et al (2008) Comparative study of the effect of traditional pigments on artificially aged oil paint systems using complementary analytical techniques. In: Bridgland J (ed) Preprints of ICOM–CC 15th triennial meeting, New Delhi, vol II. Allied Publishers Pvt. Ltd, New Delhi, pp 833–842

Mayer R (1970) The artist's handbook of materials and techniques, vol 12–16. Viking Press, New York, pp 125–222

Mills L, Burnstock A, Duarte F et al (2008) Water sensitivity of modern artists' oil paints. In: Bridgland J (ed) Preprints of ICOM–CC 15th triennial meeting, New Delhi, vol II. Allied Publishers Pvt. Ltd, New Delhi, pp 651–659

Mills JS, White R (1994) Oils and fats, Chapter 3. In: The organic chemistry of museum objects, 2nd edn. Butterworth-Heinemann, Oxford

Osmond G (2014) Zinc white and the influence of paint composition for stability in oil based media. In: Van den Berg KJ et al (eds) Issues in contemporary oil paint. Springer, Switzerland, pp 263–282

Phenix A et al (2017) Preprints of ICOM–CC 18th triennial meeting, Copenhagen. 'The might of White'1: formulations of titanium dioxide-based oil paints as evidenced in archives of two artists' colourmen, mid-twentieth century. Accepted for publication

Schilling MR, Carson DM, Khanjian HP (1999) Gas chromatographic determination of the fatty acid and glycerol content of lipids. IV. Evaporation of fatty acids and the formation of ghost images by framed oil paintings. In: Bridgland J (ed) Preprints of ICOM–CC 12th triennial meeting, Lyon, vol I. James & James, London, pp 242–247

Schumann K, Siekmann K (2000) Soaps. In: Ullmann's encyclopedia of industrial chemistry. https://doi.org/10.1002/14356007.a24_247. p. 243

Tempest H, Burnstock A, Saltmarsh P et al (2013) Sensitivity of oil paint surfaces to aqueous and other solvents. In: Mecklenburg et al (eds) New insights into the cleaning of paintings: proceedings from the cleaning 2010 international conference Universidad Politécnica de Valencia and Museum Conservation Institute. Smithsonian Institution Scholarly Press, Washington, DC, pp 107–117

Tumosa CS (2001) A brief history of aluminum stearate as a component of paint. WAAC Newsletter, Vol 23(No 3). http://cool.conservation-us.org/waac/wn/wn23/wn23-3/wn23-304.html

Turner JHW (1958) The function of aluminium complexes as structure-modifiers in paint. JOCCA 41(11):769–781

Van Dam EP, van den Beg KJ, Proaño Gaibor AN et al (2016) Analysis of triglyceride degradation products in drying oils and oil paints using LC-ESI-MS. Int J Mass Spectrometry. https://doi.org/10.1016/j.ijms.2016.09.004

Van den Berg JDJ (2002) Analytical chemical studies on traditional linseed oil paints. PhD thesis, University of Amsterdam, Chapter 4

Van den Berg KJ, van Gurp F, Bayliss S, Burnstock A, Klein Ovink B (2016) Making paint in the 20th century: the Talens archive. In: Eyb Green S et al (eds) Sources on art technology; back to basics. Proceedings of the 6th symposium of ATSR, 16 and 17 June 2014, Archetype, pp 43–50

Weiss J (1957) Organic aluminium compounds in drying oils. JOCCA 40(11):863–879; 976–981

Chapter 20
Delamination Due to Zinc Soap Formation in an Oil Painting by Piet Mondrian (1872–1944)

Conservation Issues and Possible Implications for Treatment

Laura E. Raven, Madeleine Bisschoff, Margje Leeuwestein, Muriel Geldof, Joen J. Hermans, Maartje Stols-Witlox, and Katrien Keune

Abstract The privately owned oil painting *Composition with Color Planes 4* (1917) by Piet Mondrian (1872–1944) has been the subject of ongoing investigation

L. E. Raven (✉)
Conservation Department, Rijksmuseum, Amsterdam, The Netherlands

Conservation & Restoration, University of Amsterdam, Amsterdam, The Netherlands
e-mail: L.Raven@Rijksmuseum.nl

M. Bisschoff
Independent paintings conservator, Amsterdam, The Netherlands
e-mail: madeleinebisschoff@gmail.com

M. Leeuwestein
Kröller-Müller Museum, Otterlo, The Netherlands
e-mail: MargjeLeeuwestein@Krollermuller.nl

M. Geldof
Cultural Heritage Agency of the Netherlands (RCE), Amsterdam, The Netherlands
e-mail: M.Geldof@CultureelErfgoed.nl

J. J. Hermans
Van't Hoff Institute for Molecular Sciences, University of Amsterdam, Amsterdam, The Netherlands
e-mail: j.j.hermans@uva.nl

M. Stols-Witlox
Conservation & Restoration, University of Amsterdam, Amsterdam, The Netherlands
e-mail: M.J.N.Stols-Witlox@uva.nl

K. Keune
Conservation Department, Rijksmuseum, Van't Hoff Institute for Molecular Sciences, University of Amsterdam, Amsterdam, The Netherlands
e-mail: K.Keune@rijksmuseum.nl

F. Casadio et al. (eds.), *Metal Soaps in Art*, Cultural Heritage Science,
https://doi.org/10.1007/978-3-319-90617-1_20

since 2011. The painting consists of color planes in a field of differing whites. Some of these white areas suffer from delamination issues, in combination with flaking. Previous research demonstrated a link between the presence of zinc oxide and the delamination phenomena. More recently, the formation of zinc soaps was found to play a role. In this study, cross sections from both delaminating and relatively intact white areas were investigated with light microscopy, SEM-EDX, and ATR-FTIR imaging to obtain more information about the stratigraphy and condition of the paint layers. Two stages in metal soap formation were identified in the delaminating areas. The first stage consists of noncrystalline zinc soaps or zinc ions bound to carboxylate functional groups in the polymerized oil network. Crystalline zinc soaps, which represent the second, final stage of metal soap formation, are generally linked to the development of zinc soap related deterioration phenomena. In this case, they were found at the interface between the delaminating paint layers. Possible implications for treatment and factors that might trigger further delamination will be discussed.

Keywords Zinc soaps · Crystalline · Non-crystalline · Delamination · Flaking · Piet Mondrian · SEM-EDX · ATR-FTIR · Treatment · Consolidant

20.1 Introduction

In 1917, Piet Mondrian (1872–1944) painted *Composition with Color Planes 4* (Fig. 20.1). The abstract composition consists of floating colored planes against a background of white planes in different hues. It forms part of a series of five works with a landscape format, of which four are painted on canvas (Joosten 1998).[1] All works in this series show similar compositions, but they vary slightly in color and in the arrangement of the planes. Mondrian worked on them in the same period, and all paintings are signed and dated 1917.

Composition with Color Planes 4 belongs to a private collection. It is the only canvas painting within the series that for most of its lifetime has not been kept in a museum environment. Also unique to this painting is that it is the only painting that has neither been lined nor impregnated with wax or wax-resin, but it is also the only painting in this series that is currently showing delamination problems. Two of the three shades of white used in the final stage of the composition show severe cracking and flaking. We cannot exclude the possibility that the same alterations in

[1] Private collection, *Composition with Color Planes 1*, 1917 (gouache on paper, 48 × 60 cm); Museum Boijmans Van Beuningen, *Composition with Colour Planes 2*, 1917 (oil on canvas, 48 × 61,5 cm, accession no: 1543 MK); Gemeentemuseum Den Haag, *Compositie No. 3, with Color Planes 3*, 1917 (oil on canvas, 48 × 61 cm, accession no: 0332897); Museum of Modern Art New York (MoMa), *Compositie No.5, with Color Planes 5*, 1917 (oil on canvas, 49 × 61,2 cm, accession no.: 1774.1967)

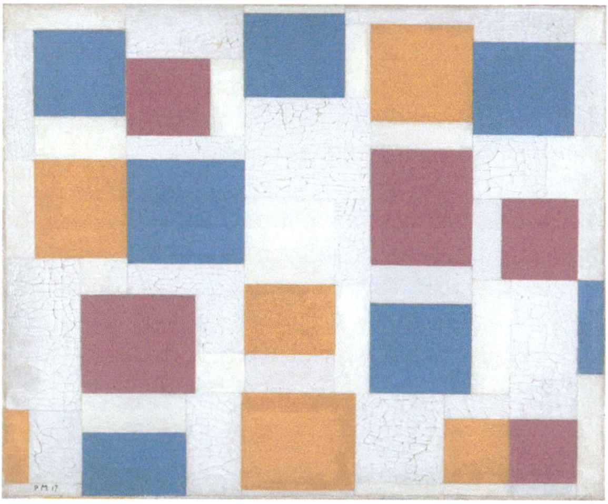

Fig. 20.1 Piet Mondrian. *Composition with Color Planes 4*, 1917. Oil on canvas. 48 × 61 cm. Private collection. (Image courtesy of the owners and the Kröller-Müller Museum)

the paint are latent in the other paintings in the series, as they were created in the same period and possibly with the same materials.

Between 2011 and 2013, the condition of the painting led to in-depth research at both the Kröller-Müller Museum and the Cultural Heritage Agency of the Netherlands (RCE). At the time, it was suggested that drying cracks, in combination with loss of cohesion in an underlying zinc oxide containing paint layer had caused the paint to delaminate (Geldof et al. 2013). Condition issues relating to the use of zinc oxide are not uncommon (Osmond 2012; Mecklenburg et al. 2013); they have been noted in paintings by abstract expressionists such as Hans Hofmann, Franz Kline, and Jackson Pollock (Rogala et al. 2009, 2010, 2016; Maines et al. 2011).

In the case of the Mondrian painting, consolidation of the paint has proven to be very complicated, and it is the subject of ongoing discussion to this day. Softening and re-adhering the brittle paint has turned out to be difficult without risking further cracking or breakage. The paint film also seems to have slightly expanded (Bisschoff 2013). All this points to other factors being involved that appear to affect the painting's condition. The recently developed hypothesis that zinc soaps could play a role in the formation of the degradation phenomena in this painting gave a new impulse to this research.

In the past few years, the formation of zinc soaps has been receiving more attention (Osmond 2012, 2014a, b; Osmond et al. 2012, 2014; Hermans et al. 2014, 2015, 2016a, b, 2019). The relation between zinc soaps and structural instability of paint layers has been reported in another study on a painting by Mondrian (Van Loon et al. 2019) and, among others, in studies on paintings by Joan Miró, Jean Paul Riopelle, Jean McEwen, and Franz Kline (O'Donoghue et al. 2006; Maor and Murray 2008; Corbeil et al. 2011; Ebert et al. 2011; Helwig et al. 2014; Rogge and Véliz-Bomford 2015). However, the role of zinc soaps in the delamination of paint is still quite underexposed.

Little is known about how paint layers affected by zinc soaps should be treated. Brittleness of the paint is a common problem encountered during consolidation treatments. The changing properties of saponified paint layers, however, pose new challenges to the conservator with regard to the effectiveness and long-term stability of consolidation measures. As stated, many twentieth century paintings already show problems related to zinc soap formation. As zinc oxide is still a common component of contemporary oil paints, a better understanding is needed to establish suitable active and preventive conservation methods.

The present study aims to investigate whether zinc soaps play a role in the delamination of the paint layers in two of the three white-colored shades used in *Composition with Color Planes 4* by Piet Mondrian. The results will be used as a basis to critically evaluate the suitability of current consolidation methods and to define parameters which conservators can take into account when dealing with similar delamination issues.

20.2 Experimental

To establish why delamination is occurring in some areas in the painting, it is important to understand how samples from affected areas differ from those taken in intact areas, and how this difference corresponds to what we see on the paint surface. Handheld X-ray fluorescence (XRF) analysis and the characteristic fluorescence of the paint surface in UV light already pointed toward a possible relation between the use of zinc oxide and the paint delamination in some of the white planes (Geldof et al. 2013). To confirm and understand the role of zinc oxide in the delaminating areas, three previously taken cross sections were reexamined: two samples (#3 and #4) taken from two different light-colored planes that show delamination and one sample taken from an intact white plane (#7).

To obtain a better understanding of the painting and its condition, visual and microscopic examinations were carried out. The paint layer buildup was investigated with optical light microscopy in dark field (DF) and ultraviolet fluorescence (UV). Scanning electron microscopy was used in combination with energy-dispersive X-ray spectrometry (SEM-EDX, both spot analysis and elemental mapping) to understand the morphology and stratigraphy of the samples and to determine the spatial distribution of the inorganic materials present. Attenuated total reflection Fourier transform infrared spectroscopy (ATR-FTIR) imaging provided more information about the internal conditions of the paint layers and the degree of change.

20.3 Results

20.3.1 Visual and Stereomicroscopic Examination of the Paint Surface

Composition with Color Planes 4 consists of planes in six different colors, of which three are more brightly colored in dark pink, yellow, and blue. In some areas, it can be seen that Mondrian toned down the brighter colors by repainting them, a technique that has also been observed in *Composition No. 3 with Color Planes* (van der Werf 1994). The white planes can be subdivided into groups according to the hue of their final layer, namely, bluish white, greyish white, and pure white.

When viewed in raking light, it becomes apparent that the delamination primarily affects the bluish and greyish white planes. In these areas the paint is cracking, cupping, lifting, and flaking, which has resulted in local paint loss. Two of the relatively intact pure white planes however also show flaking, indicating that the issue probably relates to the underlayers and not (solely) to the top layers.

Closer visual and stereomicroscopic examination of the painting shows that delamination is mainly occurring at the interface between a creamy or yellowy colored underlayer and the white top layer(s). This underlayer is not the first ground layer, but more likely a layer applied by the artist to cover up parts of an earlier composition. Piet Mondrian was known to frequently make changes in his compositions, and this painting is no exception (Blok et al. 2011; van der Werf 1994). Parts of an earlier stage of the composition are hidden by the frame rebate, and more compositional changes can also be observed in the X-radiograph. Consequently the stratigraphy of the paint is complex and may vary considerably from area to area.

20.3.2 Light Microscopy and SEM-EDX

A thin white ground consisting of chalk in a drying oil was used to fill the interstices of the canvas (Geldof et al. 2013). SEM-EDX revealed that this layer was followed by the uniform application of a second ground layer that primarily contains lead. Depending on the area where the sample was taken, a third layer is present, which also comprises mainly lead. In sample #3, this layer is followed by a really thin (ca. 5 μm), densely pigmented lead white layer (Fig. 20.2, #3: layer 3).

In samples #3 and #4, the lead layers are followed by multiple wet-in-wet applied layers, which consist primarily of zinc and lead white (Fig. 20.2, #3, layers 4a–4b; #4, layers 3–4). A small amount of an unidentified yellow pigment was found in the paint of the first layer in sample #4, hence its yellow color (Fig. 20.2, #4: layer 2). Elemental mapping reveals that the relative quantities of lead and zinc in this layer are comparable to that of the white layer above, which indicates a very similar overall chemical composition.

All samples look very similar in DF. In UV radiation however, the samples taken from delaminating planes look quite different from the sample taken in an intact paint area. Samples #3 and #4 show relatively more bright zinc oxide fluorescence than sample #7 (Fig. 20.2). This indicates a higher zinc content, which is confirmed by elemental mappings. In sample #7 the zinc content is lower than, for example,

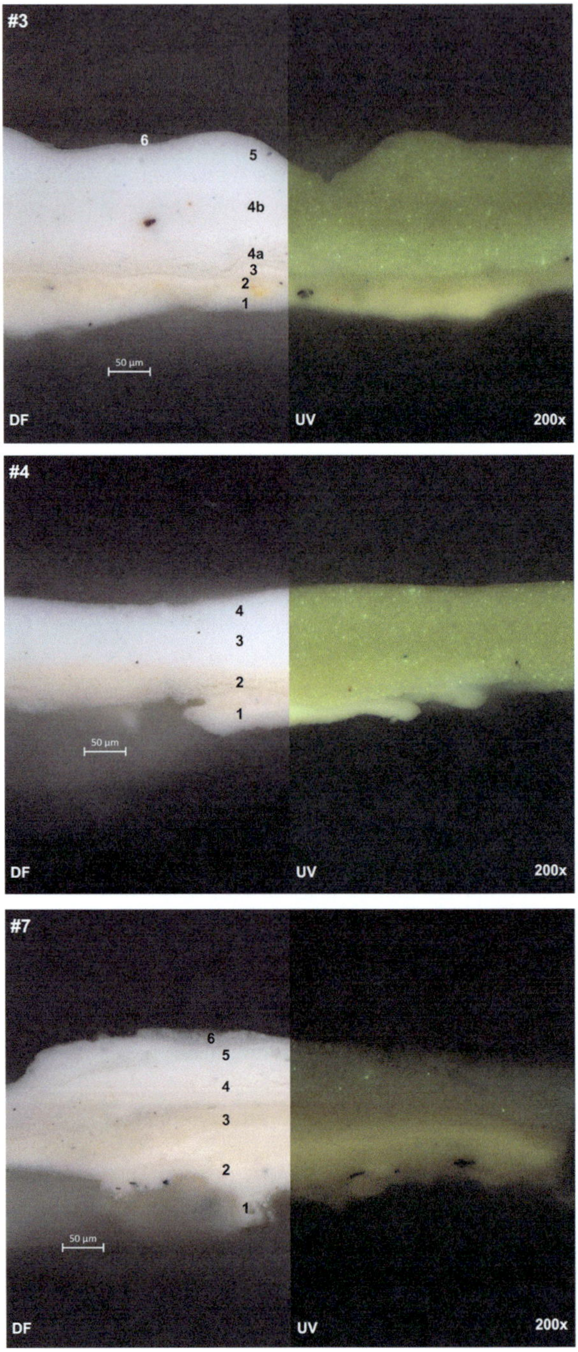

Fig. 20.2 Dark-field (DF, left) and ultraviolet (UV, 365 nm, right) light microscopy images of the cross sections. Magnification 200x. The main constituents were determined with SEM-EDX. Top: sample #3 (location, greyish white plane #18). Constituents: 1, lead white; 2, lead white, carbon black, and an unidentified yellow pigment; 3, lead white; 4a, zinc white and lead white. Part of layer 4b but is showing increased transparency; 4b, zinc white and lead; 5, lead white, zinc white, chalk, and ultramarine blue; 6, MS2A varnish layer. Center: sample #4 (location, bluish white plane #34). Constituents: 1, lead white; 2, zinc white, lead white, ultramarine blue, and an unidentified yellow pigment; 3, zinc white and lead white; 4, lead white, zinc white, chalk, and ultramarine blue; 5, MS2A varnish layer. Bottom: sample #7 (location, white plane #51). Constituents: 1, chalk; 2, lead white; 3, lead white and an unidentified yellow pigment; 4, lead white and zinc white; 5, lead white and zinc white; 6, lead white, zinc white, and ultramarine blue

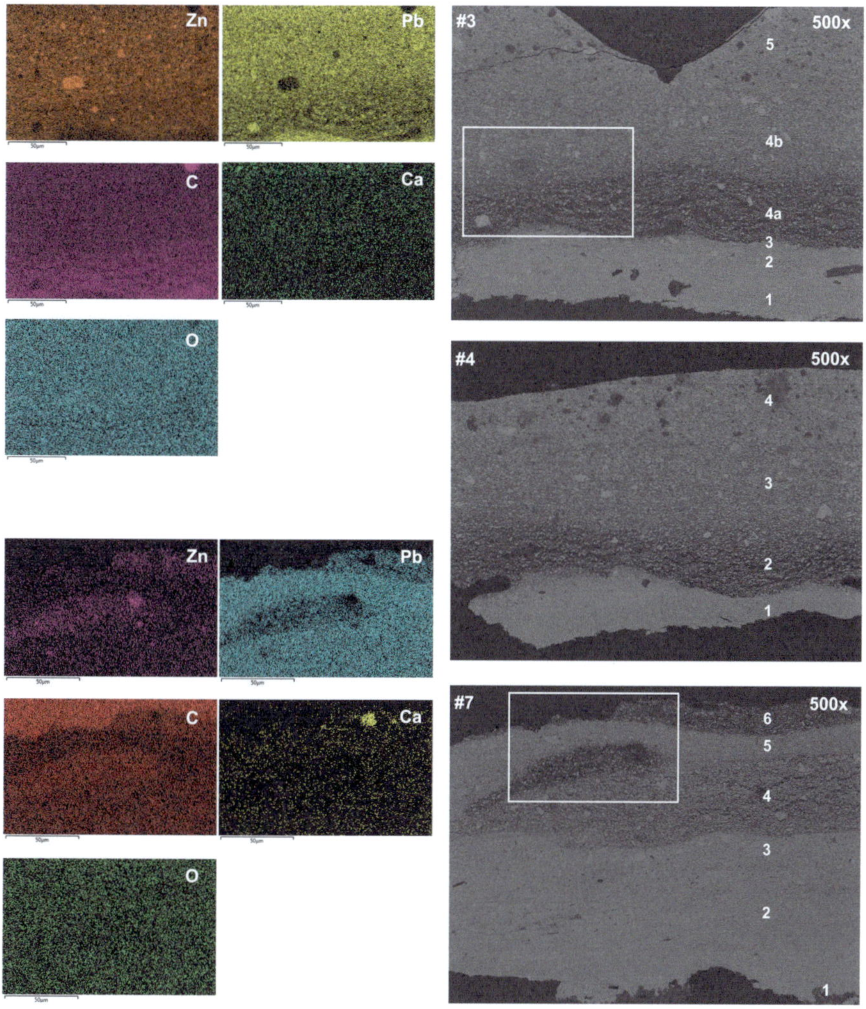

Fig. 20.3 SEM-EDX elemental mappings of cross sections #3 and #7 (left): zinc (Zn), lead (Pb), carbon (C), calcium (Ca), and oxygen (O). SEM-BSE images (right) of cross sections #3, #4, and #7. Magnification 500×

sample #3, whereas the lead content is relatively high (Fig. 20.3). UV also reveals a faint fluorescence relating to an MS2A varnish (Geldof et al. 2013).[2]

When viewed with higher magnification, it becomes apparent that in samples #3 and #4, something is going on at the interface of a lead-rich layer and the first zinc

[2]No documentation is known for the conservation history of this painting. MS2A has been in use as a varnish since 1962 and must therefore have been applied during treatment after this date.

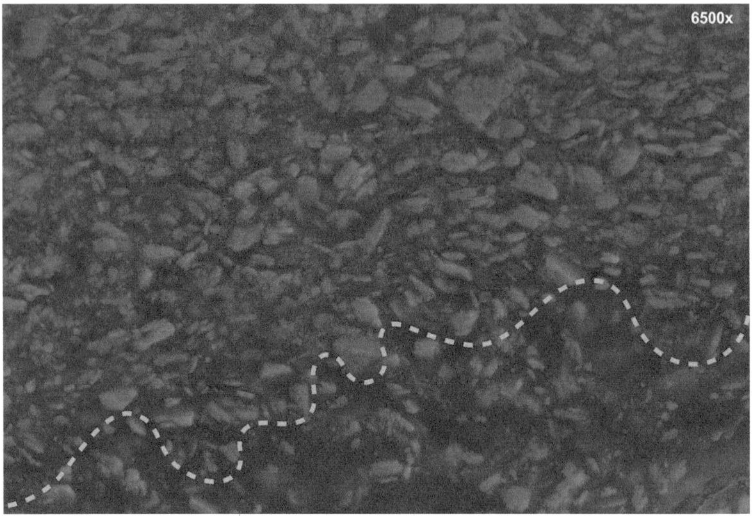

Fig. 20.4 SEM-BSE image of detail layer 4 in sample #3. Magnification 6500×. Above the green line, the zinc oxide particles are still visible in between the lead particles. In the lower area below the green line, where the fissuring takes place, there is hardly any zinc oxide left

oxide-containing layer (Fig. 20.3, #3, interface layers 4a–4b; #4, interface layers 1–2). The backscatter electron (BSE) images of these two samples reveal an area with lower electron density and lamellar fissures along the layer interface (Fig. 20.3). This phenomenon has previously been linked to the presence of zinc soaps, and it has also been noted in paintings by other artists such as Jean McEwen (Helwig et al. 2014). Elemental analysis revealed that this area is carbon rich (Fig. 20.3), which could be the consequence of saponification.

With higher magnification, a compositional change becomes visible in the area where the fissuring occurs (Fig. 20.4). Zinc oxide particles are still present in the upper part of the paint matrix, but there are hardly any detectable in the part where the fissuring is visible. The fact that there is a gradient instead of a clear border, and that there is an even distribution of lead particles in both areas, confirms that a conversion of zinc oxide particles is taking place in the lower part of the paint layer. This phenomenon can also be observed in sample #4. Similar fissuring is visible in the BSE image of sample #7, which was taken from an intact white plane (Fig. 20.3). It is remarkable that in this case, the fissuring is not occurring at the interface, but at the center part of the paint layer.

20.3.3 ATR-FTIR Imaging

Zinc oxide does not produce a spectrum in the mid-infrared range (4000–600 cm^{-1}) and is therefore not directly traceable in FTIR spectra. When mixed in oil, zinc oxide is, however, likely to produce a strong carboxylate absorption (Osmond

2012, 2014a). Recently, the existence of two different types of zinc carboxylate absorptions has been demonstrated experimentally. These have been linked to two different phases of zinc soaps (Hermans et al. 2015, 2016a, 2019).

A single sharp COO^- band around ~1536 cm^{-1} is characteristic for crystalline zinc carboxylates; in previous studies, it has been used to determine the presence of zinc soaps (van der Weerd et al. 2003; Shimadzu et al. 2008; Corbeil et al. 2011; Osmond 2012; Osmond et al. 2014). Noncrystalline zinc soaps however, as well as zinc ions bound to carboxylate moieties on the polymerized oil network (ionomer-like network), can be identified by a broadened asymmetric stretch COO^- band shifted to ~1570–1590 cm^{-1} (Hermans et al. 2016a, b).

Both zinc carboxylate vibration bands were identified in the zinc oxide rich paint layer in the samples from the delaminating white planes. ATR-FTIR mappings show that the crystalline zinc carboxylate band is present at the interface, where the paint layer is delaminating (Fig. 20.5). In samples #3 and #4, its spatial

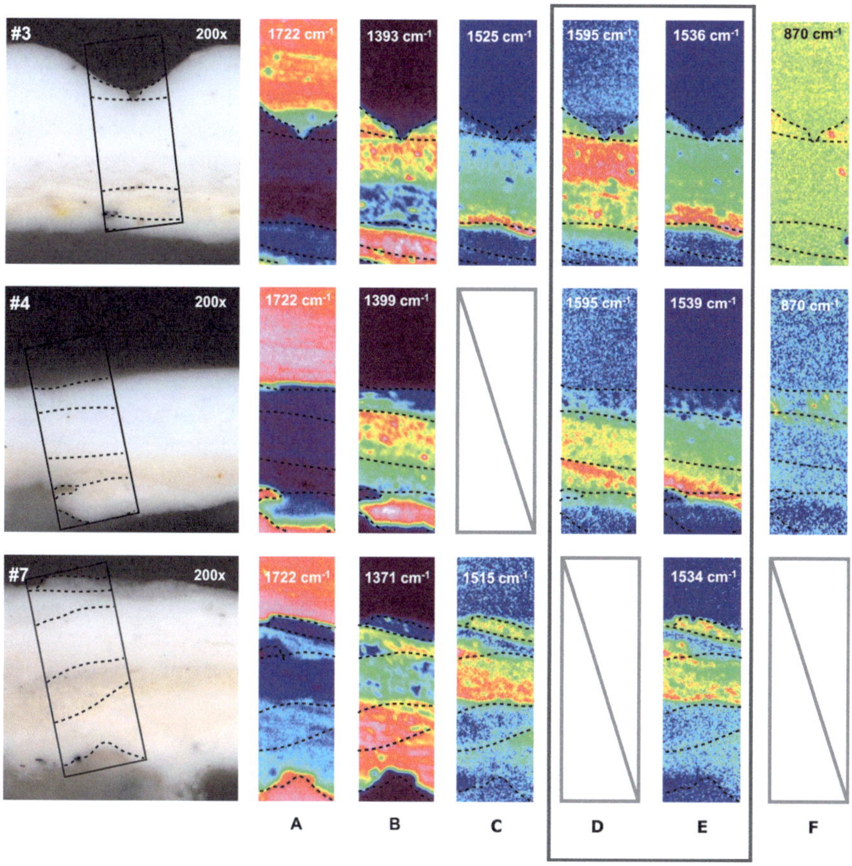

Fig. 20.5 ATR-FTIR imaging mappings of the three different samples: A, embedding medium; B, carbonate; C, lead carboxylate; D, noncrystalline zinc soaps; E, crystalline zinc soaps; F, chalk

distribution corresponds to the areas of lamellar fissuring visible in the BSE images. Interestingly, the upper part of the same layer, which shows the absorption band that is characteristic for noncrystalline zinc carboxylate (amorphous zinc soaps and the ionomer-like network), is better preserved.

20.4 Discussion

20.4.1 The Role of Zinc Soaps in the Delamination of Composition with Color Planes 4

Light microscopy and SEM-EDX examination showed that the samples from the delaminating paint areas, #3 and #4, have a relative high content of zinc white, whereas in sample #7, from the intact white plane, lead white dominates. The difference in pigmentation between the "delaminating" and intact samples, i.e., the zinc white and lead white ratio, plays a crucial role in the degree of zinc soap degradation resulting in delamination, as demonstrated in this case study. It is presumed that the free fatty acids are derived from within the layer and not from the lower lead white-containing layer (Kühn 1986). According to the BSE images, the lead white in the lower layer appears to be intact and unaffected by lead soap degradation. Therefore, we hypothesize that the fatty acids in this compact lower lead white layer are likely to be trapped by the lead and thus not able to migrate into the upper zinc white-containing layer.

The finding of vibration bands for crystalline and for noncrystalline zinc carboxylates within a single paint layer is quite remarkable, and for the first time, this can be linked to delamination issues.[3] The FTIR vibration band for noncrystalline zinc carboxylate is representative for the amorphous zinc soaps and the ionomer-like network. Unfortunately, it is not possible to distinguish between both types with ATR-FTIR.[4] However, it is likely that the ionomer-like network is homogenously distributed throughout the paint layer. As the noncrystalline zinc carboxylate dominates in the upper part of the paint layer, we can expect amorphous zinc soaps to be present.

Amorphous zinc soaps can be considered an intermediate stage and crystalline zinc soaps the final stage in zinc soap formation (Hermans et al. 2015, 2016a, b). A recent study into the crystallization process of metal soaps shows that amorphous, noncrystalline zinc soaps are inherently unstable and that they are able to crystallize spontaneously in oil (Hermans et al. 2016a). However, in a polymerized oil film,

---

[3]The presence of two different types of zinc carboxylates within oil paints has been noted before but has never been linked to delamination. *See* Corbeil et al. (2011) and Szafran et al. (2014).

[4]Surface-sensitive techniques such as secondary ion mass spectrometry would be helpful to answer this question.

their potential to crystallize decreases due to changes in viscosity and chemical composition (Hermans et al. 2015, 2016a).

For the phase transition from noncrystalline to crystalline zinc soap in a polymerized oil network to occur, favorable kinetic conditions are needed that allow structural reordering of the carboxylate groups and aligning of the fatty acid chains. Temperature, water or moisture, solvents, and open spaces in the paint structure, such as (micro)-cracks, are likely to favor the packing of the fatty acid carboxylates (Hermans et al. 2016a).

In the Mondrian painting, it is possible that the formation of noncrystalline zinc soaps first caused the paint to delaminate, as a result of which the lower part of this layer could crystallize. Another possible scenario is that the noncrystalline zinc carboxylates crystallized, which then caused the paint to delaminate. The question at which point delamination occurred and whether the zinc soaps increased in volume during either of these scenarios is uncertain and requires further research.[5]

Composition with Color Planes 4 was kept in an environment with uncontrolled climate conditions for most of its lifetime. The painting has been exposed to strong fluctuations and most likely high and low levels of relative humidity and temperature. The ability of the upper paint layers to respond to climatological changes was in all likelihood diminished by the loss of the unbound noncrystalline zinc soaps, which acted as plasticizers (Van Loon et al. 2012). One can imagine that when the flexible, mobile noncrystalline zinc soaps are converted into rigid and brittle crystallized zinc soaps, the paint film becomes less pliable and thus more prone to failure.

20.4.2 Consequences for Treatment

The existence of two types of zinc soaps and the notion that the degradation process is ongoing mean that standard approaches to consolidation have to be reconsidered, not only for the Mondrian painting but for all paintings that show zinc soap-related issues. Minimal intervention and compatibility of restoration materials to those that are original are two guiding principles for the consolidation of paint layers, especially since most consolidation treatments are irreversible by default. However, paint delamination due to zinc soap formation is most likely a continuous process. In the case of the Mondrian painting, it is feasible that new problems will arise when the noncrystalline zinc soaps in the part of the paint layer that lies above the consolidated area become crystalline. Ideally, a consolidant should be chosen that is capable of impregnating and consolidating both phases of zinc soaps. However, since the problem often concerns an underlayer, local treatment is quite difficult to achieve without impregnating adjacent layers.

---

[5]Volume expansion has been observed in paints where metal soap aggregates have formed. *See* Keune (2005), Keune and Boon (2007), and Noble and Boon (2007).

It could be interesting to compare *Composition with Color Planes 4* with the other three canvas paintings in this series, which have been wax or wax-resin impregnated. Analytical research could provide insight into the condition of the paint structure and into the degree of saponification of the zinc white-containing layers, as well as the influence of previous treatments. This could provide valuable information with regard to the development of new treatment strategies.

20.4.3 Treatment or Risk Management?

We assume that the development of zinc soap related damage is mostly dependent on the formation of crystalline zinc soaps. Much can be achieved by preventing the phase transition from noncrystalline to crystalline zinc soaps. The kinetics of metal soap crystallization are affected by several factors, such as the use of solvents, water, and heat. All have the ability to mobilize the fatty acid chains on the carboxylate groups, which can then reorder in an energetically more favorable arrangement and become crystalline (Hermans et al. 2016a). Awareness of this possibility is crucial, as all mentioned factors are regularly involved in consolidation treatments.

At this moment, to our knowledge, among the range of adhesives and consolidation methods currently available, not a single, suitable material and effective method can be found that fully negates all risk-increasing factors. In light of recent research, the question arises if the currently employed methods are still adequate, if not even potentially harmful. In order to safeguard potential high-risk paintings, it is important for conservators to be aware of the fact that their intervention might contribute to an acceleration of the degradation process.

Since an unstable climate is very likely to speed up the saponification and crystallization process, emphasis should be placed on preventive measures. High humidity levels will have a similar effect as water and allow more noncrystalline zinc soaps to form by releasing more free fatty acids through hydrolysis. It is presumed that the crystallization temperature of zinc soap drops below room temperature as the paint film becomes more polymerized (Hermans et al. 2016a). Chances for crystallization to occur are lower, if the painting is kept in a stable climate. Fluctuations will catalyze the saponification and crystallization process, as high temperatures will mobilize carboxylates, and low temperature levels will enable their crystallization.

This is of particular importance until more insight into the delamination process has been acquired, and innovative consolidation methods and materials have been developed. As previously stated, there is a need for a local impregnation method that enables the conservator to treat the zinc soap affected paint layers locally without having to affect the entire paint structure. In addition, the ideal consolidant should not contain solvents nor water and should work without heat activation.

20.5 Conclusion

With this case study, the presence of both noncrystalline and crystalline zinc soaps has for the first time been linked to serious paint delamination in an oil painting. The delamination occurs in specific paint areas that are rich in zinc oxide, and in the presence of free fatty acids, the highly reactive zinc oxide reacts to zinc soaps. The accumulation of crystalline zinc soaps at the interface where paint layers are detaching is responsible for the brittle nature of the paint. Noncrystalline zinc soaps, which have also been identified in what is chemically considered to be the same paint layer (Fig. 20.2, #3, layers 4a–4b; #4, layers 2–3), represent a first stage in the delamination process. The presence of two different phases of zinc soaps implies that further delamination is theoretically possible. This might compromise the long-term effect of consolidation treatments.

This research has helped to improve the understanding of the delamination processes due to zinc soap formation, which is a first step in finding a way to develop safer conservation strategies for paintings suffering from the described phenomena. A number of questions could be answered during this research, but many others still remain, which hopefully will be addressed by future research. In particular, whether the obtained results are representative for other paintings showing similar problems, and how this type of degradation may be influenced by treatment.

The question why in this case the zinc soaps formed at the interface with an underlayer is also still a subject of debate. In this regard, studies of other paintings with related problems, as well as comparing *Composition with Color Planes 4* to the other works in the same series, could give important clues. These matters are the subject of current research undertaken at the Rijksmuseum and the University of Amsterdam.

Reconstruction-based investigations could be useful to gain more insight into the mechanics behind the saponification process and how this can eventually lead to delamination. This could also provide more detailed information about the potentially harmful effects of agents such as water, heat, and solvents involved in consolidation treatments. This research is still ongoing, and we are optimistic that the knowledge gained can serve as a starting point for the development of safer treatments in the future.

Acknowledgments The authors would like to express their gratitude to the following people and institutions for their contributions to this research: the owners of *Composition with Color Planes 4*, H. van Keulen (RCE), L. Megens (RCE), M. van Bommel (UvA, RCE), N. de Keyser (RMA), R. Hoppe (Gemeentemuseum Den Haag), G. Osmond (Queensland Art Gallery), S. Theobald Clark (Queensland Art Gallery), C. Rogge (The Museum of Fine Arts Houston), H. Janssen (Gemeentemuseum Den Haag), K. J. van den Berg (RCE), M. de Visser (independent paintings conservator), I. Joosten (RCE), L. van Halem (RMA), and V. Blok (independent paintings conservator). This work is part of the PAinT project, supported by the Science4Arts program of the Dutch Organization for Scientific Research (NWO).

A.1 Appendix

A.1.1 Experimental Conditions

A.1.1.1 Embedding

Samples were embedded in a polyester resin (Polypol) and dry polished with SiC polishing cloths (Micro-mesh®, final step 12.000 mesh).

A.1.1.2 Light Microscopy

All paint cross sections were examined under a Zeiss Axioplan 2 microscope, with both incident polarized light and incident UV light (from a xenon lamp and a mercury short arc photo optic lamp HBO, respectively). The UV H365 filter set used for examination in UV consists of the following filters: excitation BP 365/12, beam splitter FT 395, and emission LP 397.

A.1.1.3 SEM-EDX

Scanning electron microscopy in combination with energy-dispersive X-ray analysis (SEM-EDX) studies were performed on a Verion high-vacuum electron microscope (FEI, Eindhoven, Netherlands) with an EDX system with spot analysis and elemental mapping facilities (Oxford). Backscattered electron images of the cross sections were taken at a 20 kV accelerating voltage, at a 5 mm eucentric working distance and with current density of approximately 130 pA. Samples were gold coated (3 nm thickness) in an SC7640 gold sputter coater (Quorum Technologies, Newhaven, East Sussex, UK) prior to SEM-EDX analysis to improve surface conductivity.

A.1.1.4 ATR-FTIR Imaging

FTIR spectral data were collected on a Perkin Elmer Spectrum 100 FTIR spectrometer combined with a Spectrum Spotlight 400 FTIR microscope equipped with a 16 × 1 pixel linear mercury cadmium telluride (MCT) array detector. A Perkin Elmer ATR imaging accessory consisting of a germanium crystal was used for ATR imaging.

References

Bisschoff M. *KM 132.832* 'Compositie met kleurvlakken 4', Piet Mondriaan, testfase 7/10/13-1/11/13. Kröller-Müller Museum, 2013 [unpublished treatment report]

Blok V, Bracht E, Wijnberg L (2011) Mondrian in the Stedelijk museum Amsterdam: research and conservation of five paintings. Zeitschrift für Kunsttechnologie und Konservierung (ZKK) 25(2):187–222

Corbeil M, Helwig K, Poulin J, Riopelle J-P (2011) The Artist's materials. The Getty Conservation Institute, Los Angeles
Ebert B, MacMillan Armstrong S, Singer B, Grimaldi N (2011) Analysis and conservation treatment of Vietnamese paintings. In: Bridgland J (ed) Preprints of the 16th ICOM-CC triennial conference, Lisbon. Critério-Produção Grafica, Lda, Lisbon, p 8
Geldof M, Megens L, van Keulen H, van Bommel M. Mondriaan 'Compositie met Kleurvlakken 4.' Rijksdienst voor Cultureel Erfgoed, project no. 2012-046 (14 Feb 2013) [unpublished research report]
Helwig K, Poulin J, Corbeil M, Moffatt E, Duguay D (2014) Conservation issues in several twentieth-century Canadian oil paintings: the role of zinc carboxylate reaction products. In: Van den Berg KJ, Burnstock A, de Tagle A, de Keijzer M, Heydenreich G, Krueger J, Learner T (eds) Issues in contemporary oil paints. Springer, Cham, pp 167–184
Hermans JJ, Keune K, Van Loon A, Stols-Witlox MJN, Corkery RW, Iedema PD (2014) The synthesis of new types of lead and zinc soaps: a source of information for the study of oil paint degradation. In: Bridgland J (ed) Preprints of the 17th ICOM-CC triennial conference, Melbourne. The International Council of Museums, Paris
Hermans JJ, Keune K, Van Loon A, Iedema PD (2015) An infrared spectroscopic study of the nature of zinc carboxylates in oil paintings. J Anal At Spectrom 30:1600–1608
Hermans JJ, Keune K, Van Loon A, Iedema PD (2016a) The crystallization of metal soaps and fatty acids in oil paint model systems. Phys Chem Chem Phys 18:10896–10905
Hermans JJ, Keune K, Van Loon A, Corkery RW, Iedema PD (2016b) Ionomer-like structure in mature oil paint binding media. RSC Adv 6:93363–93369
Hermans JJ, Keune K, Van Loon A, Iedema PD (2019) Toward a complete molecular model for the formation of metal soaps in oil paints. In: Casadio F, Keune K, Noble P, Van Loon A, Hendriks E, Centeno S, Osmond G (eds) Metal soaps in art: conservation and research. Springer, Cham, pp 47–64
Joosten JM (1998) Piet Mondrian Catalogue Raisonné of the work of 1911–1944, vol I–II. V+K Publishing, Blaricum
Keune K (2005) Binding medium, pigments and metal soaps characterised and localised in paint cross-sections [PhD dissertation]. Universiteit van Amsterdam, Amsterdam
Keune K, Boon JJ (2007) Analytical imaging studies of cross-sections of paintings affected by lead soap aggregate formation. Stud Conserv 52:161–176
Kühn H (1986) Zinc white. In: Feller RL (ed) Artists' pigments. A handbook of their history and characteristics, vol 1. National Gallery of Art, Washington, pp 169–186
Maines C, Rogala D, Lake S, Mecklenburg MF (2011) Deterioration in abstract expressionist paintings: analysis of zinc oxide paint layers in works from the collection of the Hirshhorn Museum and Sculpture Garden, Smithsonian Institution. In: Vandiver P, Reedy CL, Li W, Ruvalcaba Sil JL (eds) Materials issues in art and archaeology IX: symposium held 29 Nov-3 Dec 2010, Boston, MA
Maor Y, Murray A (2008) Delamination of Oil Paints on Acrylic Grounds. In: Materials issues in art and archaeology VIII, Materials Research Society, Symposia Proceedings, vol 1047, pp 127–136
Mecklenburg MF, Tumosa CS, Vicenzi EP (2013) The influence of pigments and ion migration on the durability of drying oil and alkyd paints. In: Mecklenburg MF, Charola AE, Koestler RJ (eds) New insights into the cleaning of paintings. Proceedings from the cleaning 2010 international conference universidad politécnica de valencia and museum conservation institute. Smithsonian Institution Scholarly Press, Washington, pp 59–68
Noble P, Boon JJ (2007) Metal soap degradation of oil paintings: aggregates, increased transparency and efflorescence. In: Parkin HM (ed) AIC paintings specialty group postprints, vol 19. American Institute for Conservation of Historic and Artistic Works, Washington, DC, pp 5–19
O'Donoghue E, Johnson AM, Mazurek J, Preusser F, Schilling M, Walton MS (2006) Dictated by media: conservation and technical analysis of a 1938 Joan Miró Canvas painting. In: Saunders D, Townsend JH, Woodcock S (eds) The object in context: crossing conservation

boundaries: contributions to the Munich congress, 28 Aug–1 Sept 2006. International Institute for Conservation of Historic and Artistic Works, London, pp 62–68

Osmond G (2012) Zinc white: a review of zinc oxide pigment properties and implications for stability in oil-based paintings. AICCM Bull 33:20–29

Osmond G (2014a) Zinc white and the influence of paint composition for stability in oil based media. In: Van den Berg KJ, Burnstock A, de Tagle A, de Keijzer M, Heydenreich G, Krueger J, Learner T (eds) Issues in contemporary oil paints. Springer, Cham, pp 263–281

Osmond GI (2014b) Zinc oxide-centred deterioration of modern artists' oil paint and implications for the conservation of twentieth century paintings. PhD dissertation, The University of Queensland, Queensland

Osmond G, Boon JJ, Puskar L, Drennan J (2012) Metal stearate distributions in modern artists' oil paints: surface and cross-sectional investigation of reference paint films using conventional and synchrotron infrared microspectroscopy. Appl Spectrosc 66(10):1136–1144

Osmond G, Ebert B, Drennan J (2014) Zinc oxide-centred deterioration in 20th century Vietnamese paintings by Nguyễn Trọng Kiệm (1933–1991). AICCM Bull 34:4–14

Rogala D, Lake S, Maines C, Mecklenburg M (2009) A closer look: condition issues in abstract expressionist ground layers. In: AIC paintings specialty group postprints, vol 21. American Institute for Conservation, Los Angeles, pp 41–46

Rogala D, Lake S, Maines C, Mecklenburg M (2010) Condition problems related to zinc oxide underlayers: examination of selected abstract expressionist paintings from the collection of the Hirshhorn Museum and Sculpture Garden, Smithsonian Institution. JAIC 49:96–113

Rogge C, Véliz-Bomford Z (2015) Paint never behaves the same: Franz Kline case studies. In: Abstract expressionism: time, intention, conservation, and meaning symposium organized by the Getty Conservation Institute and Clyfford Still Museum Research Center (12.11.2015). Accessed 22 Dec 2016. https://www.youtube.com/watch?v=g9DjpFHmAl0

Shimadzu Y, Keune K, van den Berg KJ, Boon JJ, Townsend JH (2008) The effects of lead and zinc white saponification on surface appearance of paint. In: Bridgland J (ed) Preprints of the 15th ICOM-CC Triennial conference, New Delhi. The International Council of Museums, Paris, pp 626–632

Szafran Y, Rivers L, Phenix A, Learner T, Landau EG, Martin S (2014) Jackson Pollock's mural: the transitional moment. Getty Publications, Los Angeles

van der Weerd J, Geldof M, Struik van der Loeff L, Heeren RMA, Boon JJ (2003) Zinc soap aggregate formation in 'falling leaves (les Alyscamps)' by Vincent van Gogh. Zeitschrift für Kunsttechnologie und Konservierung (ZKK) 17(2):407–416

van der Werf P (1994) Het zoeken naar harmonie. Over de werkwijze van Piet Mondriaan/The working methods of Piet Mondriaan. *kM*, no. 12 , pp 13–19, 31–35

Van Loon A, Noble P, Burnstock A (2012) The ageing and deterioration of traditional oil and tempera paints. In: Hill Stoner J, Rushfield R (eds) Conservation of easel paintings. Routledge, Oxon, pp 214–241

Van Loon A, Hoppe R, Keune K, Hermans J, Diependaal H, Bisschoff M, Thoury M, Van der Snickt G (2019) Paint delamination as a result of zinc soap formation in an early Mondrian painting. In: Casadio F, Keune K, Noble P, Van Loon A, Hendriks E, Centeno S, Osmond G (eds) Metal soaps in art: conservation and research. Springer, Cham, pp 361–373

Chapter 21
Paint Delamination as a Result of Zinc Soap Formation in an Early Mondrian Painting

Annelies Van Loon, Ruth Hoppe, Katrien Keune, Joen J. Hermans, Hannie Diependaal, Madeleine Bisschoff, Mathieu Thoury, and Geert van der Snickt

Abstract The Evolution triptych by Piet Mondrian (1911, oil on canvas, Gemeentemuseum Den Haag) presents a case study of a painting that is seriously affected by zinc soap formation, which has resulted in paint delamination and paint loss, particularly in the cadmium yellow paint areas. The paint is extremely fragile, which makes the paintings vulnerable with regard to handling and treatment. This paper

A. Van Loon (✉)
Conservation Department, Rijksmuseum, Amsterdam, The Netherlands
e-mail: a.vanloon@rijksmuseum.nl

R. Hoppe (✉)
Gemeentemuseum Den Haag, The Hague, The Netherlands
e-mail: rhoppe@gemeentemuseum.nl

K. Keune
Conservation Department, Rijksmuseum, Van't Hoff Institute for Molecular Sciences, University of Amsterdam, Amsterdam, The Netherlands

J. J. Hermans
Van't Hoff Institute for Molecular Sciences, University of Amsterdam, Amsterdam, The Netherlands

H. Diependaal · M. Bisschoff
Independent paintings conservator, Amsterdam, The Netherlands

M. Thoury
IPANEMA, CNRS, ministère de la Culture et de la Communication, Université de Versailles Saint-Quentin-en-Yvelines, Muséum National d'Histoire Naturelle, USR 3461, Université Paris-Saclay, Gif-sur-Yvette, France

G. van der Snickt
Department of Chemistry – AXES group, University of Antwerp, Antwerp, Belgium

Conservation Studies, University of Antwerp, Antwerp, Belgium

F. Casadio et al. (eds.), *Metal Soaps in Art*, Cultural Heritage Science,
https://doi.org/10.1007/978-3-319-90617-1_21

focuses on the analytical research of the painting using various state-of-the-art and novel macro- and micro-imaging techniques. Macro X-ray fluorescence scanning (MA-XRF) revealed the presence of cadmium (Cd) and zinc (Zn) in the affected yellow paints. Paint cross sections of both affected and intact paint areas were investigated using light microscopy, scanning electron microscopy coupled with energy dispersive X-ray analysis (SEM-EDX), attenuated total reflection Fourier transform infrared (ATR-FTIR) micro-imaging, and synchrotron photoluminescence (PL) micro-imaging. With the help of these techniques, the cadmium yellow pigment could be identified as a mixture of cadmium sulfide and cadmium oxalate. The presence of zinc white was established in areas where the yellow paint film is degraded, while the intact areas of yellow paint do not contain any zinc white. In samples of the degraded paints, it was demonstrated that high concentrations of zinc soaps have formed, accumulating at interfaces. This has caused local chemical and physical changes of the paint resulting in delamination between paint layers.

Keywords Zinc soaps · Metal soaps · Paint delamination · Cadmium yellow · Mondrian

21.1 Introduction

The *Evolution* triptych by Piet Mondrian (1911, oil on canvas, 178 × 85; 183 × 87.5; 178 × 85 cm), which belongs to the artist's early period of figurative painting, is one of the most prominent works in the collection of the Gemeentemuseum Den Haag (Fig. 21.1). The triptych consists of three separate paintings that are overall surprisingly well preserved, given the fact that the paintings have survived in their original frames and are still unvarnished and not relined. The yellow paint areas in the center and left panels, however, suffer from a peculiar form of degradation, which is only visible with higher magnification (Fig. 21.2). Local wrinkling can be observed and also delamination between the top layer and the underlying paint. This occurs in the form of blistering or "tunneling," which appears to correspond with the brushstrokes of an underlying paint application. Associated paint loss can also be seen with the microscope, indicating that the yellow paint is very vulnerable and has already suffered in the past from inadequate handling and treatment. During tests for consolidation treatment, it became clear that the lifting paint breaks upon the slightest touch.

Degradation phenomena in cadmium yellow paints are a familiar problem in works by Mondrian and his contemporaries (Kolkena 2014; Van der Snickt et al. 2009, 2012; Mass et al. 2013). In the collection of the Gemeentemuseum, these problems have been identified in some later paintings dating from 1921, *Composition with Large Red Plane, Yellow, Black, Grey and Blue*, and *Composition with Red, Blue, Black, Yellow and Grey,* and only in one early painting, *Seascape*, painted in 1909. In these paintings, however, the degradation phenomena appear differently: the paint is crumbling or disintegrating, or showing discoloration.

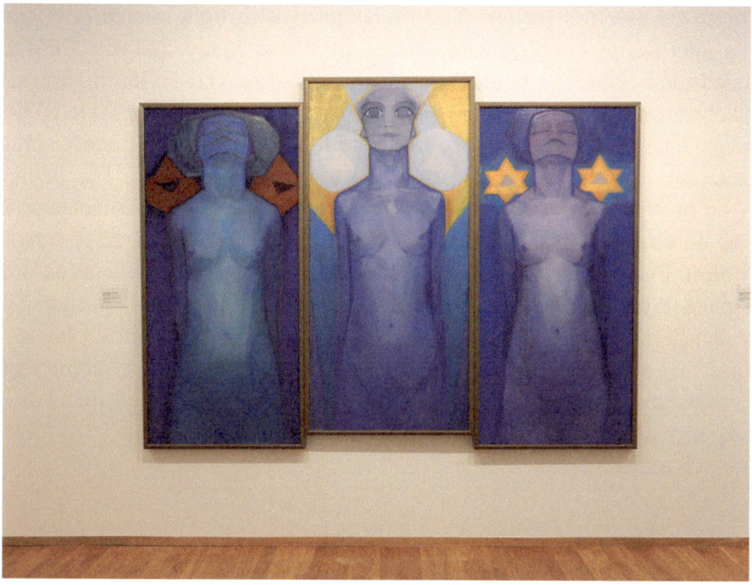

Fig. 21.1 Piet Mondrian, *Evolution*, 1911, oil on canvas, triptych: 178 × 85; 183 × 87.5; 178 × 85 cm, Gemeentemuseum Den Haag, the Netherlands. The paintings are currently presented in the galleries in three custom-made display frames

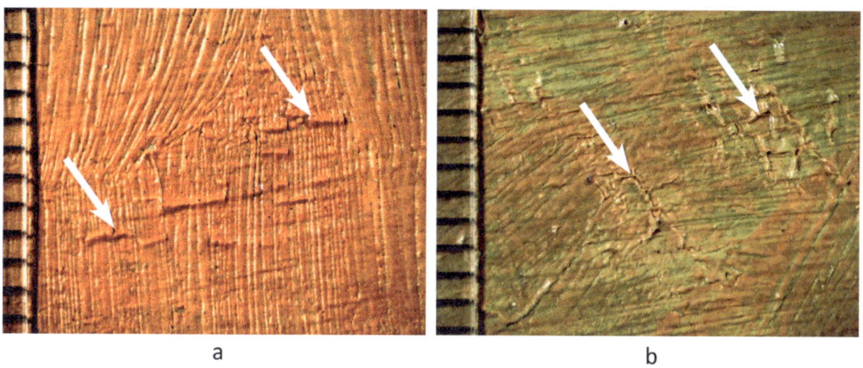

Fig. 21.2 Macrophotographs of degradation phenomena in cadmium yellow paint (**a**) tunneling and lifting across brushstroke relief; (**b**) wrinkling with associated paint loss

As part of the technical examination of the triptych, an area of 60 × 80 cm at the top right part of the center panel, including the yellow paint areas, was mapped with macroscopic X-ray fluorescence scanning (MA-XRF). This is a recent diagnostic imaging tool that reveals the elemental distributions on and below the paint surface in a non-invasive manner, thus providing information about the paint composition (Alfeld et al. 2011). Additional micro-sampling was necessary, in order to be able to fully interpret the MA-XRF data and to assess the condition of the paint. Subsequently, sample locations were carefully selected, with the aid of the XRF scans. Both affected and intact paint areas were sampled for comparison. Samples were prepared as cross sections and investigated with light microscopy, scanning electron microscopy coupled with energy dispersive X-ray analysis (SEM-EDX), attenuated total reflection Fourier transform infrared (ATR-FTIR) micro-imaging, and synchrotron photoluminescence (PL) micro-imaging. For the ATR-FTIR measurements, a new sample accessory was used in the experimental setup, containing a germanium ATR hemisphere with a flat top part (1 mm ∅). The hemisphere has to be brought in contact with the sample surface, but the advantage is that it leaves no circular depressions in the sample, as conventional ATR crystals do. Synchrotron PL was used here as a novel approach to visualize alteration products in the paint layers at high, submicron spatial resolution, based on their specific luminescence properties (see also Thoury et al. 2019). Since the most affected areas in the painting are in the yellows, the first thought was that the degradation would be related to the use of a cadmium yellow pigment. The combined chemical analyses, however, revealed that the major cause of the delamination is zinc soap formation, as discussed in this paper.

21.2 Experimental

21.2.1 MA-XRF

Macroscopic X-ray fluorescence maps were collected using a Bruker M6 Jetstream Instrument (Alfeld et al. 2013). This instrument consists of a measuring head on an XY-motorized stage that is slowly moved over the surface of the painting (no actual contact is made with the paint surface), scanning the painting pixel by pixel and line by line. The measuring head consists of a Rh-target microfocus X-ray tube (30 W, maximum voltage 50 kV, maximum current 0.6 mA) and a 30 mm^2 XFlash silicon drift detector (energy resolution <145 eV at Mn-Kα). The beam size is defined by means of a polycapillary optic and is variable, determined by the distance between the paint surface and the measuring head (varying between 150 μm at 6 mm working distance and 860 μm at 15 mm working distance). For the Mondrian painting, element distribution maps were collected over an area measuring 767 by 578 mm^2 (1096 by 826 pixels) in a single 24 h session. X-ray tube settings were 50 kV and 600 μA; a step size of 700 μm, a dwell time of 70 ms/step, and a beam size estimated to be 350 μm in diameter were used. All data were collected with the Bruker M6 Jetstream software package. The acquired spectra were then exported and processed using PyMca and the in-house developed Datamuncher software (Alfeld and Janssens 2015).

21.2.2 Microanalyses

21.2.2.1 Samples

Five micro-samples were taken in total from the center panel (Fig. 21.3a) (from the area scanned with XRF) for comparison. They were all taken along the edges of

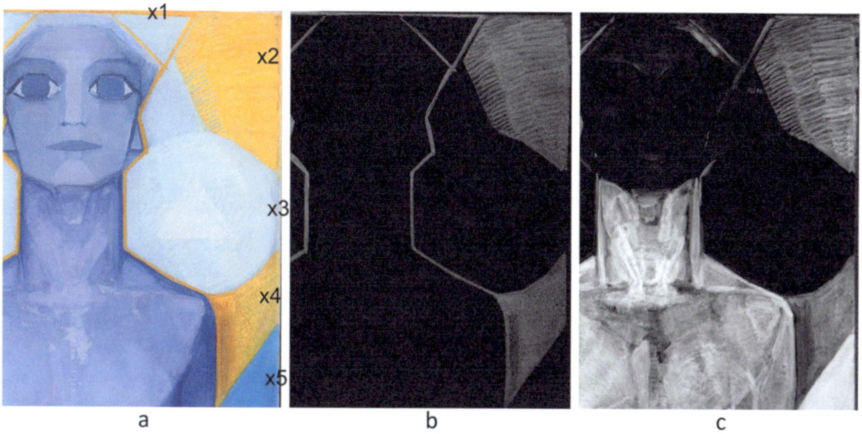

Fig. 21.3 Macro-XRF scanning of the top right part of the center panel: (**a**) area scanned with MA-XRF. With the aid of the XRF scans, sample sites (X1–5) were selected from intact and degraded areas for further analysis; (**b**) cadmium (Cd-L) map (higher brightness indicates higher signal); (**c**) zinc (Zn-K) map

the painting, with the aid of the stereo-microscope, from areas of both intact and degraded paint (Table 21.1).

21.2.2.2 Sample Preparation

The paint micro-samples were embedded in Technovit 2000 LC mounting resin, a one-component methacrylate that polymerizes under visible blue light (Heraeus Kulzer GmbH, Germany). They were wet-polished on a polishing machine to expose the complete paint layer buildup, with the assistance of a sample holder. The final polishing steps were done with Micromesh sheets up to grade 12,000 (Micro-Surface Finishing Products Inc., Wilton, Iowa, USA), using the dry polishing method (Van Loon et al. 2005).

21.2.2.3 Light Microscopy

Light microscopic (LM) studies of the paint cross sections were performed on a Leica DMRX microscope at the FOM Institute AMOLF in Amsterdam. The cross sections were examined at magnifications up to 1000×, in normal incident light, in bright-field (BF) (with and without oil immersion) and in dark-field illuminations, and with ultraviolet (UV-A) using Leica filters A (excitation 340–380 nm, emission >425 nm) and D (exc. 355–425 nm, em. >470 nm).

Table 21.1 Summary of the cross-sectional analyses

Sample no.	Sample location	XRF	Paint composition and buildup	Microanalyses
X1	Yellow border along the top edge. Paint is intact	Cd, Pb	2. Yellow paint: cadmium yellow (CdS and Cd-oxalate), very little chalk 1. White ground: lead white	LM, SEM-EDX
X2	Yellow paint area at upper right, from a pastose brushstroke. Showing wrinkling	Cd, Zn, Pb	4. Yellow paint: cadmium yellow, zinc white (*zinc soap formation*) 3. Light-blue paint: lead white, single particle of synth. ultramarine 2. Blue paint: lead white, synth. ultramarine, barium sulfate 1. White ground: lead white	LM, SEM-EDX, ATR-FTIR, PL
X3	Light-blue paint area. Paint is intact	Pb	5. Light-blue paint: lead white, few blue particles 4. Blue paint: lead white, fine blue pigment (same as in layer 2) 3. Light-blue paint: lead white, single blue particle 2. Blue paint: lead white, fine blue pigment 1. White ground: lead white	LM
X4	Pale yellow paint area, next to paint loss. Showing paint delamination and wrinkling	Cd, Zn, Pb	8. Pale yellow paint: zinc white, cadmium yellow, single particle of blue (*zinc soap formation*) 7. Yellow paint: cadmium yellow (CdS and Cd-oxalate), very little chalk 6. Blue paint: lead white, blue pigment, few pale-reddish particles 5. Light-blue paint: lead white, some blue pigment 4. Blue paint: lead white, lots of blue pigment 3. Light-blue paint: lead white, single blue particle 2. Light-blue paint: lead white, some blue pigment 1. White ground: lead white	LM, SEM-EDX, ATR-FTIR, PL
X5	Turquoise blue paint area. Slightly deformed paint	Fe, Cr, Zn	7. Blue paint: zinc white, fine blue pigment (*zinc soap formation*) 6. Green-blue paint: zinc white, cadmium yellow, few blue-green particles (*zinc soap formation*) 5. Yellow paint: cadmium yellow (CdS and Cd-oxalate), very little chalk 4. Blue paint: lead white, fine blue 3. Light-blue paint: lead white, some blue pigment 2. Blue paint: lead white, fine blue pigment 1. White ground: lead white	LM, SEM-EDX

21.2.2.4 SEM-EDX

SEM-EDX analysis of the embedded samples was performed at the FOM Institute AMOLF in Amsterdam, using a FEI Verios 460 high-pressure electron microscope equipped with an Oxford EDX system. Backscattered electron images were acquired at 20 kV acceleration voltage and 0.20 nA beam current. EDX spectral maps were recorded at the same kV and beam current, with 1024 $\times$ 640 pixels resolution and 100 μs dwell time. The samples were gold coated (3 nm) on an SC7640 sputter coater (Quorum Technologies, Newhaven, East Sussex, UK) prior to analysis to improve surface conductivity.

21.2.2.5 ATR-FTIR Micro-imaging

ATR-FTIR measurements were undertaken at the IPANEMA Laboratory, on a Bruker Hyperion 3000 FTIR microscope using a focal plane array (FPA) detector, coupled to a VERTEX 70 FTIR spectrometer. An ATR hemisphere sample accessory was used, equipped with a base germanium hemisphere crystal with a 1 mm diameter flat top that is brought in contact with the sample. The spectra were collected with 4 cm^{-1} spectral resolution in the range 4000–900 cm^{-1} averaging 128 scans. Data were processed using the OPUS 7.2 software.

21.2.2.6 Synchrotron-Photoluminescence Micro-imaging

PL experiments were accomplished on the DISCO beamline at Synchrotron SOLEIL using the full-field micro-imaging setup (TELEMOS) (Giuliani et al. 2009; Bertrand et al. 2013). The excitation wavelength was set at 280 nm. The use of a 100 $\times$ 1.25 NA ultrafluar objective allowed collecting images up to 250 nm spatial resolution. Images were acquired in nine spectral bands between 327 and 870 nm using high transmittance band-pass interference filters. The intensity of each channel of the full-field RGB image was stretched between the 2nd and 98th percentile using the ENVI 5.0 software (EXELIS) to facilitate visual comparison.

21.3 Results

21.3.1 Macro-XRF Scanning

In the context of this study, the cadmium (Cd-L) and zinc (Zn-K) distribution maps are the most informative; see Fig. 21.3b, c, respectively. The Cd map indicates the use of a cadmium-containing pigment in the yellow paint areas, in the two large areas at the right part of the XRF scan, and in the yellow lines following the contours of the face and neck and along the top edge of the painting. Zinc is also abundantly present in the degraded areas of yellow paint, particularly in the two large areas. Zinc is absent in the yellow border along the top edge, where the paint film is intact. Interestingly, the XRF scan reveals significantly high amounts of zinc in the turquoise blue paint area at the lower right corner of the scan, where the paint film shows a certain degree of deformation, whereas in the intact light blue paint area at the center right part, no zinc, only lead, was detected. This points to a causal connection between the presence of zinc and the degradation of the paint. The MA-XRF maps alone did not permit the exact zinc source to be pinpointed. In particular, it remained unclear whether the detected zinc originated from a cadmium zinc sulfide pigment or from the admixture of a separate zinc compound (pigment or drier) to the paint. Paint cross-sectional analysis was expected to shed further light on this and on the paint buildup, as well as identify possible degradation products. In this framework, the MA-XRF maps proved to be particularly helpful for selecting good sample spots and thus for keeping the number of samples to a minimum.

21.3.2 Cross-Sectional Analyses

The findings arising from all the studied micro-samples are summarized in Table 21.1. Two representative samples will be discussed here in detail: sample X1 taken in an intact yellow paint area and sample X4 originating from a degraded yellow paint area.

21.3.2.1 Intact Yellow Paint (X1)

Sample X1 is taken from the yellow border at the top edge of the painting, where the paint film is thin and appears intact (Fig. 21.4a). In this area only cadmium and lead were detected with MA-XRF; zinc is absent. The light microscopic image of the cross section shows a simple layer buildup (Fig. 21.4b): a lead white ground followed by a thin, evenly applied yellow paint layer (thickness *c.*20 μm). The yellow layer contains cadmium yellow pigment only, without any addition of zinc white or other pigments. The absence of zinc white is supported by the UV-A image, in which the typical fluorescent "sparks" of zinc white cannot be observed (Fig. 21.4c). SEM-EDX analysis revealed the presence of cadmium and sulfur in the yellow paint; no zinc was detected (Fig. 21.4d-f). Therefore it can be presumed that the cadmium yellow pigment is cadmium sulfide and not cadmium zinc sulfide. The same paint is present in sample X4 as an underlayer (layer 7) where analysis shows the cadmium sulfide pigment is mixed with larger chunks of cadmium oxalate (see section "Degraded Yellow Paint (X4)"). Although cadmium oxalate has been identified as a degradation product in other studies (Van der Snickt et al. 2009, 2012; Mass et al. 2013), in this case it is not a degradation product but an original paint component.

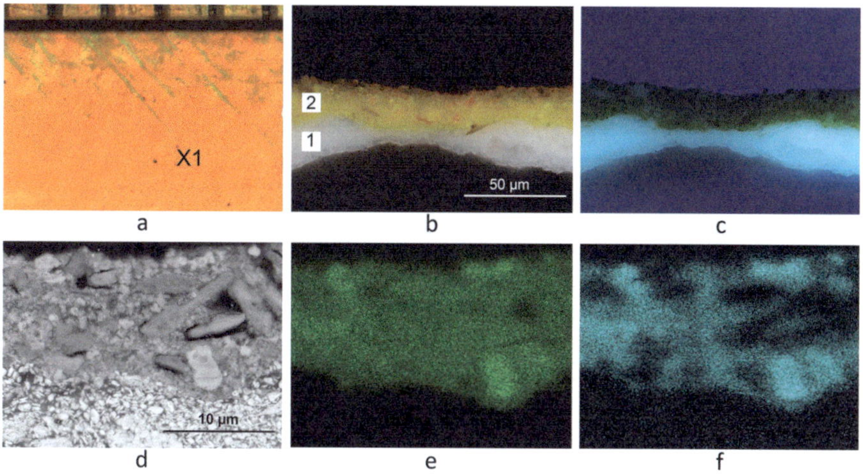

Fig. 21.4 Cross-section analyses of intact yellow paint (X1): (**a**) sample location (X1): yellow border at top edge, where the paint is intact; (**b**) light microscopic image of paint cross section in normal light (BF, oil immersion), photographed at 400×; (**c**) cross section in ultraviolet light (UV-A); (**d**) SEM backscatter image, zoomed in on the yellow paint; (**e**) EDX cadmium map of (d); (**f**) EDX sulfur map of (d)

21.3.2.2 Degraded Yellow Paint (X4)

Sample X4 represents a degraded yellow paint area. It was taken from a spot just next to an area where the top part of the paint layers had already delaminated, in order to make sure that the sample would contain all layers (Fig. 21.5a[1]).

The cross section shows a thick, multilayered package. It contains several underlayers of lead white and blue paint (these are not visible in the images). Figure 21.5b shows the top layers only, where the delamination takes place.[2] A thin yellow paint layer (7) (thickness *c.*15–20 μm) comprising purely cadmium yellow pigment is followed by a thick, pale yellow paint layer (8) (thickness up to *c.*60 μm), which contains a lot of zinc white and little cadmium yellow. Translucent areas are noticeable in layer 8, especially at the surface and at the interface with paint layer 7 below. In ultraviolet light (UV-A), the same layer shows numerous tiny, strongly fluorescent particles characteristic of zinc white (Fig. 21.5c). The fluorescent "sparks" are absent in the translucent areas, which suggests that the

[1]LM and SEM-EDX were done prior to ATR-FTIR and PL micro-imaging. When re-polishing the sample to remove the 3 nm-thick gold coating applied for the SEM-EDX analysis, the surface was slightly modified. This explains why the images obtained with the different analytical techniques cannot be 100% overlayed.

[2]During sampling, the thick paint broke into two halves, which were embedded separately. The sample containing the lower part of the layer structure was also analyzed, but the images are not shown here since those layers do not show any sign of degradation and appear stable.

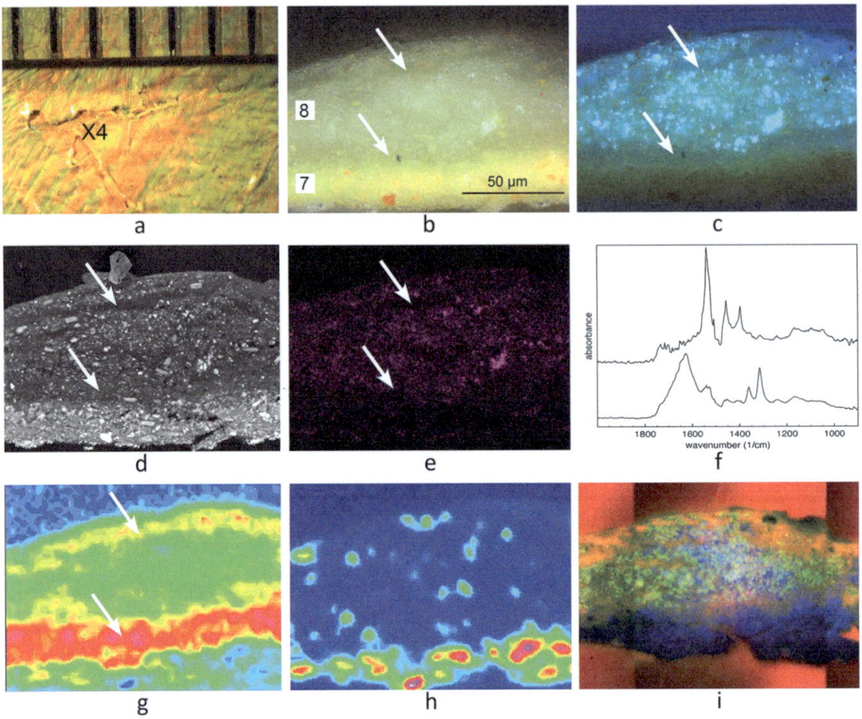

Fig. 21.5 Cross-section analyses of degraded yellow paint (X4): (**a**) sample location (X4): yellow paint area showing delamination (wrinkling); (**b**) light microscopic image of paint cross section in normal light (BF, oil immersion, photographed at 400×). The sample shows top layers only, where the delamination takes place. The white arrows point to the translucent areas in paint layer 8; (**c**) cross section in ultraviolet light (UV-A); (**d**) SEM backscatter image; (**e**) EDX zinc map; (**f**) ATR-FTIR spectra. Upper trace: spectrum from pale yellow top layer (8) showing characteristic zinc soap peaks. Lower trace: spectrum from lower yellow layer (7) showing characteristic cadmium oxalate peaks; (**g**) ATR-FTIR image of the *c.*1540 cm^{-1} band in false color, showing the distribution of zinc soaps in the pale yellow top paint layer (8) (yellow/red zones); (**h**) ATR-FTIR image of the *c.*1315 cm^{-1} band in false color, showing the distribution of cadmium oxalate particles in the yellow underlayer (7) (red zones); (**i**) photoluminescence RGB composite at 280 nm excitation (UV-C) with different emission filters. Blue channel, emission 800–870 nm, indicative of cadmium yellow pigment. Green channel, em. 370–410 nm, indicative of zinc white. Red channel, em. 327–353 nm, indicative of zinc soaps

original zinc white has dissolved or been altered in these areas. SEM backscattered electron imaging reveals that the top layer (8) is physically disrupted (Fig. 21.5d). The translucent areas are low scattering (dark) and therefore interpreted as organic-rich. What appears as small (dark) "cracks" are also filled with organic-rich material. The EDX zinc map confirms the presence of zinc in layer 8 (Fig. 21.5e). The zinc is not associated with the cadmium pigment. Thus it can be ruled out that the zinc is present in the form of cadmium zinc sulfide; it is clear that zinc white is mixed into the top paint layer (8) in high quantities. The results of light microscopic and

SEM-EDX analyses also suggest that the zinc white is in the process of reacting with the medium, converting into zinc soaps, prompting further investigation with ATR-FTIR. Zinc white is known to be very reactive toward fatty acid compounds from the oil medium resulting in the formation of zinc soaps.

ATR-FTIR micro-imaging of the cross section was carried out to characterize the translucent areas and to reveal the zinc soap distribution within the paint. The ATR-FTIR spectra demonstrate the presence of zinc soaps in the top layer (8) (Fig. 21.5f, upper trace). The spectra show an intense, sharp absorption band at *c.*1540 cm^{-1} that is characteristic of the asymmetric stretch vibration of the zinc carboxylate group $\nu_{as}(COO^{-})$ in crystalline form, together with other characteristic features: the symmetric stretch vibration $\nu_{s}(COO^{-})$ at c.1397 cm^{-1} and the C-H bend vibration δ(C-H) at *c.*1456 cm^{-1} (Hermans et al. 2014). The FTIR image of the *c.*1540 cm^{-1} band shows the distribution of zinc soaps in the paint (Fig. 21.5g). It reveals high concentrations of zinc soaps that correspond with the translucent areas in the top layer (8), especially at the paint surface and at the interface with underlying paint (layer 7). This is a very significant result, since the accumulation of zinc soaps at the interface can be directly related to the paint delamination. The conversion of the paint into zinc soaps changes its chemical and physical properties, which may cause the paint to lose adhesion with the layer below, resulting in delamination of the paint film.[3] Accumulation of zinc soaps at the lower part of a paint layer has also been found in zinc white paint reconstructions in a recent study (Osmond et al. 2012). A similar trend is apparent in other case studies of paint delamination and paint loss associated with zinc soaps in modern oil paintings, including previously published studies (Rogala et al. 2010; Helwig et al. 2014) and in other contributions to this volume (Raven et al. 2019; Osmond 2019).

ATR-FTIR micro-imaging also proved a useful tool for identifying the exact composition of the cadmium yellow pigment. Apart from cadmium sulfide, the purely yellow paint layers of X1 (layer 2) and X4 (layer 7) contain coarser particles, which are also cadmium-based but do not contain sulfur (EDX analysis). The ATR-FTIR spectra of X4 reveal absorption bands at *c.*1626 cm^{-1} (strong, broad), *c.*1361 cm^{-1} (medium, sharp) and *c.*1315 cm^{-1} (strong, sharp), which are characteristic for cadmium oxalate (Mass et al. 2013) (Fig. 21.5f. lower trace). Imaging these bands demonstrates that they correspond with the coarser particles (high cadmium/no sulfur) in the yellow underlayer (7), thus confirming the identification of cadmium oxalate (Fig. 21.5h). Since the pure cadmium yellow paint layers appear unaltered and show homogeneous mixtures of cadmium sulfide and cadmium oxalate pigment particles, it can be concluded that the latter are present as an original paint component, rather than a degradation product. At the

[3]In sample X5, which was taken from turquoise blue paint that showed some wrinkling (Table 21.1), zinc soap formation is visible at the interface between an underlying, zinc white-containing green-blue paint layer and a cadmium yellow paint layer. However, possibly due to its thickness, the turquoise blue top paint layer appears to have prevented the paint from developing more serious degradation phenomena and lifting off, at least for the time being. Due to length restrictions of this paper, sample X5 is not further discussed here.

time, cadmium oxalate was intentionally used as filler in the manufacture of paler shades of cadmium yellow paints (Fiedler and Bayard 1986).

Finally we will discuss the photoluminescence (PL) images of X4, acquired at 280 nm excitation using various emission filters ranging from 327 up to 870 nm. Three different spectral ranges are presented in the RGB composite plotted in Fig. 21.5i. Since the cadmium yellow, zinc white, and zinc soaps present in the paint layers are each shown with a different PL marker that has a specific spectral range, they can be clearly distinguished with this approach. The 800–870 nm emission image (blue channel in RGB composite) is induced by a trap state emission from cadmium sulfide (CdS), and it can be used as a characteristic signature for the cadmium yellow pigment (Thoury et al. 2011). It reveals that the cadmium yellow pigment is everywhere, including in layer 8, but not in the small "cracks" that are visible in this layer. The 370–410 nm emission image (green channel) is related to the band edge emission of zinc oxide (ZnO), and it can be used to map the presence of zinc white in layer 8 (Bertrand et al. 2013). The image shows that zinc white is present only in the top layer (8), in varying amounts. There is no indication for its presence in the lower paint layer (7). Less zinc white luminescence is visible in the top and bottom parts of layer 8 and in the "cracks," which all correspond to high-intensity areas of zinc soaps in the FTIR image. The typical fluorescent sparks of zinc white could also be recognized with conventional UV-A microscopy (see Fig. 21.5c), but the image obtained with the PL synchrotron experiment using a high numerical aperture objective centered on the band edge emission allows a more specific visualization of ZnO particles and thus reveals more details. The 327–353 nm emission image (red channel) can be indicative of zinc soaps (see also Thoury et al. 2019). The luminescence is most intense in the "cracks," where no cadmium yellow is present, showing that the "cracks" are filled with zinc soaps. In other areas, the luminescence emitted by the zinc soaps may partly be reabsorbed by the cadmium yellow pigment, resulting in lower luminescence intensity than might be expected from the ATR-FTIR imaging. This could explain why the zinc soap concentrations at the interface between layer 7 and 8 appear less bright than in the cracks. Nevertheless the emission image confirms that the zinc soaps are present throughout the whole layer. The photoluminescence data complement the FTIR imaging data. It demonstrates the specification capabilities of luminescence imaging to visualize different paint components and alteration products within a paint sample at submicron spatial resolution, which cannot be revealed with conventional UV-A microscopy.

21.4 Stages in Zinc Soap Delamination

From the chemical analyses, it can be concluded that the presence of zinc white and associated zinc soap formation must be held responsible for the paint delamination in Mondrian's *Evolution* triptych. The sample from the degraded yellow paint area X4 represents an intermediate phase in the delamination process. The paint has not yet delaminated or fallen off, but the zinc soap formation is already in an advanced state, and its distribution in the paint layer is such that we can identify the potential weak spots in the paint structure. Why zinc soaps preferentially form or concentrate at layer interfaces is the topic of ongoing research.

21.5 Conservation Measures

In 2013, the *Evolution* triptych underwent conservation treatment. The main objective of the treatment was to stabilize the fragile paint areas in order to minimize the risk of future damage or loss of original material. For consolidation, a 15% solution of Paraloid B72 dissolved in Shellsol A was locally applied in these areas using a fine sable brush. Re-adhering lifting paint was not an option, because any attempt to apply even the slightest pressure would cause the extremely delicate paint to break. The conservation treatment included the removal of surface dirt in non-affected areas, as well as filling and retouching of minor paint loss. For reasons of protection, it was decided that the works should be presented behind glass or Perspex. Since it proved impossible to fit glazing in the narrow profile of the original frames, the paintings are now presented in the galleries in three custom-made display frames constructed from aluminum with low-reflective, shatter-proof glass and rigid backing (Fig. 21.1). Great efforts have been made to match as closely as possible the profile and coloring of the original framing. A restrictive loan policy should ensure that handling and traveling are kept to a minimum.

21.6 Conclusion

The results of the combined chemical analyses discussed in this paper have made evident that the peculiar form of degradation observed in the yellow paint areas of Mondrian's *Evolution* triptych is related to the use of zinc white in the paint and to the formation of zinc soaps and not the result of cadmium yellow degradation, as was initially thought. Zinc soaps often appear to accumulate at the lower part of a layer, at the interface with an underlying paint layer or ground, which eventually can lead to paint delamination. The case study shown here is a clear example of this phenomenon.

Acknowledgments This research is part of the Paint Alterations in Time project (PAinT), which is financially supported by the Science4Arts Programme of the Netherlands Organization for Scientific Research (NWO). The synchrotron photoluminescence experiments were carried out at the DISCO beamline, at Synchrotron SOLEIL, with the help of Matthieu Réfrégiers. Macro-XRF scanning was made possible with the support from the Baillet-Latour fund. We would also like to thank Prof. Joris Dik, Delft University of Technology, for making the Bruker M6 scanner available.

References

Alfeld M, Janssens K (2015) Strategies for processing mega-pixel X-ray fluorescence hyperspectral data: a case study on a version of Caravaggio's painting *Supper at Emmaus*. J Anal At Spectrom 30:777–789. https://doi.org/10.1039/c4ja00387j

Alfeld M, Janssens K, Dik J, de Nolf W, Van der Snickt G (2011) Optimization of mobile scanning macro-XRF systems for the in situ investigation of historical paintings. J Anal At Spectrom 26:899–909. https://doi.org/10.1039/C0JA00257G

Alfeld M, Vaz Pedroso J, Van Eikema Hommes M, Van der Snickt G, Tauber G, Blaas J, Haschke M, Erler K, Dik J, Janssens K (2013) A mobile instrument for in situ scanning macro-XRF investigation of historical paintings. J Anal At Spectrom 28:760–767. https://doi.org/10.1039/c3ja30341a

Bertrand L, Réfrégiers M, Berrie B, Échard J-P, Thoury M (2013) A multiscalar photoluminescence approach to discriminate among semiconducting historical zinc white pigments. Analyst 138:4463–4469. https://doi.org/10.1039/c3an36874b

Fiedler I, Bayard MA (1986) Cadmium yellows, oranges and reds. In: Feller RL (ed) Artists' pigments: a handbook of their history and characteristics. Cambridge University Press, National Gallery of Art, Washington, pp 65–108

Giuliani A, Jamme F, Rouam V, Wien F, Giorgetta J-L, Lagarde B, Chubar O, Bac S, Yao I, Rey S, Herbeux C, Marlats JL, Zerbib D, Polack F, Réfrégiers M (2009) DISCO: a low-energy multipurpose beamline at synchrotron SOLEIL. J Synchrotron Radiat 16:835–841

Helwig K, Poulin J, Corbeil M-C, Moffatt E, Duguay D (2014) Conservation issues in several twentieth-century Canadian oil paintings: the role of zinc carboxylate reaction products. In: Issues in contemporary oil paint. Springer, Switzerland, pp 167–184

Hermans JJ, Keune K, Van Loon A, Stols-Witlox M, Corkery R, Iedema P (2014) The synthesis of new types of lead and zinc soaps: a source of information for the study of oil paint degradation. In Bridgland J (ed) ICOM committee for conservation 17th triennial conference preprints, Melbourne, 15–19 Sept 2014, pp 1604–1612

Kolkena L (2014) A phenomenological atlas of degradation of cadmium yellow paint in paintings by Piet Mondrian (1872–1944) and some of his contemporaries, Master Thesis University of Amsterdam

Mass J, Sedlmair J, Schmidt Patterson C, Carson D, Buckley B, Hirschmugl C (2013) SR-FTIR imaging of the altered cadmium sulfide yellow paints in Henri Matisse's Le Bonheur de vivre (1905–6) – examination of visually distinct degradation regions. Analyst 138:6032–6043. https://doi.org/10.1039/c3an00892d

Osmond G (2019) Zinc soaps: an overview of zinc oxide reactivity and consequences of soap formation in oil-based paintings. In: Casadio F, Keune K, Noble P, Van Loon A, Hendriks E, Centeno S, Osmond G (eds) Metal soaps in art: conservation and research. Springer, Cham, pp 25–43

Osmond G, Boon JJ, Puskar L, Drennan J (2012) Metal stearate distributions in modern artists' oil paints: surface and cross-sectional investigation of reference paint films using conventional and synchrotron infrared microspectroscopy. Appl Spectrosc 66(10):1136–1144

Raven L, Bisschoff M, Leeuwestein M, Geldof M, Hermans JJ, Stols-Witlox M, Keune K (2019) Delamination due to zinc soap formation in an oil painting by Piet Mondrian (1872–1944). In: Casadio F, Keune K, Noble P, Van Loon A, Hendriks E, Centeno S, Osmond G (eds) Metal soaps in art: conservation and research. Springer, Cham, pp 345–357

Rogala D, Lake S, Maines C, Mecklenburg M (2010) Condition problems related to zinc oxide underlayers: examination of selected abstract expressionist paintings from the collection of the Hirshhorn Museum and Sculpture Garden, Smithsonian Institution. J Am Inst Conserv 49:96–113

Thoury M, Delaney JK, de la Rie ER, Palmer M, Morales K, Krueger J (2011) Near-infrared luminescence of cadmium pigments: in situ identification and mapping in paintings. Appl Spec 68(5):939–951

Thoury M, Van Loon A, Keune K, Réfrégiers M, Hermans JJ, Berrie BH (2019) Photoluminescence micro-imaging sheds new light on the development of metal soaps in oil paintings. In: Casadio F, Keune K, Noble P, Van Loon A, Hendriks E, Centeno S, Osmond G (eds) Metal soaps in art: conservation and research. Springer, Cham, pp 213–226

Van der Snickt G, Dik D, Cotte M, Janssens K, Jaroszewicz J, De Nolf W, Groenewegen J, van der Loeff L (2009) Characterization of a degraded cadmium yellow (CdS) pigment in an oil painting by means of synchrotron radiation based X-ray techniques. Anal Chem 81(7):2600–2610. https://doi.org/10.1021/ac802518z

Van der Snickt G, Janssens J, Dik J, De Nolf W, Vanmeert F, Jaroszewicz J, Cotte M, Falkenberg G, van der Loeff L (2012) Combined use of synchrotron radiation based micro-X-ray fluorescence, micro-X-ray diffraction, micro-X-ray absorption near-edge, and micro-Fourier transform infrared spectroscopies for revealing an alternative degradation pathway of the pigment cadmium yellow in a painting by van Gogh. Anal Chem 84(23):10221–10228. https://doi.org/10.1021/ac3015627

Van Loon A, Keune K, Boon JJ (2005) Improving the surface quality of paint cross-sections for imaging analytical studies with specular reflection FTIR and static-SIMS. In: Proceedings of Art'05 conference on non-destructive testing and microanalysis for the diagnostics and conservation of the cultural and environmental heritage, Lecce, 15–19 May 2005 (CD-ROM)

Chapter 22
Photometric Stereo by UV-Induced Fluorescence to Detect Protrusions on Georgia O'Keeffe's Paintings

Johanna Salvant, Marc Walton, Dale Kronkright, Chia-Kai Yeh, Fengqiang Li, Oliver Cossairt, and Aggelos K. Katsaggelos

Abstract A significant number of oil paintings produced by Georgia O'Keeffe (1887–1986) show surface protrusions of varying width, up to several hundreds of microns. These protrusions are similar to those described in the art conservation literature as metallic soaps. Since the presence of these protrusions raises questions about the state of conservation and long-term prospects for deterioration of these artworks, a 3D-imaging technique, photometric stereo using ultraviolet illumination, was developed for the long-term monitoring of the surface shape of the protrusions and the surrounding paint. Because the UV fluorescence response of painting materials is isotropic, errors typically caused by non-Lambertian (anisotropic) specularities when using visible reflected light can be avoided providing a more accurate estimation of shape. As an added benefit, fluorescence provides additional contrast information contributing to material characterization. The developed methodology aims to detect, characterize, and quantify the distribution of micro-protrusions and their development over the surface of entire artworks. Combined with a set of analytical in situ techniques, and computational tools, this approach constitutes a novel methodology to investigate the selective distribution of protrusions in correlation with the composition of painting materials at the macroscale. While focused on O'Keeffe's paintings as a case study, we expect the proposed approach to have broader significance by providing a non-invasive protocol to the conservation community to probe topological changes for any relatively flat painted surface of an artwork, and more specifically to monitor the dynamic formation of protrusions, in

---

J. Salvant
Centre de recherche et restauration des musées de France, Paris, France

M. Walton (✉)
Materials Science and Engineering Department, Northwestern University, Evanston, IL, USA
e-mail: marc.walton@northwestern.edu

D. Kronkright
Georgia O'Keeffe Museum, Santa Fe, NM, USA

C.-K. Yeh · F. Li · O. Cossairt · A. K. Katsaggelos
Electrical Engineering and Computer Science, Northwestern University, Evanston, IL, USA

F. Casadio et al. (eds.), *Metal Soaps in Art*, Cultural Heritage Science,
https://doi.org/10.1007/978-3-319-90617-1_22

relation to paint composition and modifications of environmental conditions, loans, exhibitions, and storage over the long term.

Keywords Georgia O'Keeffe · UV-Induced fluorescence · Protrusions · Photometric stereo · Metal soaps · Computational imaging

22.1 Introduction

Metal soap protrusions are a form of deterioration that affects scores of oil-based paintings made from antiquity until the present day (Noble et al. 2002, 2005; Higgitt et al. 2003; Keune 2005; Jones et al. 2007; Ferreira et al. 2011). Understanding how metal soap protrusions form and develop over time constitutes a major challenge to painting conservation, as this widespread phenomenon is known to alter the visual appearance of the painted surface (Noble et al. 2005; Noble and Boon 2007; Shimadzu and Van den Berg 2006; Centeno and Mahon 2009) and to compromise its chemical and mechanical stability (Rogala et al. 2010; Maines et al. 2011). A growing body of literature has emerged on the investigation of metal soap protrusions in paintings over the past 20 years. While most of this research has focused on investigating metal soap protrusions at the micro- and molecular scale using a variety of techniques requiring micro-samples (Heeren et al. 1999; Noble et al. 2002; van der Weerd 2002; Higgitt et al. 2003; Plater et al. 2003; Keune 2005; Osmond et al. 2005, 2012, Keune et al. 2005; Keune and Boon 2007; Cotte et al. 2007; Spring et al. 2008; Ferreira et al. 2011, 2015; Osmond 2014), in this study, an easy-to-implement and non-invasive methodological approach is described, using artworks by Georgia O'Keeffe as models, to document the macro distribution of protrusions.

Based on a survey initiated by the Georgia O'Keeffe Museum in 2009, a subset of O'Keeffe's paintings produced between 1920 and 1950 were identified as having disfiguring micro-protrusions scattered across their surfaces. These protrusions exhibit a strong UV-induced fluorescence response, range in size from 10 to greater than 200 μm, and occasionally appear erupted with a "caldera-like" shape consistent with soap aggregates found in other modern and early modern paintings (Faubel et al. 2011; O'Donoghue et al. 2006; Osmond et al. 2005; Ferreira et al. 2011; Duffy et al. 2014; Helwig et al. 2014). The protrusion formation process must have started at an early stage in these artworks' history: in a 1947 correspondence between O'Keeffe and conservator Caroline Keck (1947), the artist mentioned that she noticed opaque, granular textures and pinpoint losses appearing in several of her oil paintings created between 1928 and 1936, some of which today exhibits particularly large protrusions.[1] The identification of these protrusions in the 2009 survey led to a number of questions about when they first developed on these artworks and whether they were still actively changing.

[1]It should be noted that the Georgia O'Keeffe Museum has a collection of paint tubes belonging to the artist as well as documentation of O'Keeffe's color reference cards from the late 1920s to mid-1930s.

The presence and potential ongoing growth of protrusions on paintings raise significant concerns for their long-term preservation. Characterizing the protrusions, their distribution over the surface of the artworks and monitoring their development in time have now become a conservation priority not only for O'Keeffe's oeuvre but for all paintings affected by protrusions. Likewise, investigating the mechanisms and factors promoting the protrusion development in artworks has become essential, especially since the collections are often used in traveling exhibitions and are subjected to environmental fluctuations that may contribute to the ongoing development of protrusions.

In this work, in order to document the dynamic development of protrusions over time, a novel photometric stereo (PS) protocol was developed to image the 3D-texture of the painted surface where fine features of $\sim$10 μm wide can be resolved. We achieve these figures of merit by exploiting the fluorescence properties of pigments and binding media when excited by ultraviolet wavelengths. This isotropic response (Rost 1995) is ideally suited to using PS, which assumes Lambertian (diffuse) reflectance from a surface (Woodham 1978). As a second step, we also describe two data processing methods for detection of the protrusions using the PS data as an input: a manual count is compared to a data-driven approach. These counts of protrusions in different regions and the measurement of their size distribution are then correlated with paint composition.

22.2 Experimental Capture and Processing Procedures

The Georgia O'Keeffe oil painting *Pedernal* (Fig. 22.1a) has been chosen as a case study to illustrate the processing and analysis of PS data. Imaging was undertaken on a region of the painting, demarcated in Fig. 22.1a that exhibits a high number of protrusions and is representative of the overall palette. The region size of approximately 7.6 $\times$ 5.0 cm (height $\times$ width) was chosen to achieve high spatial resolution of 10 μm using our camera/lens configuration. Specifically, in the conditions used, spatial sampling of 15 μm per pixel over a field of view of 4 cm was achieved. It should also be noted that exposure to these UV wavelengths was kept at a minimum, similar to that needed for a UV photograph of a painted surface, to ensure the safety of the painting. The setup utilizes a 365 nm UV lamp (fitted with a Sylvania F15 T8/BLB bulb) as the illuminant and a Canon EOS 5D Mark III camera equipped with a 50 mm prime lens.

22.2.1 Photometric Stereo Data Acquisition and Processing

22.2.1.1 Capture

At its core, our method utilizes a common camera configuration described previously for PS (Woodham 1978) and reflectance transformation imaging (RTI) measurements; the camera is fixed orthogonal to the artwork and a series of

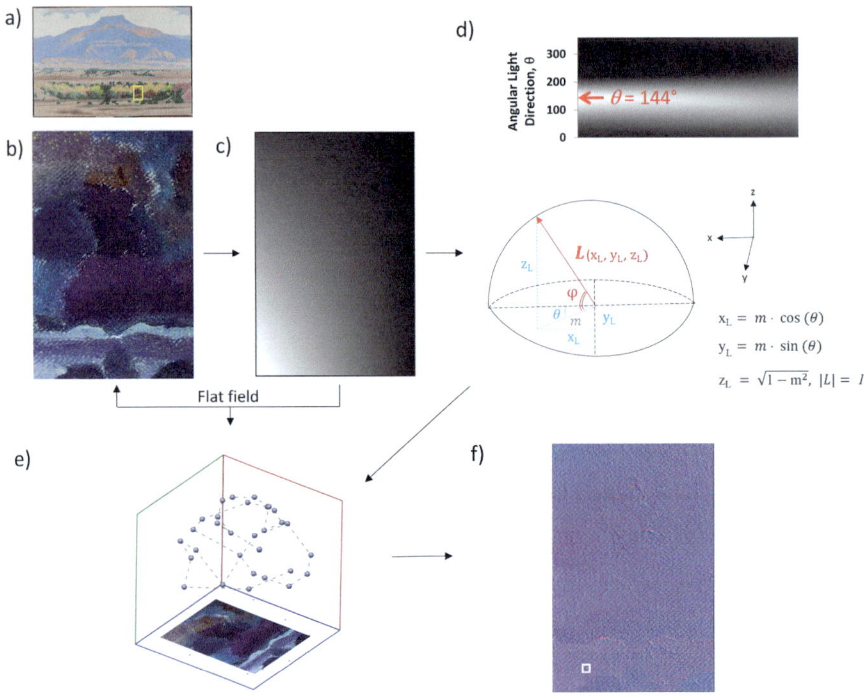

Fig. 22.1 Preprocessing pipeline of photometric stereo data illustrated for case study (**a**) Georgia O'Keeffe unvarnished oil painting *Pedernal* (1941–1942, 1997.5.12, CR1029, 51.1 × 76.8 cm, Georgia O'Keeffe museum, Santa Fe, NM). Region of ca. 7.6 × 5.0 cm, demarcated by yellow box, was selected as representative example for investigation. A single image illuminated with a directional light (**b**) was fit with an orthogonal Legendre polynomial to calculate a map of light drop-off (**c**) which shows the light originating from the direction of the lower left corner of the image. To achieve a precise measure of the lighting direction, polynomial-fit image was reshaped into polar coordinates (top d) to determine the angle θ of the illumination vector $\boldsymbol{L}$. Cartesian and polar coordinates of illumination vector $\boldsymbol{L}$ are schematically represented (bottom d). The x_L, y_L, z_L position of each illumination direction (**e**) was determined using an optimization procedure described in the text. PS was calculated using linear least squares to produce a normal vector map (**f**). The white box demarcates the detail region shown in Fig. 22.2

photographs are taken with different angles of illumination (Malzbender et al. 2001; Padfield et al. 2005; Mudge et al. 2005, 2006, 2008; Fattal et al. 2007; Earl et al. 2011, 2010a, 2010b; Malesevic et al. 2013;Duffy et al. 2013). Following the well-known photometric stereo equation, albedo (color) and normal maps (shape) of the object surface are recovered:

$$I = \tilde{\boldsymbol{N}} \cdot \boldsymbol{L}, \quad \tilde{\boldsymbol{N}} = k\boldsymbol{N}, \tag{22.1}$$

In this parameterization, the albedo (color) k is absorbed into the surface normal so that $\left|\tilde{\boldsymbol{N}}\right| = k$. Also the image intensity is given by $I(x, y)$ under illumination vector $\boldsymbol{L}$.

Shape recovery in PS requires that the surfaces being measured scatter light diffusely, often called Lambertian scattering. However, few materials exhibit ideal diffuse behavior but instead reflect light specularly. Painted surfaces often exhibit both diffuse and specular properties. These non-ideal surfaces are difficult to model accurately using PS alone (Ma et al. 2007), so recent studies have highlighted the advantages of using fluorescence-based PS for 3D-shape reconstruction compared to previous methods using reflected light (Sato et al. 2012; Treibitz et al. 2012). Stokes fluorescence emissions induced by a directional short wavelength light are isotropically radiated at longer wavelengths with ideal Lambertian characteristics where the emission intensity varies according to the cosine of the lighting angle (Kratohvil et al. 1978; Gordon et al. 1993; Glassner 1995). Therefore the use of fluorescence emission as the source of light for PS of painted surfaces can enable more accurate, precise, and repeatable shape measurements than the use of reflected light.

22.2.1.2 Preprocessing

Calculating the Illumination Vector

In a typical "free-form" PS or reflectance transformation imaging capture setup, a reflective ball is used to record the position of a far-light source $\boldsymbol{L}$ (Mudge et al. 2008). Here we have eliminated the need for using the reflective ball through an algorithm designed for near-lighting conditions (Huang et al. 2015; Cossairt et al. 2015). One of the principal benefits of removing the mirror ball and other calibration hardware from the scene is that the entire field of view may be dedicated to capturing the desired image. Also, this method where the light position is estimated directly from the object itself exploiting the inverse square law of light drop-off may be more accurate than using a mirror ball (Huang et al. 2015). This is especially true if the material being measured is a true Lambertian reflector as is the case of fluorescence measurements (Treibitz et al. 2012). Lastly, the weak intensity of most fluorescent sources necessitates near-light conditions in order to provide enough photon flux to produce an acceptable signal to noise characteristic in the captured images. In our example, the light was positioned 500 cm away from the surface.

In comparison to Huang et al. (2015), in this study, a different image processing pipeline is used which is more user-friendly for conservators, museum imaging departments, and conservation scientists interested in adopting the framework described. All of these steps were performed with either open-source and off-the-shelf software, such as ImageJ (Schneider et al. 2012, Schindelin et al. 2015). The code used in this project is available from an online repository (Github) as Python scripts that can be readily downloaded and installed into ImageJ.[2]

Light position estimation was achieved in a three-step process, as illustrated in Fig. 22.1b–f. First, each lighting direction image (Fig. 22.1b) was fit with an orthogonal polynomial in ImageJ, as exemplified in Fig. 22.1c, that is normalized to the mean grey value. While polynomial fitting of an image is typically used to compensate for uneven illumination, here we use the fit image (Fig. 22.1c) to estimate illumination direction. The fit-image can be reshaped into polar coordinates by radially slicing the image (Fig. 22.1d), from which the angle of illumination θ can be directly extracted from the point of maximum intensity. Second, the polynomial image is used to normalize or "flat-field" each lighting direction image to correct for the unevenness of illumination induced by the near light. Third, we borrow from the near-light model (Huang et al. 2015), in which the Cartesian coordinates (x_L, y_L, z_L) of the vector of illumination $\boldsymbol{L}$ are calculated from the direction and magnitude of the light drop-off. It is assumed that all light source positions are equidistant to the center of the region of investigation (Fig. 22.1d), meaning $|\boldsymbol{L}| = \mathbf{1}$.

As illustrated in Fig. 22.1e, each lighting direction $\boldsymbol{L}$ is determined by iteratively adjusting a parameter m while updating the surface normal $\tilde{\boldsymbol{N}}$. This is performed until the difference between $\boldsymbol{L}(m)\cdot \tilde{\boldsymbol{N}}$ and I becomes very small as per this cost function:

$$\operatorname{argmin}_{\tilde{\boldsymbol{N}},m} \left| \boldsymbol{L}(m)\cdot \tilde{\boldsymbol{N}} - I \right| \tag{22.2}$$

Calculating Surface Normal, Depth, and Albedo Maps

Based on the estimated vectors of illumination $\boldsymbol{L}$, the PS Eq. (22.1) is solved using least squares to recover the albedo and surface normal maps, the latter shown in Fig. 22.1f. Examples of x and y gradients, depth and albedo images are illustrated in Fig. 22.2a–d as high-resolution details from *Pedernal*. The depth map (Fig. 22.2c) is calculated by integrating the x and y gradients using standard algorithms such as described by Frankot and Chellappa (1988). Clearly apparent in the center of the image is a large protrusion (200 μm across) with other smaller protrusions (30–150 μm) peppered across the surface that have whitish to yellowish fluorescence (Fig. 22.2d). Figure 22.2e illustrates the 3D visualization of the depth map that enables characterization of the morphological features of the protrusions.

---

[2]https://github.com/NU-ACCESS/ImageJ-Photometric-Stereo-Tools

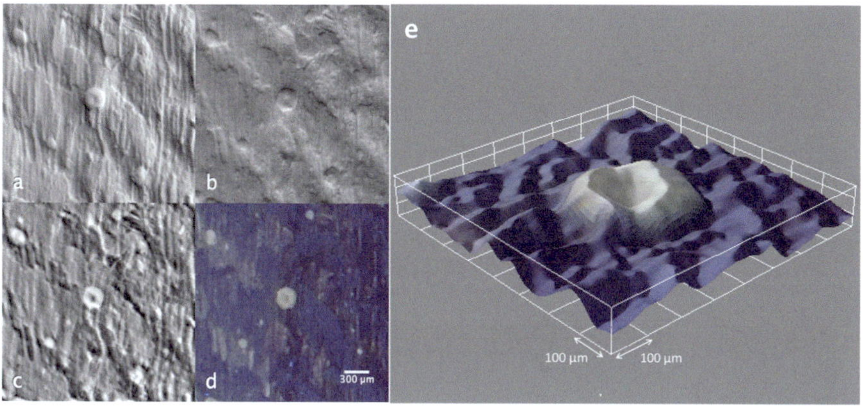

Fig. 22.2 After the flat-fielding procedure described by Fig. 22.1, the complete set of raking light images were used to calculate normal vector, depth, and albedo images. Details (**a–d**) of an area of 2.3 × 2.3 mm show the x-gradient (**a**) and y-gradient (**b**) of the surface normal vector N, the depth map (**c**) – where the grey scale intensity corresponds to relative height – and albedo (**d**) providing information on the diffuse color generated by the UV-induced fluorescence of the surface with underlying contrast from shape removed. The 3D-rendering (**e**) shows a "caldera" shape of the central protrusion

22.2.1.3 Protrusion Detection

The use of UV-induced fluorescence presents two benefits in addition to the improved accuracy of the PS measurements. First, the calculated albedo k is a perfectly diffuse and flat record of fluorescence which aids in the accurate detection and segmentation of the protrusions. Second, information on the UV-induced fluorescence response of the protrusions and of surrounding material is simultaneously collected in the visible range, which can contribute to material differentiation and characterization.

Using both the albedo and shape information, two data analysis methods were tested and compared to investigate the occurrence and distribution of protrusions in different colored areas. The first method, which is manual, uses ImageJ to detect protrusions on the albedo image with local color thresholding. The albedo image allows for a convenient visualization of the protrusions, as shown in Fig. 22.2d, due to their strong fluorescence response. This method provides quantitative information on the protrusions by detecting all visible protrusions within the investigated region. The second method is a fully automated detection scheme using an experimental algorithm implemented with Matlab and OpenCV (Bradski 2000). This algorithm has been designed for unaided detection of the protrusions, based on the combination of characteristic colors and surface features.

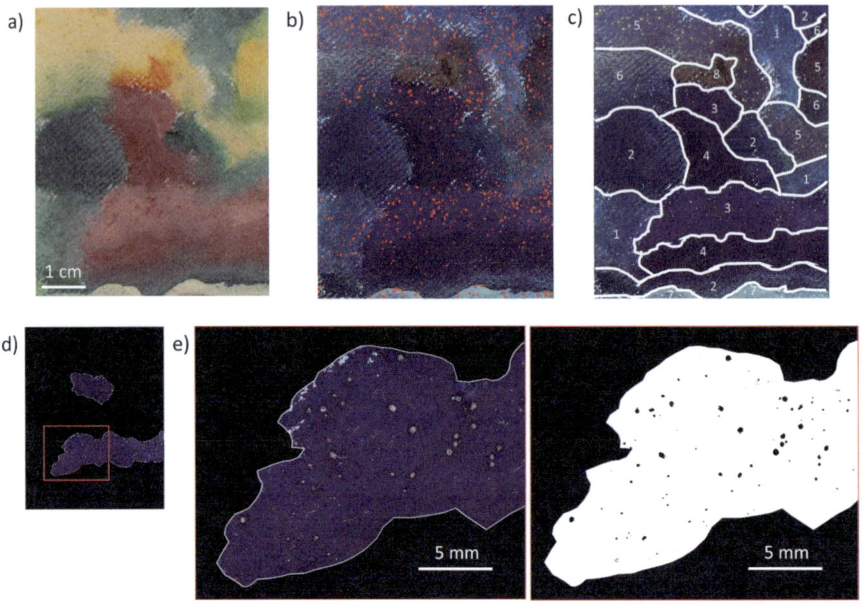

Fig. 22.3 (**a**) Visible image of the investigated region from *Pedernal* – shown in Fig. 22.1a – used to test methods for data analysis; (**b**) albedo image of the same region with overlaid binary mask of protrusions distribution (in red) manually selected using ImageJ; (**c**) albedo image with schematic indication of the eight different colored areas for segmentation; (**d**) mask from the albedo image for the light red painted area, labeled 3 in Fig. 22.3c; (**e**) steps for protrusions analysis in ImageJ: detail of the mask of the light red area (left); results for that detail from applying the mask of the light red area to the protrusion mask: distribution of protrusions (in black) for the light red area is obtained

Manual Protrusion Detection

Visible image of the investigated region is shown in Fig. 22.3a. For each observable protrusion in the albedo image, contiguous pixels of similar RGB values within an optimized adjusted range are hand-selected to fit the outline of the protrusion area (*via* the "wand tool"). All selected pixels are used to create a map of the protrusion distributions illustrated as a red overlay on the image of the fluorescence response (Fig. 22.3b).

The visible image (Fig. 22.3a) was manually segmented to isolate areas that appeared to have similar colors based on the visible and albedo images (Fig. 22.3c). Each of these areas was masked (Fig. 22.3d) to count the total number of protrusions per color. This, together with their location, size, and area were all recorded (Fig. 22.3e). The process was repeated for each segmented color area.

Despite being time-consuming, the manual method is advantageous since it is very easy to implement and accurate for detection of protrusions larger than 10 μm, considering the resolution of the PS data used in the current example.

Automated Protrusion Detection

A custom algorithm was applied to test automated quantification of the protrusions in the images acquired by PS. As for the manual method, each color area was manually segmented and masked (Fig. 22.3c–d). For the automated detection algorithm, the masks were then applied to both the albedo and normal images. Color thresholding and template matching on the albedo image were used to generate a binary mask, where a value of 1 indicates the presence of a protrusion. The normal image was integrated using the Frankot-Chellappa algorithm to produce a depth/height map of the surface of the painting. The OpenCV blob detection routine was then used to locate protrusions in the depth image, generating a second binary mask. A composite mask is then generated from the union of the two binary masks. The same procedure for analyzing protrusion statistics within each color region is then applied. The combination of information from the color albedo and surface-shaped images enables better rejection of false-positive protrusion detections, such as those that may be caused by brushwork texture or areas where the primer is apparent.

For both methods, the number of protrusions/cm^2 was calculated as a function of protrusion width for each color area (Fig. 22.4a).

22.2.2 *X-Ray Fluorescence (XRF) and Portable Fourier Transform Infrared Spectroscopy (FTIR)*

XRF imaging was carried out on the same region captured during the PS analysis (Fig. 22.1a), using an ELIO X-ray fluorescence imaging spectrometer (XGLab), equipped with a Rh tube and 1 mm spot size. Rastering was executed with a 250 μm step size and 1.6 s acquisition time for each point. The instrument was operated at 50 kV and 40 μA. The open access PyMCA software (Solé et al. 2007) was used to fit the XRF map and obtain the distribution of each element.

FTIR spectra were collected for each painted color area found in the region of interest using a Bruker Alpha small footprint portable FTIR spectrometer, in reflection mode and spectral range 6000–400 cm^{-1}, a measurement spot of 6 mm in diameter, and working distance of 15 mm. Two hundred and fifty six scans were acquired with a 4 cm^{-1} resolution.

22.3 Results and Discussion

22.3.1 Detection of Protrusions

22.3.1.1 "Ground truth" Manual Method

The red overlay on the fluorescence image (Fig. 22.3b) shows the distribution of protrusions across the painted surface. Protrusions appear most abundant in areas corresponding to light tonalities (e.g., light red, light green, beige) whereas they are almost absent in other darker tonalities (e.g., dark green, dark red). Frequency histograms (Fig. 22.4a) show the number of protrusions/cm^2 as a function of their width. From these data, the beige, light green, light red, yellow, and yellow green areas have a high number of protrusions (34 to 57 protrusions/cm^2). In contrast, the number of protrusions in the dark green, dark red, and orange areas is significantly

a)

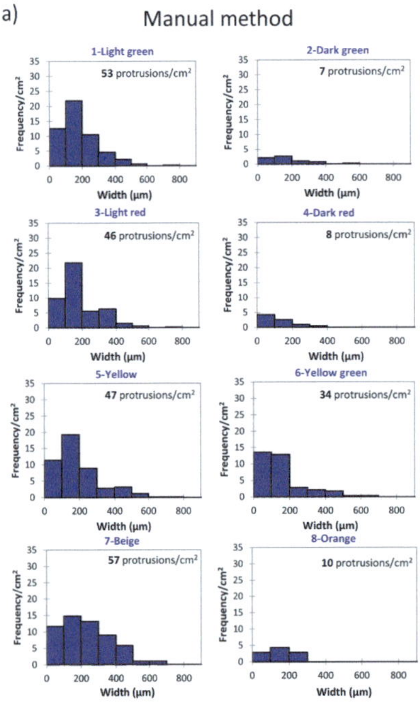

b)

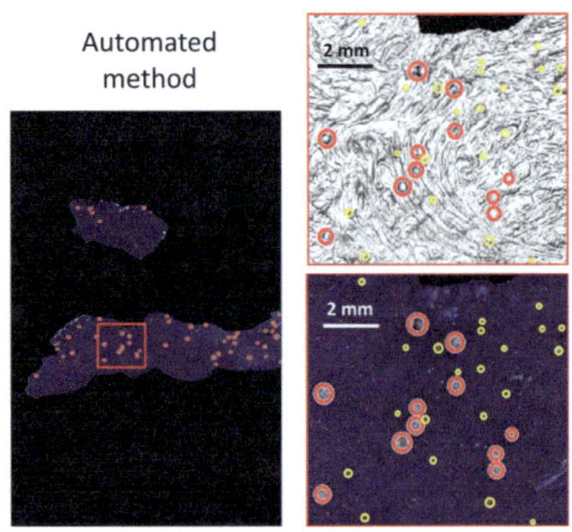

c)

Colored area (total rounded area in cm²)	Manual	Automated
	Number of protrusions detected/cm²	
1- Light green (5)	53	9
2-Dark green (9)	7	1
3- Light red (5)	46	11
4- Dark red (4)	8	2
5- Yellow (5)	47	11
6- Yellow green (3)	34	13
7- Beige (2)	57	31
8- Orange (1)	10	3

Fig. 22.4 Results of protrusion analysis and comparison of methods: (**a**) results of "ground truth" manual method of analysis. Histograms of protrusion occurrence and range of size for the eight colored areas; (**b**) results from the computer-automated method for the light red region: mask (top); details of results of protrusion detection (circled in red) overlaid onto depth map (bottom left) and albedo map (bottom right); the manually detected protrusions missed by automated processing are circled in yellow for comparison; (**c**) table comparing the occurrence of protrusions detected (number/cm^2) for each colored area in the investigated region by the two data analysis methods. Colored areas with high number of protrusions are shown in blue

lower, never exceeding 10 protrusions/cm^2. As will be discussed in more detail below, these data show that lighter and darker shades of the same colors (as in dark vs. light green – likely made from the same chromium-based pigment as determined by XRF) – clearly show dissimilar protrusion counts.

The profiles of size distributions are overall similar for the different colored areas, with a dominant proportion of protrusions in the 100–200 μm range. Only the dark red and yellow green areas exhibit a slightly higher proportion of small protrusions less than 100 μm wide. It can also be noted that the areas with abundant protrusions skewed to distributions with larger protrusions – with widths often exceeding 400 μm. It should be noted that for this proof-of-concept work, relatively large bins of 100 μm have been chosen to illustrate the capabilities of the method and reduce the number of false positives; however, in the future, smaller bins, especially in the 0–300 μm range, should be considered to better characterize the variety of sizes of the protrusions.

22.3.1.2 Evaluation of the Automated Method

Figure 22.4b shows the detected protrusions, circled in red, for the light red area using the automated algorithm overlaid onto the depth map and albedo images. For comparison, the manually detected protrusions missed by the automated method are circled in yellow. The detection of relatively big protrusions is fairly robust, while detection of small protrusions (~<200 μm) is not accurate. The challenge in detecting small protrusions lies in the fact that false positives are generated from the surface texture produced by the brush marks and from the white priming layer, contrasting in color with paint materials in areas where it remains visible. While automation significantly reduces processing time, further development of this preliminary version of the automated algorithm is however necessary to optimize detection accuracy so that it can be reliably applied to monitoring the variation of protrusion statistics over time.

22.3.2 Correlating Protrusions with Paint Composition at the Macroscale

Non-invasive FTIR and XRF analysis were performed on *Pedernal* as a case study to investigate how the distribution of protrusions correlates with paint materials. Many of O'Keeffe's artworks that exhibit protrusions have been painted on a specific type of commercial pre-primed canvas. Based on this observation, it was originally speculated by the museum conservators that the priming layer could have played a role in the protrusion formation, and this hypothesis was thus a focus of our investigation. As explained below, our analysis determined that, on the contrary, the distribution of protrusions correlates with the composition and color of paint layers, and therefore the degradation phenomena observed at the surface cannot be attributed to the priming.

Reflection FTIT analysis of the bare commercial primer (Fig. 22.5a) indicates that it is composed of lead white (mixture of hydrocerussite ($Pb_3(CO_3)_2(OH)_2$) and cerussite ($PbCO_3$)), barium sulfate ($BaSO_4$), and calcium carbonate ($CaCO_3$), based on the presence of the strong derivative band of the asymmetric CO_3^{2-} stretching at ca. 1465 cm^{-1} and the characteristic stretching and bending vibrational modes of hydrocerussite (3550, 1045, 680 cm^{-1}) and combination and overtone modes of cerussite (2413 cm^{-1}) and of calcium carbonate (4260, 2511, 1794 cm^{-1}) (Miliani et al. 2012). It should be noted that SEM/EDX analysis of a cross section from *Pedernal* clarified the relative proportions of these components, indicating that the priming is primarily composed of chalk with barium sulfate, and only minor quantities of lead compounds, which are, on the contrary, dominant in the paint layers. The sharp CH stretching at 2922 and 2846 cm^{-1} combined with the doublets in the 1540–1520 and 730–720 cm^{-1} regions, as shown in the details of Fig. 22.5a, point out toward the presence of metal carboxylates (Robinet and Corbeil 2003). Only few visible protrusions are observed on the unpainted commercial primer, and when they are detected, they tend to be small (<15 μm).

In all areas analyzed by FTIR, hydrocerussite and metal soap carboxylates were identified (Fig. 22.5a). Furthermore, the asymmetric carboxylate bands exhibit similar spectral features for the different colors suggesting the presence of the same type of metal carboxylates (Robinet and Corbeil 2003; Hermans et al. 2014, 2015) in both the protrusion-rich and protrusion-poor regions. These results highlight that while metal soaps are ubiquitous throughout the paint, they do not always aggregate into protrusions. Instead, paint layers with specific pigment compositions play a decisive role in whether or not surface protrusions form.

To investigate if protrusion formation is correlated with the presence of certain pigments or paints, the elemental composition was recorded via XRF mapping in the same region as the PS measurements. Elemental distributions of iron, chromium, and cadmium may be seen in Fig. 22.5b. These data indicate that red iron oxide pigments were employed for the red areas, chromium-based green pigments for the greens, and cadmium-based pigments for the yellow and orange. The distribution map of lead indicates that lead white was likely mixed with the colored pigments in varying proportions to produce lighter tonalities. The regions that are richest in lead correlate well with the areas with a higher occurrence of surface protrusions. These observations suggest that the specific type of lead white paint formulation in tubes employed by O'Keeffe and its proportion within the paint play a critical role in the development of surface protrusions, likely related to the aggregation of lead soaps. The latter would be also consistent with the observed UV-induced fluorescence response of the protrusions as reported for lead soap (Higgitt et al. 2003; Keune 2005; Keune and Boon 2007). It is currently being investigated whether the formulation of lead white tube paints O'Keeffe consistently used throughout several decades of her long career might be responsible for the observed abundance of soap protrusions and whether the same correlation in the distribution of lead and surface protrusions is observed on other O'Keeffe paintings.

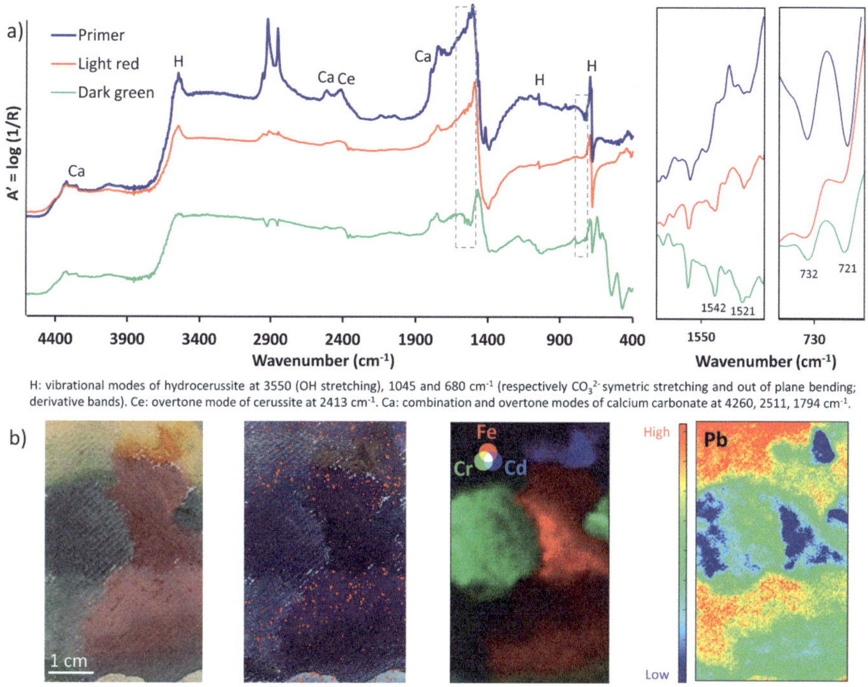

Fig. 22.5 Correlation between the distribution of protrusions and type of painting materials using complementary non-invasive analytical techniques. (**a**) FTIR spectra in reflectance of the primer (in blue), of the light red area (in red), and of the dark green area (in green); (**b**) XRF mapping of the studied area (from left to right): visible image of the area; albedo image of the area with overlays of the protrusions in red; false color K-lines elemental map distribution of iron (in red), chromium (in green), and cadmium (in blue); temperature L-lines elemental map distribution of lead where red indicates highest concentrations and blue the lowest (while only Lα peak areas are plotted here, however the Lβ distribution was found to be similar demonstrating that there is minimal contribution to the lead intensity from the ground layer)

22.4 Conclusions

Based on the case study of *Pedernal*, a Georgia O'Keeffe painting exhibiting high number of protrusions, this paper presents an innovative off-the-shelf methodology, based on imaging photometric stereo acquisition under UV illumination, to characterize the 2D-distribution and occurrence of the surface protrusions at the macroscale over an entire artwork. The two detection methods described show that the automated method of analysis produces fast protrusion distribution estimates, but with decreased precision with respect to the manual method, and substantial underestimation of protrusion numbers when parameters are adjusted to avoid false positives. The developed methodology also demonstrates that the uneven distribution of protrusions can be correlated to the paint composition: in the case study of *Pedernal*, the findings suggest that the development of protrusions, more numerous in the lead-rich areas, is primarily related to the lead white paint tube formulation in the paint layers, rather than the commercially primed canvas (containing mostly calcium and barium compounds, with only minor amounts of lead), which was previously thought to be the culprit. As an outlook, future research will focus on developing a more robust algorithm for automated protrusion counting and on providing user-friendly platforms so that the described imaging methodology can be widely applied to following the dynamic evolution of protrusions and thus directly interrogate the myriad factors that may lead to their development.

Acknowledgments Research at NU-ACCESS is supported by a generous grant from the Andrew W. Mellon Foundation. We thank Dr. Emeline Pouyet, NU-ACCESS, for her assistance with processing the XRF data; Dr. Francesca Casadio, Art Institute of Chicago, for discussing the interpretation of FTIR spectra; and Dr. Xiang Huang, Argonne National Laboratory, for our productive discussions.

References

Bradski G (2000) The openCV library. Doct Dobb's J 25(11):120–126. http://opencv.org/

Centeno S, Mahon D (2009) The chemistry of aging in oil paintings: metal soaps and visual changes. Metropolitan Museum Art Bull 67(1):12–19

Cossairt O, Huang X, Matsuda N, Stratis H, Broadway M, Tumblin J, Bearman G, Doehne E, Katsaggelos A, Walton M (2015) Surface shape studies of the art of Paul Gauguin. In: Guidi G, Scopigno R, Torres JC et al (eds) International congress on digital heritage granada, vol 2015. IEEE, Piscataway, pp 13–20

Cotte M, Checroun E, Susini J, Walter P (2007) Micro-analytical study of interactions between oil and lead compounds in paintings. Appl Phys Mater Sci Process 89(4):841–848

Duffy S, Bryan P, Earl G, Beale G, Pagi H, Kotoula E (2013) Multi-light imaging for heritage applications. English Heritage, London

Duffy M, Martins A, Boon J (2014) Metal soaps and visual changes in a painting by René Magritte – The Menaced Assassin, 1927. In: Van den Berg KJ, Burnstock A, de Tagle A, de Keijzer M, Heydenreich G, Krueger J, Learner T (eds) Issues in contemporary oil paints. Springer, Cham, pp 197–203

Earl G, Beale G, Martinez K, Pagi H (2010a) Polynomial texture mapping and related imaging technologies for the recording, analysis and presentation of archaeological materials. In: Mills J, Barber D, Miller P, Newton I (eds) Proceedings of the ISPRS commission V Midterm symposium, Newcastle upon Tyne, 21–24 June, pp 218–223

Earl G, Martinez K, Malzbender T (2010b) Archaeological applications of polynomial texture mapping: analysis, conservation and representation. J Archaeol Sci 37(8):2040–2050

Earl G, Basford P, Bischoff A, Bowman A, Crowther C, Dahl J, Hodgson M, Isaksen L, Kotoula E, Martinez K, Pagi H, Piquette K (2011) Reflectance transformation imaging systems for ancient documentary artefacts. In: Dunn S, Bowen J, Ng K (eds) Electronic visualisation and the arts (EVA 2011). BCS, The Chartered Institute for IT, London, pp 147–154

Fattal R, Agrawala M, Rusinkiewicz S (2007) Multiscale shape and detail enhancement from multi-light image collections. ACM Trans Graph 26 (3):Paper 51

Faubel W, Simon R, Heissler S, Friedrich F, Weidler PG, Becker H, Schmidt W (2011) Protrusions in a painting by max Beckmann examined with confocal μ-XRF. J Anal At Spectrom 26(5):942–948
Ferreira E, Boon J, Stampanoni M, Marone F (2011) Study of the mechanism of formation of calcium soaps in an early 20th-century easel painting with correlative 2D and 3D microscopy. In: Bridgland J (ed) ICOM committee for conservation 16th triennial meeting, Lisbon, 19–23 Sept 2011. Paper 1604
Ferreira E, Gros D, Wyss K, Scherrer N, Zumbühl S, Marone F (2015) Faded shine.... The degradation of brass powder in two nineteenth century paintings. Herit Sci 3:1–11
Frankot RT, Chellappa R (1988) A method for enforcing integrability in shape from shading algorithms. IEEE Trans Pattern Anal Mach Intell 10(4):439–451
Glassner AS (1995) The radiance equation, Chapter 17. In: Barsky B (ed) Principles of digital image synthesis, vol 2. Morgan Kaufmann, San Francisco, pp 871–882
Gordon H, Voss K, Kilpatrick K (1993) Angular distribution of fluorescence from phytoplankton. Limnol Oceanogr 38(7):1582–1586
Heeren M, Boon J, Noble P, Wadum J (1999) Integrating imaging FTIR and secondary ion mass spectrometry for the analysis of embedded paint cross-sections. In: Vontobel R (ed) ICOM committee for conservation 12th triennial meeting, Lyon, 29 Aug–3 Sept, pp 228–233
Helwig K, Poulin J, Corbeil M-C, Moffatt E, Duguay D (2014) Conservation issues in several twentieth-century canadian oil paintings: the role of zinc carboxylate reaction products. In: Van den Berg KJ, Burnstock A, de Tagle A, de Keijzer M, Heydenreich G, Krueger J, Learner T (eds) Issues in contemporary oil paints. Springer, Cham, pp 167–184
Hermans JJ, Keune K, Van Loon A, Stols-Witlox M, Corkery, R, Iedema P (2014) The synthesis of new types of lead and zinc soaps: a source of information for the study of oil paint degradation. In Bridgland J (ed) ICOM committee for conservation 17th triennial conference preprints, Melbourne, 15–19 Sept 2014, pp 1604–1612
Hermans J, Keune K, Van Loon A, Iedema P (2015) An infrared spectroscopic study of the nature of zinc carboxylates in oil paintings. J Anal At Spectrom 30(7):1600–1608
Higgitt C, Spring M, Saunders D (2003) Pigment-medium interactions in oil paint films containing red lead or lead-tin yellow. Natl Gallery Tech Bull 24:75–95
Huang X, Walton M, Bearman G, Cossairt O (2015) Near Light Correction for Image Relighting and 3D Shape Recovery. In: Guidi G, Scopigno R, Torres JC et al (eds) International congress on digital heritage granada, 28 Sept – 2 Oct 2015. IEEE, Piscataway, pp 215–222
Jones R, Townsend J, Stonor K, Duff N (2007) Lead soap aggregates in sixteenth and seventeenth century British paintings. In: Parkin HM (ed) AIC annual meeting 2006, paintings specialty group postprints, Providence, 16–19 June 2007, pp 24–32
Keck C (1947) Letter from Caroline Keck to Georgia O'Keeffe, 17 June 1947, Georgia O'Keeffe Museum Research Center. Copy, Gift of the Georgia O'Keeffe Foundation
Keune K (2005) Binding medium, pigments and metal soaps characterised and localised in paint cross-sections. PhD thesis, University of Amsterdam
Keune K, Boon J (2007) Analytical imaging studies of cross-sections of paintings affected by lead soap aggregate formation. Stud Conserv 52(3):161–176
Keune K, Ferreira E, Boon J (2005) Characterisation and localization of the oil-binding medium in paint cross-sections using imaging secondary ion mass spectrometry. In: Verger I (ed) ICOM committee for conservation 14th triennial meeting, The Hague, 12–16 Sept 2005, pp 796–802
Kratohvil J, Lee M-P, Kerker M (1978) Angular distribution of fluorescence from small particles. Appl Opt 17(13):1980–1982
Ma W-C, Hawkins T, Peers P, Chabert C-F, Weiss M, Debevec P (2007) Rapid acquisition of specular and diffuse normal maps from polarized spherical gradient illumination. In: Kautz J, Pattanaik S (eds) Proceedings of the 18th eurographics conference on rendering techniques, 25–27 June 2007. Eurographics Association, Grenoble, pp 183–194

Maines C, Rogala D, Lake S, Mecklenburg M (2011) Deterioration in abstract expressionistic paintings: analysis of zinc oxide paint layers in works from the collection of the Hirshhorn Museum and Sculpture Garden, Smithsonian Institution. Mater Res Soc Symp Proc 1319: 275–284

Malesevic B, Obradovic R, Banjac B, Jovovic I, Makragic M (2013) Application of polynomial texture mapping in process of digitalization of cultural heritage. arXiv preprint:13126935

Malzbender T, Gelb D, Wolters HJ (2001) Polynomial texture maps. In: Pocock L (ed) Proceedings of the 28th annual conference on computer graphics and interactive techniques (SIGGRAPH 2001), 12–17 Aug 2001. ACM, Los Angeles, pp 519–528

Miliani C, Rosi F, Daveri A, Brunetti BG (2012) Reflection infrared spectroscopy for the non-invasive in situ study of artists' pigments. Appl Phys Mater Sci Process 106:295–307

Mudge M, Voutaz J-P, Schroer C, Lum M (2005) Reflection transformation imaging and virtual representations of coins from the hospice of the grand St. Bernard. In: Mudge M, Ryan N, Scopigno R (eds) VAST'05 proceedings of the 6th international conference on virtual reality, archaeology and intelligent cultural heritage, Pisa, 8–11 Nov 2005. Eurographics Association, pp 29–40

Mudge M, Malzbender T, Schroer C, Lum M (2006) New reflection transformation imaging methods for rock art and multiple-viewpoint display. In: IEEE conference on visual analytics science and technology, 31 Oct–2 Nov 2006. IEEE VAST, Baltimore, pp 195–202

Mudge M, Malzbender T, Chalmers A, Scopigno R, Davis J, Wang O, Gunawardane P, Ashley M, Doerr M, Proenca A, Barbosa J (2008) Image-based empirical information acquisition, scientific reliability, and long-term digital preservation for the natural sciences and cultural heritage. Eurographics tutorials 2008, 29th annual conference of the European association for computer graphics, 14–18 Apr 2008. The Eurographics Association, Crete. http://homepages.inf.ed.ac.uk/rbf/CVonline/LOCAL_COPIES/MUDGE/EG-mudge-tutorial-notes-final.pdf

Noble P, Boon J (2007) Metal soap degradation of oil paintings: aggregates, increased transparency and efflorescence. In: Parkin HM (ed) AIC annual meeting 2006, paintings specialty group postprints, Providence, 16–19 June 2007, pp 5–19

Noble P, Boon J, Wadum J (2002) Dissolution, aggregation and protrusion. Lead soap formation in 17th century grounds and paint layers. ArtMatters 1:46–61

Noble P, Van Loon A, Boon J (2005) Chemical changes in old master paintings II: darkening due to increased transparency as a result of metal soap formation. In: Verger I (ed) ICOM committee for conservation 14th triennial meeting, The Hague, 12–16 Sept 2005, pp 496–503

O'Donoghue E, Johnson A, Mazurek J, Preusser F, Schilling M, Walton M (2006) Dictated by media: conservation and technical analysis of a 1938 Joan Miró canvas painting. Stud Conserv 51(sup2):62–68

Osmond G (2014) Zinc white and the influence of paint composition for stability in oil based media. In: Van den Berg KJ, Burnstock A, de Tagle A, de Keijzer M, Heydenreich G, Krueger J, Learner T (eds) Issues in contemporary oil paints. Springer, Cham, pp 263–281

Osmond G, Keune K, Boon J (2005) A study of zinc soap aggregates in a late 19th century painting by R.G. Rivers at the Queensland art gallery. AICCM Bull 29(1):37–46

Osmond G, Boon J, Puskar L, Drennan J (2012) Metal stearate distributions in modern artists' oil paints: Surface and cross-sectional investigation of reference paint films using conventional and synchrotron infrared microspectroscopy. Appl Spectrosc 66(10):1136–1144

Padfield J, Saunders D, Malzbender T (2005) Polynomial texture mapping: a new tool for examining the surface of paintings. In: Verger I (ed) ICOM committee for conservation 14th triennial meeting, The Hague, 12–16 Sept 2005, pp 504–510

Plater M, De Silva B, Gelbrich T, Hursthouse M, Higgitt C, Saunders D (2003) The characterisation of lead fatty acid soaps in 'protrusions' in aged traditional oil paint. Polyhedron 22(24):3171–3179

Robinet L, Corbeil M-C (2003) The characterization of metal soaps. Stud Conserv 48:23–40

Rogala D, Lake S, Maines C, Mecklenburg M (2010) Condition problems related to zinc oxide underlayers: examination of selected abstract expressionist paintings from the collection of the Hirshhorn Museum and Sculpture Garden, Smithsonian Institution. J Am Inst Conserv 49: 96–113

Rost FW (1995) Fluorescence microscopy, vol 2. Cambridge University Press, Cambridge

Sato I, Okabe T, Sato Y (2012) Bispectral photometric stereo based on fluorescence. In: 2012 IEEE conference proceedings on computer vision and pattern recognition (CVPR), 16–21 June 2012. IEEE, Providence, pp 270–277

Schindelin J, Rueden CT, Hiner MC, Eliceiri KW (2015) The ImageJ ecosystem: an open platform for biomedical image analysis. Mol Reprod Dev 82(7–8):518–529

Schneider CA, Rasband WS, Eliceiri KW (2012) NIH image to ImageJ: 25 years of image analysis. Nat Meth 9(7):671–675

Shimadzu Y, Van den Berg JD (2006) On metal soap related colour and transparency changes in a 19th C painting by Millais. In: Boon J, Ferreira E (eds) Reporting highlights of the De Mayerne programme. Netherlands Organization for Scientific Research (NWO), The Hague, pp 43–52

Solé V, Papillon E, Cotte M, Walter P, Susini J (2007) A multiplatform code for the analysis of energy-dispersive X-ray fluorescence spectra. Spectrochim Acta Part B 62(1):63–68

Spring M, Ricci C, Peggie D, Kazarian S (2008) ATR-FTIR imaging for the analysis of organic materials in paint cross sections: case studies on paint samples from the National Gallery, London. Anal Bioanal Chem 392(1–2):37–45

Treibitz T, Murez Z, Mitchell BG, Kriegman D (2012) Shape from fluorescence. In: Fitzgibbon A, Lazebnik S, Perona P, Sato Y, Schmid C (eds) 12th European conference on computer vision (ECCV 2012), 7–13 Oct 2012. Springer, Florence, pp 292–306

van der Weerd J (2002) Microspectroscopic analysis of traditional oil paint. PhD thesis, University of Amsterdam

Woodham RJ (1978) Photometric stereo: a reflectance map technique for determining surface orientation from image intensity. In: Proceedings of SPIE's 22nd annual technical symposium, international society for optics and photonics, pp 136–143

Chapter 23
A Montparnasse Disease? Severe Manifestations of Metal Soaps in Paintings by Pierre Soulages from Around 1959 to 1960 (Delaminating Oil Paint Layers, Medium Exudates, Discolorations)

Pauline Hélou-de La Grandière

Abstract Pierre Soulages (b.1919) is the most renowned French painter alive today, well-known for his use of the color black. During his long and prolific career (around 1500 paintings made to date), he has worked with various oil paint techniques and later on acrylics, making continuous use of the color black. This paper deals with paintings from the end of the 1950s, where black paint was layered on and then scraped off with tools to reveal colored underpainting (Fig. 23.1). The creation process of Pierre Soulages has often compelled him to adapt his tools (Ragon 1990; McEnroe 1991; Encrevé 1995) and to innovate in his use of materials. Yet he has always remained interested in understanding the physical phenomena of the drying process, discussing the subject with friends such as Marc Havel, chief engineer at Bourgeois and connoisseur of painting techniques (Havel 1974), or Pierre-Gilles de Gennes, winner of the Nobel Prize in Physics. Paradoxically, in spite of this constant interest, some of his paintings from an important period in the 1950s–1960s are deteriorating. This paper focuses on those paintings, their specific patterns of degradation, and the roles that technique and material composition have played in the process.

Keywords Pierre Soulages · Montparnasse · Lead white · Zinc white · Ivory black · Delamination · Consolidation · Paint exudation · Blanching · Mat

---

P. Hélou-de La Grandière (✉)
Paintings Conservator, Nîmes, France

F. Casadio et al. (eds.), *Metal Soaps in Art*, Cultural Heritage Science,
https://doi.org/10.1007/978-3-319-90617-1_23

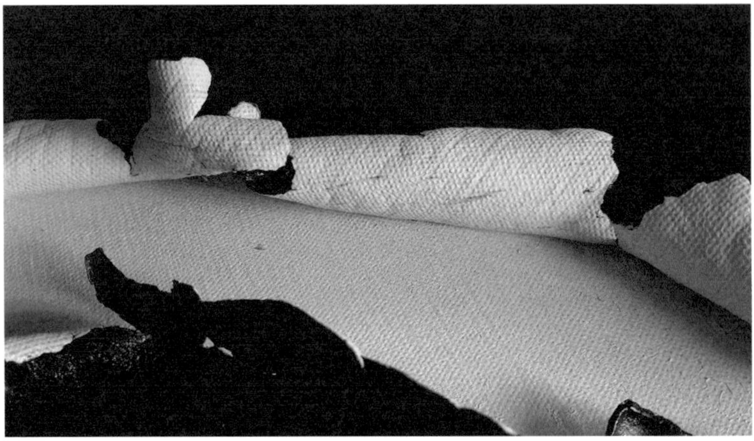

Fig. 23.1 Severe delamination occurring between ground and paint layer in *Peinture, 12 mars 1960* (Sara Hildén Foundation/Sara Hildén Art Museum). (© Pauline Hélou-de La Grandière 2016).

23.1 Soulages' Materials and Working Methods

To learn details of Soulages' working practices, we were lucky enough to be able to talk to the artist himself. This provided important insights into different aspects of his creative process, ranging from the studio conditions under which he worked to the methods and materials he employed, as discussed below.

23.1.1 Studio Conditions

Before 1957, Pierre Soulages lived and worked in a studio located on rue Schœlcher, in the Montparnasse neighborhood of Paris (Ragon 1990:37). There his practice was quite conventional; he used linseed oil paint, and he stretched and prepared his canvases himself, or with his friend Hans Hartung (1904–1989), following the Doerner recipes (Doerner 1934). The room was not very spacious, so he mostly made small or medium format pictures.

In December 1957, he moved to a larger studio on rue Galande, in the *Quartier Latin* near Notre-Dame-de-Paris. Consequently his practices changed. Located on the fifth and top floor of the building, this spacious studio now allowed him to make pictures in large formats and to store materials, so he could keep about 30 canvases on hand (de La Grandière 2005). The studio had continuous windows along two opposite sides, was very poorly insulated, and had few walls for hanging paintings. On photographs by Izis[1] dated March 1960 (Galerie de France 1960; Ragon 1990:72,77; Encrevé 1994:24; Aix-en-Provence 2016), one can see a dozen paintings placed facing the window looking southwest. Pierre Soulages remembered that the poor insulation allowed cold and moisture to enter into the studio in the winter and made for high temperatures in the summer (de La Grandière 2005). The year 1959 was the sunniest year of the century in Paris, whereas 1960 was a wet year with a heat wave in the last week of February (Séchet 2016). These fluctuations in weather condition may have had an impact on the drying of the paintings in his studio.

At this new address, Soulages continued to buy his materials from the same colormen in Montparnasse as he had done before: with some exceptions, his usual suppliers were Lucien Lefebvre-Foinet, established in rue Vavin (de La Grandière 2005; Hélou-de La Grandière 2006; Hélou-de La Grandière et al. 2007; Adam 2010; Helwig et al. 2014), and Edouard Adam, situated in Boulevard Edgar Quinet (Adam 2010).

23.1.2 *Supports and Grounds for Oil Painting*

During the 1950s, Soulages bought his pre-primed canvases on rolls, mostly from Lefebvre-Foinet[2] and sometimes from Adam.[3] He stretched the canvases himself onto quality stretchers, often adding a wooden strip to increase the bevel on the stretcher bars and thus avoid contact between the stretcher bars and the canvas when painting. This practice was copied by other artists, such as Zao Wou-Ki (1920–2013).

Soulages was very attentive to the choice of ground type. Since Lefebvre-Foinet was well-known for the high quality of his materials,[4] Soulages bought canvas with nonabsorbent ground from him, primed with two or three layers of lead white so as to avoid matte defects (de La Grandière 2005; Hélou-de La Grandière et al. 2007).

Sometimes the ground was too greasy, so he needed to degrease the surface, as his friend Roger Vailland describes he did when creating the work *Peinture, 200 × 160 cm, 27 mars 1961*:

[1]Israëlis Bidermanas, alias Izis, photographer (1911–1980).

[2]A supplier's stamp is sometimes visible on the reverse of the canvas. Oral exchange with Soulages confirmed his common use of Lefebvre-Foinet's canvas (2004, 2006, and 2017).

[3]The stamp of Adam, with a palette, is visible on some stretchers after 1959. Adam began to sell stretchers at the advice of Soulages ((Adam 2010) and oral exchange with Soulages (2017) and Adam (2004)).

[4]The shop issued an edition of postcards of the workshop with the caption (in English): "Atelier of LUCIEN LEFEBVRE-FOINET, at 19, rue Vavin – PARIS/Where the finest artists' oils are ground by hand." Pierre Soulages recalled that artists reported the grounds to be of high quality (verbal exchange 2004).

> He touches the canvas: the ground is too greasy. The canvas had been delivered by the supplier in the way Soulages had asked for it, but he degreased it with a cloth and an appropriate liquid (Vailland 1961).[5]

Pierre Soulages remembers this characteristic well:

> "For Lefebvre-Foinet's, there were 2 and 3 layers of grounds: (...) for the 3-layer ones, which appeared too glossy, I degreased the surface to promote adhesion"[6] (de La Grandière 2005).

The colourman's grandson confirmed that the specific three-layer grounds were too greasy:

> "(...) especially during the summer, when high temperatures caused exudation from the ground (!). Degreasing the surface with turpentine was then absolutely necessary"[7] (de La Grandière 2005).

Lefebvre-Foinet used poppy-seed oil to grind lead white, to limit the yellowing of the white color (O'Malley and Moffat 2001: 86; de La Grandière 2005). Analysis confirmed that poppy-seed oil was used to grind the lead white found in the tube paints (Corbeil et al. 2011:40), but for the grounds, linseed oil seems to have been used more often, as it was identified in Lefebvre-Foinet's grounds used in Jean-Paul Riopelle's paintings (Corbeil et al. 2011:40). In this type of ground, a specific mixture of plumbonacrite and hydrocerussite varieties of lead white was identified (Corbeil and Helwig 2008; Corbeil et al. 2011:54).

The use of a three-layer ground described in the sources is confirmed by observation. On the reverse of some canvases (such as *15 décembre 1959* or *12 mars 1960)*, broad lines or "tracks" are visible that measure roughly 10 to 35 centimeters wide and show variable saturation (Fig. 23.5). This is a consequence of the three-layer ground preparation, where the tracks correspond to the overlap of the layers. Another feature readily distinguishes the three-layer grounds from "traditional" ones, namely, in the case of traditional canvas, where the black impasto paint was applied with such force that the oil was pushed through the canvas, it causes corresponding stains on the reverse. The nonporous canvases with a triple ground layer, however, do not show such stains.

[5]"Il touche la toile. '–La préparation est. trop grasse'. La toile avait été livrée par le marchand, telle que Soulages l'avait demandée, mais il se met à la dégraisser avec un chiffon et un liquide approprié".

[6]Conversation with Pierre Soulages.

[7]Phone conversation with Lefebvre-Foinet's descendant.

23.1.3 Oil Paint, Medium, and Varnish

Soulages used colors from the tube such as yellow ochre, lead white, ultramarine blue, or the *Orangé de Mars*.[8] For the black paint, he used his own recipe of a black paste with a binding medium, which consisted of ivory black, drying oil boiled with lead oxide, and Flemish siccative (from Bourgeois). This black paste was often made by Edouard Adam who was very open to technological innovations. Indeed Adam made the formulation of International Klein Blue (IKB) with Yves Klein in the same period (Adam 2010). Soulages sometimes used the Venetian medium from Bourgeois and he rapidly stopped using Venetian turpentine because of drying issues. To obtain a viscous paste with his binding media, a common practice at the time was to expose it to the sun for several weeks. Soulages remembers doing this and recalls that Lefebvre-Foinet also used the technique, especially for linseed oil from Bombay and La Plata (de La Grandière 2005).

Pierre Soulages made his own dammar varnish in turpentine, or he sometimes used a retouching Vibert varnish instead (de La Grandière 2005), which consists of a dammar resin plasticized by the addition of a little poppy-seed oil and dissolved in white spirit (Mills and White 1999:107; Mayer and Myers 2013:77).

The tools used by Soulages are large brushes measuring 10 to 20 cm wide, or palette knives, often homemade, such as a knife made of a rubber slab fixed to a wooden handle (Ragon 1990; McEnroe 1991:113, 149–173; Encrevé 1994:167). But he has also had tools specially designed for him, such as an extra-large brush made by Adam (Giraudy 1996:112).

23.1.4 The Paintings

Onto the pre-primed canvas, Pierre Soulages laid down a color: blue, brown, red, yellow ochre, or white. Onto this fresh paint, he poured his black liquid paste and spread it out with brushes or spatulas. This revealed the colored underlayer and made variable mixtures in the fresh paint.

Soulages' paintings of the period 1959–1960 were quickly a great success all over the world. Today they are present in more than 50 museums and in important collections. Unfortunately though, some of them are suffering severe forms of deterioration, as the next section of this article goes on to discuss.

[8]During verbal exchange, Pierre Soulages specified that Edouard Adam was the unique supplier for this nice Mars Orange Colour in Paris.

23.2 Conservation Issues

23.2.1 Identification of a Corpus

In the second half of the 1990s, lifting paint and severe problems of delamination started to be noticed in Soulages' paintings dating from April 1959 to March 1960.

23.2.2 Descriptions of Degradation Phenomena

23.2.2.1 Delaminating Oil Paint Layers

The delamination results from loss of adhesion between the ground and the paint layer. It occurs only in the black impastos and is not influenced by the colors underneath, which can be white, red, blue, or yellow.[9] This loss of adhesion is also associated with internal stresses within the paint film. Strong retraction of the top layer (black) seems to be responsible for the lifted edges of black impasto, while an elongation or a swelling of the paint layer underneath (often white) causes lifting like a raised bubble from underneath the black impastos (Fig. 23.1). In the latter case, the flakes consist of softened paint with a "waxy" texture, curling back upon themselves without breaking.

23.2.2.2 Softening of Paint Layers

The softening of the paint propagates from the ground up toward the surface and may concern underlying or upper layers or both. When softening develops in underlayers, wave-like deformations appear that affect both the canvas and the paint layers (Fig. 23.2). When it develops in the upper layers, dust is attracted and becomes attached to the surface. In the 1990s, the paintings in question were varnished by conservators or by galleries. Unfortunately this resulted in the phenomenon becoming isolated underneath the varnish, making it worse (Barabant and Hélou-de La Grandière 2012). The most severe softening of paint was observed in *8 décembre 1959* and in *15 décembre 1959*, which had both been varnished to avoid the need for dusting. In these two cases, the delamination seems to have become worse after varnishing. After a few years, the varnish itself lost adhesion and flaked off, probably due to the solvent emanating from the paint layers underneath.

Softening can also be connected with a strong increase in the glossiness of the surface, which emits a strong fluorescence under UV light, suggesting the presence

[9]This may be different for the brown color that could have a stabilizing effect, as noticed by Jaap Boon when looking together at *Peinture 11 mars 1960* in Tampere (Finland). Work in progress.

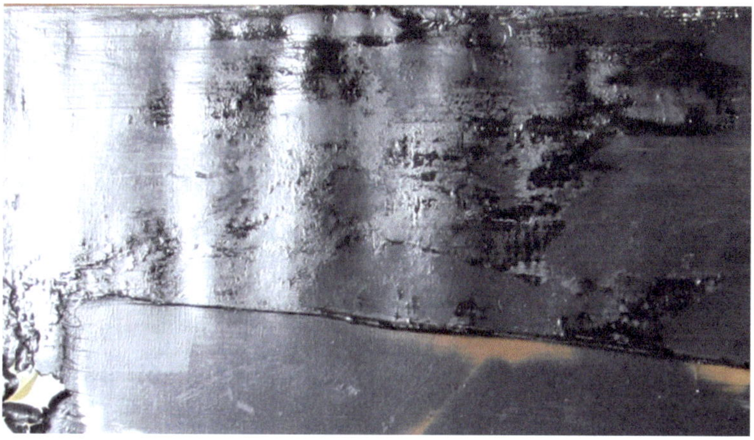

Fig. 23.2 Severe delamination occurring between ground and paint layer in *Peinture, 12 mars 1960* (Sara Hildén Foundation/Sara Hildén Art Museum). Note the deformation and glossy appearance of the paint film during the softening process. (© Pauline Hélou-de La Grandière 2016)

of lead soaps (Boon and Hoogland 2014: 229; Boon 2016). Ultimately the softening either leads to the surface becoming sticky and attracting dust or to exudation of the binding media.

As previously mentioned, softening can be concomitant with the process of lifting paint, in which case the paintings appear to have been overpainted (*Peinture, 8 décembre 1959* is an example) and the surface looks swollen.

23.2.2.3 Medium/Oil Exudates

Exudates were first noted in 1992 (see Table 23.1), that is, after 32 years of "drying" of the paint. The drips extend rapidly (a rate of 2 mm in 3 months was noted for *Peinture, 15 décembre 1959*). The drips run in long exudations from the impastos, causing strong visual disturbance. A flattened and matte impasto appears with a soft and glossy drip running from it (Fig. 23.3). Exudates show a strong fluorescence in UV and may contain color from the underlayer, as in *Peinture 12 mars* (Fig. 23.4). The phenomenon of exudation has only been noted in conjunction with the use of nonabsorbent, pre-primed canvas that is probably prepared with a three-layer ground, as may be distinguished by the priming tracks visible on the reverse and the absence of oil stains penetrated from the black impasto (see section 23.1.2) (Fig. 23.5).

Table 23.1 Paintings studied during conservation treatments

Paintings: *date*, collection, current location	Absorbency of the pre-primed canvas	Microscale properties of the ground and paint layers	Conservation issue	Picture of degradation
Peinture, 18 avril 1959, **Musée de St Etienne** (kept in storage)	Absorbent (oil stains on the reverse)	SEM showed acicular shaped lead white particles in the ground. FTIR detected lead palmitates in the ground, and in areas of lifted paint as well as in areas of paint in good condition. SEM shows rounded particles of lead white in the paint [1], [2]	Early example of delamination (restored by the artist in 1984) and rapid growth of lifting areas. Consolidated in 2008 and 2013	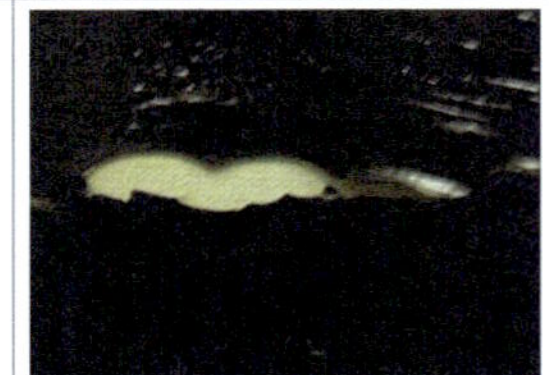
Peinture, 2 novembre 1959, **Musée Fabre, Montpellier** (consolidated in 2015, can be exhibited)	Absorbent	Round-shaped particles of lead white in the ground (with few acicular shape particles on the top of the ground). Few lead soaps [1]	Softening tendency in thicker layers. Some lifting of paint (stabilized in 2009 and after a flood in 2015)	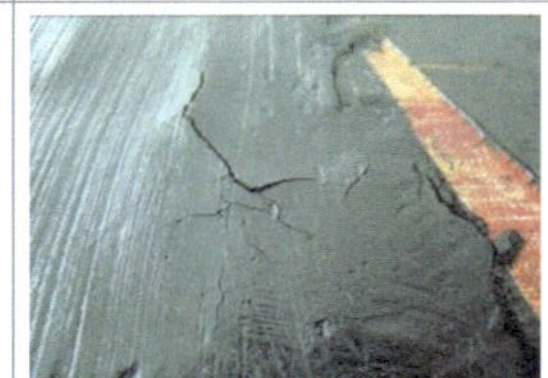
Peinture, 8 décembre 1959, **Kunstsammlung NordRhein Westfallen K20-K21, Düsseldorf** (consolidated in 2004, but not exhibited for conservation reasons)	Absorbent	Acicular shape of the lead white particles in the ground with many lead soaps detected. Round-shaped particles of lead white with very few lead soaps in the paint layer [1]	Severe paint delamination (treated in 1994, 1998 and 2004). Softened paint layer. Varnish defects. Condition unstable	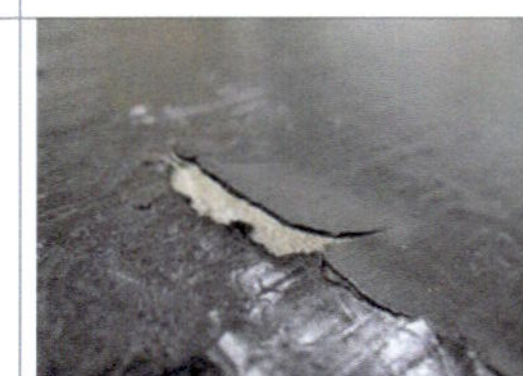
Peinture, 15 décembre 1959, **les abattoirs, Toulouse.** (consolidated in 2015 but not exhibited for conservation reasons. Stored flat because of rapid drip extension (2 mm growth in 3-month period))	Nonabsorbent (tracks visible on reverse)	Azelaic acid present in excess in the ground and in the paint layer [3]	Siccative defects with dripping medium exudates, since 1992–1994. Lifted paint areas since 1995. Varnish defects. Condition unstable	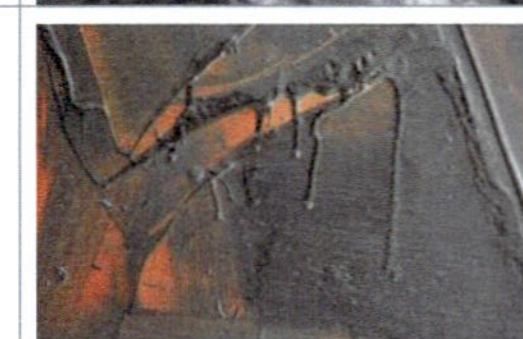

Peinture, 16 décembre 1959, **Musée d'art moderne de la Ville de Paris** (consolidated in 2005, but not exhibited for conservation reasons)	Absorbent (diffused oil visible as "clouds" on the reverse)	SEM shows acicular shape of lead white particles and many lead soaps present in the ground. Round-shaped particles of lead white and very few lead soaps in the paint layers [2], [4], [5]	Severe paint delamination. Consolidated in 1998 and 2005. Condition unstable	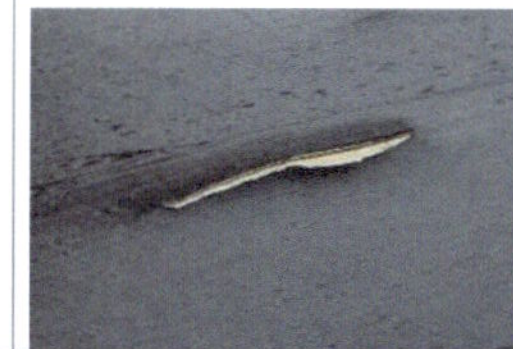
Peinture, 28 décembre 1959, **Musée Fabre, Montpellier,** (consolidated in 2015 and can be exhibited)	Unknown (reverse hidden by lining)	SEM shows "fluffy" shape of the lead white particles in the ground. Few lead soaps detected by FTIR and XRD. Plumbonacrite [1], [2]	Severe paint delamination. Lined in the 1990s. Condition unstable	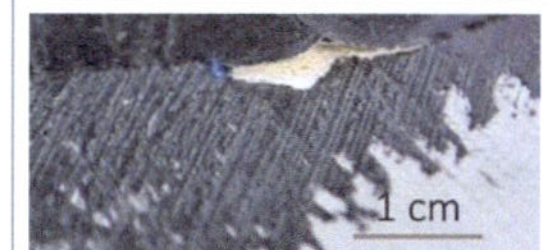
Mars 1960 **(private collection)** (consolidated in 2008 and can be exhibited)	Nonabsorbent (tracks visible on reverse)	Not analyzed	Exudates since 1990–1992. Delamination in 2008. Condition now stable	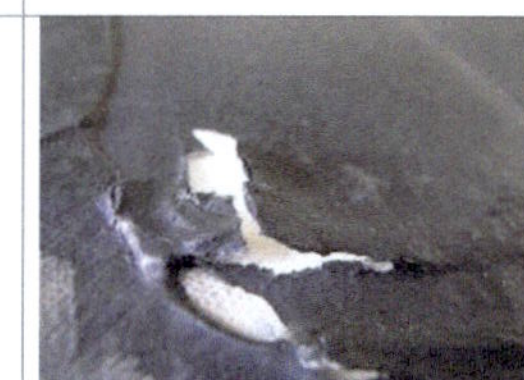
Peinture, 12 mars 1960 **Sara Hildén foundation/Sara Hildén art Museum** (consolidation in progress)	Nonabsorbent (tracks visible on reverse)	Analysis results pending	Severe delaminated paint and dripping medium exudates, removed with swab in 1995. Previously never consolidated. Consolidation in progress	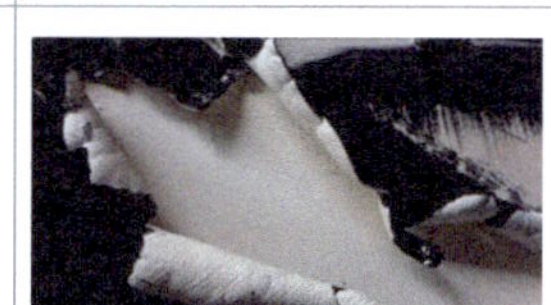

Paintings presenting conservation issues and treated. [1] Le Hô (C2RMF 2007), [2] Corbeil- Helwig (ICC 2008), [3] Balcar (C2RMF 2016), [4] Papillon (Inp 2004), [5] Le Hô and Mirambet (C2RMF 2005)

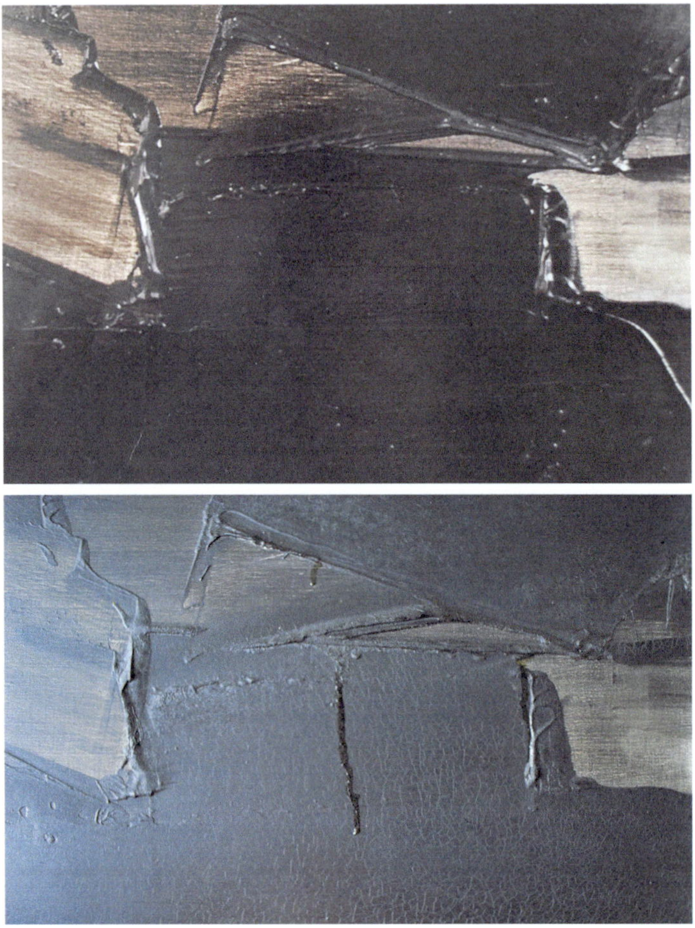

Fig. 23.3 Medium exudates occurring between 1990 (top) and 2013 (bottom) on *Peinture, mars 1960* (private collection) © Unknown (1990), and © Pauline Hélou-de La Grandière 2013

23.2.2.4 Shiny and Matte Appearance: Blanching Phenomena

Whereas the glossiness of Soulages' paint surfaces has never been objectively monitored, it can be an indicator of condition. As noted, a shinier surface is observed when the paint is in softened state, usually accompanied by other phenomena described, namely, an increase in the volume of the paint (elongation) and wavy distortions (Fig. 23.2). The rest of the paint surface is comparatively matte, which suggests a lack of binding medium.

Following a period of enclosure (as in a crate during a transport), blanching of the paint surface has also been noted, as in the case of *Peinture 15 décembre 1959*.

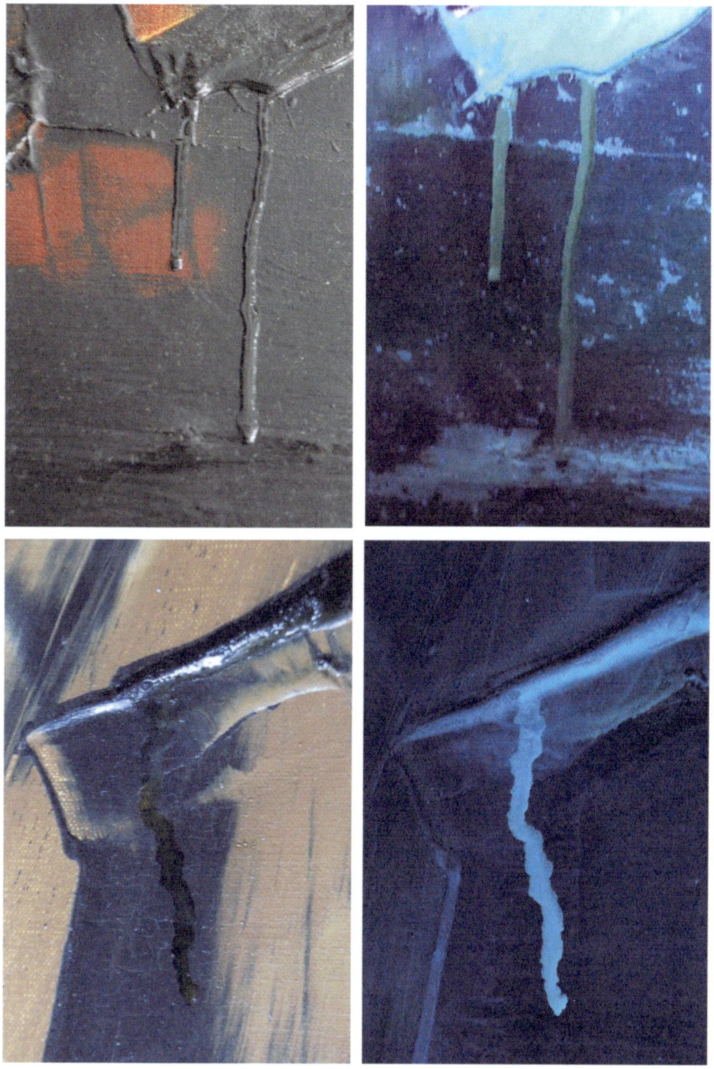

Fig. 23.4 Medium exudates in natural and UV light. Top-*Peinture, 15 décembre 1959* (Les Abattoirs, Toulouse) © Pauline Hélou-de La Grandière 2014 and Bottom-*Peinture, 12 mars 1960* (Sara Hildén Foundation/Sara Hildén Art Museum) © Pauline Hélou-de La Grandière 2016

23.2.3 Analysis

Conservation treatments have given the opportunity to conduct some analysis (see Table 23.1). When investigating the specific paint formulation used for the impastos with SEM, GC-MS, and FTIR, nothing other was found than the ivory black in

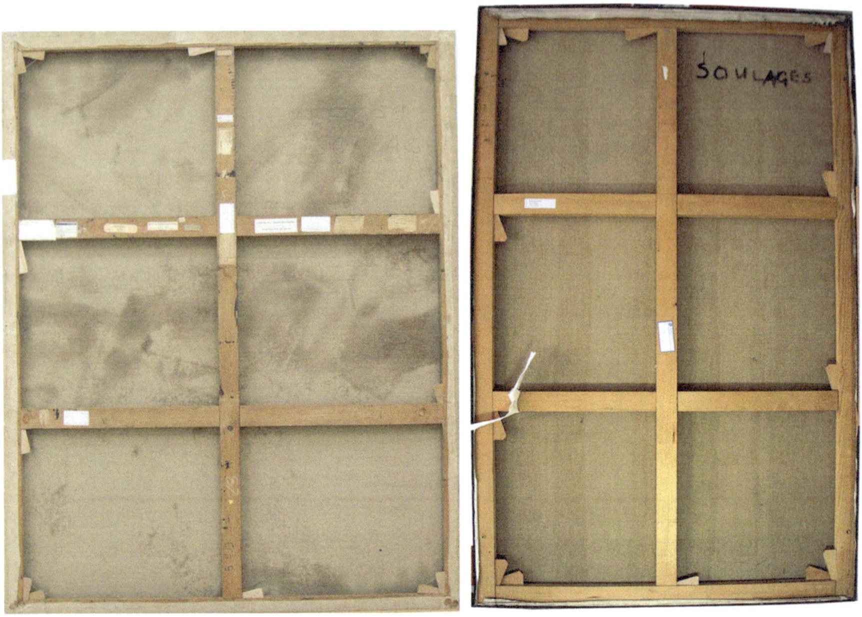

Fig. 23.5 Comparison of reverse side of a canvas showing strong absorption of the binding media (left) (*Peinture 8 décembre 1959*, with severe delamination) Kunstsammlung NordRhein Westfallen, Düsseldorf © Pauline de La Grandière 2004, and a nonabsorbent canvas (right) (*Peinture 15 décembre 1959*, with both delamination and exudates*)*, Musée des abattoirs Toulouse © Pauline Hélou-de La Grandière 2016

oil expected from study of oral and written sources. Despite rumors over the years suggesting otherwise, no ink, wax, bitumen, or Venice turpentine was detected.

In all the samples examined, the priming is seen to contain a similar type of lead white, with an unexpected needle-like shape of the pigment particles viewed under the SEM and with a high proportion of lead soaps detected by FTIR (Le Hô 2006, unpublished; Le Hô 2007, unpublished; Hélou-de La Grandière et al. 2009). The lead white was identified as plumbonacrite in *Peinture 14 avril 1959* (Unpublished report by Helwig and Corbeil, IIC, 2007; Hélou-de La Grandière et al. 2009) and could be a signature mark of the supplier, a point we will return to later on.

A higher concentration of lead soaps was noted at the interface where delamination of the ground or paint layer occurred (Hélou-de La Grandière et al. 2009).

In an area of exuding paint in *15 décembre 1959*, a high proportion of azelaic acid C_9 was identified by GC/MS analysis, both in the lead white ground and in the paint drips. To the scientist involved, this suggested the presence of a cartam oil (Balcar, C2RMF 2016, unpublished).

23.2.4 Conservation Strategies

While conservation treatments are necessary to halt the fast progression of deterioration, it is important that the chosen methods are compatible with any future treatments. To find the least invasive yet efficient option, prior tests are conducted on the ground around the edges of the painting, using adhesives that take into account the solvent sensitivity of the paint. The locations of all the treated areas are noted, ideally on a 1:1 scale transparent sheet.

Concerning the consolidation process, since the paint layers are very sensitive to heat and to solvents (both polar and apolar ones), adhesives in aqueous solution are often used. While natural glues never prove effective, synthetic adhesives may be used with success,[10] after adding surfactants to increase adhesion onto nonporous surfaces. Years later, the consolidation treatments performed are still largely successful, but unfortunately, the delamination phenomenon tends to propagate around the treated areas. This further deterioration becomes evident when comparing records made on transparent 1:1 scale sheets after a period of 2, 4, and 6 years (as was the case for *Peinture 18 avril 1959*). The consolidation only remains effective if the climate is kept stable after treatment. We experienced this in relation to both *Peinture 2 novembre 1959* and *Peinture 28 décembre 1959*, as the delamination returned in treated areas immediately after the occurrence of a flood causing elevated conditions of relative humidity. The drips from exudates may be satisfactorily removed using apolar solvents and a scalpel. However, their stabilization is more problematic. The option of hardening the liquefied oil paint by heating it to around 70–80 °C (Franken et al. 2014:344) is not appropriate for these very sensitive paintings. For the time being, we can only recommend storing the affected paintings in horizontal position (to avoid running drips) and under low temperature conditions below 17 °C (to be certain that the glass transition temperature of the paint is never reached). We also advise against confinement (which encourages blanching) and humidity variations (that are suspected to be a key factor behind the degradation).

23.3 Discussion: A Montparnasse Disease?

The conservation issues relating to paintings by Pierre Soulages became manifest in the period from the mid-1980s to the mid-1990s, that is, about 30 years after the paint was applied. This suggests a specific stage in the drying, or rather "deterioration" process. The cause of degradation appears to be a high level of stress within the thick paint layer, which contains a high proportion of oil-binding

[10] Up until now, successful adhesives have included Klucel®G in alcohol/water (CTS), Lascaux® Hydro-fund (Kremer), and Aquazol®500 (Kremer). The surfactant used is Triton DF 16, nonionic and biodegradable.

medium and was applied with strong pressure when fresh (de La Grandière 2005). The retraction of the paint layer over a distance of several mm explains the lifting of the paint along the edges of impastos, whereas elongation (swelling) of the paint is also a factor leading to delamination (see under Fig. 23.1).

Since only some of Soulages' paintings of the period are affected, while the majority are still in good condition, it is relevant to consider which technical features the problematic works have in common.

23.3.1 Role of the Commercially Primed Canvas

A shared feature of all the degraded paintings by Soulages is the commercially primed canvas. The commercially applied grounds contain metal soaps in abundance, which are thought to play a key role in determining mechanical behavior, an imbalance in pigment volume concentration, as well as affecting processes of chemical interaction.

Pierre Soulages also recollects a conflict between Lefebvre-Foinet and his employees, which led one of the best craftsmen to add some water during the preparation of the lead white, probably during the heating process (verbal exchange with Pierre Soulages, 2004 and 2017). This procedure is known to create lead soaps and to dramatically increase the speed of oil saponification (Cotte et al. 2016:12) and could therefore explain the high amount of lead carboxylate present in the priming. According to Soulage's wife, Colette, this event had a strong impact in Montparnasse.

The porosity of the commercially primed canvas is also a factor of key importance. Indeed observations reveal a correlation between the occurrence of dripping exudates and the use of nonporous, pre-primed canvases.[11] This type of canvas is easy to recognize by the eye, due to a pattern of dark lines on the reverse corresponding to where oil has penetrated from an additional layer of ground. In our experience, the paintings by Soulages made on a canvas with such "lines" on the reverse inevitably suffer from dripping paints and severe exudation. We suppose that these nonabsorbent canvases have prevented the absorption of excess oil from the paint, contrary to other canvases where stains of oil on the reverse have been wicked through from corresponding spots of impasto on the front side. Similarly, drippings and exudations on paintings by Georges Mathieu (1921–2012) were also observed on nonabsorbent types of pre-primed canvas, with lines on the reverse exactly like those observed in works by Soulages (Mathieu, *sans titre*, mai 1956, private collection). The precise correspondence of these lines makes clear that both artists used ready-primed canvas from the same consignment sold by Lefebvre-Foinet. Considering the different dates of the works they made however, Soulages

[11]The canvases are of a closely woven, coarse linen fiber (ca. 15–18 × 20–22 threads per centimeter).

must have used an old canvas still present in his studio, or an unsold canvas that had been held in stock by Lefebvre-Foinet.

It is interesting to note that the phenomenon of paint delamination does not depend upon this feature, since it occurs on both absorbent and nonabsorbent canvases, reminding us that in this case the painted composition may be an important factor instead.

23.3.2 Hypothesis: A Rancid Oil?

Recent study of exudate formation paid close attention to the quality of the oil used in the affected paintings. On paintings by Jean-Paul Riopelle (1923–2002), the degradation products from the paint appear very similar to those found in the Soulages examples that we studied. In the Soulages (*15 décembre 1959*), Nathalie Balcar (C2RMF 2016) noticed a high proportion of azelaic acid, as was similarly found in Riopelle's exudates (Corbeil et al. 2011; Bronken and Boon 2014:257). Recent studies of exudates in works by Soulages (paintings dated March 1960), Riopelle (paintings dated 1952), and Mathieu (paintings dated 1957) show that these are rich in polar fractions with stearic acids and azelaic acids present (Bronken and Boon 2015). Jaap Boon proposes that the oil in the paint undergoes an internal oxidation process as it ages, whereby the oxidized fractions of oil cannot bind with the paint and so gradually accumulate, eventually leading to the dripping exudates. These observations agree with our considerations about the making of oil at that time by Lefebvre-Foinet (Soulages' colourman and supplier), especially concerning his practice of pre-oxidizing the oil by exposing it to the sun in order to "bleach" it. This measure could have created a "rancid oil," which, as it was well-known at the time, might lead to siccative defects, as noted by Xavier de Langlais (de Langlais 1959:114–116):

> Some suppliers imagine, probably earnestly, that a ground layer can never be too flexible. They use really rancid oil for the priming of their canvas. Yet the flexibility of such grounds is due to their grinding oil having lost some of its siccative qualities. A rancid oil (that means acidified) can never fully dry. Painters cannot be wary enough of this kind of ground.[12]

This sentence from 1959 seems to be an open criticism of the colourmen who supplied grounds with lead white. Furthermore, it was clearly directed at Lefebvre-Foinet, who was well-known for his lead white grounds in Paris in the 1950s. In practice this warning seems to have been ignored by artists and colourmen, though artists' recipes specifically advise against the use of rancid oil (Ordonnez and Twilley 1998:48). Some oils become rancid more easily, such as walnut oil (Gettens and Stout 1966:77), while others may become rancid only after sunlight exposure Lefebvre-Foinet exposed oils to the sun on the rooftop of his rue vavin studio (de La Grandière 2005). The rancidification of oil leads to an excess of acidic

[12]Translation by Juliette Jacqmin.

compounds that precipitate with basic compounds as lead white (Doerner 1934), explaining the high level of lead soaps. This is consistent with what has been found by analysis.

23.3.3 Epidemic Defect

Other paintings from the 1950s show very similar delamination and corresponding grounds with an abundance of metal soaps. Examples include works painted by Paul-Emile Borduas (1905–1960) around 1957–1959 (Moffat and Miller 1994; O'Malley and Moffat 2001; Corbeil et al. 2004), paintings by Jean-Paul Riopelle dated 1952–1957 (Corbeil and Sirois 2007; Corbeil et al. 2011; Bronken and Boon 2014; Soldano 2015) and works by Georges Mathieu dated 1957 (Götz 2003; de Ségogne 2014). Paintings by these artists and by Soulages are currently being studied and analyzed by Ida Bronken and Jaap Boon (Boon and Lister 2014, Bronken and Boon 2015).

In French museums and private collections, we have also observed similar severe delaminations and exudates on paintings by Joan Mitchell (1925–1992) dated 1958, Zao Wou Ki (1920–2013) dated 1964, Mario Prassinos (1916–1985) dated 1959, and Georges Mathieu dated 1956. No analyses were performed, but similar visual signs of degradation suggest their use of common materials. Indeed these artists are linked by a common factor, since they all lived in the Parisian area of Montparnasse and are known or supposed to have been customers of Lefebvre-Foinet (Hélou-de La Grandière 2006, 2008). To this list we can also add Karel Appel (1921–2006), who was living in Montparnasse (rue Brézin) and whose paintings are suffering the same defects (Wijnberg 2014:28; Boon 2016). "Everybody went to Lefebvre-Foinet" as Pierre Soulages remembers, especially foreign artists since Lefebvre-Foinet covered the customs costs for transporting works abroad. This is a very worrying realization, since, in that case, all paintings from the second school of Paris may be involved.

23.4 Conclusion

The observed pandemic of delaminating and exudating paint is due to a combination of technical features specific to each painter, and the common use of materials obtained from the most widely used supplier in Paris at that time.

Today these defects are more widely analyzed and hence better understood, but two aspects remain worrisome. While effective treatments to consolidate or to remove the drips exist, these are still very much open for improvement. Also, we do not know how to stabilize the phenomenon, let alone prevent it from occurring.

We investigated the creation process for paintings by Soulages and understood that the drying conditions during the first three decades after they were made are essential, as was also demonstrated during the Metal soaps in Art conference

(Catalano et al. 2019; Bomford et al. 2019; Van Loon et al. 2016; Hermans et al. 2019; Cotte et al. 2019; Boon 2016). Due to the especially harsh climate in Paris in 1959 and 1960, Pierre Soulages' paintings made in the Rue Galande studio were exposed to poor conditions early on. This may well have served as a catalyst for the degradation of the corpus of works considered here. Other painters are involved in this issue too, all of them working in the Montparnasse neighborhood of Paris. They have in common their supplier of pre-primed canvas, and comparisons between the works of Soulages and those of other artists have led to a plausible hypothesis about the deterioration phenomena. We need to worry about this "Montparnasse disease," since it is likely that in the coming decades, more defects will appear on works by other painters after 50 years of "normal drying." We hope that research will keep ahead of this potentially expanding problem.

List of paintings

Peinture, 18 avril 1959, Musée de St Etienne (France)
Peinture, 2 novembre 1959, Musée Fabre, Montpellier (France)
Peinture, 8 décembre 1959, Kunstsammlung NordRhein Westfallen K20-K21, Düsseldorf (Germany)
Peinture, 15 décembre 1959, Les Abattoirs, Toulouse (France)
Peinture, 16 décembre 1959, Musée d'art moderne de la Ville de Paris (France)
Peinture, 28 décembre 1959, Musée Fabre, Montpellier (France)
Peinture, Mars 1960 (Private collection)
Peinture, 12 mars 1960, Sara Hildén Foundation/Sara Hildén Art Museum (Finland)

Acknowledgements Special and extensive thanks are expressed to Pierre and Colette Soulages, who kindly shared their memories and stories, and to Dan McEnroe and Pierre Encrevé.

We would like to thank the museums and collectors who agreed to share information (Les Abattoirs (Toulouse), Musée Fabre (Montpellier) Musée Soulages (Rodez), KNRW (Düsseldorf), Musée d'Art moderne (Saint-Etienne), Musée d'Art Moderne de la Ville de Paris, Sara Hildén Foundation/Sara Hildén Art Museum (Tampere)), the scientists who collaborated (Anne-Solenn Le Hô, François Mirambet, and Nathalie Balcar (C2RMF), Marie-Claude Corbeil and Kate Helwig (ICC), and Marie-Christine Papillon (INP)), and the conservators (Anne Craveia, Amalia Ramanankirahina, Cecilia Winter, Marie Zdyb, Pauline Fossier, Elisa Paolini, and Agata Graczyk). For the corrections, thank you to Maggie Montanaro, Dan McEnroe, Juliette Jacqmin, and Dave Mailman.

References

Adam E (2010) Itinéraire d'un marchand de couleurs à Montparnasse, Chêne, Paris

Aix-en-Provence (2016) Photographies d'ateliers d'artistes – Willy Maywald à l'atelier de Cézanne, Exhibition by Office du Tourisme d'Aix-en-Provence, Atelier de Cézanne, July–Sept 2016

Barabant G, Hélou-de La Grandière P (2012) Soulages, technique originale, restaurations. In: La restauration des peintures et des sculptures. Connaissance et reconnaissance de l'œuvre. Armand Colin, Paris, pp 313–330

Boon JJ (2016) Polar medium exudates in oil paintings and their disastrous consequences. Presented at the conference Metal soaps in Art, Rijksmuseum, Amsterdam, 15–16 Mar 2016

Boon JJ, Hoogland FG (2014) Investigating fluidizing dripping pink commercial paint on van Hermert's seven-series works from 1990–1995. In: Van den Berg KJ, Burnstock A, de Tagle A, de Keijzer M, Heydenreich G, Krueger J, Learner T (eds) Issues in contemporary oil paints. Springer, Cham, pp 227–246

Boon JJ, Lister K (2014) Investigating aged oil-medium phase separating and exuding as drips from a 1960s painting by Pierre Soulages. In: Bridgland J (ed) ICOM-CC 17th Triennial conference preprints, Melbourne, 15–19 Sept 2014. International Council of Museums

Bronken I, Boon JJ (2014) Hard dry paint, softening tacky paint, and exuding drips on composition (1952) by Jean-Paul Riopelle. In: Van den Berg KJ, Burnstock A, de Tagle A, de Keijzer M, Heydenreich G, Krueger J, Learner T (eds) Issues in contemporary oil paints. Springer, Cham, pp 247–262

Bronken I, Boon JJ (2015) Investigating softening and dripping paints in oil paintings made between 1952 and 2007. http://aics43rdannualmeeting2015.sched.org. Accessed 5 Sept 2016

Catalano J, Murphy A, Yao Y, Zumbulyadis N, Centeno SA, Dybowski C (2019) Understanding the dynamics and structure of lead soaps in oil paintings using multinuclear NMR. In: Casadio F, Keune K, Noble P, Van Loon A, Hendriks E, Centeno S, Osmond G (eds) Metal soaps in art: conservation and research. Springer, Cham, pp 69–82

Corbeil M-C, Sirois PJ (2007) Note on a modern lead white, also known as 'Synthetic Plumbonacrite. Stud Conserv 52(4):281–288

Corbeil M-C, Helwig K, Poulin J (2004) Analysis of the painted oeuvre of Jean-Paul Riopelle: from oil to mixed media. In: Modern art, new museums, contributions to the Bilbao Congress, Bilbao, 13–17 Sept 2004. International Institute for Conservation, Londres, pp 170–173

Corbeil MC, Helwig K (2008) Canadian Conservation Institute, "Analyse d'échantillons de tableaux de Pierre Soulages pour Pauline de La Grandière-Hélou", 8th Feb 2008, Report no LRA 4508

Corbeil MC, Helwig K, Poulin J (2011). Jean Paul Riopelle: The Artist's Materials, Getty publication, Los Angeles

Corbeil MC, Helwig K, Poulin J (2004) Analysis of the painted oeuvre of Jean-Paul Riopelle: from oil to mixed media. In: Modern art, new museums, contributions to the Bilbao Congress, Bilbao, 13–17 Sept 2004. International Institute for Conservation, Londres, pp 170–173

Cotte M, Checroun E, De Nolf W, Taniguchi Y, (de) Viguerie L, Burghammer M, Walter P, Rivard C, Salomé M, Janssens K, Susini J (2016) Lead soaps in paintings: friends or foes? Stud Conserv 62:2–23

Cotte M, De Viguerie L, Checroun E, Susini J, Walter P (2019) Historical evolutions of lead-fat/oil formula from Antiquity to modern times in a set of European pharmaceutical and painting treatises. In: Casadio F, Keune K, Noble P, Van Loon A, Hendriks E, Centeno S, Osmond G (eds) Metal soaps in art: conservation and research. Springer, Cham, pp 85–103

de La Grandière P (2004) Conservation Report on paintings by Pierre Soulages, 'Peinture 8 décembre 1959', Kunstsammlung NordRhein Westfallen in K20-K21, Düsseldorf, INP

de La Grandière P (2005) Restauration de Peinture, 16 décembre 1959, Musée d'Art Moderne de la Ville de Paris, Dissertation (painting conservation), Institut national du Patrimoine, Paris

de Langlais X (1959) La technique de la peinture à l'huile, Histoire du procédé à l'huile, de Van Eyck à nos jours. Eléments, recettes et manipulations. Pratique du métier, suivie d'une étude sur la peinture acrylique. Flammarion, Paris

de Ségogne H (2014) Théorème de Gödel by Georges Mathieu, 1957. Study and restoration: consolidation through cohesive regeneration using a solvent. In: Van den Berg KJ et al (eds) Issues in contemporary oil paint. Springer, Dordrecht, pp 149–157

Doerner M (1934) The materials of the artist and their use in painting with notes on the techniques of the old masters, Harcourt, brace and company (ed), New York. First edited (1921) Malmaterial und seine Verwendung im Bilde, Neuhaus Ed. Munich

Encrevé P (1994) Soulages: L'œuvre complet. Peintures I. 1946–1959, Editions du Seuil, Paris

Encrevé P (1995) Soulages: L'œuvre complet. Peintures II. 1959–1978. Editions du Seuil, Paris

Franken V, Heydenreich G, Jägers E, Müller W, Schulz J, Zumbühl S (2014) Set back the race: treatment strategies for running oil paint. In: Van den Berg KJ, Burnstock A, de Tagle A, de Keijzer M, Heydenreich G, Krueger J, Learner T (eds) Issues in contemporary oil paints. Springer, Cham, pp 333–349

Galerie de France (1960) Reportage photographique Izis, Latelier de Soulages-mars 1960, exhibition of recent paintings by Soulages (17 May–12 June 1960), Galerie de France ed. Paris

Giraudy D (1996) Soulages/Outrenoir. Dent Tech 4:108–113

Gettens R, Stout G (1966) Painting Materials, A short Encylopaedia. Dover Publication, New-York, p 175

Götz E (2003) Zur Trocknungsproblematik pastoser Leinölfarben, Dissertation (painting conservation) Institut für Konservierungs- und Restaurierungswissenschaft der Fachhochschule, Cologne

Havel M (1974) La technique du tableau. Dessain et Tolra. Paris

Hélou-de La Grandière P (2006) La restauration de 'Peinture, 114 × 165 cm, 16 décembre 1959'_ de Pierre Soulages. Patrimoines 2(2):124–129

Hélou-de La Grandière P (2008) Recherche sur le rôle des savons métalliques dans le développement des clivages des peintures de Pierre Soulages de 1959 Centre National des Arts Plastiques. https://doi.org/10.13140/2.1.1961.5528. Available on researchgate.net. Accessed 19 Sept 2016

Hélou-de La Grandière P, Le Hô A-S, Mirambet F (2007) Delaminating paint film at the end of 1950s: a case study on Pierre Soulages. Preparation for painting: the Artist's choice and its consequences. Archetypes Publications, London, pp 156–162

Hélou-de La Grandière P, Corbeil M-C, Le Hô A-S, Mirambet F (2009) Quand l'étude des matériaux constitutifs mène au diagnostic, cas de la problématique soulevée par quelques peintures de Pierre Soulages. In: Art d'aujourd'hui, Patrimoine de demain, Conservation et restauration des œuvres contemporaines/art today, cultural properties of tomorrow, conservation of contemporary artwork. SFIIC, Champs-sur-Marne, pp 340–341

Hélou-de La Grandière P (2014) Conservation Report on paintings by Pierre Soulages in Saint-Etienne, C2RMF

Hélou-de La Grandière P (2016) Hélou-de La Grandière P, Conservation Report on paintings by Pierre Soulages in Les Abattoirs, C2RMF

Helwig K, Poulin J, Corbeil M-C, Moffatt E, Duguay D (2014) Conservation issues in several twentieth-century Canadian oil paintings: the role of zinc carboxylate reaction products. Issues in contemporary oil paint. Springer, Dordrecht, 2014:167–184

Hermans JJ, Keune K, Van Loon A, Iedema PD (2019) Toward a complete molecular model for the formation of metal soaps in oil paints. In: Casadio F, Keune K, Noble P, Van Loon A, Hendriks E, Centeno S, Osmond G (eds) Metal soaps in art: conservation and research. Springer, Cham, pp 47–65

Le Hô AS (2006) C2RMF, "Analyse d'échantillons de peintures de Pierre Soulages pour le Musée Fabre, Montpellier, pour Pauline Hélou-de La Grandière, recherche CNAP"

Le Hô AS (2007) C2RMF, "Analyse d'échantillons de peintures de Pierre Soulages pour le Musée Fabre, Montpellier, Saint-Etienne et Düsseldorf, pour Pauline Hélou-de La Grandière, recherche CNAP"

Mayer L, Myers G (2013) American painters on technique: 1860–1945. Getty Publications, Los Angeles

McEnroe D (1991) Les outils et le matériel dans la peinture de Soulages, Dissertation (Art Histoty). Université Paris-IV Sorbonne, Paris

Mills JS, White R (1999) The organic chemistry of museum objects. Butterworth-Heinemann, Boston

Moffat E, Miller D (1994) Spectroscopic and chromatographic analysis of selected paintings from the Parisian period of Paul-Émile Borduas. Journal de l'ACCR 19, pp 3–31

O'Malley M, Moffat E (2001) Paul-Emile Borduas: the blacksand whites of the later paintings (1955–1960). In: Phenix A (ed) Deterioration of artists' paints: effects and analysis. UKIC and British Museum, London, pp 84–87

Ordonnez E, Twilley J (1998) Clarifying the haze, efflorescence on works of art. WAAC Newsletter, Vol 20(No 1), p 48

Ragon M (1990) Les ateliers de Pierre Soulages. Albin Michel, Paris

Rogge CE, Bomford ZV, Leal M, (2019) Seldom black and white: the works of Franz Kline. In: Casadio F, Keune K, Noble P, Van Loon A, Hendriks E, Centeno S, Osmond G (eds) Metal soaps in art: conservation and research. Springer, Cham, pp 415–425.

Séchet G (2016) Les chroniques météo de l'année 1959. http://wwwmeteo-pariscom/chronique/annee/1959. Accessed 6 Sept 2016

Soldano A (2015) Etude de la dégradation des œuvres de Jean Paul Riopelle des années 1952–1956, Dissertation (Painting conservation), Université Paris I-Sorbonne

Vailland R (1961) Comment travaille Pierre SOULAGES, reportage photographique de Claude Michaelides. In L'Œil (77):40–47. Reedited (1998) Comment travaille Pierre Soulages, Le Temps des Cerises, Paris

Van Loon A, Keune K, Hermans J,Sawicka A, Thoury M,Bertrand L, Sandt C, Reguer S, Tokarski C, Corkery R, Iedema P (2016) Investigation of the dynamic processes underlying lead soap-related degradation phenomena in multi-layer model systems. Presented at the conference Metal soaps in Art, Rijksmuseum, Amsterdam, 15–16 Mar 2016

Wijnberg L (2014) Do we see what we know or do we know what we see? Conservation of oil paintings in the Stedelijk museum Amsterdam. In: Van den Berg KJ et al (eds) Issues in contemporary oil paint. Springer, Dordrecht, pp 21–32

Chapter 24
Seldom Black and White: The Works of Franz Kline

Corina E. Rogge, Zahira Velíz Bomford, and Maite Leal

Abstract Franz Kline (1910–1962) was one of the defining artistic personalities of Abstract Expressionism, and like others in this movement, he embraced modern, inexpensive alternatives to conventional oil paints. This break with tradition often adversely impacted the long-term stability of the materials, and indeed, three of the four paintings by Kline in The Museum of Fine Arts, Houston, are manifesting condition issues that pose challenges to their aesthetic and physical integrity. *Wotan* (1950–1) exhibits fragile and persistently cracking and lifting paint layers, while *Orange and Black Wall* (1959) and *Red Brass* (1959) suffer from extreme interlayer cleavage and active paint loss. In contrast, *Corinthian II* (1961) is in stable, nearly pristine condition. Zinc and lead soaps have been identified in FTIR spectra of all four works, including *Corinthian II*; therefore their presence alone cannot explain the cohesive failures observed in the other three paintings. Other causes include under-bound paint, heavy sizing, poor quality canvas, and, for *Wotan*, a problematic complex support. Comparison of these paintings elucidates how Kline's choices of materials and paint application result in subtle but critical differences that significantly influence the relative stability of his works.

Keywords Franz Kline · Abstract expressionism · Zinc soaps · Lead soaps · Metal carboxylates · Interlayer cleavage · Under-bound paint · SEM · FTIR · GC-MS · py-GC-MS

24.1 Introduction

> *For all its impact, Kline's most celebrated art – his black and white painting – is among the most fragile of our century. In that fateful contrast of appearance* vs. *physical condition, it truly mirrors the artist* (Gaugh 1994).

C. E. Rogge (✉) · Z. V. Bomford · M. Leal
The Museum of Fine Arts Houston, Houston, TX, USA
e-mail: crogge@mfah.org

F. Casadio et al. (eds.), *Metal Soaps in Art*, Cultural Heritage Science,
https://doi.org/10.1007/978-3-319-90617-1_24

Franz Kline (1910–1962), one of the most significant artists in the American Abstract Expressionist movement, died at the height of his career from a rheumatic heart caused by a childhood infection (Gaugh 1979). Unfortunately, many of the paintings he created during his iconic and most productive years (1950–1961) are exhibiting their own condition issues – yellowing, cracking, lifting, cupping, and interlayer cleavage – that distort the impact of the works on the viewer and indicate the threat of mechanical failure. Identifying the causes of these problems may lead to treatments that prolong the survival of these paintings or at least allow the long-term prognosis for these works to be better understood.

Although Kline was said to favor Winsor & Newton oil colors during his pre-abstract expressionist years, the move to larger canvases, the need for personal economy, and the desire for a more liquid paint with specific handling characteristics caused Kline to adopt house paints (Gaugh 1994; Goodnough 1952). He supposedly favored Behlen's zinc white (a commercial house paint), modifying it with titanium to alter the handling properties, and a black oil enamel used to tint house paints. The use of these nontraditional paints evidently caused early problems; in particular, the whites were prone to yellowing (Sylvester 2001). Although Kline himself claimed not to view the yellowing negatively, explaining that the whites, no matter how yellow, retained their proper relationship with the blacks, collectors of his works were not so sanguine. When Kline joined the Sidney Janis Gallery in 1956, he was encouraged to switch to more stable artists' paints and to bill the gallery if he could not afford them (Gaugh 1979). He evidently struggled with this switch (de Kooning 1962), and the presence of alkyd paint in the relatively late painting *Delaware Gap* (1958) (Lake 1999) suggests that difficulty in adapting to the different handling properties of artists' paint may have encouraged Kline to continue using nonartists' paints for a considerable period of time.[1]

Although the use of nonstandard paints can account for the yellowing observed in many Kline paintings, the mechanical issues are more commonly attributed to the use of paints containing zinc oxide and their resultant formation of zinc soaps. Studies on abstract expressionist paintings, including two works by Franz Kline, held in the collection of the Hirshhorn Museum and Sculpture Garden suggest that intralayer cleavage of zinc white paints and cracking of paint layers applied over those paints are characteristic markers of mechanical failure caused by zinc soaps and embrittled paint, which result from the reaction of zinc oxide and the oil binder (Maines et al. 2001; Rogala et al. 2010). These studies also found high levels of oleic acid in *Palladio* (1961) by Franz Kline and zinc white reference paints. The presence of significant amounts of oleic acid in oil paints indicates an incomplete polymerization and oxidation of the paint film and was ascribed by the authors to the formation of zinc soap lamellar structures that limit the oxidation and cross-linking of the oil medium.

While the presence of oleic acid may be an indicator of zinc soaps, FTIR analysis is one of the best ways to determine the presence of crystalline metal stearates and palmitates using the sharp, strong, and distinctive COO^{-} stretching

[1]Artists' alkyds were introduced only in 1970 (Learner 2004).

absorptions between ~1540 and 1400 cm^{-1} (Robinet and Corbeil 2003; Otero et al. 2014). While crystalline soaps are often detected in paintings, a more common phenomenon is a broad band centered at ~1580 cm^{-1} ascribed to an asymmetric COO^- stretch. Recent work on model systems suggests that this band may be due to amorphous zinc carboxylates, perhaps including both noncrystalline zinc soaps and zinc bound to carboxylate moieties within the polymeric network of the oil medium, essentially a zinc ionomer (Hermans et al. 2015). Both types of carboxylates may be present within a given sample of a painting, but the detection of crystalline zinc soaps may herald greater structural instability as they indicate phase separation, aggregation, and lamellar structure formation that can result in embrittlement and inter- and intralayer cleavage (Maor and Murray 2008; Osmond et al. 2014).

The Museum of Fine Arts Houston (MFAH) holds four oil paintings by Kline: *Wotan* (1950–1), *Red Brass* (1959), *Orange and Black Wall* (1959), and *Corinthian II* (1961) (Fig. 24.1a). While the last of these is in good condition, the three earlier paintings are exhibiting condition issues including cracking, interlayer cleavage, and active paint loss (Fig. 24.1b). As all four are unvarnished and retain their original structure (none are wax lined), they offer a valuable opportunity to explore how the role of Kline's choices of supports, materials, and method of paint application contribute to the stability of his aging works and to better understand the role metal soaps and carboxylates play in the degradation of abstract expressionist works.

24.2 Methods of Analysis

FTIR spectroscopy: ATR-FTIR spectra were collected using a Lumos FTIR microscope equipped with a motorized germanium ATR crystal with a 100 μm tip (Bruker). The spectra are an average of 64 or 128 scans at 4 cm^{-1} spectral resolution, and an ATR correction was automatically applied by the Opus 7.0 instrument control and data collection software (Bruker). Transmission spectra were collected on a Continuum microscope coupled to a Nicolet Nexus 670 FTIR spectrometer (Thermo Scientific). Samples were prepared by flattening them in a diamond compression cell (Thermo Spectra Tech), removing the top diamond window, and analyzing the thin film in transmission mode on the bottom diamond window (2 × 2 mm surface area). An approximately 100 × 100 μm square microscope aperture was used to isolate the sample area for analysis. The spectra are an average of 64 scans at 4 cm^{-1} spectral resolution.

Dispersive Raman spectroscopy: Dispersive Raman spectra were collected on an inVia Raman microscope (Renishaw) using a 785 nm excitation laser operating at powers of 384 μW to7.51 mW at the sample as measured using a PM100D laser power meter (Thorlabs) equipped with a S120C photodiode power sensor. A 50x objective was used to focus the excitation beam on the sample supported on a glass microscope slide. The resulting Raman spectra are the average of 1 to 15 scans of 10 s duration. Spectral resolution was 3–5 cm^{-1} across the spectral range analyzed.

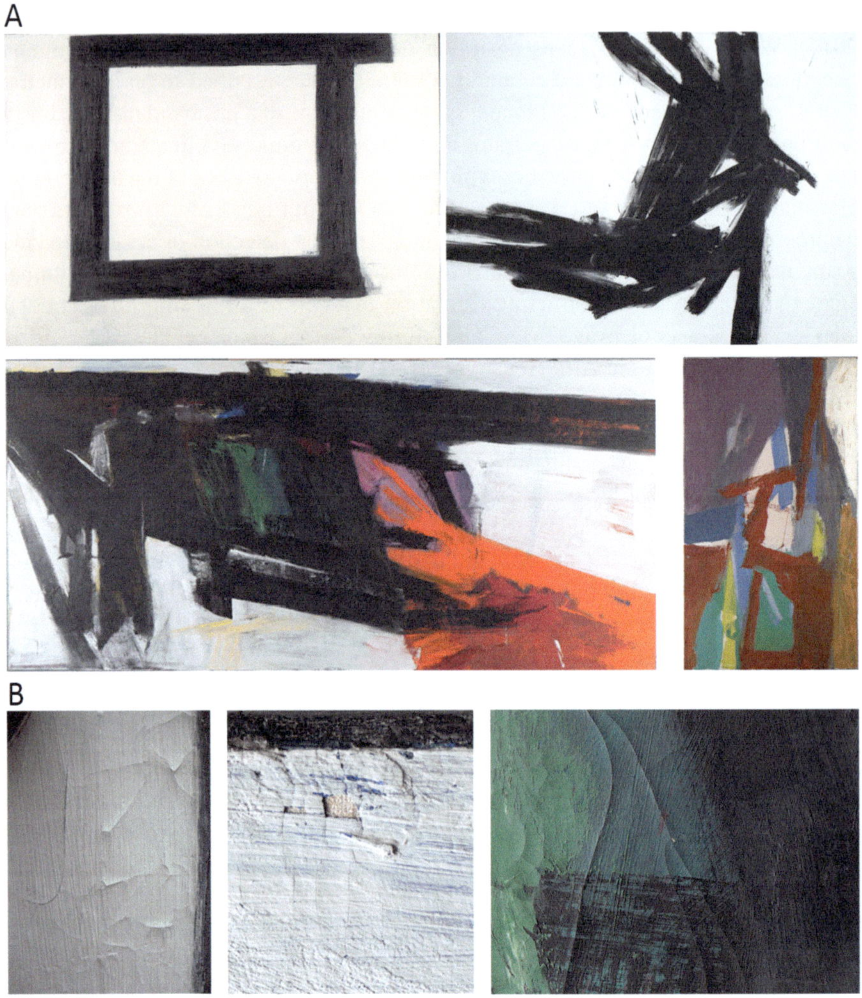

Fig. 24.1 (**a**) Franz Kline paintings in the MFAH collection. Clockwise from upper left: Wotan (1950-1), oil on canvas mounted on Masonite, 139.7 × 201.5, The Museum of Fine Arts, Houston, Museum purchase by exchange, 80.120 © The Franz Kline Estate/Artists Rights Society (ARS), New York; Corinthian II (1961), oil on canvas, 202.2 × 272.4 cm, The Museum of Fine Art, Houston, Bequest of Caroline Wiess Law, 2004.26 © The Franz Kline Estate/Artists Rights Society (ARS), New York; Orange and Black Wall (1959), oil on canvas, 169.5 × 367 cm, The Museum of Fine Arts, Houston, Bequest of Caroline Wiess Law, 2004.27 © The Franz Kline Estate/Artists Rights Society (ARS), New York; Red Brass (1959), oil on canvas, 173.4 × 99.3 cm, The Museum of Fine Arts, Houston, Bequest of Caroline Wiess Law, 2004.28 © The Franz Kline Estate/Artists Rights Society (ARS), New York. (**b**) Condition issues shown by the paintings, from left to right: detail of cracking and lifting paint in Wotan; detail of interlayer cleavage and paint loss in Orange and Black Wall; detail of cracking, lifting and paint loss in Orange and Black Wall. (Photo credits: Matthew Golden and Maite Leal, © The Museum of Fine Arts, Houston.)

XRF spectroscopy: X-ray fluorescence spectra were collected using either a Bruker Artax 400 energy dispersive X-ray spectrometer system or a portable Bruker Tracer III-V instrument. The Artax 400 excitation source was a Rhodium (Rh) target X-ray tube with a 0.2 mm thick beryllium (Be) window, operated at 40 kV and 400 mA current. The X-ray beam was directed at the artifact through a polycapillary tube. X-ray signals were detected using Peltier-cooled XFlash silicon drift detector (SDD) with a resolution of 146.4 eV. Helium purging was used to enhance sensitivity to light elements. Spectra were collected over 180 s (live time). The Bruker Tracer III-V handheld energy dispersive X-ray spectrometer equipped with a Si-Pin detector with a resolution of 190 eV. The excitation source was a Rhodium (Rh) target X-ray tube, operated at 40 kV and 10 μA current, and spectra were collected over 120 s (live time). Spectral interpretation was performed using the Bruker Artax Spectra 7.4.0.0 Software.

SEM-EDX: Analysis was performed on a JEOL 6010LA/Plus Intouch Scope operating at 15 kV in low vacuum mode and equipped with backscatter and EDS detectors.

GC-MS: GC-MS data were acquired by J. Mazurek using the instrumentation, conditions, and procedures outlined in Mazurek (2014) and Mazurek et al. (2014). Py-GC-MS data were acquired using the instrumentation, conditions, and procedures described in Heginbotham and Schilling (2011).

24.3 Results and Discussion

XRF analysis of all four paintings confirmed the presence of zinc (Fig. 24.2). The high levels of zinc and relatively low levels of barium suggest that while the presence of lithopone (a mixture of barium sulfate and zinc sulfide) cannot be excluded, zinc oxide is likely the predominant source of the zinc signal. In addition to zinc pigments, XRF, FTIR, and Raman spectroscopy show that all four paintings also contain titanium dioxide and lead white, and cross-sectional samples (Fig. 24.3) confirm the presence of multiple white paints in the paintings. Visual inspection of *Corinthian II* indicates the presence of at least three different white paints based upon their gloss and color. XRF and Raman analysis of the various whites confirms that they have different elemental signatures, revealing that in this painting, Kline was deliberately choosing to use different white paints. However, in *Wotan*, the surface white paint in all of the cross-sectional samples was the same, although at least five different subsurface white paints were present. Therefore it is unclear if the different subsurface white paints are signifying that Kline used whatever material was at hand or if in an unfinished state *Wotan* had modulated whites that Kline deliberately painted over to achieve the homogeneous surface of the finished work.

Wotan (1950–1) is a painting with a complicated history. Listed in Kline's first one man show in 1950 at the Egan gallery as an oil painting on canvas, the canvas was later modified by adhering it to two layers of fiberboard support with poly(vinyl acetate). As Kline signed the verso, and was known to have carpentry

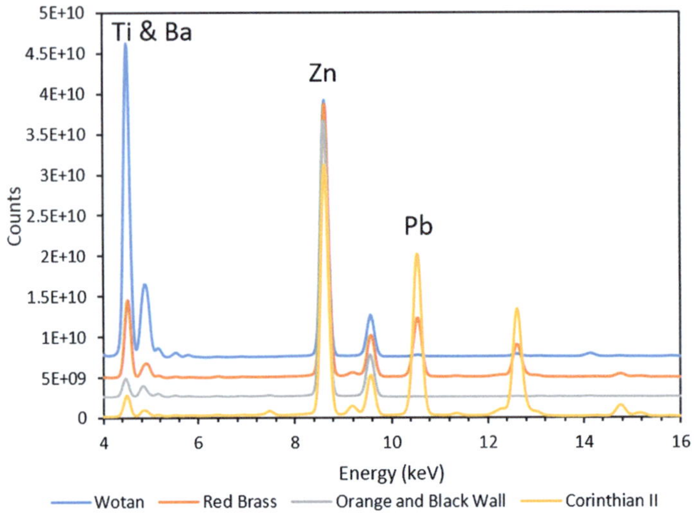

Fig. 24.2 XRF spectra taken from white areas of the four paintings. The Ti K_α, Ba L_α, Zn K_α, and Pb Ba L_α peaks are labeled. For ease of comparison, the Zn K_β counts have been normalized. © The Museum of Fine Arts, Houston

skills and tools, he may have done this himself (Gaugh 1994). The reasons for the remounting are not known, but could include early displays of inherent vice or damage caused by impact, which is evident in current X-radiographs. Since entering the MFAH collection in 1980, *Wotan* has undergone repeated consolidation due to extensive cracking, lifting, and cupping in the white regions (Fig. 24.1b), although the blacks are in relatively good condition with only a few drying cracks visible. The detection of zinc in the white areas and fragility of those areas suggested that zinc soaps may be the prime contributor to the current condition. FTIR-ATR mapping of the broad amorphous carboxylate peak centered near 1580 cm^{-1} shows that the ground of *Wotan* contains metal carboxylates (Fig. 24.3a). Of the seven cross-sectional samples taken from *Wotan*, only one has the sharp asymmetric COO^- peak at 1540 cm^{-1} characteristic of crystalline zinc soaps. Mapping of the narrow peak showed that crystalline metal soaps are localized to an area between a coarse calcium carbonate layer and overlapping zinc-containing layers (Fig. 24.3b), suggesting the formation of zinc soap lamellae at the layer interface. While seven samples taken from a painting with an area of 3 m^2 cannot be said to be representative, these results show that crystalline zinc soaps are not present in all areas and that additional factors (inherent and environmental) have contributed to the current condition of *Wotan*. Inspection of the cross-sectional samples by light microscopy and SEM-EDX reveals the presence of at least six different white paints, one of which, present in two samples, is a very coarse and under-bound calcium carbonate layer, visible in Fig. 24.4a, b. The two samples are incomplete and missing the lower layers as the coarse calcium carbonate layer has cleaved during sample collection. This fragile layer, evidently unable to bear mechanical stress, may be the origin of the cracking observed whenever the painting was moved

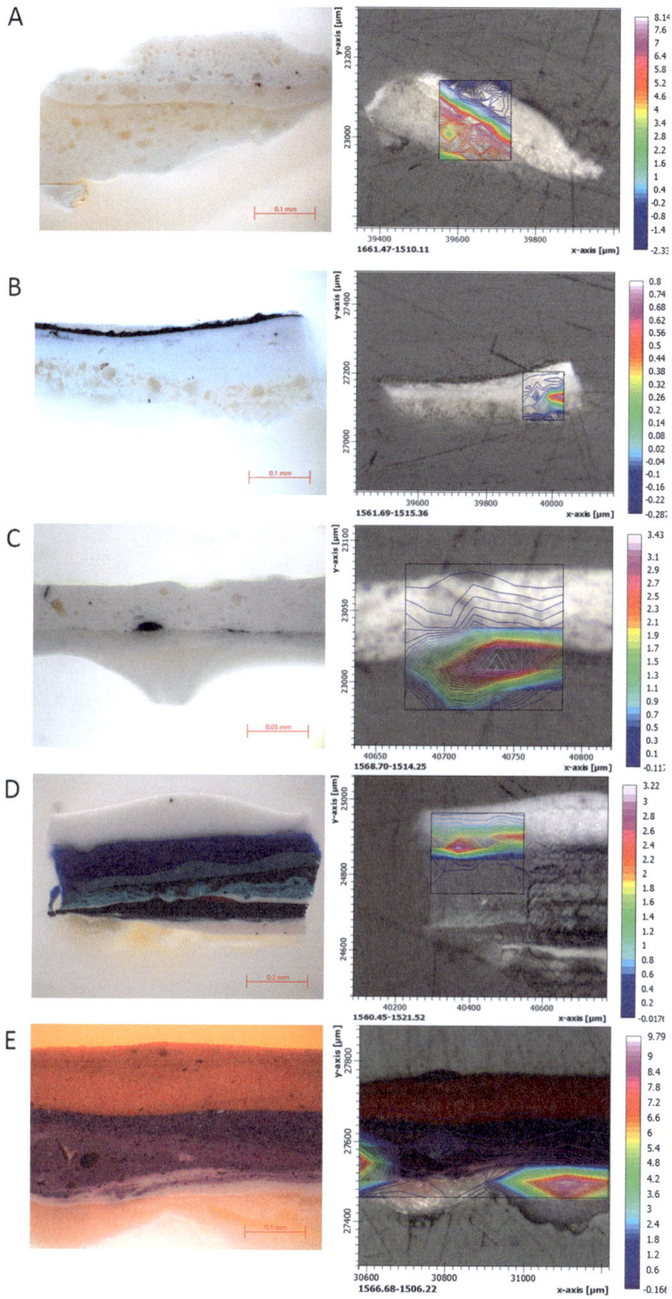

Fig. 24.3 Visible light images and ATR maps of cross-sectional samples taken from Franz Kline paintings. (**a**) sample from *Wotan* showing the presence of metal carboxylates in the ground; (**b**) sample from *Wotan* with a high concentration of crystalline zinc stearate in an intermediate paint layer; (**c**) sample from *Red Brass* showing the presence of crystalline zinc stearate in the ground layer; (**d**) sample from *Orange and Black Wall* with high levels of crystalline zinc stearate in between the surface white and underlying blue paint in an area exhibiting interlayer cleavage; (**e**) sample from *Orange and Black Wall*, crystalline lead stearate is present in high amounts at the ground/paint interface. All images © The Museum of Fine Arts, Houston

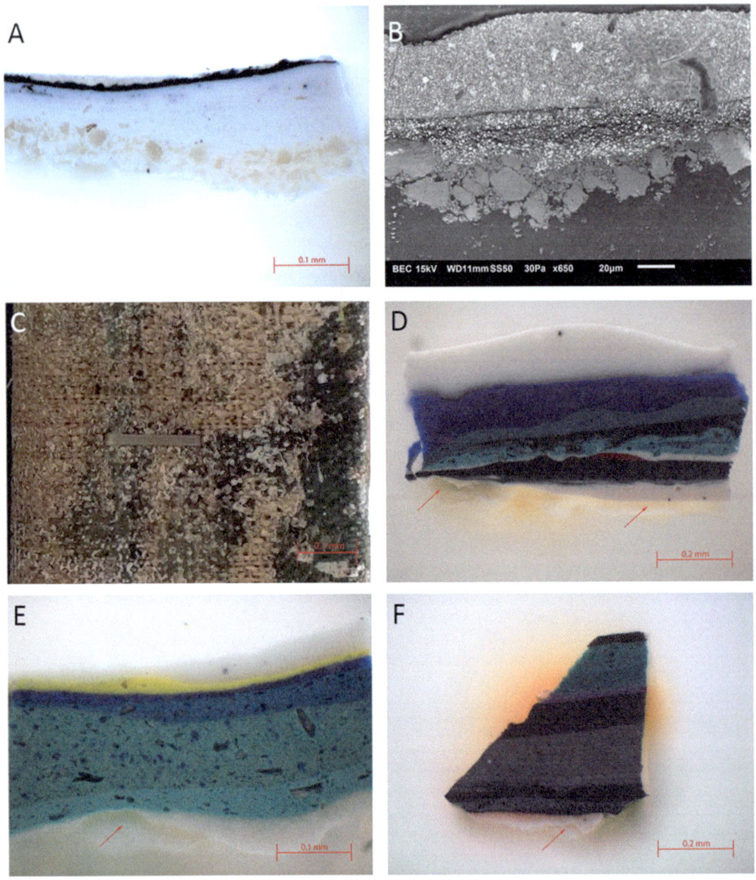

Fig. 24.4 Additional causes of deterioration present in Franz Kline paintings. (**a**) Light microscope image a cross section of *Wotan* showing the coarse, under-bound and fragile calcium carbonate layer at the bottom of the sample; (**b**) Backscatter SEM image of a cross section of *Wotan* showing the coarse, under-bound, and fragile calcium carbonate layer at the bottom of the sample; (**c**) normal light image showing the irregularly woven canvas present on *Orange and Black Wall*; (**d-f**) cross-sectional samples from *Orange and Black Wall* showing the very thick paint application, many paint layers, the presence of both wet-on-wet and wet-on-dry mixing, and intralayer cleavage in the ground layer, the latter indicated by arrows. All images © The Museum of Fine Arts, Houston

and the fiberboard support torqued. Embrittled paints caused by the formation of zinc soaps may exacerbate the fragility of the white passages, but zinc soaps do not appear to be the primary cause of the observed condition issues.

Red Brass (1959) is in generally good condition, except for a region of interlayer cleavage, lifting, and loss that occurs all along the proper right side of the painting, approximately 2 cm from the edge of the canvas, in both the orange and white paints. XRF, FTIR, and Raman microscopy indicate that the ground is a mixture of lead white and zinc white, suggesting that metal soaps may be the origin of the observed issues. A cross-sectional sample taken from the region of cracking and lifting (Fig. 24.3c) preserved a portion of the ground, and ATR-FTIR mapping suggests that it contains high levels of crystalline zinc soaps. The detection of these species, combined with the classic presentation of blind cleavage seen in many zinc soap-afflicted abstract expressionist works (Rogala et al. 2010), suggest that in this painting zinc soaps may be the primary cause of deterioration. However, the reason for the highly localized area of observed damage is unknown and suggests that additional factors, perhaps related to the mechanics of the painting as the zone of damage is close to the turnover edge, influence where the cleavage is occurring.

Orange and Black Wall (1959) displays a variety of condition issues, including cracking, lifting, interlayer cleavage, and loss (Fig. 24.1b), and the areas of damage do not appear to be localized to particular surface paint colors. XRF, FTIR, and Raman spectroscopy indicate that the ground contains lead white and calcium sulfate. ATR-FTIR mapping reveals that some samples contain high concentrations of lead soaps at the ground/paint interface (Fig. 24.3e), which may account for the interlayer cleavage seen between the ground and paint layers (Fig. 24.1b). However, zinc white paints are also present in some areas, including the cross section shown in Fig. 24.3d. In the visible light image of the cross section in question, interlayer cleavage can be seen between the surface zinc white paint and the underlying cobalt blue paint, exactly where ATR-FTIR mapping indicates the highest concentration of crystalline zinc soaps is located; macroscopic examples of this cleavage can also be seen in Fig. 24.1b. Thus, zinc soaps may be the cause of the interlayer cleavage occurring in the non-ground white paint. While the presence of metal soaps is decidedly a factor in the condition issues seen in *Orange and Black Wall*, other factors also play important roles. The canvas has an irregular weave (Fig. 24.4c) and has been heavily sized with animal glue, a material identified by FTIR and py-GC-MS. To the canvas, size, and lead white ground structure, Kline applied very thick layers of paint: some cross sections have more than 14 layers (Fig. 24.4d–f). Such a layer structure, applied to over 6 m^2, is very heavy and causes high strain, which is particularly problematic for an irregularly woven canvas. This force is aggravated by the dimensional response of the heavy glue sizing layer to variations in relative humidity: this painting spent 8 years in a private home in Houston, Texas, where seasonal relative humidity varies between 30 and 100% (interior RH records are not available for the private location). Dimensional changes induced by the sorption of water by the sizing layer and the resulted stress applied to the paint are likely the cause of the failure of the paint layer, which is visible as long cracks over much of the paintings surface (Fig. 24.1b). Thus, the structure Kline created has strongly influenced the current condition of this painting, although his choice to utilize lead and zinc-rich paints and the past environment have also played roles.

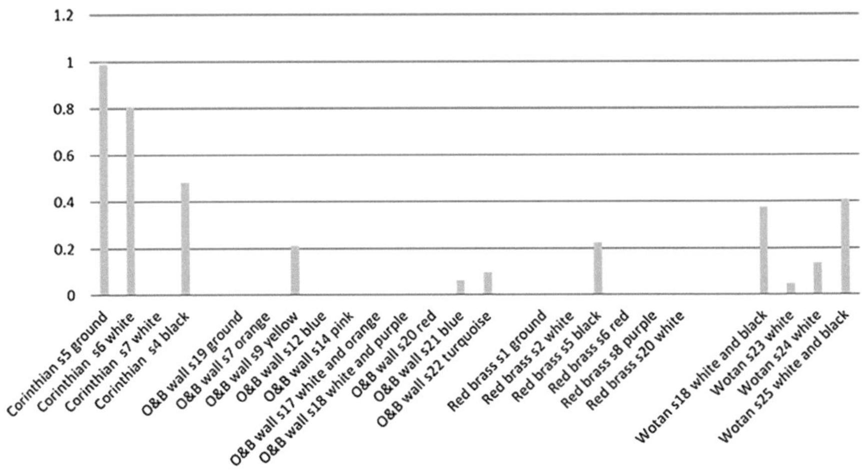

Fig. 24.5 Oleic acid to palmitic acid ratio of samples of Franz Kline paintings; the title of *Orange and Black Wall* has been abbreviated to "*O & B wall*." The absence of a bar indicates that no oleic acid was detected in a sample, © The Museum of Fine Arts, Houston

Corinthian II (1961) displays virtually no condition issues, and so sampling opportunities were limited, and no ATR-FTIR mapping was performed. XRF, FTIR, and Raman spectroscopy indicate the ground, which appears to be commercially applied, contains lead white and zinc oxide. The surface artist-applied paints are titanium white-based. Transmission FTIR microscopy of samples taken from the ground and surface white paints show the presence of high levels of metal carboxylates, as indicated by the broad band at 1580 cm^{-1}, and while one sample displayed the characteristic COO^- peak indicative of lead soaps, no crystalline zinc soaps were detected, unlike in *Red Brass*, which has a similar ground. The structure of *Corinthian II* is quite simple, with limited layering, thinly applied paint layers and a limited number of paints or paint mixtures. This may be because it is a recreation of *Corinthian I*, a painting Kline made in 1957 that was destroyed in the 1961 fire at the Governor's mansion in Albany, NY (Gaugh 1979). As Kline had already finalized the composition, the creation of this work involved less struggle and fewer revisions, resulting in a simpler, lighter structure that – combined with the commercially prepared canvas and apparent absence of crystalline zinc soaps – is less prone to degradation despite the presence of metal carboxylates and lead soaps.

As all four of the paintings display the presence of metal carboxylates and soaps, GC-MS analysis of the oil binders provides an ideal way to determine if high oleic acid levels are related to the presence of zinc soaps as detected by FTIR and to the relative condition of the works. Figure 24.5 displays the oleic acid to palmitic acid ratio of samples taken from the four paintings. Palmitic acid rather than stearic acid levels were chosen to avoid the issue of added sources of stearate, such as aluminum stearate, a material commonly included in many modern oil paint formulations (Tumosa 2001). The black paints in *Wotan*, *Red Brass*, and *Corinthian II* contain oleic acid, consistent with the poor drying often observed in black paints (van den et al. 1999; Sutherland 2001). However, the paint samples with the highest levels of oleic acid are from *Corinthian II*. The reason behind the incomplete oxidation of oleic acid in *Corinthian II* is not understood at this time, but as the FTIR spectra of samples from *Corinthian II* do not display peaks indicative of crystalline zinc soaps and as this painting is in impeccable condition, the presence of high levels of oleic acid may not always be directly correlated with degradation caused by zinc soap lamellae, as has been previously suggested (Maines et al. 2001; Rogala et al. 2010).

24.4 Conclusions

Lawrence Alloway described New York painters as having "a simplicity not so simple," (Alloway 1960) and Franz Kline's work might be the quintessential example of this. His striking paintings seem at first glance to be paragons of simplicity, but our ongoing project at the MFAH has proven that his paintings are complex and nuanced, each one a testimony of the artist's struggle to achieve the desired balance between equilibrium and tension. And just as his paintings are not simple, neither are their causes of deterioration: while zinc and lead soaps play an undoubted role, so do the choices Kline made regarding support, canvas preparation, and types and layering of paint. These factors can work synergistically, as in the case of *Corinthian II* where the presence of metal soaps may be offset by good quality canvas, sizing and ground, thin paint layers, and the absence of crystalline zinc soaps. The other three paintings show that it is possible for metal soaps and other factors to exacerbate condition issues. Caretakers of abstract expressionist works should be aware that sources of inherent vice other than metal soaps may play important roles in the "life" of an artwork and not oversimplify the causes of deterioration. But they should also take heart, as these other causes can often be more readily addressed than the inevitable formation of metal soaps.

Acknowledgments We thank Joy Mazurek and Michael Schilling of the Getty Conservation Institute for GC-MS and py-GC-MS analysis, Drs. Cruz Hinojos and Tom Tague of Bruker Optics for loan of the Lumos FTIR microscope, Mark Wypyski and Julie Arslanoglu of the Metropolitan Museum of Art for SEM-EDX analysis and sharing of analytical results, Zane Mařek and coworkers of JEOL USA Inc. for SEM-EDX analysis, Rice University Shared Equipment Authority for access to their FTIR, Raman, and SEM instruments. We thank MFAH colleagues Matthew Golden, Alison de Lima Green, Steve Pine, Ivan Reyes-García, and Bert Samples for their contributions to this project, and the Andrew W. Mellon Foundation for supporting conservation science in Houston.

References

Alloway L (1960) Sign and surface: notes on black and white painting in New York. Quadrum 9:49–62

de Kooning E (1962) Franz Kline memorial exhibition. Washington Gallery of Modern Art, Washington, DC

Gaugh HF (1979) Franz Kline: the color abstractions. The Phillips Collections, Washington, DC

Gaugh HF (1994) Franz Kline. Abbeville Press, New York

Goodnough R (1952) Franz Kline paints a picture. Art News 51(8):36–39, 63–64

Heginbotham A, Schilling M (2011) New evidence for the use of southeast Asian raw materials in seventeenth-century Japanese export lacquer. In: Rivers S, Faulkner R, Pretzel B (eds) East Asian lacquer: material culture, science, and conservation. Archetype Publications, London, pp 92–106

Hermans JJ, Keune K, Van Loon A, Iedema PD (2015) An infrared spectroscopic study of the nature of zinc carboxylates in oil paintings. J Anal At Spectrom 30(7):1600–1608

Lake S (1999) The relationship between style and technical procedure: Willem de Kooning's paintings of the late 1940s and 1960s. University of Delaware, Newark

Learner TJS (2004) Analysis of modern paints. The Getty conservation institute, Los Angeles

Maines CA, Rogala D, Lake S, Mecklenburg M (2001) Deterioration in abstract expressionist paintings: analysis of zinc oxide paint layers in works from the collection of the Hirshhorn Museum and Sculpture Garden, Smithsonian Institution. In: Vandiver PB, Li W, Sil JLR, Reedy CL, Frame LD (eds) Materials issues in art and archaeology IX. Cambridge University Press, Cambridge, pp 275–284

Maor Y, Murray A (2008) Delamination of oil paints on acrylic grounds. MRS Symp Proc 1047:127–136

Mazurek J (2014) Free fatty acid profiles in water sensitive oil paints: a comparison of modern and 15th century oil paints. In: Research and technical studies specialty group postprints from the 42nd annual meeting of the American Institute for Conservation, San Francisco. The American Institute for Conservation of Historic and Artistic Works, Washington, DC, pp 214–238

Mazurek J, Svoboda M, Maish J, Kawahara K, Fukakusa S, Nakazawa T, Tanigushi Y (2014) Characterization of binding media in Egyptian Romano portraits using enzyme-linked immunosorbent assay and mass spectrometry. e-Preserv Sci 11:76–83

Osmond G, Ebert B, Drennan J (2014) Zinc oxide-centered deterioration in 20th century Vietnamese paintings by Nguyễn Trọng Kiệm (1933–1991). AICCM Bull 34(1):4–14

Otero V, Sanches D, Montagner C, Vilariques M, Carlyle L, Lopes JA, Melo MJ (2014) Characterization of metal carboxylates by Raman and infrared spectroscopy in works of art. J Raman Spectrosc 45(11–12):1197–1206

Robinet L, Corbeil MC (2003) The characterization of metal soaps. Stud Conserv 48(1):23–40

Rogala D, Lake S, Maines C, Mecklenburg M (2010) Condition problems related to zinc oxide underlayers: examination of selected abstract expressionist paintings from the collection of the Hirshhorn Museum and Sculpture Garden, Smithsonian Institution. J Am Inst Conserv 49(2):96–113

Sutherland K (2001) Solvent extractable components of oil paint films. Ph.D. Thesis, University of Amsterdam, Amsterdam

Sylvester D (2001) Interviews with American arists. Chatto & Windus, London

Tumosa CS (2001) A brief history of aluminum stearate as a component of paint. WAAC Newsletter, Vol 23(No 3), pp 10–11

van den Berg JDJ, van den Berg KJ, Boon JJ (1999) Chemical changes in curing and aging oil paints. 12th triennial meeting of the ICOM committee for conservation, Lyon, vol 1, pp 248–253